Current Topics in Behavioral Neurosciences

Series Editors:
Mark Geyer, La Jolla, CA, USA
Bart Ellenbroek, Wellington, New Zealand
Charles Marsden, Nottingham, UK

About this series

Current Topics in Behavioral Neurosciences provides critical and comprehensive discussions of the most significant areas of behavioral neuroscience research, written by leading international authorities. Each volume offers an informative and contemporary account of its subject, making it an unrivalled reference source. Titles in this series are available in both print and electronic formats.

With the development of new methodologies for brain imaging, genetic and genomic analyses, molecular engineering of mutant animals, novel routes for drug delivery, and sophisticated cross-species behavioral assessments, it is now possible to study behavior relevant to psychiatric and neurological diseases and disorders on the physiological level. The *Behavioral Neurosciences* series focuses on "translational medicine" and cutting-edge technologies. Preclinical and clinical trials for the development of new diagnostics and therapeutics as well as prevention efforts are covered whenever possible.

Clare Stanford • Rosemary Tannock
Editors

Behavioral Neuroscience of Attention Deficit Hyperactivity Disorder and Its Treatment

Editors
Dr. S. Clare Stanford
Department of Neuroscience
Physiology & Pharmacology
University College London
Gower Street
WC1E 6BT London
United Kingdom
c.stanford@ucl.ac.uk

Dr. Rosemary Tannock
Applied Psychology & Human Development
The Ontario Institute for Studies in Education
University of Toronto
252 Bloor Street West
Toronto, Ontario M5S 1V6
Canada
rosemary.tannock@utoronto.ca

ISSN 1866-3370
ISBN 978-3-642-24611-1
DOI 10.1007/978-3-642-24612-8
Springer Heidelberg Dordrecht London New York

e-ISSN 1866-3389
e-ISBN 978-3-642-24612-8

Library of Congress Control Number: 2011943172

© Springer-Verlag Berlin Heidelberg 2012
This work is subject to copyright. All rights are reserved, whether the whole or part of the material is concerned, specifically the rights of translation, reprinting, reuse of illustrations, recitation, broadcasting, reproduction on microfilm or in any other way, and storage in data banks. Duplication of this publication or parts thereof is permitted only under the provisions of the German Copyright Law of September 9, 1965, in its current version, and permission for use must always be obtained from Springer. Violations are liable to prosecution under the German Copyright Law.
The use of general descriptive names, registered names, trademarks, etc. in this publication does not imply, even in the absence of a specific statement, that such names are exempt from the relevant protective laws and regulations and therefore free for general use.

Printed on acid-free paper

Springer is part of Springer Science+Business Media (www.springer.com)

A Tribute to Terje Sagvolden (Feb 12, 1945 to Jan 12, 2011)

We dedicate this edited volume on Attention-Deficit Hyperactivity Disorder (ADHD) to Terje Sagvolden, a highly accomplished neuroscientist of international repute, a wonderful caring man, a kind considerate friend, and an inspiring researcher. His seminal work was to demonstrate the relevance of a specific animal model, the Spontaneously Hypertensive Rat, for human ADHD, which he describes in a chapter in this volume, completed shortly before his sudden death. As a founding member of the European Network for Hyperkinetic Disorder (Eunethydis), he emphasized with legendary enthusiasm, the importance of reinforcement and learning as a major focus for human ADHD research. Terje worked tirelessly and passionately to forge links between basic and clinical researchers from different disciplines, across different countries, to better understand the neurobiology of ADHD. He leaves a rich legacy of ideas to be pursued by the next generation of ADHD researchers, some of whom are represented in this volume.

Preface

As one in the series of *Current Topics in Behavioral Neuroscience*, this book draws together the latest developments in both preclinical and clinical research of attention-deficit hyperactivity disorder (ADHD). We have tried to highlight the science that is common to both, as well as the chasms that separate them. This appraisal is timely in view of the forthcoming publication of DSM-5 in 2013, which aims to update the diagnostic criteria for ADHD, based on its current neuroscientific understanding.

ADHD has a worldwide incidence of about 5% in children but it is not a benign disorder found only in the young, as was once believed. As the opening chapter of this book makes clear, ADHD persists into adulthood in more than half of the cases. These patients often experience serious comorbidity, such as substance misuse (especially alcohol), anxiety and emotional lability; bipolar disorder and criminality. Any one of these problems can ruin social function, employability, and quality of life, and they all illustrate why it is so important that we find ways of understanding the neurobiology of this disorder and develop effective treatment approaches.

There is an obvious and justifiable emphasis on the latest research that points to ADHD as a neurodevelopmental disorder involving alterations in prefrontal brain regions, but the role of other brain regions, such as those coupled to the periphery, is covered too. At last, there is real progress in understanding the consequences of disrupting these complex feed-back and feed-forward loops and their functional connections with forebrain neuronal circuits.

After that, the theme of the chapters switches to comorbid problems, especially drug misuse and obesity. It is not at all obvious why these debilitating conditions are so prominent with ADHD, compared with other psychiatric disorders, but that anomaly must give clues to their underlying neurobiology. More clues are emerging from genetic studies, which are gradually identifying candidates that are more certain, or more common, than others. However, as is now evident, all these factors can be confounded by the impact of early life experience on gene expression and factors that govern brain development, including those that determine gender.

As in other fields, the development of an animal model of ADHD is seen as an essential step in translational research. Validating these animal models, even as mere drug screens, is a constant and challenging process. Several mouse contenders are reviewed here but whether a strain of mouse will ever be developed that replicates the benchmark rodent model of ADHD, the Spontaneously Hypertensive Rat (SHR), remains to be seen.

Apart from the SHR, which has been studied in exceptional detail, there is a striking dearth of preclinical research of ADHD compared with other CNS disorders. That could well explain why, despite the prevalence of ADHD, only five drugs are licensed to treat ADHD in the USA – and only three of these are available in the UK. Yet, given that even the first-line treatments for ADHD (psychostimulants) are ineffective in about 20–25% of cases, there is a pressing need for new approaches to pharmacotherapy of ADHD, particularly in adults, based on a strong scientific rationale.

The last chapter offers a novel framework in which to view the various accounts and explanations of ADHD presented herein and in the broader ADHD literature. It presents an updated version of the theoretical framework initially presented by Peter Killeen at the international multidisciplinary research group led by Terje Sagvolden at the Centre for Advanced Studies at the Norwegian Academy for Science and Letters (2004–2005). This "think-tank" on ADHD acted as a catalyst in generating new ideas and new lines of international collaborative research between basic and clinical scientists and the establishment of the journal *Behavioral and Brain Functions*, of which Terje was Editor-in-Chief.

Contents

ADHD in Children and Adults: Diagnosis and Prognosis 1
Renata Kieling and Luis A. Rohde

ADHD: Volumetry, Motor, and Oculomotor Functions 17
E. Mark Mahone

Neurodevelopmental Abnormalities in ADHD 49
Chandan J. Vaidya

Intraindividual Variability in ADHD and Its Implications
for Research of Causal Links 67
Jonna Kuntsi and Christoph Klein

Hypothalamic–Pituitary–Adrenocortical Axis Function
in Attention-Deficit Hyperactivity Disorder 93
Graeme Fairchild

Brain Processes in Discounting: Consequences of Adolescent
Methylphenidate Exposure 113
Walter Adriani, Francesca Zoratto, and Giovanni Laviola

Attention Deficit Hyperactivity Disorder and Substance
Use Disorders 145
Oscar G. Bukstein

Linking ADHD, Impulsivity, and Drug Abuse: A Neuropsychological
Perspective 173
Gonzalo P. Urcelay and Jeffrey W. Dalley

Obesity and ADHD: Clinical and Neurobiological Implications 199
Samuele Cortese and Brenda Vincenzi

Face Processing in Attention Deficit/Hyperactivity Disorder 219
Daniel P. Dickstein and F. Xavier Castellanos

Quantitative and Molecular Genetics of ADHD 239
Philip Asherson and Hugh Gurling

Rodent Models of ADHD 273
Xueliang Fan, Kristy J. Bruno, and Ellen J. Hess

Rat Models of ADHD 301
Terje Sagvolden and Espen Borgå Johansen

Epigenetics: Genetics Versus Life Experiences 317
Josephine Elia, Seth Laracy, Jeremy Allen, Jenelle Nissley-Tsiopinis, and Karin Borgmann-Winter

Sexual Differentiation of the Brain and ADHD: What Is a Sex Difference in Prevalence Telling Us? 341
Jaylyn Waddell and Margaret M. McCarthy

ADHD: Current and Future Therapeutics 361
David J. Heal, Sharon L. Smith, and Robert L. Findling

The Four Causes of ADHD: A Framework 391
Peter R. Killeen, Rosemary Tannock, and Terje Sagvolden

Index 427

Contributors

Walter Adriani
Section of Behavioural Neuroscience, Department of Cell Biology & Neurosciences, Istituto Superiore di Sanità, Rome, Italy

Jeremy Allen
The Children's Hospital of Philadelphia, Science Center, Philadelphia, PA, USA

Philip Asherson
MRC Social Genetic and Developmental Psychiatry, Institute of Psychiatry, Kings College London, London, UK

Karin Borgmann-Winter
The Children's Hospital of Philadelphia, Science Center, Philadelphia, PA, USA

Kristy J. Bruno
Departments of Pharmacology and Neurology, Emory University School of Medicine, Atlanta, GA, USA

Oscar G. Bukstein
University of Texas Health Science Center at Houston, Houston, Texas, USA

F. Xavier Castellanos
Neidich Professor of Child and Adolescent Psychiatry, Director of the Phyllis Green and Randolph Cöwen Institute for Pediatric Neuroscience, Director of Research, Nathan Kline Institute for Psychiatric Research, University Child Study Center. Research Psychiatrist, Orangeburg, NY, USA

Samuele Cortese
Institute for Pediatric Neuroscience, New York University Child Study Center, New York, NY, USA

Jeffrey W. Dalley
Behavioural and Clinical Neuroscience Institute, Department of Experimental Psychology & Department of Psychiatry, University of Cambridge, Cambridge, UK

Daniel P. Dickstein
PediMIND Program, E.P. Bradley Hospital, Alpert Medical School of Brown University, Providence, Rhode Island, USA

Josephine Elia
The Children's Hospital of Philadelphia, Science Center, Philadelphia, PA, USA

Graeme Fairchild
Developmental Psychiatry Section, Department of Psychiatry, Cambridge University, Cambridge, UK

Xueliang Fan
Departments of Pharmacology and Neurology, Emory University School of Medicine, Atlanta, GA, USA

Robert L. Findling
University Hospitals Case Medical Center, Case Western Reserve University School of Medicine, Cleveland, OH, USA

Hugh Gurling
Mental Health Sciences Unit, Faculty of Brain Sciences, University College London, UK

David J. Heal
RenaSci Consultancy Ltd, BioCity, Nottingham, UK

Ellen J. Hess
Departments of Pharmacology and Neurology, Emory University School of Medicine, Atlanta, GA, USA

Espen Borgå Johansen
Akershus University College, Lillestrøm, Norway

Renata Kieling
ADHD Outpatient Clinic, Hospital de Clínicas de Porto Alegre, Ramiro Barcelos, Porto Alegre, Brazil

Peter R. Killeen
Psychology Department, Arizona State University, Tempe, AZ, USA

Christoph Klein
School of Psychology, Bangor University School of Psychology, Adeilad Brigantia, Gwynedd, UK; Department of Child and Adolescent Psychiatry and Psychotherapy, University of Freiburg, Freiburg im Breisgau, Germany

Jonna Kuntsi
King's College London, MRC Social, Genetic and Developmental Psychiatry Centre, Institute of Psychiatry, London, UK

Seth Laracy
The Children's Hospital of Philadelphia, Science Center, Philadelphia, PA, USA

Giovanni Laviola
Section of Behavioural Neuroscience, Department of Cell Biology & Neurosciences, Istituto Superiore di Sanità, Rome, Italy

E. Mark Mahone
Kennedy Krieger Institute, Johns Hopkins University School of Medicine, Baltimore, MD, USA

Margaret M. McCarthy
Department of Physiology, School of Medicine, University of Maryland, Baltimore, MD, USA

Jenelle Nissley-Tsiopinis
The Children's Hospital of Philadelphia, Science Center, Philadelphia, PA, USA

Luis A. Rohde
ADHD Outpatient Clinic, Hospital de Clínicas de Porto Alegre, Ramiro Barcelos, Porto Alegre, Brazil

Terje Sagvolden
Department of Physiology, Institute of Basic Medical Sciences, University of Oslo, Oslo, Norway

Sharon L. Smith
RenaSci Consultancy Ltd, BioCity, Nottingham, UK

S. Clare Stanford
Department of Neuroscience, Physiology & Pharmacology, University College London, London, UK

Rosemary Tannock
Neuroscience & Mental Health Research Program, The Hospital for Sick Children, Toronto, ON, Canada

Gonzalo P. Urcelay
Behavioural and Clinical Neuroscience Institute, Department of Experimental Psychology, University of Cambridge, Cambridge, UK

Chandan J. Vaidya
Department of Psychology, Georgetown University, Washington, DC, USA; Children's Research Institute, Children's National Medical Center, Washington, DC, USA

Brenda Vincenzi
Ellern Mede Centre, London, UK

Jaylyn Waddell
Department of Physiology, School of Medicine, University of Maryland, Baltimore, MD, USA

Francesca Zoratto
Section of Behavioural Neuroscience, Department of Cell Biology & Neurosciences, Istituto Superiore di Sanità, Rome, Italy

ADHD in Children and Adults: Diagnosis and Prognosis

Renata Kieling and Luis A. Rohde

Contents

1. Introduction 2
2. Diagnostic Criteria 3
3. Controversial Issues in the Diagnosis 5
4. ADHD in Preschool Children 7
5. ADHD in School-Age Children 8
6. ADHD in Adolescents 9
7. ADHD in Adults 11
8. Future Perspectives 12
9. Conclusion 12
References 12

Abstract Attention-deficit hyperactivity disorder (ADHD) is one of the most common neuropsychiatric disorders with childhood onset, having a chronic course associated with high dysfunction and morbidity throughout life. Despite significant advances in our understanding of the neurobiological underpinnings of the disorder, diagnosis of ADHD remains strictly clinical and is based on behavioral symptoms of inattention, impulsivity, and hyperactivity. In this chapter, we review the diagnostic process and current controversies in the diagnosis of ADHD, discuss the clinical presentation of the disorder across the lifespan, and examine the patterns of comorbidity and the longitudinal predictors of outcomes. We conclude by pointing out some of the challenges that need to be addressed in future classifications systems to improve the characterization and validity of the diagnosis of ADHD.

Keywords Age of onset · Attention-deficit hyperactivity disorder · Clinical profile · Comorbidity · Diagnostic criteria

R. Kieling (✉) and L.A. Rohde
ADHD Outpatient Clinic, Hospital de Clínicas de Porto Alegre, Ramiro Barcelos, 2350 – room 2201A, Porto Alegre, Brazil
e-mail: renatakieling@uol.com.br

Abbreviations

ADHD   Attention-deficit hyperactivity disorder
APA   American Psychiatric Association
CD   Conduct disorder
DSM   Diagnostic and Statistical Manual (of Mental Disorders)
ICD   International Classification of Diseases
ODD   Oppositional defiant disorder

1 Introduction

Attention-deficit hyperactivity disorder (ADHD) is characterized by pervasive and impairing symptoms of inattention, hyperactivity, and impulsivity, being one of the most common neuropsychiatric disorders with childhood onset. As one of the most extensively investigated disorders in medicine (Goldman et al. 1998), ADHD is considered a multifactorial developmental neuropsychiatric disorder, based on genetic predisposition and neurobiological dysregulation. Epidemiological data indicate a worldwide prevalence of 5.29% among individuals under the age of 18 (Polanczyk et al. 2007), and a recent meta-analysis documented a 2.5% prevalence in adults (Simon et al. 2009).

Although previously thought to remit before or during adolescence, long-term follow-up studies of children and retrospective studies of adults with ADHD have shown that the disorder often persists into adulthood (Biederman et al. 2010). ADHD may be associated with a number of comorbid psychiatric disorders, including oppositional defiant disorder (ODD), conduct disorder (CD), learning disabilities, substance use disorder, and mood and anxiety disorders. It is usually accompanied by impaired academic and social skills, leading to significant emotional distress for both patients and families.

In each developmental stage, ADHD is associated with various negative outcomes, which may be exacerbated by the presence of comorbid disorders. Overall, ADHD-related problems include the following: (for children) greater risk of poor school performance and low academic achievement, grade retention, school suspensions and expulsions; (for adolescents) poor peer and family relations, school drop-out, aggression, conduct problems and delinquency, early substance experimentation and abuse; (for adults) driving accidents and speeding violations, difficulties in social relationships, marriage, and employment (Biederman et al. 2010; Fischer et al. 2007; Kessler et al. 2006; Nijmeijer et al. 2008). Thus, impairments go beyond the individual's own attainments to affect the various domains of home life, occupational functioning, dealings in the community, marital or dating relationships, money management, leisure and recreational activities (Barkley and Murphy 2010a).

Here, we review the current knowledge on diagnostic criteria and clinical presentation of ADHD in children, adolescents, and adults exploring changes in symptomatology across the lifespan. We also examine the comorbidities most frequently associated with ADHD and discuss their impact on the prognosis of child and adult patients.

2 Diagnostic Criteria

The current version of the Diagnostic and Statistical Manual of Mental Disorders – DSM-IV–TR (American Psychiatric Association 2000) lists 18 symptoms of ADHD (Table 1). DSM-IV criteria specify two dimensions of symptoms, namely

Table 1 DSM-IV-TR diagnostic criteria for ADHD

A. Either (1) or (2):
Inattention
1. Six (or more) of the following symptoms of *inattention* have persisted for at least 6 months to a degree that is maladaptive and inconsistent with developmental level:
    (a) Often fails to give close attention to details or makes careless mistakes in schoolwork, work, or other activities
    (b) Often has difficulty sustaining attention in tasks or play activities
    (c) Often does not seem to listen when spoken to directly
    (d) Often does not follow through on instructions and fails to finish schoolwork, chores, or duties in the workplace (not due to oppositional behavior or failure to understand instructions)
    (e) Often has difficulty organizing tasks and activities
    (f) Often avoids, dislikes, or is reluctant to engage in tasks that require sustained mental effort (such as schoolwork or homework)
    (g) Often loses things necessary for tasks or activities (e.g., toys, school assignments, pencils, books, or tools)
    (h) Is often easily distracted by extraneous stimuli
    (i) Is often forgetful in daily activities
2. Six (or more) of the following symptoms of *hyperactivity–impulsivity* have persisted for at least 6 months to a degree that is maladaptive and inconsistent with developmental level:
Hyperactivity
    (a) Often fidgets with hands or feet or squirms in seat
    (b) Often leaves seat in classroom or in other situations in which remaining seated is expected
    (c) Often runs about or climbs excessively in situations in which it is inappropriate (in adolescents or adults, may be limited to subjective feelings of restlessness)
    (d) Often has difficulty playing or engaging in leisure activities quietly
    (e) Is often "on the go" or often acts as if "driven by a motor"
    (f) Often talks excessively
Impulsivity
    (g) Often blurts out answers before questions have been completed
    (h) Often has difficulty awaiting turn
    (i) Often interrupts or intrudes on others (e.g., butts into conversations or games)
B. Some hyperactive–impulsive or inattentive symptoms that caused impairment were present before age 7 years
C. Some impairment from the symptoms is present in two or more settings [e.g., at school (or work) and at home]
D. There must be clear evidence of clinically significant impairment in social, academic, or occupational functioning
E. The symptoms do not occur exclusively during the course of a pervasive developmental disorder, schizophrenia, or other psychotic disorder and are not better accounted for by another mental disorder (e.g., mood disorder, anxiety disorder, dissociative disorder, or a personality disorder)

inattention and hyperactivity/impulsivity, which define three subtypes of the disorder: predominantly hyperactive–impulsive type, predominantly inattentive type, and combined type. The diagnosis is based on the presence of at least six symptoms of inattention and/or six symptoms of hyperactivity/impulsivity (American Psychiatric Association 2000). The International Classification of Diseases, tenth revision (ICD-10) (WHO 1993) uses a different nomenclature – hyperkinetic disorder – and requires the presence of symptoms in both dimensions (at least six of inattention, three of hyperactivity, and one of impulsivity), but lists similar operational criteria for the disorder.

The full diagnosis of ADHD currently requires four additional criteria: symptoms causing impairment must be present before the age of 7 years; they need to be observed in two or more settings (i.e., home, school, workplace); they must cause significant impairment in social, academic, or occupational functioning; they do not occur exclusively during the course of autism, schizophrenia, or other psychotic disorder; and they are not better accounted for by another mental disorder (e.g., mood disorder, anxiety disorder, dissociative disorder, or personality disorder) (American Psychiatric Association 2000).

Thus, ADHD is diagnosed by clinical history, using diagnostic criteria to check for the presence of characteristic symptoms, lasting for at least 6 months and observed in two or more settings. Evaluation for ADHD should consist of clinical interviews with the patient and parents/family members, obtaining information about the patient's school/work functioning, evaluation for comorbid psychiatric disorders, and review of medical, social, and family histories (Pliszka et al. 2007). The clinician should make a detailed interview to check for the presence of each of the 18 ADHD symptoms listed in DSM-IV. For each symptom, emphasis should be given to determine its presence, duration, severity, and frequency. In addition to having the required number of symptoms, each occurring more often than not, they must show a stable, chronic course (oscillating symptoms with asymptomatic periods are not characteristic of ADHD), and with onset of symptoms during childhood. The level of impairment and the settings in which it occurs must also be assessed. DSM-IV requires impairment in at least two settings (home, school, or job) to meet the criteria for the disorder, but the clinical consensus is that severe impairment in one setting warrants treatment (Parens and Johnston 2009).

Attention deficits are particularly observed when individuals with ADHD are assigned boring, tedious, or repetitive tasks, especially when lacking intrinsic appeal. Also, inattentive symptoms increase while working on tasks that exceed the individual's cognitive processing abilities (e.g., that require fast processing speed or place high demands on working memory) (Kofler et al. 2010). Poor sustained attention or sustained effort to complete tasks often results in difficulty following through on instructions and organizing tasks; being easily distracted; shifting from one unfinished activity to another; and failing to give close attention to details or avoiding careless mistakes (American Psychiatric Association 2000). Hyperactivity can be observed in behaviors such as fidgeting with hands or feet, squirming in seat, leaving seat in situations in which remaining seated is expected, and acting as if driven by a motor. Notably, some of the currently listed diagnostic criteria for hyperactivity seem to apply mainly for children (such as running about or climbing excessively, for example). In adults with ADHD, the gross motor overactivity seen in children is believed to switch toward a greater difficulty with fidgeting and a more subjective sense of restlessness (Asherson et al. 2007).

It should be underscored that inattention, hyperactivity, and impulsivity as isolated symptoms may result from problems in relationships with parents and/or colleagues and friends, inappropriate educational systems, or inadequate work environments. Thus, symptoms of hyperactivity or impulsivity with no effect on the individual's daily activities may result from different functioning or temperament styles other than a psychiatric disorder. Likewise, symptoms that occur only at home, or only at school or work, should warn clinicians of the possibility that inattention, hyperactivity, or impulsivity may reflect a chaotic marital or family situation or negative school experiences. Therefore, for the diagnosis of ADHD it is always necessary to contextualize the symptoms in the patient's history (Rohde and Halpern 2004).

In the assessment of children, because many patients with ADHD show academic impairment, it is important to ask specific questions about school performance. Clinicians should verify the patient's academic or intellectual progress, keeping in mind the possibility of a concurrent learning disorder. Although parents are thought to be the best information sources for the diagnosis of ADHD, because children and adolescents tend to underreport symptoms, teacher report is recommended as a complement to parent information in the assessment (Frazier and Youngstrom 2006). Parent reports about symptoms at school moderately correlate with teacher ratings (Sayal and Goodman 2009) and are influenced by the child's behavior at both home and school as well as parental mental health (Sayal and Taylor 2005). Several difficulties may arise in combining data across informants and measures. It is important to note that we still lack adequate models regarding how discrepant information across informants is best integrated into the diagnostic decision-making process (Sowerby and Tripp 2009). Rating scales, symptom inventories, and neuropsychological batteries – although not diagnostic – may help to provide additional evidence for both the disorder and comorbid conditions (e.g., Conners Rating Scales; Swanson, Nolan, and Pelham version IV scale, SNAP-IV; Child Behavior Checklist; Vanderbilt ADHD Diagnostic Parent and Teacher Scales).

3 Controversial Issues in the Diagnosis

Over the last decade, an enormous amount of new data has been published suggesting that ADHD is best conceptualized as a neurobiological disorder (Kieling et al. 2008). The increasingly sophisticated tools of modern research in genetics, molecular biology, neuropharmacology, and neuroimaging have significantly advanced our understanding of the biological underpinnings of ADHD. Although much debate exists as to whether it would be worth including any molecular genetic, neuropsychologic, neurophysiologic, or neuroimaging marker in future diagnostic classification systems for ADHD, currently no biological marker is deemed ready due to low sensitivity and specificity (Rohde 2008). Thus, it is important to reiterate that the diagnostic process for ADHD is essentially clinical.

Another key point for future diagnostic systems involves the age-of-onset criterion. The increased recognition of ADHD as a life-long disorder is now challenging the onset criterion of before 7 years, as problems with retrospective recall make the age-of-onset criterion particularly problematic for the adult population. Prospective and retrospective data indicate that only half of adults assessed for ADHD symptoms report their presence by the age of 7 (Kieling et al. 2010). However, it should be emphasized that, although the specific age requirement is under debate, it remains unquestionable that the diagnosis of adult ADHD requires evidence of childhood onset of several symptoms.

The validity of the subtypes has also recently been put to test. In a comprehensive literature review and meta-analysis of 431 studies, strong evidence was found for the symptom dimensions (hyperactivity/impulsivity and inattention), but findings in regard to subtypes were more nuanced (Willcutt et al. submitted). The authors reported that the hyperactive subtype is markedly understudied and only validated in young children. The combined type is more strongly associated with externalizing disorders and aspects of social impairment, whereas the inattentive subtype is more strongly associated with shy and passive social behavior. Moreover, longitudinal data indicate that all three subtypes are highly unstable, with some individuals shifting between them from year to year.

Diagnosis of ADHD in females is also an issue of debate. Studies in children consistently suggest that the prevalence of ADHD is higher in boys than in girls, with male/female ratios varying from 3:1 to 9:1, depending on the origin of the sample (Polanczyk and Rohde 2007). In a systematic review and meta-regression, the pooled ADHD prevalence was 2.45 times higher for boys than for girls (only non-referred samples were included) (Polanczyk et al. 2007), and recent evidence suggests that the male/female ratio remains relatively stable across the lifetime (Ramtekkar et al. 2010).

The lower prevalence of ADHD in females may be the result of higher liability toward ADHD in boys than girls, the expression of different genes for ADHD in boys versus girls or, alternatively, more noticeable symptoms in males, who tend to have higher levels of disruptive behaviors. Taken together, these issues suggest a possible barrier to diagnosis and treatment of ADHD in girls, either because the condition goes unrecognized or because it is not seen as severe enough to warrant intervention. Most studies suggest that the clinical symptoms of ADHD vary little as a function of gender, except for the intensity of symptoms, with girls being less symptomatic than boys, particularly regarding hyperactivity and externalizing behaviors, although impulsivity levels appear comparable (Graetz et al. 2005; Newcorn et al. 2001).

In the following sections, we explore in detail the clinical presentation and correlates of ADHD across the lifespan. We discuss the most prevalent patterns of comorbidity with ADHD in children, adolescents, and adults; examine the correlates and longitudinal predictors of outcomes; and conclude by summarizing research findings and discussing the relevance of a diagnosis of ADHD.

4 ADHD in Preschool Children

Although the majority of ADHD referrals are for school-age children, early evaluation is critical. Behavioral, social, familial, and academic difficulties have been reported in preschool children with ADHD. Reports of parents and clinicians support the existence of a relation between preschool behavioral difficulty and later adjustment problems in school-age children (Spira and Fischel 2005).

Highly deviant behavior at a very young age may be observed in children with ADHD, expressed by poor concentration, overactivity, restlessness, disobedience, poor social relations, and antisocial behavior. Aberrant maternal–child interactions are seen in preschool children with ADHD and may contribute to the persistence of these disruptive behaviors (DuPaul et al. 2001). In addition to core symptoms of ADHD, multiple developmental delays, including sensory, motor, language, and intellectual functioning deficits, are often found in these children (Yochman et al. 2006). Thus, even as early as the preschool period, young children with ADHD are multiply handicapped and often exhibit sensorimotor difficulties, leading to higher rates of utilization of remedial services, such as speech and language, occupational, and physical therapy, and they are more likely to be placed in special education programs (Marks et al. 2009).

Preschool children with ADHD, despite their significant younger age, show striking similarities to school-age ADHD children, both in terms of impairment and psychiatric comorbidity. A study by Wilens et al. (2002) showed that these children already had prominent school, social, and overall dysfunction at the same level of the older, school-age group. Mood disorders occurred in approximately half of both age groups with ADHD, with a trend to higher rates of comorbid bipolar disorders in the preschool children (Wilens et al. 2002). In a large study with 303 preschoolers diagnosed with ADHD, 69.6% experienced comorbid disorders, with ODD, communication disorders, and anxiety disorders being the most common (Posner et al. 2007).

It has been hypothesized that difficult mother–child interactions and family stress can interfere with the acquisition of appropriate means of coping with the initially problematic behavior of these children, thereby preventing the decline of symptoms (Chi and Hinshaw 2002). Parents of 3–5-year-old children with ADHD experience greater stress, are more likely to display negative behavior toward their children, and cope less adaptively than parents of non-ADHD children (DuPaul et al. 2001). Moreover, mothers of preschool-age children with ADHD were found to report more symptoms of depression and a reduced sense of parental competence (Byrne et al. 1998).

In the 2–5-year-old range, the core symptoms of ADHD are often seen in daily activities; the level of impairment is often difficult to assess, and behaviors are less likely to be considered developmentally inappropriate (Connor 2002; Greenhill et al. 2008). Research shows that activity level, as well as squirming, is not a reliable phenotype to distinguish between preschoolers with and without ADHD (DuPaul et al. 2001). Similarly, up to 40% of children exhibit inattention problems by the age of 4 years, which are of concern to their parents and preschool teachers (Greenhill et al. 2008), without necessarily having ADHD.

A pattern of inability to cooperate and be productive in the preschool setting is often described in preschoolers with ADHD. Young children with ADHD exhibit more noncompliant and inappropriate behavior, particularly during task situations, and are less socially skilled than their normally developing classmates (McGoey et al. 2002). They typically demonstrate difficulty modulating their behavior in response to different situational demands and function best in highly structured environments (Fabiano et al. 2009).

The clinical diagnosis of ADHD in preschool children is challenging, and the validity of the diagnosis in the preschool years has often been questioned (Greenhill et al. 2008; Posner et al. 2007). The absence of developmentally appropriate assessment protocols has contributed to an understandable reluctance to diagnose ADHD in preschoolers. The combined type and the hyperactive/impulsive type are the most common but, as pointed earlier, children rarely remain in the hyperactive classification over time; rather, they most often shift to combined type in later years (Lahey et al. 2005). Moreover, the changing pattern of ADHD across ages means that, even among those children whose symptoms are frequent and severe enough to warrant a diagnosis of ADHD in the preschool years, only 48% will have this same diagnosis by later childhood or adolescence (Connor 2002). Nevertheless, as hyperactive–impulsive symptoms are quite common in preschoolers, for the purpose of early identification, inattentive behavior at an early age may be indicative of psychopathology (Smidts and Oosterlaan 2007).

5 ADHD in School-Age Children

Most school-age children with ADHD have significant difficulties with academic performance and/or peer relationships (American Psychiatric Association 2000). These academic, social, and behavioral problems tend to vary according to subtype, with relatively independent areas of impairment for each diagnostic group. Children in the combined subtype are significantly more impaired than the other two subtypes on measures, such as global impairment, overall social functioning, and tendency to be disliked by peers, whereas inattentive and hyperactive children do not differ on measures of these domains (Willcutt et al. submitted). Both combined type and inattentive individuals with ADHD are significantly more impaired than those presenting with only hyperactivity on measures of academic functioning and are more likely to be ignored by peers (Gaub and Carlson 1997).

Upon school entry, children with ADHD are likely to lag behind in basic math concepts, pre-reading skills, and fine motor abilities (DuPaul et al. 2001).

Approximately 30% of children with ADHD have a learning disability, and a large majority underachieve academically (DuPaul and Stoner 2003). These academic difficulties cut across academic subject domains, including both reading and math, and occur despite average or above average IQ estimates (Biederman et al. 1996). Moreover, children with ADHD are more likely to be expelled, suspended, or repeat a grade, compared with controls (Loe and Feldman 2007). They also have a greater chance of obtaining a lower-level diploma and failing to graduate from secondary school (Galera et al. 2009).

Several studies have shown that the poor academic outcomes seen in these children are the result of ADHD itself and not the associated comorbid disorders (Daley and Birchwood 2010): both neuropsychological and performance deficits are thought to underlie the low scholastic outcome associated with the disorder. Thus, neuropsychological impairments typically seen in ADHD patients, such as poor motor inhibition and planning, may contribute to early intellectual problems, which may subsequently be exacerbated by disruptive behavior or other comorbid conditions (Hinshaw 2003). Academic difficulties may also be secondary to inherent problems these children have with engagement in classroom activities, failure to follow through on instructions, and to complete assigned tasks and tests (DuPaul and Stoner 2003).

In addition to the academic problems, school-age children with ADHD, particularly those with the combined or hyperactive–impulsive subtypes, typically show behavioral issues that interfere with their relationships at school and at home. Social impairments and difficulty with peer relations are often described as difficulty in attaining and keeping friends, resolving conflicts, and managing anger and frustration (Nijmeijer et al. 2008). Moreover, children with ADHD are often rejected by peers even after only brief interactions; have fewer friends than their non-ADHD peers; tend to choose other ADHD youths as playmates; and have difficulty regulating their emotions and sustaining associative play (Bagwell et al. 2001). Dysfunctional behavior seen in ADHD children also includes excessive talkativeness, noisiness, noncompliance, and physical aggression toward peers.

Finally, as a result of impaired social relationships and school performance, the development of low self-esteem is of particular concern in this age group. The persistent inability to concentrate, multiple failures, disapproval, and demoralization may all contribute to self-esteem issues. Furthermore, poor self-esteem is thought to be a mediating factor of future adverse outcomes, such as depression, deviant peer choices, and substance abuse in children with ADHD (Bussing et al. 2000).

6 ADHD in Adolescents

The majority of children diagnosed with ADHD while in elementary school continue to have significant manifestations of the disorder throughout adolescence. Persistence rates vary in different studies, according to how persistence is defined (Faraone et al. 2006). Yet, regardless of definition, ADHD seems to lessen with age, as studies show that salient features of ADHD change during adolescence. Hyperactivity, especially, although still present, becomes much less visible during this age period (Ramtekkar et al. 2010).

As the cognitive demands increase during middle and high school, academic problems that might have been less problematic, or effectively managed during elementary school, may become a significant issue in adolescence (Bussing et al. 2010). Additionally, peer problems, such as fewer close friendships and greater peer rejection, become more evident as social interactions assume greater relevance during teenage years (Bagwell et al. 2001).

Adolescents with ADHD are often emotionally immature and tend to feel more comfortable interacting with younger children. Affect is poorly regulated, and they often display both negative and positive responses in excess for the situation, becoming easily frustrated, with sudden outbursts of anger and irritability (Barkley and Fischer 2010). In extreme, these symptoms may be a manifestation of comorbid ODD and CD.

Approximately 50% of youngsters with ADHD also meet diagnostic criteria for ODD and/or CD, adding significant impairment to the clinical picture (Kutcher et al. 2004). ODD is defined as a pattern of negativistic, hostile, and defiant behaviors, with persistent and developmentally inappropriate levels of irritability and oppositionality (American Psychiatric Association 2000). The combination of ODD symptoms with those of ADHD is associated with greater than normal conflicts, anger, poor communication, unreasonable beliefs, and negative interactive styles often seen in ADHD adolescents (Edwards et al. 2001). CD is characterized by persistent antisocial behaviors that include acts of aggression, destruction of property, deceitfulness, theft, and rule violation. CD is more common among adolescents and more prevalent among boys than girls (Biederman et al. 2008). Childhood ODD is significantly associated with adolescent CD (Whitinger et al. 2007) and roughly two-thirds of all adolescents with CD also have ADHD (Kutcher et al. 2004).

Both comorbidities predict the persistence of ADHD into adulthood and mediate the risk for the development of other problems, such as substance use, antisocial personality disorder, and major depression (Biederman et al. 2008; Daviss 2008; Fischer and Barkley 2003). Several studies have demonstrated that adolescents with ADHD are at risk for early initiation of cigarette smoking (Huizink et al. 2009) and illicit drug use (Langley et al. 2010). ADHD has also been associated with higher levels of alcohol, tobacco, and illicit drug use compared to controls (Molina and Pelham 2003; Szobot and Bukstein 2008). There is also a greater likelihood to transition from an alcohol use disorder to a drug use disorder and to continue to abuse substances (Biederman et al. 1998; Wilson and Levin 2001), particularly when there is an association between conduct and attentional problems (Fergusson et al. 2007).

Although the use of stimulants in young children has generated considerable controversy, an extensive review found no evidence that stimulant treatment of children with ADHD leads to an increased risk for substance experimentation, use, dependence, or abuse by adulthood (Barkley et al. 2003). Moreover, there is evidence of an age-related benefit for stimulant treatment – the later the treatment, the greater the chances of developing substance use disorder (Mannuzza et al. 2008).

7 ADHD in Adults

Longitudinal studies have shown that ADHD persists into adulthood in approximately 65% of cases (Barkley et al. 2002; Faraone et al. 2006), with differences in remission rates being attributed both to the different definitions of ADHD and to the natural history of the disorder. Although they may no longer meet the full symptom criteria, young adults with a history of lifetime ADHD maintain higher levels of ADHD symptoms compared with the general population (Ramtekkar et al. 2010). Childhood ADHD severity and childhood treatment are thought to significantly predict persistence (Kessler et al. 2006).

ADHD remains a highly comorbid disorder in adults. The National Comorbidity Survey Replication found that 38.3% of adults with ADHD had a comorbid mood disorder; 47.1% had a comorbid anxiety disorder; 15.2% had a substance use disorder; and 19.6% had other impulse-control disorders (Kessler et al. 2006). In fact, impulsive emotions are strikingly evident in ADHD, including impatience, low frustration tolerance, hot-temperedness, quickness to anger or volatility, irritability, and a general propensity for being easily emotionally excitable. In this sense, Barkley has argued that emotional impulsiveness should be explicitly included in the conceptualization of ADHD, as it explains the high comorbidity of the ADHD with other psychiatric disorders, particularly with ODD. Furthermore, it may better account for the various social, occupational, and familial impairments that are not as easily explained by the two traditional dimensions of inattention and hyperactivity/impulsivity (Barkley and Fischer 2010).

Overall, symptoms of ADHD in adults tend to be more heterogeneous and subtle than in children or adolescents. In adults, the hyperactivity observed as children may translate into constant activity, overscheduling, or choosing a busy job. Similarly, impulsivity may manifest through problems such as ending relationships prematurely, quitting jobs, being unwilling to wait in line, or losing temper with a child. Inattention in adults may be seen in dialogical situations, in tasks that require organization and sustained attention over time, as well as in difficulties with time management and memory. Impairments of executive functioning clinically translated as procrastination, among others, are extremely frequent in adults with the disorder. Other adult variants of these symptoms include being a workaholic, feeling uncomfortable sitting through meetings, and driving impulsively (Weiss and Weiss 2004).

Not surprisingly, several negative driving outcomes have been reported for young adults with ADHD, including more traffic citations, citations for speeding, involvement in crashes, and license suspensions (Barkley and Cox 2007; Kieling et al. in press). In terms of social functioning, investigations of the relationship difficulties of adults with ADHD show that they have significantly higher rates of marital dissatisfaction and discord, higher divorce rates, and parenting difficulties (Barkley and Fischer 2010; Eakin et al. 2004). Work-related problems also negatively affect the long-term outcome of the disorder, as adults with ADHD are found more often to perform poorly, quit, or to have been fired from jobs; have a history of poorer educational performance; achieve significantly less formal training; and have lower-ranking occupational positions than controls (Barkley and Murphy 2010b; Stein 2008).

8 Future Perspectives

Several issues clearly need to be addressed through more extensive investigation to improve characterization and, ultimately, the validity of the ADHD diagnosis in future classifications systems. Some of the aspects that deserve special attention are refining diagnostic criteria for ADHD in adults, possibly with the inclusion of new criteria, such as impulsivity items; making DSM and ICD classification for ADHD more alike; proposing a way to integrate different information sources (teachers, parents, family members, friends, and coworkers); and excluding the age of onset of impairment criteria, considered one of the most fragile DSM criterion for ADHD, with low validity and reliability (Rohde 2008).

9 Conclusion

Work to date has established ADHD as a valid disorder in children and adults. Research clearly shows that childhood ADHD tends to continue in adulthood, causing significant impairment in educational, social, and occupational contexts throughout life. As it becomes increasingly recognized that this is a fairly common, impairing, but treatable disorder, it is important to emphasize that diagnosis and subsequent treatment may not only provide significant relief to patients and families, but also reduce social costs associated with unemployment, criminality, smoking, substance abuse, and driving accidents.

References

American Psychiatric Association (2000) Diagnostic and statistical manual of mental disorders (4th edn, Text Revision) (DSM-IV-TR). American Psychiatric Association, Washington, DC
Asherson P, Chen W, Craddock B, Taylor E (2007) Adult attention-deficit hyperactivity disorder: recognition and treatment in general adult psychiatry. Br J Psychiatry 190:4–5

Bagwell CL, Molina BSG, Pelham WE, Hoza B (2001) Attention-deficit hyperactivity disorder and problems in peer relations: Predictions from childhood to adolescence. J Am Acad Child Adolesc Psychiatry 40:1285–1292

Barkley RA, Cox D (2007) A review of driving risks and impairments associated with attention-deficit/hyperactivity disorder and the effects of stimulant medication on driving performance. J Safety Res 38:113–128

Barkley RA, Fischer M (2010) The unique contribution of emotional impulsiveness to impairment in major life activities in hyperactive children as adults. J Am Acad Child Adolesc Psychiatry 49:503–513

Barkley RA, Murphy KR (2010a) Deficient emotional self-regulation in adults with attention-deficit/hyperactivity disorder (ADHD): the relative contributions of emotional impulsiveness and ADHD symptoms to adaptive impairments in major life activities. Journal of ADHD and Related Disorders 1:5–28

Barkley RA, Murphy KR (2010b) Impairment in occupational functioning and adult ADHD: the predictive utility of executive function (EF) ratings versus EF tests. Arch Clin Neuropsychol 25:157–173

Barkley RA, Fischer M, Smallish L, Fletcher K (2002) The persistence of attention-deficit/hyperactivity disorder into young adulthood as a function of reporting source and definition of disorder. J Abnorm Psychol 111:279–289

Barkley RA, Fischer M, Smallish L, Fletcher K (2003) Does the treatment of attention-deficit/hyperactivity disorder with stimulants contribute to drug use/abuse? A 13-year prospective study. Pediatrics 111:97–109

Biederman J, Faraone S, Milberger S, Guite J, Mick E, Chen L, Mennin D, Marrs A, Oullette C, Moore P, Spencer T, Norman D, Wilens T, Kraus I, Perrin J (1996) A prospective 4-year follow-up study of attention-deficit hyperactivity and related disorders. Arch Gen Psychiatry 53:437–446

Biederman J, Wilens TE, Mick E, Faraone SV, Spencer T (1998) Does attention-deficit hyperactivity disorder impact the developmental course of drug and alcohol abuse and dependence? Biol Psychiatry 44:269–273

Biederman J, Petty CR, Dolan C, Hughes S, Mick E, Monuteaux MC, Faraone SV (2008) The long-term longitudinal course of oppositional defiant disorder and conduct disorder in ADHD boys: findings from a controlled 10-year prospective longitudinal follow-up study. Psychol Med 38:1027–1036

Biederman J, Petty CR, Evans M, Small J, Faraone SV (2010) How persistent is ADHD? A controlled 10-year follow-up study of boys with ADHD. Psychiatry Res 177:299–304

Bussing R, Zima BT, Perwien AR (2000) Self-esteem in special education children with ADHD: relationship to disorder characteristics and medication use. J Am Acad Child Adolesc Psychiatry 39:1260–1269

Bussing R, Mason DM, Bell L, Porter P, Garvan C (2010) Adolescent outcomes of childhood attention-deficit/hyperactivity disorder in a diverse community sample. J Am Acad Child Adolesc Psychiatry 49:595–605

Byrne JM, Bawden HN, DeWolfe NA, Beattie TL (1998) Clinical assessment of psychopharmacological treatment of preschoolers with ADHD. J Clin Exp Neuropsychol 20:613–627

Chi TC, Hinshaw SP (2002) Mother-child relationships of children with ADHD: the role of maternal depressive symptoms and depression-related distortions. J Abnorm Child Psychol 30:387–400

Connor DF (2002) Preschool attention deficit hyperactivity disorder: a review of prevalence, diagnosis, neurobiology, and stimulant treatment. J Dev Behav Pediatr 23:S1–S9

Daley D, Birchwood J (2010) ADHD and academic performance: why does ADHD impact on academic performance and what can be done to support ADHD children in the classroom? Child Care Health Dev 36:455–464

Daviss WB (2008) A review of co-morbid depression in pediatric ADHD: etiologies, phenomenology, and treatment. J Child Adolesc Psychopharmacol 18:565–571

DuPaul GJ, Stoner GD (2003) ADHD in the schools: assessment and intervention strategies, 2nd edn. Guilford Press, New York

DuPaul GJ, McGoey KE, Eckert TL, VanBrakle J (2001) Preschool children with attention-deficit/hyperactivity disorder: impairments in behavioral, social, and school functioning. J Am Acad Child Adolesc Psychiatry 40:508–515

Eakin L, Minde K, Hechtman L, Ochs E, Krane E, Bouffard R, Greenfield B, Looper K (2004) The marital and family functioning of adults with ADHD and their spouses. J Atten Disord 8:1–10

Edwards G, Barkley RA, Laneri M, Fletcher K, Metevia L (2001) Parent-adolescent conflict in teenagers with ADHD and ODD. J Abnorm Child Psychol 29:557–572

Fabiano GA, Pelham WE, Coles EK, Gnagy EM, Chronis-Tuscano A, O'Connor BC (2009) A meta-analysis of behavioral treatments for attention-deficit/hyperactivity disorder. Clin Psychol Rev 29:129–140

Faraone SV, Biederman J, Mick E (2006) The age-dependent decline of attention deficit hyperactivity disorder: a meta-analysis of follow-up studies. Psychol Med 36:159–165

Fergusson DM, Horwood LJ, Ridder EM (2007) Conduct and attentional problems in childhood and adolescence and later substance use, abuse and dependence: Results of a 25-year longitudinal study. Drug Alcohol Depend 88:S14–S26

Fischer M, Barkley RA (2003) Young adult follow-up of hyperactive children: self-reported psychiatric disorders, comorbidity, and the role of childhood conduct problems and teen CD (vol 30, pg 463, 2002). J Abnorm Child Psychol 31:563–563

Fischer M, Barkley RA, Smallish L, Fletcher K (2007) Hyperactive children as young adults: driving abilities, safe driving behavior, and adverse driving outcomes. Accid Anal Prev 39:94–105

Frazier TW, Youngstrom EA (2006) Evidence-based assessment of attention-deficit/hyperactivity disorder: using multiple sources of information. J Am Acad Child Adolesc Psychiatry 45:614–620

Galera C, Melchior M, Chastang JF, Bouvard MP, Fombonne E (2009) Childhood and adolescent hyperactivity-inattention symptoms and academic achievement 8 years later: the GAZEL Youth study. Psychol Med 39:1895–1906

Gaub M, Carlson CL (1997) Behavioral characteristics of DSM-IV ADHD subtypes in a school-based population. J Abnorm Child Psychol 25:103–111

Goldman LS, Genel M, Bezman RJ, Slanetz PJ, Amer Med A (1998) Diagnosis and treatment of attention-deficit/hyperactivity disorder in children and adolescents. JAMA 279:1100–1107

Graetz BW, Sawyer MG, Baghurst P (2005) Gender differences among children with DSM-IV ADHD in Australia. J Am Acad Child Adolesc Psychiatry 44:159–168

Greenhill LL, Posner K, Vaughan BS, Kratochvil CJ (2008) Attention deficit hyperactivity disorder in preschool children. Child Adolesc Psychiatr Clin N Am 17:347

Hinshaw SP (2003) Impulsivity, emotion regulation, and developmental psychopathology: specificity versus generality of linkages. Roots of Mental Illness in Children 1008:149–159

Huizink AC, van Lier PAC, Crijnen AAM (2009) Attention deficit hyperactivity disorder symptoms mediate early-onset smoking. Eur Addict Res 15:1–9

Kessler RC, Adler L, Barkley R, Biederman J, Conners CK, Demler O, Faraone SV, Greenhill LL, Howes MJ, Secnik K, Spencer T, Ustun TB, Walters EE, Zaslavsky AM (2006) The prevalence and correlates of adult ADHD in the United States: results from the National Comorbidity Survey Replication. Am J Psychiatry 163:716–723

Kieling C, Goncalves RRF, Tannock R, Castellanos FX (2008) Neurobiology of attention deficit hyperactivity disorder. Child Adolesc Psychiatr Clin N Am 17:285

Kieling C, Kieling RR, Rohde LA, Frick PJ, Moffitt T, Nigg JT, Tannock R, Castellanos FX (2010) The age at onset of attention deficit hyperactivity disorder. Am J Psychiatry 167:14–16

Kieling RR, Szobot CM, Matte B, Coelho R, Kieling C, Pechansky F, Rohde LA (2011) Mental disorders and delivery motorcycle drivers (motoboys): a dangerous association. Eur Psychiatry (Paris) 26:23–27

Kofler MJ, Rapport MD, Bolden J, Sarver DE, Raiker JS (2010) ADHD and working memory: the impact of central executive deficits and exceeding storage/rehearsal capacity on observed inattentive behavior. J Abnorm Child Psychol 38:149–161

Kutcher S, Aman M, Brooks SJ, Buitelaar J, van Daalen E, Fegert J, Findling RL, Fisman S, Greenhill LL, Huss M, Kusumakar V, Pine D, Taylor E, Tyano S (2004) International consensus statement on attention-deficit/hyperactivity disorder (ADHD) and disruptive behaviour disorders (DBDs): clinical implications and treatment practice suggestions. Eur Neuropsychopharmacol 14:11–28

Lahey BB, Pelham WE, Loney J, Lee SS, Willcutt E (2005) Instability of the DSM-IV subtypes of ADHD from preschool through elementary school. Arch Gen Psychiatry 62:896–902

Langley K, Fowler T, Ford T, Thapar AK, van den Bree M, Harold G, Owen MJ, O'Donovan MC, Thapar A (2010) Adolescent clinical outcomes for young people with attention-deficit hyperactivity disorder. Br J Psychiatry 196:235–240

Loe IM, Feldman HM (2007) Academic and educational outcomes of children with ADHD. J Pediatr Psychology 32:643–654

Mannuzza S, Klein RG, Truong NL, Moulton JL, Roizen ER, Howell KH (2008) Age of methylphenidate treatment initiation in children with ADHD and later substance abuse: prospective follow-up into adulthood. Am J Psychiatry 165:604–609

Marks DJ, Mlodnicka A, Bernstein M, Chacko A, Rose S, Halperin JM (2009) Profiles of service utilization and the resultant economic impact in preschoolers with attention deficit/hyperactivity disorder. J Pediatr Psychology 34:681–689

McGoey KE, Eckert TL, DuPaul GJ (2002) Early intervention for preschool-age children with ADHD: a literature review. J Emot Behav Disord 10:14–28

Molina BSG, Pelham WE (2003) Childhood predictors of adolescent substance use in a longitudinal study of children with ADHD. J Abnorm Psychol 112:497–507

Newcorn JH, Halperin JM, Jensen PS, Abikoff HB, Arnold LE, Cantwell DP, Conners CK, Elliott GR, Epstein JN, Greenhill LL, Hechtman L, Hinshaw SP, Hoza B, Kraemer HC, Pelham WE, Severe JB, Swanson JM, Wells KC, Wigal T, Vitiello B (2001) Symptom profiles in children with ADHD: effects of comorbidity and gender. J Am Acad Child Adolesc Psychiatry 40:137–146

Nijmeijer JS, Minderaa RB, Buitelaar JK, Mulligan A, Hartman CA, Hoekstra PJ (2008) Attention-deficit/hyperactivity disorder and social dysfunctioning. Clin Psychol Rev 28:692–708

Parens E, Johnston J (2009) Facts, values, and attention-deficit hyperactivity disorder (ADHD): an update on the controversies. Child Adolesc Psychiatry Ment Health 3:1

Pliszka S, Bernet W, Bukstein O, Walter HJ, Arnold V, Beitchman J, Benson RS, Chrisman A, Farchione T, Hamilton J, Keable H, Kinlan J, McClellan J, Rue D, Schoettle U, Shaw JA, Stock S, Ptakowski KK, Medicus J (2007) Practice parameter for the assessment and treatment of children and adolescents with attention-deficit/hyperactivity disorder. J Am Acad Child Adolesc Psychiatry 46:894–921

Polanczyk G, Rohde LA (2007) Epidemiology of attention-deficit/hyperactivity disorder across the lifespan. Curr Opin Psychiatry 20:386–392

Polanczyk G, de Lima MS, Horta BL, Biederman J, Rohde LA (2007) The worldwide prevalence of ADHD: a systematic review and metaregression analysis. Am J Psychiatry 164:942–948

Posner K, Melvin GA, Murray DW, Gugga SS, Fisher P, Skrobala A, Cunningham C, Vitiello B, Abikoff HB, Ghuman JK, Kollins S, Wigal SB, Wigal T, McCracken JT, McGough JJ, Kastelic E, Boorady R, Davies M, Chuang SZ, Swanson JM, Riddle MA, Greenhill LL (2007) Clinical presentation of attention-deficit/hyperactivity disorder in preschool children: The preschoolers with attention-deficit/hyperactivity treatment study (PATS). J Child Adolesc Psychopharmacol 17:547–562

Ramtekkar UP, Reiersen AM, Todorov AA, Todd RD (2010) Sex and age differences in attention-deficit/hyperactivity disorder symptoms and diagnoses: implications for DSM-V and ICD-11. J Am Acad Child Adolesc Psychiatry 49:217–228

Rohde LA (2008) Is there a need to reformulate attention deficit hyperactivity disorder criteria in future nosologic classifications? Child Adolesc Psychiatr Clin N Am 17:405

Rohde LA, Halpern R (2004) Recent advances on attention deficit/hyperactivity disorder. J Pediatr (Rio J) 80:S61–S70

Sayal K, Goodman R (2009) Do parental reports of child hyperkinetic disorder symptoms at school predict teacher ratings? Eur Child Adolesc Psychiatry 18:336–344

Sayal K, Taylor E (2005) Parent ratings of school behaviour in children at risk of attention deficit/ hyperactivity disorder. Acta Psychiatr Scand 111:460–465

Simon V, Czobor P, Balint S, Meszaros A, Bitter I (2009) Prevalence and correlates of adult attention-deficit hyperactivity disorder: meta-analysis. Br J Psychiatry 194:204–211

Smidts DP, Oosterlaan J (2007) How common are symptoms of ADHD in typically developing preschoolers? a study on prevalence rates and prenatal/demographic risk factors. Cortex 43:710–717

Sowerby P, Tripp G (2009) Evidence-based assessment of attention-deficit/hyperactivity disorder (ADHD). In: Matson JL, Andrasik F, Matson ML (eds) Assessing childhood psychopathology and developmental disabilities. Springer, New York, pp 209–239

Spira EG, Fischel JE (2005) The impact of preschool inattention, hyperactivity, and impulsivity on social and academic development: a review. J Child Psychol Psychiatry 46:755–773

Stein MA (2008) Impairment associated with adult ADHD. CNS Spectr 13:9–11

Szobot CM, Bukstein O (2008) Attention deficit hyperactivity disorder and substance use disorders. Child Adolesc Psychiatr Clin N Am 17:309

Weiss MD, Weiss JR (2004) A guide to the treatment of adults with ADHD. J Clin Psychiatry 65:27–37

Whitinger NS, Langley K, Fowler TA, Thomas HV, Thapar A (2007) Clinical precursors of adolescent conduct disorder in children with attention-deficit/hyperactivity disorder. J Am Acad Child Adolesc Psychiatry 46:179–187

WHO (1993) The ICD-10 classification of mental and behavioural disorders: diagnostic criteria for research. WHO, Geneva

Wilens TE, Biederman J, Brown S, Tanguay S, Monuteaux MC, Blake C, Spencer TJ (2002) Psychiatric comorbidity and functioning in clinically referred preschool children and school-age youths with ADHD. J Am Acad Child Adolesc Psychiatry 41:262–268

Willcutt EG, Nigg JT, Pennington B, Solanto MV, Rohde LA, Tannock R, Loo SK, Carlson CL, McBurnett K, Lahey BB. Validity of DSM-IV attention–deficit/hyperactivity disorder symptom dimensions and subtypes. Under review edn

Wilson JJ, Levin FR (2001) Attention deficit hyperactivity disorder (ADHD) and substance use disorders. Curr Psychiatry Rep 3:497–506

Yochman A, Ornoy A, Parush S (2006) Co-occurrence of developmental delays among preschool children with attention-deficit-hyperactivity disorder. Dev Med Child Neurol 48:483–488

ADHD: Volumetry, Motor, and Oculomotor Functions

E. Mark Mahone

Contents

1 Introduction 18
2 Volumetry in ADHD 19
   2.1 Frontostriatal Anomalies in ADHD 19
   2.2 Caudate Anomalies in ADHD 21
   2.3 Cerebellar Anomalies in ADHD 23
   2.4 Nonfrontal Cortical Anomalies in ADHD 24
   2.5 Cortical Morphology in ADHD 24
   2.6 Longitudinal Volumetric Studies of ADHD 26
   2.7 Shape Analysis Applied to ADHD 27
   2.8 Treatment with Stimulant Medication and Volumetry 27
   2.9 Genes, ADHD, and Volumetry 28
   2.10 Summary: Volumetry in ADHD 29
3 Motor Functions in ADHD 30
   3.1 Subtle Signs and ADHD 31
   3.2 Overflow 31
   3.3 Approaches to Motor Assessment in ADHD 33
   3.4 Summary: Motor Skills in ADHD 34
4 Oculomotor Functions and ADHD 34
   4.1 Experimental Assessment of Eye Movements and Oculomotor Control 35
   4.2 Assessment of Oculomotor Skills in ADHD 36
   4.3 Summary: Oculomotor Functions in ADHD 38
5 Conclusions 38
References 39

Abstract The use of quantitative neuroimaging (volumetry), motor, and oculomotor assessments for studying children with attention-deficit/hyperactivity disorder (ADHD) has grown dramatically in the past 20 years. Most evidence to date suggests

E.M. Mahone
Kennedy Krieger Institute, Johns Hopkins University School of Medicine, 1750 East Fairmount Avenue, Baltimore, MD 21231, USA
e-mail: mahone@kennedykrieger.org

C. Stanford and R. Tannock (eds.), *Behavioral Neuroscience of Attention Deficit Hyperactivity Disorder and Its Treatment,* Current Topics in Behavioral Neurosciences 9,
DOI 10.1007/7854_2011_146, © Springer-Verlag Berlin Heidelberg 2011,
published online 5 July 2011

that anomalous basal ganglia development plays an important role in early manifestation of ADHD; however, widespread cerebellar and cortical delays are also observed and are associated with the behavioral (cognitive, motor, oculomotor) phenotype in children with ADHD. These motor and "executive" control systems appear to develop in parallel, such that both systems display a similar protracted developmental trajectory, with periods of rapid growth in elementary years and continued maturation into young adulthood. Development of each system is dependent on the functional integrity and maturation of related brain regions, suggesting a shared neural circuitry that includes frontostriatal systems and the cerebellum (i.e., those identified as anomalous in studies of volumetry in ADHD). Motor and oculomotor paradigms provide unique opportunities to examine executive control processes that exist at the interface between movement and cognition in children with ADHD, also linking cognition and neurological development. The observed pattern of volumetric differences, together with the known parallel development of motor and executive control systems, appears to predict motor and oculomotor anomalies in ADHD, which are highly relevant, yet commonly overlooked in clinical settings.

Keywords Attention · Childhood · Executive function · MRI · Saccade · Sensorimotor · Volume

Abbreviations

ADHD	Attention-deficit/hyperactivity disorder
aMRI	Anatomic (MRI)
DAMP	Deficits in attention motor control, and perception
DAT	Dopamine transporter
DCD	Developmental coordination disorder
DRD1/DRD4	Dopamine receptor (D1 or D4 subtype)
LDDMM	Large deformation diffeomorphic metric mapping
MABC	Motor Assessment Battery for Children
MFNA	Motor Function Neurological Assessment (MFNU)
MGS	Memory-guided saccades
MRI	Magnetic resonance imaging
ODD	Oppositional defiant disorder
SMC	Supplementary motor complex
TMS	Transcranial magnetic stimulation

1 Introduction

Attention-deficit/hyperactivity disorder (ADHD) is a neurodevelopmental disorder involving motor intentional systems that are mediated in part by frontostriatal and fronto-cerebellar circuitry (Durston et al. 2010). These neuroanatomic anomalies (delays), observed in children with ADHD, set the stage for deficits in motor and oculomotor coordination and speed, which, when carefully assessed, are ubiquitous to the disorder and contribute to persistent cognitive and academic dysfunction. Children with ADHD commonly exhibit deficits in controlled behavior including difficulties with inhibition, delay aversion, and temporal processing (Sonuga-Barke and Halperin 2010) that are supported by a distributed neural network with cortical and subcortical components, including the frontal cortex and its striatal–thalamic–cerebellar connections (Durston et al. 2003, 2010) – those identified as most anomalous in ADHD. Indeed, current neurological models of frontal lobe structure and function have their basis in a well-described series of at least five parallel frontal–subcortical circuits (Lichter and Cummings 2001), of which two are related to motor function, originating in skeletomotor and oculomotor regions of the cortex. The other three, originating in dorsolateral prefrontal, orbitofrontal, and anterior cingulate regions, are thought to be crucial in cognitive ("executive") and socioemotional control. Frontal projections to the basal ganglia and cerebellum form a series of frontal–striatal–thalamo–frontal and frontal cerebello (dentato)-frontal circuits (Krause et al. 2000). These circuits link specific regions of the frontal lobes to subcortical structures, supply modality-specific mechanisms for interaction with the environment, and provide the framework for understanding the neurobiological substrate of ADHD.

2 Volumetry in ADHD

The use of quantitative neuroimaging (volumetry) for studying whole brain, as well as regional development in children with ADHD, is now over 20 years old. Compared with the earliest volumetric investigations, researchers now benefit from the availability of MRI scanners with higher field strength (up to 7.0 Tesla) and increased computational power that has enabled the development of more sophisticated analytic methods (Castellanos and Proal 2009). These newer methodologies now allow for the examination of morphology, as well as volume, including thickness of various regions of the cortical mantle, shape analysis of surface changes, and higher resolution of imagery – thereby allowing better demarcation of gray matter, white matter, and cerebrospinal fluid in measurement.

2.1 Frontostriatal Anomalies in ADHD

Among children with ADHD, the early structural MRI evidence showed consistent reductions in total cerebral volume (3–8%) compared to typically developing children without ADHD (Hill et al. 2003). It soon became clear that the convergence of findings revealed robust abnormalities in frontostriatal systems, including reduced size of the left caudate (Hynd et al. 1993), right caudate (Castellanos et al. 1994), right globus pallidus (Castellanos et al. 1996), and smaller left globus pallidus (Aylward et al. 1996). Decreased size of frontal regions was also a consistent early finding in ADHD, including bilateral frontal volumes (Hynd et al. 1990), right anterior frontal volumes (Castellanos et al. 1996), right anterior superior white matter (Filipek et al. 1997), and right dorsolateral prefrontal cortex (Hill et al. 2003). Children with ADHD were also found to have decreased area of the rostral body corpus callosum relative to controls (Baumgardner et al. 1996).

Early anatomic MRI (aMRI) studies of the cerebral cortex in ADHD relied on a volumetric approach, in which gray and white matter volumes were measured within parcellated subregions, initially defined by callosal landmarks (Castellanos et al. 1996; Filipek et al. 1997). Multiple aMRI studies of ADHD have continued to reveal abnormalities in frontal areas (Castellanos et al. 2000, 2002; Hesslinger et al. 2002; Kates et al. 2002; Mostofsky et al. 2002). More recently, investigators have used cortical landmarks to define functionally relevant lobar (frontal, parietal, temporal, occipital) and sublobar (e.g., prefrontal, premotor, anterior cingulate) regions. These studies have revealed decreased volumes in both prefrontal and premotor regions (Kates et al. 2002; Mostofsky et al. 2002). Subcortical anomalies in children with ADHD also continue to be identified, and include: the caudate nucleus (Mataro et al. 1997), putamen (Wellington et al. 2006), globus pallidus (Basser and Pierpaoli 1996), and cerebellum (Castellanos et al. 1996; Berquin et al. 1998; Mostofsky et al. 1998). There is an emerging convergence of findings suggesting ADHD-related reductions in medial prefrontal and anterior and posterior cingulate and precuneus (Castellanos and Proal 2009).

At the cerebral cortical level, the decreased volume of several frontal (Castellanos et al. 1996; Filipek et al. 1997; Hesslinger et al. 2002; Mostofsky et al. 2002) regions suggests that abnormalities are not localized to a specific area: rather, they appear widespread throughout the brain, including superior prefrontal cortex, reduced midsaggital area of the cerebellar vermis, and smaller splenium of the corpus callosum (Hill et al. 2003). Among school-aged children with ADHD (but not controls), prefrontal volume (especially right superior prefrontal) predicted performance on measures of sustained attention (Hill et al. 2003).

One of the challenges to MRI studies of ADHD is the presence of comorbidities that complicate the interpretation of anatomic findings, particularly among diagnostic groups for whom pathological *increases* in regional brain volume may mask the ADHD-related reductions. Further, there have been few studies that carefully contrast the neuroanatomic anomalies associated with ADHD with those of other childhood disorders that present with anomalous brain development while controlling for comorbidities. For example, while children with ADHD and autism spectrum disorders (ASD) both show regional reductions in cortical development, those with ASD show atypical *increases* (compared to controls) in gray matter volume in the right supramarginal gyrus (Brieber et al. 2007) and in frontal white matter (Mostofsky et al. 2007). Further, in studies in which oppositional defiant disorder (ODD) and conduct disorder (CD) comorbidities are taken into account, there are more widespread regions of reduced cerebral volume in children with ADHD, suggesting that these comorbidities may serve to mask some of the volumetric reductions, compared with samples of children with "pure" ADHD

(Sasayama et al. 2010). The same pattern is observed in children with "pure" Tourette syndrome (TS), for whom disproportionate *increases* in frontal white matter and rostral corpus callosum volume were observed (compared to controls) – a pattern that contrasted with that observed in children with "pure" ADHD, in whom reductions in these regions were observed, and in children with comorbid TS and ADHD, in whom no differences with controls were observed in these regions (Fredericksen et al. 2002).

The few meta-analyses of volumetry in ADHD have highlighted frontal anomalies as well as other cortical and subcortical anomalies, including differences in total right and left cerebral volume, cerebellar regions, splenium of the corpus callosum (Valera et al. 2006), and gray matter reduction in right putamen/globus pallidus (Ellison-Wright et al. 2008). However, these meta-analyses highlighted the fact that females have been underrepresented in the ADHD neuroimaging literature (Valera et al. 2006). Many of these early studies reported findings on ADHD samples that were predominantly (or exclusively) male, and so conclusions could not be drawn about female-specific anomalies in ADHD. In what may be the only volumetric study specific to girls with ADHD, Castellanos (Castellanos et al. 2001) reported ADHD-related reductions in left caudate and posterior–inferior cerebellar vermis, with equivocal frontal lobe findings. Because this study was completed separately from earlier studies of boys with ADHD, the authors remarked: "conclusions about sex differences in ADHD must remain tentative until verified in contemporaneously collected and analyzed longitudinal scans" (p 293).

More recently, Mahone et al. (2009b) examined functionally defined (and manually delineated) frontal lobe subdivisions in contemporaneously recruited samples of boys and girls with and without ADHD. Compared to age-matched controls, children with ADHD showed reduced tissue volumes involving prefrontal and premotor regions, with largest volume reductions (in both boys and girls) observed in the supplementary motor complex (SMC). Further, girls (but not boys) with ADHD showed reduced lateral premotor cortex and increased primary motor cortex volumes. Across groups, however, reduced SMC volume was associated with ADHD symptom severity, suggesting dysfunction in circuits important for motor response selection and inhibition (Mahone et al. 2009b).

2.2 Caudate Anomalies in ADHD

Although early empirical evidence highlighted anomalous frontal brain development in ADHD (Castellanos et al. 1996; Mostofsky et al. 2002; Casey et al. 1997; Durston et al. 2004), more recent research has argued that other brain regions (particularly subcortical) may provide answers to the early neurobiological unfolding of ADHD (Valera et al. 2006; Halperin and Schulz 2006). For example, among children with early prefrontal lesions, functional impairment often does not manifest until later childhood (Anderson et al. 1999) and then tends to get worse upon entry into adolescence (Denckla and Cutting 2004). In contrast, symptoms of

ADHD are almost always evident during the preschool years (Barkley 2006), and the severity of symptoms (most notably hyperactive/impulsive symptoms) tends to diminish with age (Hinshaw et al. 2006). This pattern has led some researchers to hypothesize that the prefrontal cortex may be more involved in recovery from ADHD, rather than the cause of the disorder, and that early disruption to other regions, notably the basal ganglia, may be involved in the early development of ADHD (Halperin and Schulz 2006; Soliva et al. 2010), perhaps through underuse of these brain regions, or via poor neural connectivity, giving rise to later delays in the development of cortical volumes (Shaw et al. 2006).

The preceding hypothesis has considerable appeal in explaining the parallel development of regional brain volumes and unfolding of patterns of executive and motor dysfunction in ADHD. The basal ganglia serve as the nexus through which prefrontal, premotor, and motor signals inhibit competing motor programs and disinhibit intended behaviors (Mink 1996; Nachev et al. 2008). In particular, the number of cells that project from the basal ganglia to the supplementary motor cortex (a region critical for response control, which is reduced in ADHD (Mahone et al. 2009b; Ranta et al. 2009) is three to four times the number that project from the cerebellum, suggesting a critical early link between basal ganglia development and response control (Akkal et al. 2007).

Consistent with the hypothesis of early basal ganglia dysfunction is a recent meta-analysis in which the right caudate was among the most frequently assessed and showed the largest ADHD-related reductions, compared to controls (Valera et al. 2006). Further, longitudinal studies of animal models of ADHD (using spontaneously hypertensive rats) highlight the importance of early striatal reductions that stabilize by 6 weeks of age (human equivalent $= 7$–9 years) (Hsu et al. 2010). Similarly, a longitudinal study of children with ADHD suggests that normalization of reduced caudate volume occurs by puberty (Castellanos et al. 2002). Thus, the developmental trajectory of caudate anomalies in ADHD appears to parallel the pattern of development of hyperactive–impulsive symptoms (Biederman et al. 2000), such that early caudate reduction is associated with increased symptoms that tend to resolve with accelerated growth (normalization in volume) by adolescence. For example, Carmona and colleagues reported bilateral reductions in the volume of the ventral striatum that correlated with maternal ratings of hyperactivity (Carmona et al. 2009).

Indeed, investigations in humans continue to highlight the importance of early caudate anomalies in children with ADHD. In a study of nine monozygotic twin pairs, discordant for ADHD, Castellanos et al. reported that the affected children had smaller total caudate volumes (Castellanos et al. 2003). There is also accumulating evidence that atypical caudate *asymmetry* in ADHD may be an important biomarker in the development of the disorder. For example, among boys with ADHD, there is evidence of reversed asymmetry of the head of the caudate, such that the ADHD group had a smaller volume of left caudate head, while controls had a smaller volume of right caudate head (Semrud-Clikeman et al. 2000). Similarly, among boys with ADHD in residential treatment, reversed caudate asymmetry (right $>$ left) was observed, compared to the more typical (left > right) asymmetry that was present in controls and no group differences observed for putamen volumes (Wellington et al. 2006). Other investigators have observed that atypical caudate asymmetry is a strong predictor of ADHD symptoms, such that a greater degree of right > left asymmetry predicted inattention symptoms, but not hyperactive/impulsive symptoms (Schrimsher et al. 2002). This pattern of asymmetry may hold promise as a biomarker specific to ADHD, especially among boys. Soliva and colleagues examined patterns of asymmetry of caudate volumes in ADHD in 39 children with ADHD (35 boys). They found that a ratio of right caudate volume to total bilateral caudate volume of 0.48 or lower had 95% specificity in predicting ADHD diagnosis (versus typically developing controls) (Soliva et al. 2010).

To identify early patterns of brain anomalies in ADHD, Mahone and colleagues used volumetric imaging to study 26 preschoolers, aged 4–5 years, who presented with and without symptoms of ADHD. After controlling for total cerebral volume, total caudate (but not frontal lobe) volumes were reduced in the ADHD group. Further, across groups, reduced caudate (but not frontal lobe) volume was associated with increased parent ratings of hyperactive/impulsive symptoms. These preliminary findings suggest that early anomalous caudate (perhaps more than frontal lobe) development is associated with early onset of ADHD symptoms (Mahone et al. 2011).

2.3 Cerebellar Anomalies in ADHD

The cerebellum is among the most vulnerable regions to early insult (Volpe 1995) and, like the basal ganglia, may influence the cognitive operations normally thought to be subserved by the frontal lobes (Middleton and Strick 2001), especially given the prevalence of children with ADHD who have motor control problems (Diamond 2000; Pitcher et al. 2003). While phylogenetically older regions (e.g., basal ganglia) mature earlier than newer regions (e.g., cortex), the cerebellum may also continue to develop into the 20s, a finding that has implications for how cerebellar maturation may relate to the symptom onset and developmental change in ADHD, especially because the cerebellum appears less influenced by genetics and more sensitive to environmental variables (Durston et al. 2010; Fassbender and Schweitzer 2006; Lantieri et al. 2010; Tiemeier et al. 2010).

Indeed, one of the most consistent findings in ADHD brain imaging studies is a decrease in posterior inferior cerebellar vermis (lobes VIII–X) volume (Hill et al. 2003; Castellanos et al. 1996, 2002; Berquin et al. 1998; Mostofsky et al. 1998; Valera et al. 2006; Castellanos et al. 2001; Durston et al. 2004; Bussing et al. 2002). However, unlike the pattern observed in basal ganglia reductions in ADHD, longitudinal studies suggest that volumetric reductions in posterior inferior cerebellar vermis tend to persist over time; they also remain associated with ADHD symptom severity over time (Castellanos et al. 2002). Furthermore, in a study that included unaffected siblings, total cerebral and prefrontal volumes were reduced (compared to controls) in boys with ADHD *and* their unaffected siblings. Right cerebellar volume was the only measure reduced in the ADHD group but not in the unaffected siblings (Durston et al. 2004), suggesting that the cerebellum may play an important role in the pathophysiology of ADHD that is more associated with environmental (rather than genetic) influences. Anomalies in fronto-cerebellar circuitry are thus also considered key to the development of ADHD, given the efferent outputs from the cerebellum to both the frontal cortex and the basal ganglia (Durston et al. 2010), and the links between anomalous cerebellar development and deficits in motor response control (Suskauer et al. 2008), motor timing (Van Meel et al. 2005), classical conditioning (Chess and Green 2008), and oculomotor control (Voogd et al. 2010).

2.4 Nonfrontal Cortical Anomalies in ADHD

Although the evidence from individual studies and meta-analyses has pointed to frontostriatal regions as being anomalous (reduced) in ADHD, there is also substantial evidence of structural anomalies outside frontostriatal circuitry (Cherkasova and Hechtman 2009) including reductions in cortical temporal lobes (Castellanos et al. 2002; Sowell et al. 2003; Carmona et al. 2005), parietal lobes (Filipek et al. 1997; Castellanos et al. 2002; Carmona et al. 2005), and occipital lobes (Filipek et al. 1997; Castellanos et al. 2002; Durston et al. 2004). Reductions in medial temporal volumes, as well as striatal and anterior cingulate volumes, are correlated with performance on measures of response inhibition (i.e., stop-signal reaction time) in boys with ADHD (McAlonan et al. 2009). Using voxel-based morphometry, regions that are reduced in ADHD include bilateral temporal poles, occipital cortex, and left amygdala. These regions were even greater when analyses controlled for the presence of oppositional defiant disorder (ODD) (Sasayama et al. 2010). Among boys with ADHD ($n = 19$), decreased callosal thickness was identified in anterior and posterior corpus callosum regions, with the largest reduction observed in isthmus region (which projects to parietal cortex and may be critical for sustaining attentional control) (Luders et al. 2009).

2.5 Cortical Morphology in ADHD

While development of the human nervous system begins 2–3 weeks after conception (Black et al. 1990) and continues at least into early adulthood, the trajectory of development is nonlinear and progresses in a region-specific manner that coincides with functional maturation (Halperin and Schulz 2006; Gogtay et al. 2004). By age 2 years, the brain is approximately 80% of its adult size (Giedd et al. 1999). Synapse formation (Huttenlocher and Dabholkar 1997) and myelination (Kinney et al. 1988) proceed rapidly up to age 2 years, followed by a relative plateau phase, during which neurons begin to form complex dendritic trees (Mrzljak et al. 1990). Maximum synaptic density (i.e., synaptic overproduction) is observed at age 3 months in the primary auditory cortex and at age 15 months in the prefrontal cortex (Huttenlocher and Dabholkar 1997). After age 5 years, brain development is marked by continued neuronal growth, pruning, and cortical organization. Onset of puberty accelerates the experience-dependent pruning of inefficient synapses (Gogtay et al. 2004) and eventually reduces synaptic density to 60% of maximum (Huttenlocher and Dabholkar 1997). Longitudinal studies suggest that cortical gray matter maturation progresses from primary sensorimotor areas and spreads rostrally over the frontal cortex and caudally and laterally over the parietal, occipital, and finally to the temporal cortex (Halperin and Schulz 2006; Gogtay et al. 2004; Giedd et al. 1999). Prominent volumetric reduction of frontal and parietal cortices occurs during adolescence and is considered to be attributable to synaptic pruning or, more likely, the combination of synaptic pruning and increasing myelination (Sowell et al. 2004). Regionally specific, protracted, age-related changes in white matter have been also been described. For example, myelination of optic radiations and occipital white matter begins 1–2 months before birth and extends to frontal lobes by 9 months postnatal age (Marsh et al. 2008; Paus et al. 2001); cortical myelination also follows a posterior-to-anterior direction, with sensory pathways myelinating first, followed by motor pathways, and finally by association areas (Huttenlocher and Dabholkar 1997). In normal development, this dynamic pattern of myelination and associated cortical volume reduction is associated with improvement in cognitive performance; timing of peak cortical volume is considered to be a marker for maturation (Shaw et al. 2007a). Conversely, patterns of pathological pruning may contribute to the genesis of psychiatric disorders (Marsh et al. 2008; Kotrla and Weinberger 2000) and can contribute to onset and progression of ADHD symptoms, delayed cortical maturation and worse clinical outcome (Sowell et al. 2003; Marsh et al. 2008; Shaw et al. 2007a, b).

While school-aged children with ADHD show widespread decreases in cortical volume, the precise morphologic contributions to these reductions are less clear. For example, decreased cortical volume in ADHD can potentially represent a "thinning" of the cortex or a decrease in the total surface area of the cortex, or a combination of both. As such, the study of volumetry in ADHD has begun to emphasize cortical morphology (including thickness and surface area), as well as patterns of cortical folding and shape (Wolosin et al. 2009). In this context, cortical *thickness* refers to linear *distance* from the gray/white boundary to the pial surface (Fischl and Dale 2000), and *thinning* refers to a reduction in this distance compared to controls (or a reduction over time, compared to one's own cortical thickness).

Sowell and colleagues were among the first to examine cortical morphology in children with ADHD. They identified reduced surface area in inferior portions of dorsal prefrontal cortices bilaterally (Brodmann areas 44, 45, 46), with reductions correlated with symptoms of hyperactivity (Sowell et al. 2003). Similarly, among medication-naïve children and adults with ADHD, reduced cortical thickness in the right superior frontal gyrus was predictive of ADHD symptom severity (Almeida et al. 2010). In a series of large-scale studies, Shaw et al. found that children with

ADHD had a mean thinning of cortex of 0.09 mm globally. The greatest reduction was in medial/superior prefrontal and precentral regions, and the reduction in medial prefrontal cortex was associated with worse clinical outcomes (Shaw et al. 2007a, b). Narr and colleagues also reported ADHD-related cortical thinning over large areas of frontal, temporal, parietal, and occipital association cortices and aspects of motor cortex (although not within the primary sensory regions) (Narr et al. 2009). Conversely, Wolosin (Wolosin et al. 2009) found reduced volume and surface area (but not thickness) in all four lobes bilaterally in children with ADHD.

2.6 Longitudinal Volumetric Studies of ADHD

Critical insights into the development of ADHD have emerged from longitudinal studies investigating trajectories of development in children. Castellanos et al. (2002) reported growth curves highlighting the different developmental trajectories of regional brain volumes. For most regions of interest, the growth curves of children with ADHD (relative to controls) were parallel, but on a lower track. More recently, Shaw et al. (Shaw et al. 2007a; Shaw 2010) reported a series of longitudinal studies of children with ADHD using measures of cortical thickness (i.e., distance between the pial surface and white/gray boundary in the cortex). They found that children with ADHD showed *delay* in cortical maturation (i.e., age of attaining peak thickness) throughout the cerebrum. The most prominent area of delay was the lateral prefrontal cortex, with "delay" approaching 5 years in middle prefrontal cortex in children with ADHD. In a follow-up longitudinal study, the same authors demonstrated that this delay in cortical thinning, especially in prefrontal and premotor cortex, was associated with both the severity of ADHD symptoms and the categorical presence of the disorder itself (Shaw 2010).

In a separate longitudinal study of cerebellar development and clinical outcome, Mackie et al. (2007) examined MRI scans from 36 children with ADHD and found a progressive loss of volume (compared to controls) in the superior cerebellar vermis. Moreover, those in the ADHD group who had *worse* clinical outcome had more rapid loss of inferior cerebellar lobes bilaterally, compared with controls or children with ADHD with better outcome. Different longitudinal patterns of change with regard to cortical asymmetry have also been identified in ADHD (Shaw et al. 2009a). For example, in children with ADHD, the (earlier developing) posterior components of cortical asymmetry were observed to be intact, whereas the (later developing) prefrontal components were lost, suggesting a developmental disruption in the prefrontal function in ADHD (Shaw and Rabin 2009).

While most longitudinal studies of ADHD have shown delayed patterns of development throughout the cortex, Shaw et al. (Shaw 2010) also reported that the primary motor cortex showed earlier maturation (i.e., age of attaining peak thickness) in children with ADHD. One hypothesis for this pattern may be that early, excessive motor activity among young children with ADHD activates (and thus facilitates) connections within the primary motor cortex. In other words, the development of connections in primary motor cortex outpaces the development of connections in other frontal regions that function to restrict motor hyperactivity. The result of this pattern of uneven functional maturation of the frontal cortex gives rise to hyperactivity and deficits in response control in children with ADHD.

2.7 Shape Analysis Applied to ADHD

Shape analysis represents an emerging, powerful computational method that enables more detailed definition of local surface changes, based on the degree to which a structure has to be warped to meet a template (Shaw 2010). Several studies have begun to specify more precisely the localization of ADHD-related abnormalities using shape analysis that allows for detection of anomalies beyond regional volume. For example, Plessen et al. investigated the morphology of regional hippocampal volumes in 51 children and adolescents (aged 6–18) with combined subtype ADHD. They found that the head of the hippocampus was *enlarged* in children with ADHD, with enlargement related to ADHD symptom severity (albeit weakly) (Plessen et al. 2006).

More recently, Qiu et al. used large deformation diffeomorphic metric mapping (LDDMM) to examine ADHD-related differences in basal ganglia shapes in 47 children with ADHD and 66 controls, aged 8–12 years. While boys with ADHD showed smaller overall basal ganglia volumes compared with control boys, LDDMM analysis identified markedly different basal ganglia shapes in boys with ADHD (compared with control boys), including bilateral volume *compression* in the caudate head and body, anterior putamen, left anterior globus pallidus, and right ventral putamen, but volume *expansion* in posterior putamen. In contrast, no basal ganglia volume or shape differences were revealed in girls with ADHD, compared with control girls in this age range (Qiu et al. 2009).

2.8 Treatment with Stimulant Medication and Volumetry

While the majority of volumetry research in ADHD has emphasized group differences without characterization of medication use, interest has risen in the effects of stimulant medication treatment on brain volumes. Current research findings suggest that the reductions or delays in brain development among children with ADHD do not appear to be a result of chronic stimulant medication treatment (Schnoebelen et al. 2010). In fact, the preponderance of evidence suggests a protective or even "normalizing" effect of stimulants on brain development in ADHD (Shaw et al. 2009b). Castellanos et al. demonstrated that, among children with ADHD, those treated with stimulants did not differ from controls in white matter volumes, whereas in medication-naïve children with ADHD, white matter volume was reduced compared to controls (Castellanos et al. 2002). More recently, stimulant-related "normalization" has been reported for children with ADHD in right anterior cingulate and bilateral caudate volumes (i.e., volumes closer to controls compared to those untreated children with ADHD) (Pliszka et al. 2006), and for cross-sectional area of posterior interior cerebellar vermis (Bledsoe et al. 2009). In a similar investigation, Sobel et al. examined basal ganglia surface morphology and the effects of stimulant medication treatment in children with ADHD. Among medication-naïve participants, there were volume reductions in putamen that were primarily driven by *inward* deformations of structures, including those nuclei within the putamen that are components of limbic, sensorimotor, and associative pathways. In contrast, those treated with stimulants had *outward* deviations of putamen nuclei, suggesting that stimulants may affect, and perhaps facilitate, development of basal ganglia morphology in ADHD (Sobel et al. 2010) in a manner that serves to ameliorate ADHD-related deficits in executive and motor control. Admittedly, comparison of treated versus nontreated children with ADHD is complicated, in part because it is possible that groups may have differed in some systematic way before onset of treatment. For instance, those with more severe and pervasive symptomatology or with comorbidity may opt for stimulant treatment. Nevertheless, in this context, it is even more impressive that those treated with stimulants show greater "normalization" than those not treated.

2.9 Genes, ADHD, and Volumetry

ADHD is one of the most heritable childhood neuropsychiatric disorders. Polymorphisms within the dopamine transporter genotype (DAT) and dopamine D4 receptor (DRD4) gene have been frequently implicated in its pathogenesis (Shaw et al. 2007b). As such, investigators have begun to link anatomic MRI with investigation of genetic differences hypothesized to be associated with ADHD. For example, Shook et al. (2010) examined the association between ADHD symptoms and the volume of the head of caudate. This striatal structure has a high dopamine transporter (DAT) expression that is important for inhibitory function and differs in children with ADHD. Overall caudate volumes were smaller in children who were carriers of two copies of the 10-repeat DAT allele than those with one copy, suggesting that altered caudate development, associated with 10-repeat homozygosity of DAT1, may contribute susceptibility to ADHD (Shook et al). In a related study, Shaw et al. (2007b) examined the effects of the 7-repeat microsatellite in the DRD4 gene on clinical outcome and cortical development in ADHD. Possession of the DRD4 7-repeat allele was associated with a thinner right orbitofrontal/inferior prefrontal and posterior parietal cortex, with overlap in regions found to be thinner in children with ADHD (compared with controls). Participants with ADHD carrying the DRD4 7-repeat allele had a better clinical outcome and a distinct trajectory of cortical development, with this group showing normalization of the right parietal cortical region, which is important for the development of attentional control. By contrast, there were no effects of the

DRD1 or DAT1 polymorphisms on clinical outcome or cortical development, suggesting that the DRD4 7-repeat allele confers a protective effect on volumetric changes and ADHD symptoms. While not a volumetric MRI study, Lantieri et al. (2010) examined the top single-nucleotide proteins (SNPs) that had been identified in a genome-wide association study of ADHD (Neale et al. 2008) using an independent cohort. The two SNPs identified as significant (XKR4 in 8q12.1, and FAM190A in 4q22.1) were both located in genes coding for uncharacterized proteins expressed most prominently in the cerebellum, a region that has shown robust decreases in children with ADHD (Valera et al. 2006).

2.10 Summary: Volumetry in ADHD

The preponderance of evidence to date highlights widespread "delays" in gray and white matter development that are associated with onset and progression of ADHD symptomatology and do not appear to be associated with chronic stimulant medication use. To the contrary, treatment with stimulants appears to have a protective, or even "normalizing," effect on brain development in children with ADHD. From a developmental perspective, anomalous basal ganglia development (especially involving the caudate) may play an important role in the early onset of ADHD, while atypical cerebellar development (potentially associated with environmental influences) may be linked with the persistence of symptoms over time. These anatomic anomalies provide the underlying neural substrate for ADHD-related behavioral deficits in motor and executive control, particularly during the elementary school years. In particular, the parallel development of motor and executive control systems suggests a shared neural circuitry that includes frontostriatal systems and the cerebellum (i.e., those identified as anomalous in ADHD), and thus the study of neuroanatomic development in ADHD is inextricably linked with the study of motor and response control skills.

Nevertheless, despite the proliferation of studies, conclusions from existing literature have been complicated by reliance on cross-sectional designs, samples including largely (or exclusively) males, differences related to medication treatment history, and inconsistencies in samples due to comorbidities and ADHD subtypes. The future of volumetric-imaging ADHD is likely to involve more emphasis on shape analysis, genes, and greater attempts to link volumetric imaging developmentally with emergence of (and recovery from) symptoms. Emerging methodologies may also enable investigators to differentiate ADHD subtypes according to empirically supported neuropsychological differences, such as the triple pathway model (inhibitory control, delay aversion, temporal processing) proposed by Sonuga-Barke (Sonuga-Barke and Halperin 2010), or linking these functions more directly to neural circuitry, such as defining a "dorsal fronto-striatal" (cognitive control), "orbitofronto-striatal" (reward processing), or "fronto-cerebellar" (timing) subtypes (Durston et al. 2010).

3 Motor Functions in ADHD

When carefully assessed, motor skill deficits are ubiquitous in children with ADHD, likely as a function of the known anomalous development of frontostriatal-cerebellar circuitry associated with the disorder. Unfortunately, the presence of poor motor skills in ADHD is associated with slow and effortful completion of routine tasks, and can increase attentional demands on other processes, such as working memory, thereby creating a bottleneck. Those with ADHD are often left to trade accuracy for speed (Klimkeit et al. 2004), or use more controlled attentional resources to maintain what should be otherwise automatic postural control (Roebers and Kauer 2009), thereby creating an even greater state of "inattention," or otherwise exacerbating other skill deficits.

Historically, motor assessment in children has been used to delineate behavioral development patterns that distinguish clinical groups, and to examine the effects of treatment – especially medication. Cognitive and emotional control, however, can be considered "mental" neighbors of motor control, as their respective circuitries are organized in proximity to circuits supporting motor systems. In fact, motor and executive control systems appear to develop in parallel. Both systems display a similar protracted developmental trajectory, with periods of rapid growth in elementary years and continued maturation into young adulthood (Diamond 2000). Furthermore, development of each system is dependent on the functional integrity and maturation of related brain regions, suggesting a shared neural circuitry that includes frontostriatal systems and the cerebellum (i.e., those identified as anomalous *via* volumetry) (Diamond 2000; Pennington and Ozonoff 1996; Rubia et al. 2001). Age and sex are also important mediating factors that need to be considered in motor examination with children (Denckla 1974).

Because the systems that support motor and higher order "cognitive" control both have protracted periods of development, they are vulnerable to disruption via a variety of etiologies – which is likely why so many children with neurodevelopmental disorders (such as ADHD) present with both motor and executive dysfunction. For example, among 5–6-year-old children, motor behavior was associated with executive control (i.e., performance on a Stroop task) as well as with severity of externalizing behaviors (Livesey et al. 2006). A strong association between attention and motor coordination in 7-year-old children has also been identified (Piek et al. 2004). Moreover, deficits in either executive or motor control systems frequently present with coexisting deficits in the other: for example, approximately half of the children with ADHD demonstrate problems with motor coordination (Pitcher et al. 2003; Carte et al. 1996; Denckla and Rudel 1978; Kadesjo and Gillberg 1998; Steger et al. 2001), while approximately half of the children with developmental coordination disorder (DCD) manifest problems with attention (Kaplan et al. 1998), although the presence of both ADHD and DCD is associated with greater risk for poor psychosocial outcome than ADHD alone (Rasmussen and Gillberg 2000). In Sweden, the overlap between attention and motor control has long been recognized and is characterized as part of the syndrome of deficits in attention, motor control, and perception (DAMP) (Gillberg et al. 1992). Additionally, prominent theories of ADHD, such as the cognitive-energetic model of information processing, link motor behavior to executive control (Sergeant 2000). Thus, assessment of motor function can be critical to understanding both the biological substrates and cognitive phenotypes associated with ADHD (Denckla 2005).

3.1 Subtle Signs and ADHD

A variety of standardized tests of motor function (speed, coordination, strength) with published normative data are available for school-aged children. Beyond these commercially available tests, careful clinical assessment of basic motor function in children can reveal subtle motor dysfunction. Such neurological subtle signs include *overflow* (also called "associated" or "extraneous") movements, *involuntary movements* (i.e., limb tremor, odd posturing, choreiform), and *dysrhythmia*. These subtle signs can be reliably assessed (Gustafsson et al. 2010; Stray et al. 2009a) and can serve as markers for inefficiency in neighboring parallel brain systems that are important for control of cognition and behavior. It is common to observe subtle signs in typically developing younger children (Largo et al. 2001). However, there is evidence that persistence of subtle signs into later childhood can be a marker for atypical neurological function – often associated with disorders such as ADHD (Morris et al. 2001; Mostofsky et al. 2003; Cole et al. 2008). In fact (before the institution of the ADHD diagnosis in DSM-III), motor exam results (including speed, overflow, dysrhythmia) correctly classified 89% of boys who scored highly for the "hyperactivity" syndrome (Denckla and Rudel 1978). Similarly, early research in dyslexia typically did not screen for, or measure, ADHD symptoms in their dyslexic samples. When directly screened, children with dyslexia *without* attention problems performed better than children with dyslexia *with* attention problems on five of six rapid movements. The screened dyslexic group also had fewer signs of dysrhythmia or overflow than the unscreened group (Denckla et al. 1985). Although subtle signs can be important biomarkers, they can be variable and their presence alone should be neither considered diagnostic nor the sole basis for explaining complex behavioral and neurological disorders (Touwen and Sporrel 1979; Touwen 1987) such as ADHD.

3.2 Overflow

Overflow is defined as comovement of body parts not specifically needed to efficiently complete a task. As typically developing children mature, they manifest fewer overflow movements (Largo et al. 2003). The presence of age-inappropriate overflow may reflect immaturity of cortical systems involved with automatic inhibition (Mostofsky et al. 2003). In particular, observed overflow after 10 years of age may be a strong indicator of developmental dysfunction (Denckla and Rudel 1978). Motor-skill development, particularly the developmental pattern of asymmetries in left-, versus right-, sided performance, may be a marker for maturation of the corpus callosum or for the different rates of development of the cerebral hemispheres – known to be anomalous in ADHD (Roeder et al. 2008).

Among overflow movements, the most studied are mirror movements (i.e., synkinesis), which refers to involuntary movements that accompany voluntary movements on the contralateral side of the body. The presence of mirror overflow movements in adolescents and adults in disorders of both the motor cortex and the corpus callosum suggests that the ability to perform unilateral fine motor movements is dependent upon intact interhemispheric and corticospinal connections (Nass 1985; Knyazeva et al. 1997; Meyer et al. 1998). The supplementary motor complex (SMC) (and in particular the pre-SMC) has dense callosal and interhemispheric connectivity, arguing for a role in voluntary and involuntary movements. In particular, the SMC may have a role in suppressing default movements (Addamo et al. 2007). Mirror overflow may also be due to abnormally active ipsilateral corticospinal tract (reflecting more severe early neurodevelopmental abnormalities), as well as bilaterally active corticospinal tracts, which may occur in normal individuals under conditions of fatigue (Hoy et al. 2007). Mirror overflow movements appear to develop in a U-shaped relationship with age, decreasing rapidly during childhood (suggesting increased inhibition), then increasing with age in late adulthood (suggesting age-related loss of inhibition with aging) (Koerte et al. 2010). Thus, when cortical inhibitory and excitatory systems are immature, overflow movements in children are at their peak. As these cortical systems mature, overflow movements are more difficult to elicit and their presence into adolescence is thought to be associated with delayed cortical maturation (Cole et al. 2008).

As a disorder involving delay in development of brain systems supporting motor inhibition (Shaw 2010), overflow movements may represent important neurobehavioral markers in ADHD, especially as they appear to have linear associations with ADHD symptoms in early years. For example, in 5–6-year-old children, qualitative aspects of motor problems (subtle signs/overflow) as well as dynamic balance and manual dexterity were predictive of ADHD symptoms 18 months later, but not predictive of symptoms of oppositional defiant disorder or conduct disorder (Kroes et al. 2002). In a separate cohort, subtle signs at age 4–6 years were significant predictors of ADHD symptoms, in both low birth weight and normal birth weight children (Sato et al. 2004).

Among school-aged children, there also appears to be an association between motor overflow and performance on tasks of attentional control (Waber et al. 1985), as well as effortful motor response inhibition (Mostofsky et al. 2003; Cole et al. 2008). Children aged 5–11 years, with more evidence of minor neuromotor dysfunction, perform more poorly in school and have more signs of attention deficit (Batstra et al. 2003). Conversely, reduction in overflow movements appears to parallel reduction in ADHD symptomatology. In a large sample of children aged

7–14, controls and girls with ADHD showed steady age-related reduction of overflow and dysrhythmia, whereas boys with ADHD had little improvement in these signs through age 14 years (Cole et al. 2008).

3.3 Approaches to Motor Assessment in ADHD

Motor deficits may represent an important endophenotype in ADHD. For example, on a computerized tracking task, both children with ADHD and their unaffected siblings expressed deficits, relative to controls (Rommelse et al. 2007). Motor-control difficulties also have specificity for ADHD, compared with other disorders. Among children with Tourette syndrome, for example, those with ADHD have motor slowing, whereas those with Tourette syndrome alone (without ADHD) do not (Schuerholz et al. 1997). Poor motor coordination has been associated with oppositional defiant disorder, but not with conduct disorder (Martin et al. 2010). Iranian boys with ADHD showed deficits (relative to controls) on eight of nine fine motor tasks (cutting, threading buttons, grooved pegboard), despite showing no differences in IQ (Lavasani and Stagnitti 2010). Motor problems in ADHD are readily observed when performing nonautomated (Carte et al. 1996) or skilled movements (e.g., Grooved Pegboard) more than when performing simple repetitive movements (finger tapping) (Meyer and Sagvolden 2006).

Caregiver ratings of motor skills are also important diagnostically. Parents and teachers rate motor skill problems in as many as one-third of children with ADHD (Fliers et al. 2008), although children with ADHD, who show deficits on performance-based tests of motor control, do not tend to rate themselves as having motor problems (Fliers et al. 2010). In a large study of 7–19-year-old children, those with parent ratings of ADHD and motor coordination problems were also rated as having elevated levels of autistic symptoms (Reiersen et al. 2008).

Given their effects on dopamine transmission throughout the brain, stimulant medications can improve not only attentional control but also motor control. For example, unmedicated boys with ADHD showed deficits in postural control (measured via motion analysis of head stability), characterized by increased low levels of baseline head movement, punctuated by higher-amplitude spikes. Following administration of methylphenidate, however, both the baseline and spike amplitudes of the ADHD group were suppressed to levels at or below that of controls (Ohashi et al. 2010). Similarly, treatment with stimulant medication improved handwriting (Flapper et al. 2006) and overall motor performance in children with ADHD and/or DCD on the Motor Assessment Battery for Children (MABC) (Flapper et al. 2006; Bart et al. 2010) and the Motor Function Neurological Assessment (MFNU) (Stray et al. 2009b), although balance was not improved by stimulants (Bart et al. 2006). Other studies similarly found that larger motor skills (as measured by the Test of Gross Motor Development-2) were less affected by stimulant medication treatment (Harvey et al. 2007).

3.4 Summary: Motor Skills in ADHD

The preponderance of evidence from structural imaging studies of ADHD suggests anomalous and/or delayed development of brain systems that are critical to the development of motor skills, including prefrontal and premotor cortex, corpus callosum, basal ganglia, and cerebellar vermis. When carefully assessed, children with ADHD manifest a variety of deficits in motor skills, with co-occurrence of developmental coordination disorder in nearly 50%. These motor skill weaknesses contribute to slowed processing speed, poor automatization of skills, and inefficiency of task completion. Improvements in motor speed and coordination, and reductions in "abnormal-for-age" subtle signs among children with ADHD appear to occur in parallel with the (delayed) development of cortical brain systems supporting motor control. This process of "normalization" of motor functioning occurs earlier in girls with ADHD than in boys with ADHD. While stimulant medication may help with both attention and motor control, treatment plans for children with ADHD (especially boys) should carefully consider the (likely) coexisting motor deficits and plan for intervention and accommodations accordingly.

4 Oculomotor Functions and ADHD

Motor skill deficits are ubiquitous in children with ADHD, likely as a function of the anomalous development of frontostriatal-cerebellar circuitry associated with the disorder. Like the motor system, the oculomotor system is highly relevant in children with ADHD and includes a widely distributed network, incorporating regions of the cerebral cortex as well as the basal ganglia, thalamus, cerebellum, superior colliculus, and brainstem reticular formation. Oculomotor paradigms can provide unique opportunities to examine executive control processes that exist at the interface between movement and cognition (Leigh and Zee 2006). These paradigms can complement other diagnostic and exploratory studies, because they afford a degree of quantification about information processing and timing not found in methods such as MRI or neuropsychological testing. This is because it is virtually impossible to operationalize visual attention without documenting "looking," which means "where the eyes are fixed" (Lasker et al. 2007). As such, oculomotor paradigms may offer a more precise means of assessing components of visual attention that link cognition and neurological development.

The oculomotor system includes a widely distributed network with regions in the cerebral cortex (frontal eye fields, posterior parietal cortex, supplementary eye fields, presupplementary motor area, dorsolateral prefrontal cortex), as well as the basal ganglia, thalamus, cerebellum, superior colliculus, and brainstem reticular formation (Munoz and Everling 2004). Frontal regions project to the superior colliculus, where cortical and subcortical signals converge and are integrated

(Munoz and Everling 2004). Within the frontal cortex, the frontal eye fields have a critical role in voluntary saccades (Schall 1997). The supplementary eye fields are important for sequencing of saccades and error monitoring (Stuphorn et al. 2000). The DLPFC is involved in spatial working memory and in suppressing reflexive saccades (Pierrot-Deseilligny et al. 1991). These frontal cortical oculomotor regions also project directly and indirectly to the caudate nucleus (Hikosaka et al. 2000), and directly to the paramedian pontine reticular formation, which provides input to the saccadic premotor circuit. Both projections support the initiation and suppression of visual saccades (Munoz and Everling 2004). Of importance is the observation that the same regions involved in oculomotor control are also those in which anomalous or delayed development is observed in children with ADHD. Not surprisingly, children with ADHD are found to have deficits (inefficiency) in both the initiation and suppression of controlled eye movements. These oculomotor deficits in ADHD are associated with slowed processing speed, poor automaticity of skills, and more effortful processing on routine tasks.

4.1 Experimental Assessment of Eye Movements and Oculomotor Control

Two types of oculomotor skills are examined under experimental conditions: those involving *fixation* of the eyes and those involving *movement* of the eyes. Fixation paradigms require the individual to maintain fixation on a stimulus – often in the face of varying types of distracters. Visual fixation paradigms are associated with activation of premotor, prefrontal, and striatal structures (Paus 1991; Brown et al. 2007; Curtis and Connolly 2008) and, as a result, can be useful in examining oculomotor persistence in ADHD, particularly as poor postural stability of head (Ohashi et al. 2010) is likely to have deleterious effect on eye fixation and oculomotor control.

Experimental examination of eye movements also involves assessment of *saccades,* which are fast eye movements made by the oculomotor system. Their purpose is to bring some part of the visual field onto the fovea where it can be seen and acted upon. Saccades are usually classified into various categories depending on what initiates them. *Reflexive* saccades are usually involuntary and can be triggered by a novel stimulus in the immediate environment. The saccade is usually accompanied by a quick head move in order to bring the eyes quickly on target. In laboratory assessments, reflexive saccade can be approximated by asking individuals to immediately move their eyes to a target light as soon as it comes on (usually with the head movement constrained). In contrast, *volitional* saccades are those eye movements that are made with intent, thus invoking greater executive control demand (so that the eyes go where the individual wishes).

Included within the category of volitional saccades are: delayed saccades, memory-guided saccades, antisaccades, anticipatory saccades, and predictive saccades. *Delayed* saccades are those saccades that are initiated some time after a target is perceived. The individual knows the target, but is required to inhibit responding until an external signal is given to respond. *Memory-guided* saccades are made some time after the stimulus has disappeared and the individual looks toward his/her last seen position. The *antisaccade* is a saccade that is made in the opposite direction of a presenting target, requiring inhibition of a prepotent or reflexive response. *Anticipatory* saccades are those eye movements made with the idea that a stimulus will occur sometime in the future, but without the knowledge of where or when it will occur. *Predictive* saccades are made to a regularly occurring target and have the distinction of being elicited before the stimulus has occurred. Unlike anticipatory saccades, which can occur any time before the stimulus is presented, predictive saccades occur as a specific response to the repetitiveness of the stimulus.

The types of variables obtained in oculomotor paradigms tend to capture three important components of executive control: *response preparation* (saccade latency and variability), *response inhibition* (antisaccade directional errors; anticipatory errors, go/no-go commission errors), and *working memory* (memory-guided saccade accuracy). Interpretation of saccades in children should take into account age-related change (Munoz et al. 1998; Fukushima et al. 2000), whereby saccadic performance improves rapidly during early childhood and then stabilizes after adolescence, with inhibition and response latency stabilizing at age 14 and 15, respectively, and spatial working memory stabilizing latest (age 19) (Luna et al. 2004). Frontal and parietal cortex and anterior cingulate appear to be involved in a "state of preparedness" for oculomotor tasks. Increased intraindividual variability (observable on oculomotor tasks) has been linked with frontal circuits important for motor response selection and inhibition, particularly those involving the rostral supplementary motor complex (pre-SMC) (Mostofsky and Simmonds 2008).

4.2 Assessment of Oculomotor Skills in ADHD

Studies examining oculomotor skills have contributed to the understanding of the neurobiological basis of ADHD. In general, they have supported the "motor intentional deficit" hypotheses, with ADHD groups demonstrating robust abnormalities related to response inhibition (i.e., increased commission errors on antisaccade and go/no-go tasks; greater anticipatory errors) and response preparation (i.e., increased response latency and intraindividual variability). Oculomotor deficits in smooth pursuit eye movements (Castellanos et al. 2000) and working memory have been found less consistently in studies of ADHD (Mahone et al. 2009a).

4.2.1 Response Preparation

Children with ADHD have longer and more variable response latency on even the most basic "reflexive" prosaccade tasks (Mostofsky et al. 2001) although, in a later study with a larger sample, the impairment was observed in girls (but not boys) with

ADHD (Mahone et al. 2009a). Goto (Goto et al. 2010) also found increased response latency on prosaccade and antisaccade tasks in children with ADHD, but only among younger age groups (6–8 years); similar deficits were not observed among older children with ADHD. There is also emerging evidence that (like motor skills) oculomotor skills may represent an important endophenotype for ADHD. Van der Stigchel (Van der Stigchel et al. 2007) found that boys with ADHD and their unaffected brothers had slower oculomotor responses across tasks than controls. Across groups, oculomotor response latency correlated with a dimensional rating of ADHD inattentive symptoms, such that greater symptoms were associated with increased response latency.

4.2.2 Response Inhibition

Increased response inhibition errors on oculomotor tasks have been a robust and consistent finding across different studies and paradigms, including children (Mostofsky et al. 2001; Goto et al. 2010; Mahone et al. 2009a) and adults with ADHD (Armstrong and Munoz 2003; Nigg et al. 2002). Compared to typically developing controls, children with ADHD show increased commission errors, more intrusion errors on go/no-go tasks (Ross et al. 1994), and a higher proportion of anticipatory and directional errors than controls (Goto et al. 2010; Hanisch et al. 2006). Similar results are obtained among samples comprising primarily boys (Mostofsky et al. 2001) and those with exclusively girls (Castellanos et al. 2000). In the study of only girls, those with ADHD made twice as many commission errors and three times as many intrusion errors on a go/no-go task than controls (Castellanos et al. 2000), while smooth pursuit performance was equivalent across groups.

Increased inhibition and anticipatory errors have also been observed on antisaccade and delays tasks (Van der Stigchel et al. 2007), with the proportion of intrusive saccades correlating with ADHD symptom severity. Mahone et al. (2009a) found that both boys and girls with ADHD (aged 8–12 years) had increased inhibitory errors (i.e., commissions on antisaccade and go/no-go tasks; anticipatory errors on memory-guided saccades). Among older children (aged 11–14), those with ADHD combined (but not inattentive) subtype had increased antisaccade errors (relative to male controls) but showed significant improvement in these errors following treatment with methylphenidate (O'Driscoll et al. 2005).

4.2.3 Working Memory

ADHD groups (especially those with predominantly or exclusively male samples) have shown inconsistent deficits in the accuracy of memory-guided saccades (MGS), which are thought to assess spatial working memory (Mostofsky et al. 2001; Ross et al. 1994). In one study that included only girls, there was a strong trend ($p = 0.07$) for reduced MGS accuracy (Castellanos et al. 2000) and, in a mixed sample, children with ADHD (aged 8–13) had poorer resistance to peripheral distracters (fixation), inhibition on antisaccades, and poorer spatial working memory (memory-guided saccades), compared to controls (Loe et al. 2009). However, in a more recent study including a sample of boys and girls with ADHD, matched on subtype, neither boys nor girls with ADHD (as a group) were deficient on the MGS task relative to controls although, across sexes, those with inattentive subtype were deficient on the MGS task (Mahone et al. 2009a).

4.3 Summary: Oculomotor Functions in ADHD

Models of frontal lobe structure and function describe at least five parallel frontal-subcortical circuits (Alexander et al. 1986). The most posterior are related to motor and oculomotor function (originating in skeletomotor and oculomotor regions of the cortex); the more anterior are thought to be crucial in control of higher-order behavior (e.g., cognitive "executive" and socioemotional control). These regions have been identified as anomalous in volumetry. Studies examining oculomotor skills have contributed to the understanding of the neurobiological basis of ADHD, particularly given the utility of oculomotor paradigms to examine the "interface" between motor skills and cognition in ADHD and other neurodevelopmental disorders.

5 Conclusions

ADHD is a disorder involving delayed, anomalous development of the cerebral cortex and subcortical regions, including the basal ganglia (particularly the caudate nucleus), corpus callosum, and cerebellum. From a developmental perspective, brain systems affected in ADHD, including the cortex, subcortical white matter, basal ganglia, and cerebellum, share a pattern of reciprocal influence ("crossed trophic effect"), such that early injury to subcortical structures impairs not only the cerebral cortical development but also the development of the more remote-developing cerebellum, and vice versa. It remains unclear, however, whether behaviors associated with emergence of ADHD are associated with anomalous development of the basal ganglia and/or cerebellum, with subsequent reduction in growth of the cerebral cortex, or vice versa. Volumetric MRI studies of younger children suggest that structures of the basal ganglia, which mature earlier than the cerebellum and cortex, may play a crucial role in the early development of the disorder. Nevertheless, it is clear that when the "normal" timing and trajectory of brain development is altered and that behavioral and cognitive symptoms can persist, even after the brain has "caught up" with regard to size and shape.

While the preponderance of volumetric imaging studies of childhood ADHD highlight anomalous development of frontostriatal regions (especially prefrontal and premotor cortex and caudate) and cerebellar vermis, there have been inconsistencies in the findings as a result of reliance on cross-sectional studies, samples with disproportionate numbers of boys, inconsistent screening for comorbidities, and different parcellation protocols. More recent research is taking advantage of advanced computational methods to assess more closely the shape and morphology of regions of interest, providing insights beyond those obtained through examination of volumes alone.

The developmental brain anomalies observed in ADHD occur in parallel with developmental anomalies (delays) in motor and oculomotor development. This is not surprising, given the overlap between the brain systems associated with the behavioral symptoms of ADHD and those systems involved in the development of motor and oculomotor control. While motor skill deficits are not presently considered part of the DSM-IV ADHD diagnostic criteria, motor skill deficits commonly co-occur and are often overlooked in assessment and treatment of children with ADHD. Many teachers, however, report that children with ADHD, particularly boys, can be clumsy or have poor handwriting. While these could be by-products of hyperactivity, closer examination shows that children with ADHD often have trouble coordinating motor skills of all types. Thus, as children with ADHD progress through school, increasing demands that include more and more writing can contribute to fatigue, difficulty with sustained performance, less than optimal alertness, and frustration. All these issues can contribute to impressions of increased "distractibility" and to marked difficulties under the demands for simultaneous writing and listening. Similarly, just as children with ADHD have difficulty with arm, finger, and leg movements, they are also slower and less precise when making eye movements. These findings add to the evidence that children with ADHD need more time to complete tasks. Perhaps more importantly, the results suggest that children with ADHD are likely working much harder than their peers whenever they manage to keep pace with others. Thus, they are far more likely to experience fatigue (cognitive, physical) that could adversely affect their availability for learning. Thus, the research findings from volumetric imaging, motor, and oculomotor assessment, suggest that those working with children with ADHD should consider the impact of these neurobiological factors that contribute to increased cognitive load, including demands for multitasking, speeded performance, and simultaneous writing and listening – all of which can adversely affect a variety of life functions.

References

Addamo PK, Farrow M, Hoy KE, Bradshaw JL, Georgiou-Karistianis N (2007) The effects of age and attention on motor overflow production–A review. Brain Res Rev 54:189–204

Akkal D, Dum RP, Strick PL (2007) Supplementary motor area and presupplementary motor area: targets of basal ganglia and cerebellar output. J Neurosci 27:10659–10673

Alexander GE, DeLong MR, Strick PL (1986) Parallel organization of functionally segregated circuits linking basal ganglia and cortex. Annu Rev Neurosci 9:357–381

Almeida LG, Ricardo-Garcell J, Prado H, Barajas L, Fernandez-Bouzas A, Avila D, Martinez RB (2010) Reduced right frontal cortical thickness in children, adolescents and adults with ADHD and its correlation to clinical variables: a cross-sectional study. J Psychiatr Res 44:1214–1223

Anderson SW, Bechara A, Damasio H, Tranel D, Damasio AR (1999) Impairment of social and moral behavior related to early damage in human prefrontal cortex. Nat Neurosci 2:1032–1037

Armstrong IT, Munoz DP (2003) Inhibitory control of eye movements during oculomotor countermanding in adults with attention-deficit hyperactivity disorder. Exp Brain Res 152:444–452

Aylward EH, Reiss AL, Reader MJ, Singer HS, Brown JE, Denckla MB (1996) Basal ganglia volumes in children with attention-deficit hyperactivity disorder. J Child Neurol 11:112–115

Barkley RA (2006) Attention deficit hyperactivity disorder: a handbook for diagnosis and treatment, 3rd edn. Guilford Press, New York, NY

Bart O, Podoly T, Bar-Haim Y (2006) A preliminary study on the effect of methylphenidate on motor performance in children with comorbid DCD and ADHD. Res Dev Disabil 31:1443–1447

Bart O, Podoly T, Bar-Haim Y (2010) A preliminary study on the effect of methylphenidate on motor performance in children with comorbid DCD and ADHD. Res Dev Disabil 31:1443–1447

Basser PJ, Pierpaoli C (1996) Microstructural and physiological features of tissues elucidated by quantitative-diffusion-tensor MRI. J Magn Reson B 111:209–219

Batstra L, Neeleman J, Hadders-Algra M (2003) The neurology of learning and behavioural problems in pre-adolescent children. Acta Psychiatr Scan 108:92–100

Baumgardner T, Singer HS, Denckla MB, Rubin MA, Abrams MT, Colli MJ et al (1996) Morphology of the corpus callosum in children with Tourette syndrome and attention deficit hyperactivity disorder. Neurology 47:477–482

Berquin PC, Giedd JN, Jacobsen LK, Hamburger SD, Krain AL, Rapoport JL et al (1998) Cerebellum in attention deficit hyperactivity disorder: a morphometric MRI study. Neurology 50:1087–1093

Biederman J, Mick E, Faraone SV (2000) Age-dependent decline of symptoms of attention deficit hyperactivity disorder: impact of remission definition and symptom type. Am J Psychiatry 157:816–818

Black IB, DiCicco-Bloom E, Dreyfus CF (1990) Nerve growth factor and the issue of mitosis in the nervous system. Curr Top Dev Biol 24:161–192

Bledsoe J, Semrud-Clikeman M, Pliszka SR (2009) A magnetic resonance imaging study of the cerebellar vermis in chronically treated and treatment-naïve children with attention-deficit/hyperactivity disorder combined type. Biol Psychiatry 65:620–624

Brieber S, Neufang S, Bruning N, Kamp-Becker I, Remschmidt H, Herpertz-Dahlmann B, Fink GR, Konrad K (2007) Structural brain abnormalities in adolescents with autism spectrum disorder and patients with attention deficit/hyperactivity disorder. J Child Psych Psychiatry 48:1251–1258

Brown MR, Vilis T, Everling S (2007) Frontoparietal activation with preparation for antisaccades. J Neurophys 98:1751–1762

Bussing R, Grudnik J, Mason D, Wasiak M, Leonard C (2002) ADHD and conduct disorder: an MRI study in a community sample. World J Biol Psychiatry 3:216–220

Carmona S, Vilarroya O, Bielsa A, Tremols V, Soliva JC, Rovira M et al (2005) Global and regional gray matter reductions in ADHD: a voxel-based morphometric study. Neurosci Lett 389:88–93

Carmona S, Proal E, Haekzema E, Gispert J, Picado M, Moreno I et al (2009) Ventro-striatal reductions underpin symptoms of hyperactivity and impulsivity in attention-deficit/hyperactivity disorder. Biol Psychiatry 66:972–977

Carte ET, Nigg JT, Hinshaw SP (1996) Neuropsychological functioning, motor speed, and language processing in boys with and without ADHD. J Abnorm Child Psych 24:481–498

Casey BJ, Castellanos FX, Giedd JN, Marsh WL, Hamburger SD, Schubert AB et al (1997) Implication of right frontostriatal circuitry in response inhibition and attention-deficit/hyperactivity disorder. J Am Acad Child Adolesc Psychiatry 36:374–383

Castellanos F, Proal E (2009) Location, location, and thickness: volumetric neuroimaging of attention-deficit/hyperactivity disorder comes of age. J Am Acad Child Adolesc Psychiatry 48:979–981

Castellanos FX, Giedd JN, Eckburg P, Marsh WL, Vaituzis AC, Kaysen D et al (1994) Quantitative morphology of the caudate nucleus in attention deficit hyperactivity disorder. Am J Psychiatry 151:1791–1796

Castellanos FX, Geidd JN, Marsh WL, Hamburger SD, Vaituzis AC, Dickstein DP et al (1996) Quantitative brain magnetic resonance imaging in attention-deficit hyperactivity disorder. Arch Gen Psychiatry 53:607–616

Castellanos FX, Marvasti FF, Ducharme JL, Walter JM, Israel ME, Krain A et al (2000) Executive function oculomotor tasks in girls with ADHD. J Am Acad Child Adolesc Psychiatry 39:644–650

Castellanos FX, Giedd JN, Berquin PC, Walter JM, Sharp W, Tran T, Vaituzis AC, Blumenthal JD, Nelson J, Bastain TM, Zijdenbos A, Evans AC, Rapoport JL (2001) Quantitative brain magnetic resonance imaging in girls with attention-deficit/hyperactivity disorder. Arch Gen Psychiatry 58:289–295

Castellanos FX, Lee PP, Sharp W, Jeffries NO, Greenstein DK, Clasen LS et al (2002) Developmental trajectories of brain volume abnormalities in children and adolescents with attention-deficit/hyperactivity disorder. JAMA 288:1740–1748

Castellanos FX, Sharp WS, Gottesman RF, Greenstein DK, Giedd JN, Rapoport JL (2003) Anatomic brain abnormalities in monozygotic twins discordant for attention deficit hyperactivity disorder. Am J Psychiatry 160:1693–1696

Cherkasova MV, Hechtman L (2009) Neuroimaging in attention-deficit hyperactivity disorder: beyond the frontostriatal circuitry. Can J Psychiatry 54:651–663

Chess AC, Green JT (2008) Abnormal topography and altered acquisition of conditioned eyeblink responses in a rodent model of attention-deficit/hyperactivity disorder. Behav Neurosci 122:63–74

Cole W, Mostofsky S, Larson J, Denckla M, Mahone E (2008) Age-related changes in motor subtle signs among girls and boys with ADHD. Neurology 71:1514–1520

Curtis CE, Connolly JD (2008) Saccade preparation signals in the human frontal and parietal cortices. J Neurophysiol 99:133–145

Denckla MB (1974) Development of motor co-ordination in normal children. Dev Med Child Neurol 16:729–741

Denckla M (2005) Why assess motor functions "early and often?". Ment Retard Dev Disabil Res Rev 11:3

Denckla MB, Cutting LE (2004) Genetic disorders with a high incidence of learning disabilities. Learn Disabil Res & Pract 19:131–132

Denckla MB, Rudel RG (1978) Anomalies of motor development in hyperactive boys. Ann Neurol 3:231–233

Denckla MB, Rudel RG, Chapman C, Krieger J (1985) Motor proficiency in dyslexic children with and without attentional disorders. Arch Neurol 42:228–231

Diamond A (2000) Close interrelation of motor development and cognitive development and of the cerebellum and prefrontal cortex. Child Dev 71:44–56

Durston S, Tottenham NT, Thomas KM, Davidson MC, Eigsti IM, Yang Y et al (2003) Differential patterns of striatal activation in young children with and without ADHD. Biol Psychiatry 15:871–878

Durston S, Hulshoff Pol HE, Schnack HG, Buitelaar JK, Steenhuis MP, Minderaa RB et al (2004) Magnetic resonance imaging of boys with attention-deficit/hyperactivity disorder and their unaffected siblings. J Am Acad Child Adolesc Psychiatry 43:332–340

Durston S, van Belle J, de Zeeuw P (2010) Differentiating frontostriatal and fronto-cerebellar circuits in attention-deficit/hyperactivity disorder. Biol Psychiatry. doi:10.1016/j.biopsych.2010.07.037

Ellison-Wright I, Ellison-Wright Z, Bullmore E (2008) Structural brain change in attention deficit hyperactivity disorder identified by meta-analysis. BMC Psychiatry 8:51

Fassbender C, Schweitzer JB (2006) Is there evidence for neural compensation in attention deficit hyperactivity disorder? A review of the functional neuroimaging literature. Clin Psychol Rev 26:445–465

Filipek PA, Semrud-Clikeman M, Steingard RJ, Renshaw PF, Kennedy DN, Biederman J (1997) Volumetric MRI analysis comparing subjects having attention-deficit hyperactivity disorder with normal controls. Neurology 48:589–601

Fischl B, Dale AM (2000) Measuring the thickness of the human cerebral cortex using magnetic resonance images. Proc Natl Acad Sci 97:11044–11049

Flapper B, Houwen S, Schoemaker M (2006) Fine motor skills and effects of methlphenidate in children with attention-deficit-hyperactivity disorder and developmental coordination disorder. Dev Med Child Neurol 48:165–169

Fliers E, Rommelse N, Vermeulen SH, Altink M, Buschgens CJ, Faraone SV et al (2008) Motor coordination problems in children and adolescents with ADHD rated by parents and teachers: Effects of age and gender. J Neural Transm 115:211–220

Fliers E, deHoog M, Franke B, Faraone S, Rommelse N, Buitelaar J et al (2010) Actual motor performance and self perceived motor competence in children with attention-deficit hyperactivity disorder compared with healthy siblings and peers. J Dev Beh Ped 31:35–40

Fredericksen KA, Cutting LE, Kates WR, Mostofsky SH, Singer HS, Cooper KL, Lanham DC, Denckla MB, Kaufmann WE (2002) Disproportionate increases of white matter in right frontal lobe in Tourette syndrome. Neurology 58:85–89

Fukushima J, Hatta T, Fukushima K (2000) Development of voluntary control of saccadic eye movements. I. Age-related changes in normal children. Brain Dev 22:173–180

Giedd JN, Blumenthal J, Jeffries NO, Castellanos FX, Liu H, Zijdenbos A et al (1999) Brain development during childhood and adolescence: a longitudinal MRI study. Nat Neurosci 2:861–863

Gillberg C, Gillberg IC, Steffenburg S (1992) Siblings and parents of children with autism: a controlled population-based study. Dev Med Child Neurol 34:389–398

Gogtay N, Giedd JN, Lusk L, Hayashi KM, Greenstein D, Vaituzis AC, Nugent TF, Herman DH, Clasen LS, Toga AW, Rapoport JL, Thompson PM (2004) Dynamic mapping of human cortical development during childhood through early adulthood. Proc Natl Acad Sci 101:8174–8179

Goto Y, Hatakeyama K, Kitama T, Sato T, Kanemura H, Aoyagi K et al (2010) Saccade eye movements as a quantitative measure of frontostriatal network in children with ADHD. Brain Dev 32:347–355

Gustafsson P, Svedin C, Ericsson I, Linden C, Karlsson M, Thernlund G (2010) Reliability and validity of the assessment of neurological soft-signs in children with and without attention-deficit-hyperactivity disorder. Dev Med Child Neurol 52:364–370

Halperin JM, Schulz KP (2006) Revisiting the role of the prefrontal cortex in the pathophysiology of attention-deficit/hyperactivity disorder. Psychol Bull 132:560–581

Hanisch C, Radach R, Holtkamp K, Herpertz-Dahlmann B, Konrad K (2006) Oculomotor inhibition in children with and without attention-deficit hyperactivity disorder (ADHD). J Neural Trans 113:671–684

Harvey W, Reid G, Grizenko N, Mbekou V, Ter-Stepanian M, Joober R (2007) Fundamental movement skills and children with attention-deficit hyperactivity disorder: Peer comparisions and stimulant effects. J Abnorm Psychol 35:871–882

Hesslinger B, Tebartz van Elst L, Thiel T, Haegele K, Hennig J, Ebert D (2002) Frontoorbital volume reductions in adult patients with attention deficit hyperactivity disorder. Neurosci Lett 328:319–321

Hikosaka O, Takikawa Y, Kawagoe R (2000) Role of the basal ganglia in the control of purposive saccadic eye movements. Physiol Rev 80:953–978

Hill D, Yeo R, Campbell R, Hart B et al (2003) Magnetic resonance imaging correlates of attention-deficit/hyperactivity disorder in children. Neuropsychology 17:496–506

Hinshaw SP, Owens EB, Sami N, Fargeon S (2006) Prospective follow-up of girls with attention deficit/hyperactivity disorder into adolescence: evidence for continuing cross-domain impairment. J Consult Clin Psychol 74:489–499

Hoy KE, Georgiou-Karistianis N, Laycock R, Fitzgerald PB (2007) Using transcranial magnetic stimulation to investigate the cortical origins of motor overflow: a study in schizophrenia and healthy controls. Psychol Med 37:583–594

Hsu J-W, Lee L-C, Chen R-F, Yen C-T, Chen Y-S, Tsai M-L (2010) Striatal volume changes in a rat model of childhood attention-deficit/hyperactivity disorder. Psychiatry Res 179:338–341

Huttenlocher PR, Dabholkar AS (1997) Regional differences in synaptogenesis in human cerebral cortex. J Comp Neurol 387:167–178

Hynd GW, Semrud-Clikeman M, Lorys A, Novey ES, Eliopulos D (1990) Brain morphology in developmental dyslexia and attention deficit disorder/hyperactivity. Arch Neurol 47:919–926

Hynd GW, Hern KL, Novey ES, Eliopulos D, Marshall R, Gonzalez JJ et al (1993) Attention-deficit hyperactivity disorder and asymmetry of the caudate nucleus. J Child Neurol 8:339–347

Kadesjo B, Gillberg C (1998) Attention deficits and clumsiness in Swedish 7-year-old children. Dev Med Child Neurol 40:796–804

Kaplan B, Wilson B, Dewey D, Crawford S (1998) DCD may not be a discrete disorder. Hum Mov Sci 17:471–490

Kates WR, Folley BS, Lanham DC, Capone GT, Kaufmann WE (2002) Cerebral growth in Fragile X syndrome: review and comparison with Down syndrome. Microsc Res Tech 57:159–167

Kinney HC, Brody BA, Kloman AS, Gilles FH (1988) Sequence of central nervous system myelination in human infancy. II. Patterns of myelination in autopsied infants. J Neuropathol Exp Neurol 47:217–234

Klimkeit E, Sheppard D, Lee P, Bradshaw J (2004) Bimanual coordination deficits in attention deficit/hyperactivity disorder (ADHD). J Clin Exper Neuropsych 26:999–1010

Knyazeva M, Koeda T, Njiokiktjien C, Jonkman EJ, Kurganskaya M, de Sonneville L et al (1997) EEG coherence changes during finger tapping in acallosal and normal children: a study of inter- and intrahemispheric connectivity. Behav Brain Res 89:243–258

Koerte I, eftimov L, Laubender R, Esslinger O, Schroeder A, Ertl-Wagner B et al (2010) Mirror movements in healthy humans across the lifespan: effects of development and ageing. Dev Med Child Neurol 52:1106–1112

Kotrla KJ, Weinberger DR (2000) Developmental neurobiology. In: Sadock BJ, Sadock VA (eds) Comprehensive textbook of psychiatry, 7th edn. Lippincott Williams & Wilkins, Philadelphia, PA, pp 32–40

Krause KH, Dresel SH, Krause J, Kung HF, Tatsch K (2000) Increased striatal dopamine transporter in adult patients with attention deficit hyperactivity disorder: effects of methylphenidate as measured by single photon emission computed tomography. Neurosci Lett 285:107–110

Kroes M, Kessels AG, Kalff AC, Feron FJ, Vissers YL, Jolles J et al (2002) Quality of movement as predictor of ADHD: Results from a prospective population study in 5- and 6-year-old children. Dev Med & Child Neurol 44:753–760

Lantieri F, Glessner JT, Hakonarson H, Elia J, Devoto M (2010) Analysis of GWAS top hits in ADHD suggests association to two polymorphisms located in genes expressed in the cerebellum. Am J Med Genet B Neuropsychiatri Genet 153B:1127–1133

Largo RH, Caflisch JA, Hug F, Muggli K, Molnar AA, Molinari L et al (2001) Neuromotor development from 5 to 18 years. Part 1: timed performance. Dev Med Child Neurol 43:436–443

Largo RH, Fischer JE, Rousson V (2003) Neuromotor development from kindergarten age to adolescence: developmental course and variability. Swiss Med Wkly 133:193–199

Lasker AG, Mazzocco MM, Zee DS (2007) Ocular motor indicators of executive dysfunction in fragile X and Turner syndromes. Brain Cogn 63:203–220

Lavasani N, Stagnitti K (2010) A study on fine motor skills of Iranian children with attention deficit/hyper activity disorder aged from 6 to 11 years. Occup Ther Inter. doi:10.1002/oti.306

Leigh JS, Zee DS (2006) The neurology of eye movements, 4th edn. Oxford University Press, New York, NY

Lichter DG, Cummings JL (2001) Introduction and overview. In: Lichter DG, Cummings JL (eds) Frontal-subcortical circuits in psychiatric and neurological disorders. Guilford Press, New York, pp 1–43

Livesey D, Keen J, Rouse J, White F (2006) The relationship between measures of executive function, motor performance and externalising behaviour in 5- and 6-year-old children. Hum Mov Sci 25:50–64

Loe IM, Feldman HM, Yasui E, Luna B (2009) Oculomotor performance identifies underlying cognitive deficits in attention-deficit-hyperactivity disorder. J Am Acad Child Adolesc Psychiatry 48:431–440

Luders E, Narr KL, Hamilton LS, Phillips OR, Thompson PM, Valle JS et al (2009) Decreased callosal thickness in attention-deficit/hyperactivity disorder. Biol Psychiatry 65:84–88

Luna B, Garver KE, Urban TA, Lazar NA, Sweeney JA (2004) Maturation of cognitive processes from late childhood to adulthood. Child Dev 75:1357–1372

Mackie S, Shaw P, Lenroot R, Pierson R, Greenstein DK, Nugent TF, Sharp WS, Giedd JN, Rapoport JL (2007) Cerebellar development and clinical outcome in attention deficit hyperactivity disorder. Am J Psychiatry 164:647–655

Mahone EM, Mostofsky SH, Lasker AG, Zee D, Denckla MB (2009a) Oculomotor anomalies in attention-deficit/hyperactivity disorder: evidence for deficits in response preparation and inhibition. J Am Acad Child Adolesc Psychiatry 48:749–756

Mahone EM, Richardson ME, Crocetti D, O'Brien J, Kaufmann WE, Denckla MB, Mostofsky SH (2009b) Manual MRI parcellation of frontal lobe in boys and girls with ADHD [Abstract]. NeuroImage 47:S70

Mahone EM, Crocetti D, Ranta M, Gaddis A, Cataldo M, Slifer KJ, Denckla MB, Mostofsky SH (2011) A preliminary neuroimaging study of preschool children with ADHD. Clini Neuropsychol Doi: 10.1080/13854046.2011.580784

Marsh R, Gerber AJ, Peterson BS (2008) Neuroimaging studies of normal brain development and their relevance for understanding childhood neuropsychiatric disorders. J Am Acad Child Adolesc Psychiatry 47:1233–1251

Martin N, Piek J, Bayman G, Levy F, Hay D (2010) An examination of the relationship between movement problems and four common development disorders. Hum Mov Sci 29:799–808

Mataro M, Garcia-Sanchez C, Junque C, Estevez-Gonzalez A, Pujol J (1997) Magnetic resonance imaging measurement of the caudate nucleus in adolescents with attention-deficit hyperactivity disorder and its relationship with neuropsychological and behavioral measures. Arch Neurol 54:963–968

McAlonan G, Cheung V, Chua S, Oosterlaan J, Hung S, Tang C et al (2009) Age-related grey matter volume correlates of repsonse inhibition and shifting in attention-deficit hyperactivity disorder. Br J Psychiatry 194:123–129

Meyer A, Sagvolden T (2006) Fine motor skills in South African children with symptoms of ADHD: influence of subtype, gender, age, and hand dominance. Behav Brain Funct 2:33

Meyer BU, Röricht S, Woiciechowsky C (1998) Topography of fibers in the human corpus callosum mediating interhemispheric inhibition between the motor cortices. Ann Neurol 43:360–369

Middleton FA, Strick PL (2001) Cerebellar projections to the prefrontal cortex of the primate. J Neurosci 21:700–712

Mink JW (1996) The basal ganglia: focused selection and inhibition of competing motor programs. Prog Neurobiol 50:381–425

Morris M, Inscore A, Mahone EM (2001) Overflow movement on motor examination in children with ADHD [Abstract]. Arch Clin Neuropsych 16:782

Mostofsky SH, Simmonds DJ (2008) Response inhibition and response selection: two sides of the same coin. J Cogn Neurosci 20:751–761

Mostofsky SH, Reiss AL, Lockhart P, Denckla MB (1998) Evaluation of cerebellar size in attention deficit hyperactivity disorder. J Child Neurol 13:434–439

Mostofsky SH, Lasker AG, Cutting L, Denckla MB, Zee DS (2001) Oculomotor abnormalities in attention deficit hyperactivity disorder: a preliminary study. Neurology 57:423–430

Mostofsky SH, Cooper K, Kates W, Denckla M, Kaufmann W (2002) Smaller prefrontal and premotor volumes in boys with ADHD. Biol Psychiatry 52:785–794

Mostofsky SH, Newschaffer CJ, Denckla MB (2003) Overflow movements predict impaired response inhibition in children with ADHD. Percept Mot Skills 97:1315–1331

Mostofsky SH, Burgess MP, Gidley Larson JC (2007) Increased motor cortex white matter volume predicts motor impairment in autism. Brain 130:2117–2122

Mrzljak L, Uylings HB, Van Eden CG, Judas M (1990) Neuronal development in human prefrontal cortex in prenatal and postnatal stages. Prog Brain Res 85:185–222

Munoz DP, Everling S (2004) Look away: the anti-saccade task and the voluntary control of eye movement. Nat Rev Neurosci 5:218–228

Munoz DP, Broughton JR, Goldring JE, Armstrong IT (1998) Age-related performance of human subjects on saccadic eye movement tasks. Exper Brain Res 121:391–400

Nachev P, Kennard C, Husain M (2008) Functional role of the supplementary and pre-supplementary motor areas. Nat Rev Neurosci 9:856–869

Narr K, Woods R, Lin J, Kim J, Phillips O, Del'Homme M et al (2009) Widespread cortical thinning is a robust anatomical marker for attention-deficit/hyperactivity disorder. J Am Acad Child Adolesc Psychiatry 48:1014–1022

Nass R (1985) Mirror movement asymmetries in congenital hemiparesis: the inhibition hypothesis revisited. Neurology 35:1059–1062

Neale BM, Lasky-Su J, Anney R, Franke B, Zhou K, Maller JB, Vasquez AA, Asherson P, Chen W, Banaschewski T, Buitelaar J, Ebstein R, Gill M, Miranda A, Oades R, Roeyers H, Rothenberger A, Sergeant J, Steinhausen HC, Sonuga-Barke E, Mulas F, Taylor E, Laird N, Lange C, Daly M, Faraone S (2008) Genome-wide association scan of attention deficit hyperactivity disorder. Am J Med Genet Part B 147B:1337–1344

Nigg JT, Butler KM, Huang-Pollock CL, Henderson JM (2002) Inhibitory processes in adults with persistent childhood onset ADHD. J Consult Clin Psychol 70:153–157

O'Driscoll GA, Depatie L, Holahan AL, Savion-Lemieux T, Barr RG, Jolicoeur C et al (2005) Executive functions and methylphenidate response in subtypes of attention-deficit/hyperactivity disorder. Biol Psychiatry 57:1452–1460

Ohashi K, Vitaliano G, Polcari A, Teicher MH (2010) Unraveling the nature of hyperactivity in children with attention-deficit/hyperactivity disorder. Arch Gen Psychiatry 67:388–396

Paus T (1991) Two modes of central gaze fixation maintenance and oculomotor distractibility in schizophrenics. Schizphr Res 5:145–152

Paus T, Collins DL, Evans AC, Leonard G, Pike B, Zijdenbos A (2001) Maturation of white matter in the human brain: a review of magnetic resonance studies. Brain Res Bull 54:255–266

Pennington BF, Ozonoff S (1996) Executive functions and developmental psychopathology. J Child Psychol Psychiatry 37:51–87

Piek JP, Dyck MJ, Nieman A, Anderson M, Hay D, Smith LM et al (2004) The relationship between motor coordination, executive functioning and attention in school aged children. Arch Clin Neuropsychol 19:1063–1076

Pierrot-Deseilligny C, Rivaud S, Gaymard B, Agid Y (1991) Cortical control of reflexive visually-guided saccades. Brain 114:1473–1485

Pitcher TM, Piek JP, Hay DA (2003) Fine and gross motor ability in boys with attention deficit hyperactivity disorder. Dev Med Child Neurol 45:525–535

Plessen K, Bansal R, Zhu H, Whiteman R, Amat J, Quakenbush G et al (2006) Hippocampus and amygdala morphology in attention-deficit/hyperactivity disorder. Arch Gen Psychiatry 63:795–807

Pliszka SR, Glahn DC, Semrud-Clikeman M, Franklin C, Perez R III, Xiong J et al (2006) Neuroimaging of inhibitory control areas in children with attention deficit hyperactivity disorder who were treatment naive or in long-term treatment. Am J Psychiatry 163:1052–1060

Qiu A, Crocetti D, Adler M, Mahone EM, Denckla MB, Miller M et al (2009) Basal ganglia volume and shape in children with ADHD. Am J Psychiatry 166:74–82

Ranta ME, Crocetti D, Clauss JA, Kraut MA, Mostofsky SH, Kaufmann WE (2009) Manual MRI parcellation of the frontal lobe. Psychiatry Res 172:147–154

Rasmussen R, Gillberg C (2000) Natural outcome of ADHD with developmental coordination disorder at age 22 years: Acontrolled, longitudinal, community-based study. J Am Acad Child Adolesc Psychiatry 39:1424–1431

Reiersen A, Constantino J, Todd R (2008) Co-occurrence of motor problems and autistic symptoms in attention-deficit/hyperactivity disorder. J Am Acad Child Adolesc Psychiatry 47:662–672

Roebers CM, Kauer M (2009) Motor and cognitive control in a normative sample of 7-year-olds. Dev Sci 12:175–181

Roeder MB, Mahone EM, Gidley Larson J, Mostofsky SH, Cutting LE, Goldberg MC et al (2008) Left-right differences on timed motor examination in children. Child Neuropsychol 14:249–262

Rommelse N, Altink M, Oosterlaan J, Buschgens C, Buitelaar J, DeSonneville L et al (2007) Motor control in children with ADHD and non-affected siblings: deficits most pronounced using the left hand. J Child Psychol Psychiatry 48:1071–1079

Ross RG, Hommer D, Breiger D, Varley C, Radant A (1994) Eye movement task related to frontal lobe functioning in children with attention deficit disorder. J Am Acad Child Adolesc Psychiatry 33:869–874

Rubia K, Russell T, Overmeyer S, Brammer MJ, Bullmore ET, Sharma T et al (2001) Mapping motor inhibition: conjunctive brain activations across different versions of go/no-go and stop tasks. Neuroimage 13:250–261

Sasayama D, Hayashida A, Yamasue H, Harada Y, Kaneko T, Kasai K et al (2010) Neuroanatomical correlates of attention-deficit-hyperactivity disorder accounting for comorbid oppositinal defiant disorder and conduct disorder. Psychiatry Clin Neurosci 64:394–402

Sato M, Aotani H, Hattori R, Funato M (2004) Behavioral outcome including attention deficit hyperactivity disorder/hyperactivity disorder and minor neurological signs in perinatal high-risk newborns at 4–6 years of age with relation to risk factors. Ped Inter 46:346–352

Schall JD (1997) Visuomotor areas of the frontal lobe. Cereb Cortex 12:527–638

Schnoebelen S, Semrud-Clikeman M, Pliszka SR (2010) Corpus callosum anatomy in chronically treated and stimulant naïve ADHD. J Atten Disord 14:256–266

Schrimsher GW, Billingsley RL, Jackson EF, Moore BD III (2002) Caudate nucleus volume asymmetry predicts attention-deficit hyperactivity disorder (ADHD) symptomatology in children. J Child Neurol 17:877–884

Schuerholz LJ, Cutting L, Mazzocco MM, Singer HS, Denckla MB (1997) Neuromotor functioning in children with Tourette syndrome with and without attention deficit hyperactivity disorder. J Child Neurol 12:438–442

Semrud-Clikeman M, Steingard RJ, Filipek P, Biederman J, Bekken K, Renshaw PF (2000) Using MRI to examine brain-behavior relationships in males with attention deficit disorder with hyperactivity. J Am Acad Child Adolesc Psychiatry 39:477–484

Sergeant J (2000) The cognitive-energetic model: an empirical approach to attention-deficit hyperactivity disorder. Neurosci Biobehav Rev 24:7–12

Shaw P (2010) The shape of things to come in attention deficit hyperactivity disorder. Am J Psychiatry 167:363–365

Shaw P, Rabin C (2009) New insights into attention-deficit/hyperactivity disorder using structural neuroimaging. Curr Psychiatry Rep 11:393–398

Shaw P, Lerch J, Greenstein D, Sharp W, Clasen L, Evans A et al (2006) Longitudinal mapping of cortical thickness and clinical outcome in children and adolescents with attention-deficit/hyperactivity disorder. Arch Gen Psychiatry 63:540–549

Shaw P, Eckstrand K, Sharp W, Blumenthal J, Lerch JP, Greenstein D, Clasen L, Evans A, Giedd J, Rapoport JL (2007a) Attention-deficit/hyperactivity disorder is characterized by a delay in cortical maturation. PNAS 104:19649–19654

Shaw P, Gornick M, Lerch J, Addington A, Seal J, Greenstein D et al (2007b) Polymorphisms of the dopamine D4 receptor, clinical outcome, and cortical structure in attention-deficit/hyperactivity disorder. Arch Gen Psychiatry 64:921–931

Shaw P, Lalonde F, Lepage C, Rabin C, Eckstrand K, Sharp W, Greenstein D, Evans A, Giedd JN, Raporport J (2009a) Development of cortical asymmetry in typically developing children and its disruption in attention-deficit/hyperactivity disorder. Arch Gen Psychiatry 66:888–896

Shaw P, Sharp WS, Morrison M, Eckstrand K, Greenstein DK, Clasen LS, Evans AC, Rapoport JL (2009b) Psychostimulant treatment and the developing cortex in attention deficit hyperactivity disorder. Am J Psychiatry 166:58–63

Shook D, Brady C, Lee P, Kenealy L, Murphy E, Gaillard W et al (2010) Effect of dopamine transporter genotype on caudate volume in childhood ADHD and controls. Am J Med Gen Part B 156B:28–35

Sobel L, Bansal R, Maia T, Sanchez J, Mazzone L, Durkin K et al (2010) Basal ganglia surface morphology and the effects of stimulant medications in youth with attention deficit hyperactivity disorder. AJP in Advance 167:977–986

Soliva JC, Fauquet J, Bielsa A, Rovira M, Carmona S, Ramos-quiroga JA et al (2010) Quantitative MR analysis of caudate abnormalities in pediatric ADHD: proposal for a diagnostic test. Psychiatry Res 182:238–243

Sonuga-Barke EJS, Halperin JM (2010) Developmental phenotypes and causal pathways in attention deficit/hyperactivity disorder: potential targets for early intervention? J Child Psych Psychiatry 51:368–389

Sowell ER, Thompson PM, Welcome SE, Henkenius AL, Toga AW, Peterson BS (2003) Cortical abnormalities in children and adolescents with attention-deficit hyperactivity disorder. Lancet 362:1699–1707

Sowell ER, Thompson PM, Leonard CM, Welcome SE, Kan E, Toga AW (2004) Longitudinal mapping of cortical thickness and brain growth in normal children. J Neurosci 24:8223–8231

Steger J, Imhof K, Coutts E, Gundelfinger R, Steinhausen HC, Brandeis D (2001) Attentional and neuromotor deficits in ADHD. Dev Med Child Neurol 43:172–179

Stray L, Stray T, Iversen S, Ruud A, Ellertsen B (2009a) Methylphenidate improves motor functions in children diagnosed with hyperkinetic disorder. Behav Brain Funct 5:21

Stray L, Stray T, Iversen S, Ruud A, Ellertsen B, Tonnessen F (2009b) The motor function neurological assessment (MFNU) as an indicator of motor function problems in boys with ADHD. Behav Brain Funct 5:22

Stuphorn V, Taylor TL, Schall JD (2000) Performance monitoring by the supplementary eye field. Nat Rev Neurosci 408:857–860

Suskauer SJ, Simmonds DJ, Fotedar S, Blankner JG, Pekar JJ, Denckla MB et al (2008) Functional magnetic resonance imaging evidence for abnormalities in response selection in attention deficit hyperactivity disorder: differences in activation associated with response inhibition but not habitual motor response. J Cogn Neurosci 20:478–493

Tiemeier H, Lenroot RK, Greenstein DK, Tran L, Pierson R, Giedd JN (2010) Cerebellum develoopment during childhood and adolescence: a longitudinal morphometric MRI study. NeuroImage 49:63–70

Touwen B (1987) The meaning and value of soft signs in neurology. Grune & Stratton, Inc., New York

Touwen B, Sporrel T (1979) Soft signs and MBD. Dev Med Child Neurol 21:528–538

Valera EM, Faraone SV, Murray KE, Seidman LJ (2006) Meta-analysis of structural imaging findings in attention-deficit/hyperactivity disorder. Biol Psychiatry 61:1361–1369

Van der Stigchel S, Rommelse N, Geldof C, Witlox J, Oosterlaan J, Sergeant J et al (2007) Oculomotor capture in ADHD. Cogn Neuropsychol 24:535–549

Van Meel CS, Oosterlaan J, Heslenfeld DJ, Sergeant JA (2005) Motivational effects on motor timing in attention-deficit/hyperactivity disorder. J Am Acad Child Adolesc Psychiatry 44:451–460

Volpe JJ (1995) Neurology of the newborn, 3rd edn. WB Saunders, Philadelphia, PA

Voogd J, Schraa-Tam CK, van der Geest JN, De Zeeuw CI (2010) Visuomotor cerebellum in human and nonhuman primates. Cerebellum. doi:10.1007/s12311-010-0204-7

Waber DP, Mann MB, Merola J (1985) Motor overflow and attentional processes in normal school-age children. Dev Med Child Neurol 27:491–497

Wellington TM, Semrud-Clikeman M, Gregory AL, Murphy JM, Lancaster JL (2006) Magnetic resonance imaging volumetric analysis of the putamen in children with ADHD: combined type versus control. J Atten Disord 10:171–180

Wolosin SM, Richardson ME, Hennessy J, Denckla MB, Mostofsky SH (2009) Abnormal cerebral cortex structure in children with ADHD. Hum Brain Mapp 30:175–184

Neurodevelopmental Abnormalities in ADHD

Chandan J. Vaidya

Contents

1 Introduction 50
2 Structural Imaging in ADHD 51
   2.1 Key Findings 52
   2.2 Conclusion 55
3 Functional Imaging in ADHD 55
   3.1 Executive Function: Frontal–Striatal–Cerebellar Circuitry 56
   3.2 Reward-Related Decision Making: Mesolimbic Circuitry 57
   3.3 Visual–Spatial Attention: Parietal–Temporal Regions 58
   3.4 Motor-Execution: Motor-Premotor Regions 58
   3.5 Conclusion 59
4 Resting-State Imaging in ADHD 60
   4.1 Key Findings 61
   4.2 Conclusion 61
5 Future Directions 61
References 62

Abstract Structural and functional imaging studies in subjects with attention deficit hyperactivity disorder (ADHD) are reviewed with the goal of gleaning information about neurodevelopmental abnormalities characterizing the disorder. Structural imaging studies, particularly those with longitudinal designs, suggest that brain maturation is delayed by a few years in ADHD. However, a maturational delay model alone is incomplete: alternate courses are suggested by differences associated with phenotypic factors, such as symptom remission/persistence and exposure to stimulant treatment. Findings from functional imaging studies point to multiple loci of abnormalities that are not limited to frontal–striatal circuitry, which is important for executive and motivational function, but also include parietal,

---

C.J. Vaidya
Department of Psychology, Georgetown University, Washington, DC, USA
and
Children's Research Institute, Children's National Medical Center, Washington, DC, USA
e-mail: cjv2@georgetown.edu

temporal and motor cortices, and the cerebellum. However, a definitive conclusion about maturational delays or alternate trajectories cannot be drawn from this work as activation patterns are influenced by task-specific factors that may induce variable performance levels and strategies across development. In addition, no studies have implemented cross-sectional or longitudinal designs, without which the developmental origin of differences in activation cannot be inferred. Thus, current task-evoked functional imaging provides information about dynamic or state-dependent differences rather than fixed or trait-related differences. In the future, task-free functional imaging holds promise for revealing neurodevelopmental information that is minimally influenced by performance/strategic differences. Further, studies using longitudinal designs that identify sources of phenotypic heterogeneity in brain maturation and characterize the relationship between brain function and underlying structural properties are needed to provide a comprehensive view of neurodevelopmental abnormalities in ADHD.

Keywords ADHD · Brain development · Functional magnetic resonance imaging · Magnetic resonance imaging · Neuroimaging

Abbreviations

ADHD	Attention Deficit/Hyperactivity Disorder
DSM-IV	Diagnostic and Statistical Manual for Mental Disorders – 4th Edition
DTI	Diffusion tensor imaging
EEG	Electroencephalography
FA	Fractional anisotropy
fMRI	Functional magnetic resonance imaging
MEG	Magnetoencephalography
MRI	Magnetic resonance imaging

1 Introduction

Attention Deficit Hyperactivity Disorder (ADHD) is a childhood disorder of developmental origin that persists into adulthood, with deleterious effects on educational and vocational achievement, and social adaptation at each developmental stage. Symptoms of the disorder first become apparent in preschool years and persist into adulthood in approximately 60% of the cases (Biederman et al. 1996; Weiss and Hechtman 1993). Current diagnostic criteria require that symptoms appear prior to the age of 7 years and are expressed in at least two settings for at least 6 months

(American Psychiatric Association 2000). A primary challenge for developmental cognitive neuroscience is to characterize the neurodevelopmental origins and course of the disorder.

Broadly, the developmental progression of symptoms parallels the emergence of control processes mediated by the maturation of the prefrontal cortex. Thus, age-inappropriate levels of hyperactivity/impulsivity and inattention in ADHD children could reflect a maturational course that is atypical or typical but delayed. The primary source of evidence supporting the neurodevelopmental basis of ADHD has come from the application of noninvasive brain imaging methods. Magnetic resonance imaging (MRI) visualizes differences in tissue types, without ionizing radiation, to provide estimates of morphological volume, cortical thickness, and fiber architecture. Further, functional MRI (fMRI) visualizes differences in properties of blood oxygenation to provide estimates of involvement of brain structures during cognitive activity (termed "activation") and of temporal synchrony across regions (termed "functional connectivity"). Other methods with poorer spatial resolution but superior temporal resolution relative to MRI, such as scalp-recordings of voltage (electroencephalography, EEG) or magnetic field (magnetoencephalography, MEG) changes produced by electric currents associated with neural activity, have been used in ADHD. This work has been reviewed elsewhere (Barry et al. 2003a, b). This chapter parses the structural and functional MRI findings with the goal of evaluating the amount of support for a model of ADHD as delayed or aberrant brain maturation, or both.

A major shortcoming of the current application of MRI methods in ADHD is that they cannot disambiguate whether observed structural and functional brain differences are causal or reflect a consequence of the disorder. This shortcoming could be addressed by imaging in infancy and determining which brain differences persist in children who are subsequently diagnosed with ADHD. Currently, no study has used such a design.

2 Structural Imaging in ADHD

Structural imaging measures of the brain include volumetric measurements of gray or white matter of the whole brain (including or excluding the cerebellum) and its lobes, and fine-grained measures, such as cortical thickness or the density of gray matter acquired from individual voxels in the brain or across the cortical surface. Typical development comprises an overall increase in gray matter volume prior to puberty followed by reductions in adolescence (Giedd et al. 1999). Further, these changes are regionally diverse such that the peak volume was attained at 12 years in the frontal and parietal lobes, at 16 years in the temporal lobes and, even later, at 20 years in the occipital lobes. White matter volume continued to increase linearly from 4 to 22 years. The gray matter volumetric findings parallel the time-course of neuronal maturation (i.e., synapse proliferation followed by pruning) reported by postmortem histological studies of normal development. Fine-grained measures indicate that primary visual and sensorimotor regions mature first, followed by other cortical regions in a back-to-front sequence (e.g., parietal before frontal), with higher-order association cortices such as the superior temporal cortex maturing last (Gogtay et al. 2004). Gray matter measures are thought to reflect glial, vascular, and neuronal architecture, with age-related changes reflecting maturational changes in synaptic density; these neuronal processes cannot be measured directly in the living brain at present. Further, the optimal volume or thickness/gray matter density for a certain age is not known and so interpretation of findings in ADHD children has to be in relative terms, as departures from typical development, rather than in absolute terms.

2.1 Key Findings

ADHD as a delay in brain maturation is supported primarily by structural brain imaging studies that have constructed growth trajectories using a mixed cross-sectional and longitudinal design with large samples ranging in age from 5 to 18 years. Volumes of each lobe, of gray and white matter within each lobe, and overall cerebral and cerebellar volume were approximately 4% smaller in ADHD relative to control subjects, despite controlling for differences in exposure to stimulant medication, vocabulary scores, and height (Castellanos et al. 2002). Differences were also observed in the overall thickness of the cortical mantle (Shaw et al. 2007). In both ADHD and control groups, peak cortical thickness was attained earlier in sensory cortices than in association cortices. However, control children attained peak thickness earlier, between 7 and 8 years, relative to ADHD children who attained it later, between 10 and 11 years. This evidence suggested a similar course of the sequence of regional development in the ADHD and control subjects but with cortical maturation delayed by a few years in ADHD.

More evidence in support of widespread volumetric reductions in ADHD subjects comes from cross-sectional studies comparing ADHD and control subjects in smaller samples than in the above studies [see reviews (Seidman et al. 2005; Shaw and Rabin 2009)]. While there are many mixed findings in this body of work, the majority indicated that volumes were reduced in ADHD subjects relative to age-matched controls. The loci of the reported reductions are in multimodal association cortices such as the frontal lobes and its subregions, premotor cortex, posterior cingulate, anterior and medial temporal lobes, cerebellar lobules, and basal ganglia structures (caudate, globus pallidus, putamen, ventral striatum). One study reported larger inferior parietal and posterior temporal regions in ADHD adolescents than in controls (Sowell et al. 2003). Mixed findings across studies likely reflect differences in the composition of samples in ages, subtypes, medication history, and size. Specifically for basal ganglia structures, in addition to volumetric reductions, surface shape showed differences such that some regions were compressed while others had bulges (Qiu et al. 2009). Further probing into the locations of these surface deformations suggests that they included portions of basal ganglia structures that connect with premotor, oculomotor, and association areas of prefrontal cortices that are known to be smaller in ADHD.

Findings of some longitudinal studies also provide evidence for an altered maturational course in ADHD in selected brain regions. These findings come from the same studies discussed above. First, volumetric growth trajectories for the caudate converged across age such that ADHD children had smaller volumes relative to controls in childhood. However, these differences did not persist into adolescence as caudate volumes decreased with age in both groups (Castellanos et al. 2002). Thus, it appears that the caudate has a less rapid reduction in volume across age in ADHD children than in typical development. Volumetric reductions in typical development are thought to reflect the process of synapse pruning and, thus, this finding suggests an altered time course of that process in the caudate in ADHD. Second, development of hemispheric asymmetry differed selectively in the frontal cortex in ADHD subjects. Cortical thickness increased in right orbital and inferior frontal lobes and left occipital cortex in adolescence during typical development; ADHD children, in contrast, showed increased thickness in the occipital but not frontal cortices (Shaw et al. 2009a). Third, while the cortical mantle was thinner overall in ADHD subjects, some specific regions showed more pronounced differences including superior, medial, and precentral gyri in the frontal lobes (Shaw et al. 2006). A thinner cortical mantle, overall or in specific regions, cannot be interpreted as accelerated synaptic pruning as gray matter density reflects a variety of processes (glia, vasculature), which cannot be resolved with current imaging methods. Differences relative to control children can only be interpreted as an alteration of maturational processes. Therefore, these findings suggest that frontal–striatal and occipital regions show altered maturational courses in ADHD relative to typical development.

Alterations of brain maturation in ADHD are associated with phenotypic individual differences such as clinical outcome and exposure to stimulant treatment. ADHD children whose symptoms persisted into adolescence had thinner medial prefrontal cortex at an average age of 8.7 years compared to both ADHD children whose symptoms remitted and to controls (Shaw et al. 2006). Further, the thickness of the right parietal cortex was reduced through childhood in ADHD children relative to controls; however, this difference disappeared in adolescence for ADHD children whose symptoms resolved with increasing age. Cortical maturation between 12.5 and 16.4 years was influenced by exposure to stimulant medication such that unmedicated children showed greater cortical thinning than age-matched control children (Shaw et al. 2009b). Further, comparison between medicated and unmedicated ADHD children showed that the right motor cortex, left ventrolateral prefrontal cortex, and right parietooccipital cortex were thinner in unmedicated ADHD children; clinical outcome did not differ between the two groups. Thus, ADHD children who were not medicated between 12.5 and 16.4 years showed more rapid cortical thinning. A "thinner" cortical mantle relative to control children reflects reduced amounts of glial, neuronal, vascular, and synaptic processes that comprise the cortex, and is interpreted as reflecting less cortical maturation. Together, these results suggest that frontal–parietal cortical maturation in ADHD children differed depending upon the status of symptom expression and exposure to stimulant treatment.

In addition to the gray matter findings discussed earlier, white matter is affected in ADHD [see review (D'Agati et al. 2010). Although a reduction in overall white matter volume has been reported in ADHD children when studied using both longitudinal and cross-sectional designs, there are only two studies that examined the integrity of white matter using diffusion tensor imaging (DTI), which provides microstructural information sensitive to myelination, axonal thickness, and the axis of fiber direction. DTI measures the diffusion of water, which diffuses 3–7 times faster along fiber tracts than perpendicular to the tracts (Basser and Pierpaoli 1996; Pierpaoli and Basser 1996). It provides an index of this directional coherence of water diffusion (fractional anisotropy: "FA"), with higher FA values reflecting tracts with thicker, more myelinated, and more consistently organized fibers; currently, no methods are available to measure these properties separately. In a study comparing 7- to 11-year-old ADHD children with age-matched controls, FA was reduced in ADHD children anteriorly in right premotor and striatal regions, as well as posteriorly in parieto–occipital and cerebellar areas (Ashtari et al. 2005). In a study examining specific fiber tracts, adults with ADHD had reduced FA relative to controls, specifically of the cingulum bundle and superior fasciculus, which are medial and lateral tracts, respectively, that connect the frontal and parietal lobes (Makris et al. 2008). While there are no data on the developmental course of white matter microstructure in ADHD, these cross-sectional studies suggest that anterior and posterior white matter and connecting tracts are less mature in ADHD. Less mature white matter tracts are thought to slow down neural conduction, thereby reducing the processing efficiency of functions subserved by regions connected by those tracts.

The largest band of white matter fibers in the brain, the corpus callosum, which connects the left and right hemispheres, has been the focus of a large volume of studies. Across studies, children with ADHD had reduced volume of the splenium, the posterior corpus callosum that connects bilateral parieto–temporal cortices [see meta-analysis (Valera et al. 2007)]. Further, another meta-analysis reported that boys with ADHD also showed reductions in the volume of the rostral portion of the corpus callosum (Hutchinson et al. 2008). Both these findings, differences in anterior and posterior portions, were confirmed in a study that used a finer-grained approach for dividing the corpus callosum (100 segments versus 5 in past work) (Luders et al. 2009). However, the anterior differences did not persist after controlling for brain volume and conditions that are often comorbid with ADHD (e.g., oppositional defiant disorder). The corpus callosum is important for efficient communication between the two hemispheres, and reduced volume of those fibers ought to affect the efficiency of interhemispheric communication, and therefore, cognitive functions that depend upon bilateral collaboration. Indeed, the volume of the rostral portion was positively associated with the speed of response control in boys with ADHD (McNally et al. 2010). Findings in posterior portions may influence contributions of parietal–temporal cortices to attentional function. Specifically, in an EEG study, right-hemispheric contributions over inferior parietal cortex were increased in ADHD adults during the performance of a task that required sustained attention (Hale et al. 2010). However, reduced right parietal function has also been suggested by behavioral studies showing a rightward bias in spatial attention (similar to that in visual neglect) in ADHD children (Sheppard et al. 1999).

2.2 Conclusion

The current body of structural imaging findings in ADHD provides evidence for a global maturational delay based on reduced gray and white matter volumes and cortical thickness in ADHD relative to controls through childhood and adolescence. However, there is also evidence for altered maturational courses of selected regions such as the caudate nucleus and frontal cortical mantle, relative to typical development. The frontal cortex and caudate form a network that is important for behavior requiring executive control, an area of impairment in ADHD.

There are individual differences in cortical maturational trajectories of frontal and parietal cortices depending upon whether symptoms persist into adolescence and whether they are treated with stimulant medication. ADHD children with remitted or treated symptoms did not differ from controls in adolescence. Those whose symptoms persisted into adolescence had a thinner cortical mantle in medial prefrontal cortex in childhood and in parietal cortex in adolescence. Further, those who discontinued stimulant medication during adolescence had thinner frontal, motor, and parietal–occipital cortices relative to medicated children, despite similar clinical outcome. Thus, maturational courses were influenced by symptom status as well as stimulus medication exposure. While the direction of the influence (i.e., whether gray matter differences caused symptoms to persist or vice versa) cannot be determined based on these data, these findings point to the dynamic nature of brain maturation. They also have important implications for formulating neurodevelopmental models of ADHD by suggesting that there is likely to be high heterogeneity in the nature of maturational differences, whether delayed or aberrant, or both, in ADHD relative to typical development depending upon symptom expression and treatment choices. Specifically, an early maturational delay could resolve or continue through development, depending upon environmental and genetic factors that shape the child's behavioral experience, as well as brain structure and function. As those factors vary across individuals, maturational courses are likely to be heterogeneous across ADHD children.

3 Functional Imaging in ADHD

Knowledge about developmental functional characteristics of the brain in ADHD is based upon cross-sectional studies: no study has used longitudinal or mixed designs that are needed to characterize growth trajectories of functional activation in ADHD. Further, no cross-sectional studies have compared ADHD children of different ages. Functional imaging studies fall into three classes based upon sample ages, preadolescent (7–13 years), mixed children and adolescents (7–18 years), and adults. The majority of studies has focused primarily on males. One principled way to parse these findings is in terms of neural networks that subserve a functional domain that is affected in ADHD [see recent reviews (Makris et al. 2009; Vaidya and Stollstorff 2008)]. If activation is reduced/increased in one or more regions in ADHD relative to control children, it is interpreted as reflecting reduced/greater engagement of that region, which mediates cognitive processes involved in performance of the task at hand.

3.1 Executive Function: Frontal–Striatal–Cerebellar Circuitry

Executive function, the ability to control attention and action in the service of goals, has been the focus of the bulk of functional imaging work in ADHD subjects. It has been probed using tasks that draw upon component processes such as response inhibition, interference suppression, and working memory. These processes, to varying degrees, draw upon a circuit comprising regions in the frontal cortex (e.g., lateral prefrontal, premotor, anterior cingulate), dorsal striatum (e.g., caudate), and the cerebellum via thalamic projections (see: Fig. 1). Striatal activation has been evoked, using Go/No-go and Stop Signal Response Inhibition tasks in children (Durston et al. 2003; Vaidya et al. 1998, 2005) and adolescents (Rubia et al. 1999) and has been found to be consistently reduced in ADHD [confirmed by meta-analysis (Dickstein et al. 2006)]. Further, lateral inferior frontal regions associated with inhibitory control also showed reduced activation in ADHD children (Durston et al. 2003; Vaidya et al. 2005) and adolescents (Rubia et al. 1999, 2005). Studies of inhibitory control in ADHD adults using a Stroop task showed reduced dorsal anterior cingulate activation (Bush et al. 1999). Further, a study of working memory function using a mental rotation task found reduced activation in multiple frontal regions in ADHD adolescents (Silk et al. 2005). Together, these findings showing reduced activation in ADHD subjects indicate less engagement of regions mediating inhibitory and working memory processes. In contrast, ADHD adolescents had greater activation of inferior frontal and anterior

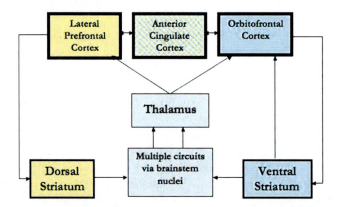

Fig. 1 Regions depicted in the figure show consistent differences in activation across studies in ADHD. *Yellow* depicts a circuit important for executive function and *blue* depicts a circuit important for motivational function

cingulate cortex during inhibitory control (Schulz et al. 2004), suggesting greater engagement of regions mediating inhibitory processes.

In addition to frontal–striatal differences, cerebellar activation was reduced in ADHD children, particularly during executive tasks relying on temporal processing (Durston et al. 2007). Thus, the bulk of the findings, which are from children and adults, indicate that ADHD is associated with reduced activation of frontal cortex and associated striatal and cerebellar structures during tasks drawing upon executive function.

Subject and performance-related factors influence the nature of frontal involvement in some studies. Inferior frontal activation was greater in ADHD and control children with better interference control (Vaidya et al. 2005), suggesting that poor inhibitory functioning is associated with inadequate inferior frontal engagement. However, studies with ADHD adolescents suggested the opposite: that is poor inhibitory function was associated with greater frontal engagement. Specifically, higher frontal activation was associated with lower inhibitory performance in adolescents with ADHD (Schulz et al. 2005a) and in adolescents with persisting symptoms and worse performance relative to those with remitted symptoms and better performance (Schulz et al. 2005b). With just a handful of studies, it cannot be discerned whether these mixed findings reflect age (children versus adolescents) or task differences (Flanker interference versus stimulus and response conflict) or both. Furthermore, as these findings are correlational, it is not possible to determine the direction of the relationship between activation and performance levels.

3.2 Reward-Related Decision Making: Mesolimbic Circuitry

In light of behavioral findings documenting atypical sensitivity to rewards in ADHD children (Luman et al. 2005), functional imaging of motivational function in ADHD has targeted reward-related brain circuitry using decision-making and choice tasks that manipulate reward contingencies. Reward processing in the brain has been primarily associated with mesolimbic dopaminergic projections encompassing ventromedial prefrontal cortex, anterior cingulate gyrus, ventral striatal nuclei, amygdala, and hippocampus (see: Fig. 1). During reward anticipation, ventral striatum was less activated in adolescents (Scheres et al. 2006) and adults (Strohle et al. 2008) with ADHD relative to controls. Further, the orbitofrontal cortex was also less activated in ADHD adults (Strohle et al. 2008). In another study in adults using a gambling task, activation was reduced in the hippocampus but increased in the anterior cingulate in ADHD relative to controls (Ernst et al. 2003). These regions, ventral striatum, hippocampus, and orbitofrontal cortex, are important for encoding the salience of a stimulus and evaluating it in the context of making decisions. Reduced activation of these regions suggests that these processes may be evoked to a lesser degree or less efficiently in ADHD than in control subjects. As these studies did not include children, it is impossible to evaluate neurodevelopmental differences in activation during motivational function in ADHD.

3.3 Visual–Spatial Attention: Parietal–Temporal Regions

Parietal–temporal involvement is apparent during executive function tasks that draw upon visual–spatial processes in working memory and inhibitory control. It is also evident during attention tasks such as selective attention to visual space and involuntary attention to auditory/visual oddballs. Spatial working memory tasks showed reduced activation of right inferior parietal cortex in ADHD children (Vance et al. 2007) and superior parietal and temporal regions in ADHD adolescents (Silk et al. 2005); medial parietal regions were more activated in ADHD adolescents than controls. Parietal cortices contribute to spatial representations associated with working memory and their reduced involvement parallels observations in the frontal lobes. Greater medial parietal activation is thought to reflect increased attentional resources necessary to perform difficult tasks such as spatial working memory.

Studies using the oddball involuntary attention task showed reduced activation in bilateral and medial–parietal cortex (Tamm et al. 2006), bilateral superior temporal gyri and posterior cingulate (Rubia et al. 2007) and parahippocampal gyrus and amygdala (Stevens et al. 2007) in adolescents with ADHD. Further, selective attention tasks also showed reduced activation of the right superior parietal cortex in ADHD children (Booth et al. 2005) and left posterior middle temporal gyrus in ADHD adolescents (Shafritz et al. 2004). Thus, reduced posterior engagement during both voluntary and involuntary attentional tasks in ADHD suggests a general reduction of attentional resources in ADHD.

In contrast to these reports of reductions in ADHD, a variety of inhibitory tasks showed increased activation of parietal or temporal regions on tasks that typically do not engage those regions in controls (Durston et al. 2003; Rubia et al. 1999; Schweitzer et al. 2000; Vaidya et al. 2005); these findings extend to all three ages (children, adolescents, and adults). This increased activation of posterior regions that are normally not engaged by these inhibitory tasks has been interpreted as reflecting compensatory strategies, which may or may not be effective in maintaining performance.

3.4 Motor-Execution: Motor-Premotor Regions

Children with ADHD often show subtle motor signs such as more variable trial-to-trial response latencies (Leth-Steensen et al. 2000) and motor overflow (Denckla and Rudel 1978) suggesting immature motor circuitry. Indeed, evidence supporting immaturity comes from a transcranial magnetic stimulation study showing reduced neural inhibition in the corticospinal tract in ADHD children (Moll et al. 2000). Motor immaturity is deleterious for executive functioning as greater motor overflow is associated with reduced response inhibition performance (Mostofsky et al. 2003). During self-paced finger-to-thumb movements, ADHD children showed reduced activation in contralateral motor cortex and right superior parietal cortex, relative to control children, despite similar levels of performance (Mostofsky et al. 2006). Such lower level motor abnormalities in ADHD are likely to contribute to higher-order executive function deficits reviewed earlier.

3.5 Conclusion

The review above indicates that functional activation differences between ADHD and control subjects are quite widespread. They include frontal-striatal-cerebellar networks, which are important for the integrity of executive function, a core domain of dysfunction in ADHD. In addition, limbic-frontal networks that are important for mediating motivational function and parietal-temporal networks that are important for mediating attentional function also show atypicalities in ADHD. Finally, differences between ADHD and control children also extend to lower level functions such as motor execution.

In the absence of studies with longitudinal or cross-sectional designs that enable evaluation of maturational courses, it is difficult to draw definitive conclusions about whether the observed functional activation differences in ADHD suggest delayed or aberrant development. Nevertheless, a tentative conclusion can be attempted in light of what is known about typical functional development. The largest body of developmental fMRI studies to date has focused on executive control processes, such as working memory and inhibitory control [see recent reviews (Bunge and Wright 2007; Luna et al. 2010)]. These findings indicate that greater prefrontal, parietal, and striatal involvement with age supports age-related improvements in executive control. Such age-related increases have also been observed for visual–spatial attentional function (Rubia et al. 2010). However, there are discontinuities such that in some studies adolescents showed greater prefrontal activation than adults despite similar levels of performance (Luna et al. 2001).

In studies that controlled for performance differences by visualizing activation associated only with correct performance, younger children showed greater prefrontal cortical engagement than adolescents or adults (Velanova et al. 2008). Thus, both reduced and increased activation signifies immaturity but yields different interpretation when considered in the context of differences in performance levels. Specifically, reduced activation accompanied by lower performance reflects processing capacity limitations, whereas greater activation paralleled by similar performance reflects more effortful processing. In findings from ADHD subjects, children tend to show reduced performance relative to controls, whereas adolescents and adults perform similarly to controls. Therefore, the observed pattern of reduced activation in ADHD children and greater activation in adolescents or adults relative to age-matched controls suggests capacity limitations and increased effort, respectively. Both patterns of differences are consistent with immature functional engagement and, furthermore, support a model of delayed maturation in ADHD as the engaged regions are the same as those in control subjects.

In drawing further conclusions about functional engagement in ADHD, it is important to note that, in typical development, fMRI studies also reveal a mixed regional pattern of reductions and increases in activation for the function at hand. For example, during response inhibition, age-related increases were observed in medial frontal gyrus and decreases in inferior frontal gyrus from 8 to 20 years (Tamm et al. 2002). These findings were interpreted as reflecting age-related improvements of inhibitory processing mediated by medial frontal gyrus and reduced effort mediated by inferior frontal gyrus. An alternate interpretation is that increases/decreases may be associated with specific strategies that draw differentially upon the two regions. Thus, the interpretation of regional group differences is not straightforward even in typical development.

In light of the above developmental fMRI findings, it would be simplistic to interpret the mixed pattern of differences between ADHD and control subjects as either developmental delay or aberrant maturation. While the structural findings suggest a developmental lag in brain maturation in ADHD, the maturational course is likely moderated by phenotypic factors affecting symptom progression. Further, in terms of functional engagement, performance effort and strategies strongly influence the observed differences. Performance is moderated by extrinsic (e.g., situational/contextual factors – with or without time pressure) and intrinsic (e.g., reward, interest) factors, making it difficult to predict what the developmentally appropriate pattern or level of activation is for a function at hand.

4 Resting-State Imaging in ADHD

One dependent measure that is minimally influenced by performance-related factors is functional connectivity during the resting state. A large body of evidence supports the view that when one is not engaged in task-specified cognition (termed resting-state), neural activity of different regions fluctuates spontaneously at slow rates (<0.1 Hz) (Biswal et al. 2010). These fluctuations are synchronized across regions forming distinct networks that are termed intrinsic connectivity networks. Most importantly, these intrinsic networks are spatially similar to those identified in task-evoked cognitive states (Smith et al. 2009). Task-evoked networks include a frontoparietal network that is observed during tasks of executive function, an insular-cingulate network evoked during monitoring and maintaining task sets, auditory and visual networks evoked by sensory tasks, and a medial prefrontal–posterior cingulate network evoked by self-referential cognition (termed the default-mode network). These same networks can be delineated when subjects are resting and not engaged in an experimenter-directed task. In fMRI studies of task states, most of these networks are activated when subjects are engaged in tasks directing attention to external stimuli, whereas the default-mode network is deactivated during those tasks. The consensus among researchers is that these networks form a fundamental functional organization of the brain, as it is independent of the subjects' current cognitive state.

4.1 Key Findings

Findings from intrinsic connectivity studies comparing ADHD adults and children to controls reveal atypicalities in specific networks, as well as the interrelationships across networks. First, functional connectivity among regions involved in executive control, such as frontal cortex, striatum, and cerebellum, was weaker in children with ADHD (Cao et al. 2006): amplitude reduced in some regions (right inferior frontal, left sensorimotor, and cerebellum) but increased in others (right anterior cingulate and brainstem) (Zang et al. 2007). Many of these regions are the same ones implicated by task-evoked studies discussed earlier. Thus, evidence from resting-state studies complements that from task-based studies by showing altered communication between the same regions that were atypically activated in ADHD. Second, studies of the default-mode network suggest that the posterior cingulate cortex is an important site of functional abnormality in ADHD. It was weakly connected functionally to other nodes within the network [e.g., medial prefrontal (Uddin et al. 2008)], but strongly connected to nodes of other networks [e.g., dorsal anterior cingulate cortex (Castellanos et al. 2008)]. This pattern of weak default-mode within-network connectivity and strong across-network connectivity was also observed in an EEG study with ADHD children (Helps et al. 2010). This pattern is thought to signify deficits in sustaining a task-oriented set due to intrusions from the default-mode network, which have been observed during attention lapses [e.g., mind-wandering (Mason et al. 2007)]. Thus, default-mode network abnormalities revealed by resting-state studies of ADHD provide new knowledge about ADHD.

4.2 Conclusion

Findings of fMRI studies of the resting-state in adults and children diagnosed with ADHD reveal atypical functional connectivity across regions, which is independent of task-directed cognitive activity. They suggest that disruptions of network-level temporal properties, specifically associated with the default-mode network, comprising medial prefrontal and posterior cingulate cortices, play an important role in attentional dysfunction in ADHD. Currently, these findings do not provide developmental information about ADHD as most studies are of adults. The findings, however, are important in opening up a new area for future neurodevelopmental investigations of ADHD.

5 Future Directions

Future studies ought to address at least five significant gaps in current research. First, current studies do not control adequately for comorbid conditions. Further, very few studies include an adequate number of females to examine sex differences.

As a result, the extent to which observed brain differences reflect ADHD, comorbidity, or general psychopathology cannot be discerned. Thus, studies including relatively homogenous samples balanced for gender are needed. Second, similar to studies of structural brain development in ADHD, functional imaging studies implementing longitudinal designs are needed. Imaging the same children at different ages is necessary to determine the stability of activation and the differences in performance between ADHD and controls at different developmental stages. Further, such within-subjects designs enable visualization of age-related performance improvement and its concomitant activation changes within the same children, regardless of absolute differences between ADHD and controls. Such data are important to reveal the neural signatures of developmental change in ADHD that are not confounded by differences in levels of performance relative to typical development. Third, it is important to investigate heterogeneity in brain development in ADHD. Currently, sources of heterogeneity in ADHD, such as symptom severity, treatment (e.g., stimulants versus nonstimulants), outcome in adolescence (e.g., symptom persistence versus remission) are controlled in the design of fMRI studies by including as homogenous a group as possible. However, as illustrated in findings from Schulz et al. (2005b), it is necessary to manipulate these as independent variables to reveal how phenotypic factors are associated with structural and functional development in ADHD. Fourth, there are few studies of motivational function in ADHD. Studies including tasks that parse apart executive components of evaluation/decision making from encoding of incentive/reward information are needed to reveal abnormalities that are not confounded by known executive deficits in ADHD. Finally, how structural and functional abnormalities relate to one another is relatively unknown in ADHD. More insight about the sources of functional atypicalities in ADHD can be gained by characterizing how they relate to underlying white matter structural properties within and across functional networks. Similarly, studies that elucidate how regional differences in cortical thickness impact upon activation or deactivation of regions during task states are important in the interpretation of functional atypicalities in ADHD.

The last 15 years have produced a wealth of information about brain development in ADHD. This forms a solid foundation for the next generation of studies with multimodal imaging methods and targeted experimental designs that will allow definitive conclusions about the developmental neuropathology underlying ADHD.

Acknowledgments Preparation of this manuscript was supported by grants MH084961 and MH086709 from the National Institutes of Mental Health.

References

American Psychiatric Association (2000) Diagnostic and statistical manual for mental disorders – 4th edition – text revision. American Psychiatric Association, Washington, DC

Ashtari M, Kumra S, Bhaskar SL, Clarke T, Thaden E, Cervellione KL et al (2005) Attention-deficit/hyperactivity disorder: a preliminary diffusion tensor imaging study. Biol Psychiatry 57:448–455

Barry RJ, Clarke AR, Johnstone SJ (2003a) A review of electrophysiology in attention-deficit/ hyperactivity disorder: I. Qualitative and quantitative electroencephalography. Clin Neurophysiol 114:171–183

Barry RJ, Johnstone SJ, Clark AR (2003b) A review of electrophysiology in attention-deficit/ hyperactivity disorder: II. Event-related potentials. Clin Neurophysiol 114:184–198

Basser PJ, Pierpaoli C (1996) Microstructural and physiological features of tissues elucidated by quantitative-diffusion-tensor MRI. J Magn Reson B 111:209–219

Biederman J, Faraone S, Milberger S, Curtis S, Chen L, Marrs A et al (1996) Predictors of persistence and remission of ADHD into adolescence: results from a four-year prospective follow-up study. J Am Acad Child Adolesc Psychiatry 35:343–351

Biswal BB, Mennes M, Zuo XN, Gohel S, Kelly C, Smith SM et al (2010) Toward discovery science of human brain function. Proc Natl Acad Sci USA 107:4734–4739

Booth JR, Burman DD, Meyer JR, Lei Z, Trommer BL, Davenport ND et al (2005) Larger deficits in brain networks for response inhibition than for visual selective attention in attention deficit hyperactivity disorder (ADHD). J Child Psychol Psychiatry 46:94–111

Bunge SA, Wright SB (2007) Neurodevelopmental changes in working memory and cognitive control. Curr Opin Neurobiol 17:243–250

Bush G, Frazier JA, Rauch SL, Seidman LJ, Whalen PJ, Jenike MA et al (1999) Anterior cingulate cortex dysfunction in attention-deficit/hyperactivity disorder revealed by fMRI and the Counting Stroop. Biol Psychiatry 45:1542–1552

Cao Q, Zang Y, Sun L, Sui M, Long X, Zou Q et al (2006) Abnormal neural activity in children with attention deficit hyperactivity disorder: a resting-state functional magnetic resonance imaging study. Neuroreport 17:1033–1036

Castellanos FX, Lee PP, Sharp W, Jeffries NO, Greenstein DK, Clasen LS et al (2002) Developmental trajectories of brain volume abnormalities in children and adolescents with attention-deficit/hyperactivity disorder. JAMA 288:1740–1748

Castellanos FX, Margulies DS, Kelly C, Uddin LQ, Ghaffari M, Kirsch A et al (2008) Cingulate-precuneus interactions: a new locus of dysfunction in adult attention-deficit/hyperactivity disorder. Biol Psychiatry 63:332–337

D'Agati E, Casarelli L, Pitzianti MB, Pasini A (2010) Overflow movements and white matter abnormalities in ADHD. Prog Neuropsychopharmacol Biol Psychiatry 34:441–445

Denckla MB, Rudel RG (1978) Anomalies of motor development in hyperactive boys. Ann Neurol 3:231–233

Dickstein SG, Bannon K, Castellanos FX, Milham MP (2006) The neural correlates of attention deficit hyperactivity disorder: an ALE meta-analysis. J Child Psychol Psychiatry 47:1051–1062

Durston S, Tottenham NT, Thomas KM, Davidson MC, Eigsti IM, Yang Y et al (2003) Differential patterns of striatal activation in young children with and without ADHD. Biol Psychiatry 53:871–878

Durston S, Davidson MC, Mulder MJ, Spicer JA, Galvan A, Tottenham N et al (2007) Neural and behavioral correlates of expectancy violations in attention-deficit hyperactivity disorder. J Child Psychol Psychiatry 48:881–889

Ernst M, Kimes AS, London ED, Matochik JA, Eldreth D, Tata S et al (2003) Neural substrates of decision making in adults with attention deficit hyperactivity disorder. Am J Psychiatry 160:1061–1070

Giedd JN, Blumenthal J, Jeffries NO, Castellanos FX, Liu H, Zijdenbos A et al (1999) Brain development during childhood and adolescence: a longitudinal MRI study. Nat Neurosci 2:861–863

Gogtay N, Giedd JN, Lusk L, Hayashi KM, Greenstein D, Vaituzis AC et al (2004) Dynamic mapping of human cortical development during childhood through early adulthood. Proc Natl Acad Sci USA 101:8174–8179

Hale TS, Smalley SL, Walshaw PD, Hanada G, Macion J, McCracken JT et al (2010) Atypical EEG beta asymmetry in adults with ADHD. Neuropsychologia 48:3532–3539

Helps SK, Broyd SJ, James CJ, Karl A, Chen W, Sonuga-Barke EJ (2010) Altered spontaneous low frequency brain activity in attention deficit/hyperactivity disorder. Brain Res 1322:134–143

Hutchinson AD, Mathias JL, Banich MT (2008) Corpus callosum morphology in children and adolescents with attention deficit hyperactivity disorder: a meta-analytic review. Neuropsychology 22:341–349

Leth-Steensen C, Elbaz ZK, Douglas VI (2000) Mean response times, variability, and skew in the responding of ADHD children: a response time distributional approach. Acta Psychol (Amst) 104:167–190

Luders E, Narr KL, Hamilton LS, Phillips OR, Thompson PM, Valle JS et al (2009) Decreased callosal thickness in attention-deficit/hyperactivity disorder. Biol Psychiatry 65:84–88

Luman M, Oosterlaan J, Sergeant JA (2005) The impact of reinforcement contingencies on AD/HD: a review and theoretical appraisal. Clin Psychol Rev 25:183–213

Luna B, Thulborn KR, Munoz DP, Merriam EP, Garver KE, Minshew NJ et al (2001) Maturation of widely distributed brain function subserves cognitive development. Neuroimage 13:786–793

Luna B, Padmanabhan A, O'Hearn K (2010) What has fMRI told us about the development of cognitive control through adolescence? Brain Cogn 72:101–113

Makris N, Buka SL, Biederman J, Papadimitriou GM, Hodge SM, Valera EM et al (2008) Attention and executive systems abnormalities in adults with childhood ADHD: A DT-MRI study of connections. Cereb Cortex 18:1210–1220

Makris N, Biederman J, Monuteaux MC, Seidman LJ (2009) Towards conceptualizing a neural-systems based anatomy of Attention Deficit/Hyperactivity Disorder. Dev Neurosci 31:36–49

Mason MF, Norton MI, Van Horn JD, Wegner DM, Grafton ST, Macrae CN (2007) Wandering minds: the default network and stimulus-independent thought. Science 315:393–395

McNally MA, Crocetti D, Mahone EM, Denckla MB, Suskauer SJ, Mostofsky SH (2010) Corpus callosum segment circumference is associated with response control in children with attention-deficit hyperactivity disorder (ADHD). J Child Neurol 25:453–462

Moll GH, Heinrich H, Trott G, Wirth S, Rothenberger A (2000) Deficient intracortical inhibition in drug-naive children with attention-deficit hyperactivity disorder is enhanced by methylphenidate. Neurosci Lett 284:121–125

Mostofsky SH, Newschaffer CJ, Denckla MB (2003) Overflow movements predict impaired response inhibition in children with ADHD. Percept Mot Skills 97:1315–1331

Mostofsky SH, Rimrodt SL, Schafer JG, Boyce A, Goldberg MC, Pekar JJ et al (2006) Atypical motor and sensory cortex activation in attention-deficit/hyperactivity disorder: a functional magnetic resonance imaging study of simple sequential finger tapping. Biol Psychiatry 59:48–56

Pierpaoli C, Basser PJ (1996) Toward a quantitative assessment of diffusion anisotropy. Magn Reson Med 36:893–906

Qiu A, Crocetti D, Adler M, Mahone EM, Denckla MB, Miller MI et al (2009) Basal ganglia volume and shape in children with attention deficit hyperactivity disorder. Am J Psychiatry 166:74–82

Rubia K, Overmeyer S, Taylor E, Brammer M, Williams SC, Simmons A et al (1999) Hypofrontality in attention deficit hyperactivity disorder during higher-order motor control: a study with functional MRI. Am J Psychiatry 156:891–896

Rubia K, Smith AB, Brammer MJ, Toone B, Taylor E (2005) Abnormal brain activation during inhibition and error detection in medication-naive adolescents with ADHD. Am J Psychiatry 162:1067–1075

Rubia K, Smith AB, Brammer MJ, Taylor E (2007) Temporal lobe dysfunction in medication-naive boys with attention-deficit/hyperactivity disorder during attention allocation and its relation to response variability. Biol Psychiatry 62:999–1006

Rubia K, Hyde Z, Halari R, Giampietro V, Smith A (2010) Effects of age and sex on developmental neural networks of visual-spatial attention allocation. Neuroimage 51:817–827

Scheres A, Milham MP, Knutson B, Castellanos FX (2006) Ventral striatal hyporesponsiveness during reward anticipation in attention-deficit/hyperactivity disorder. Biol Psychiatry 61:720–724

Schulz KP, Fan J, Tang CY, Newcorn JH, Buchsbaum MS, Cheung AM et al (2004) Response inhibition in adolescents diagnosed with attention deficit hyperactivity disorder during childhood: an event-related FMRI study. Am J Psychiatry 161:1650–1657

Schulz KP, Newcorn JH, Fan J, Tang CY, Halperin JM (2005a) Brain activation gradients in ventrolateral prefrontal cortex related to persistence of ADHD in adolescent boys. J Am Acad Child Adolesc Psychiatry 44:47–54

Schulz KP, Tang CY, Fan J, Marks DJ, Newcorn JH, Cheung AM et al (2005b) Differential prefrontal cortex activation during inhibitory control in adolescents with and without childhood attention-deficit/hyperactivity disorder. Neuropsychology 19:390–402

Schweitzer JB, Faber TL, Grafton ST, Tune LE, Hoffman JM, Kilts CD (2000) Alterations in the functional anatomy of working memory in adult attention deficit hyperactivity disorder. Am J Psychiatry 157:278–280

Seidman LJ, Valera EM, Makris N (2005) Structural brain imaging of attention-deficit/hyperactivity disorder. Biol Psychiatry 57:1263–1272

Shafritz KM, Marchione KE, Gore JC, Shaywitz SE, Shaywitz BA (2004) The effects of methylphenidate on neural systems of attention in attention deficit hyperactivity disorder. Am J Psychiatry 161:1990–1997

Shaw P, Rabin C (2009) New insights into attention-deficit/hyperactivity disorder using structural neuroimaging. Curr Psychiatry Rep 11:393–398

Shaw P, Lerch J, Greenstein D, Sharp W, Clasen L, Evans A et al (2006) Longitudinal mapping of cortical thickness and clinical outcome in children and adolescents with attention-deficit/hyperactivity disorder. Arch Gen Psychiatry 63:540–549

Shaw P, Eckstrand K, Sharp W, Blumenthal J, Lerch JP, Greenstein D et al (2007) Attention-deficit/hyperactivity disorder is characterized by a delay in cortical maturation. Proc Natl Acad Sci USA 104:19649–19654

Shaw P, Lalonde F, Lepage C, Rabin C, Eckstrand K, Sharp W et al (2009a) Development of cortical asymmetry in typically developing children and its disruption in attention-deficit/hyperactivity disorder. Arch Gen Psychiatry 66:888–896

Shaw P, Sharp WS, Morrison M, Eckstrand K, Greenstein DK, Clasen LS et al (2009b) Psychostimulant treatment and the developing cortex in attention deficit hyperactivity disorder. Am J Psychiatry 166:58–63

Sheppard DM, Bradshaw JL, Mattingley JB, Lee P (1999) Effects of stimulant medication on the lateralisation of line bisection judgements of children with attention deficit hyperactivity disorder. J Neurol Neurosurg Psychiatry 66:57–63

Silk T, Vance A, Rinehart N, Egan G, O'Boyle M, Bradshaw JL et al (2005) Fronto-parietal activation in attention-deficit hyperactivity disorder, combined type: functional magnetic resonance imaging study. Br J Psychiatry 187:282–283

Smith SM, Fox PT, Miller KL, Glahn DC, Fox PM, Mackay CE et al (2009) Correspondence of the brain's functional architecture during activation and rest. Proc Natl Acad Sci USA 106:13040–13045

Sowell ER, Thompson PM, Welcome SE, Henkenius AL, Toga AW, Peterson BS (2003) Cortical abnormalities in children and adolescents with attention-deficit/hyperactivity disorder. Lancet 362:1699–1707

Stevens MC, Pearlson GD, Kiehl KA (2007) An FMRI auditory oddball study of combined-subtype attention deficit hyperactivity disorder. Am J Psychiatry 164:1737–1749

Strohle A, Stoy M, Wrase J, Schwarzer S, Schlagenhauf F, Huss M et al (2008) Reward anticipation and outcomes in adult males with attention-deficit/hyperactivity disorder. Neuroimage 39:966–972

Tamm L, Menon V, Reiss AL (2002) Maturation of brain function associated with response inhibition. J Am Acad Child Adolesc Psychiatry 41:1231–1238

Tamm L, Menon V, Reiss AL (2006) Parietal attentional system aberrations during target detection in adolescents with attention deficit hyperactivity disorder: event-related fMRI evidence. Am J Psychiatry 163:1033–1043

Uddin LQ, Kelly AM, Biswal BB, Margulies DS, Shehzad Z, Shaw D et al (2008) Network homogeneity reveals decreased integrity of default-mode network in ADHD. J Neurosci Methods 169:249–254

Vaidya CJ, Stollstorff M (2008) Cognitive neuroscience of Attention Deficit Hyperactivity Disorder: current status and working hypotheses. Dev Disabil Res Rev 14:261–267

Vaidya CJ, Austin G, Kirkorian G, Ridlehuber HW, Desmond JE, Glover GH et al (1998) Selective effects of methylphenidate in attention deficit hyperactivity disorder: a functional magnetic resonance study. Proc Natl Acad Sci USA 95:14494–14499

Vaidya CJ, Bunge SA, Dudukovic NM, Zalecki CA, Elliott GR, Gabrieli JD (2005) Altered neural substrates of cognitive control in childhood ADHD: evidence from functional magnetic resonance imaging. Am J Psychiatry 162:1605–1613

Valera EM, Faraone SV, Murray KE, Seidman LJ (2007) Meta-analysis of structural imaging findings in attention-deficit/hyperactivity disorder. Biol Psychiatry 61:1361–1369

Vance A, Silk TJ, Casey M, Rinehart NJ, Bradshaw JL, Bellgrove MA et al (2007) Right parietal dysfunction in children with attention deficit hyperactivity disorder, combined type: a functional MRI study. Mol Psychiatry 12:826–832

Velanova K, Wheeler ME, Luna B (2008) Maturational changes in anterior cingulate and frontoparietal recruitment support the development of error processing and inhibitory control. Cereb Cortex 18:2505–2522

Weiss G, Hechtman L (1993) Hyperactive children grown up. Guilford Press, New York

Zang YF, He Y, Zhu CZ, Cao QJ, Sui MQ, Liang M et al (2007) Altered baseline brain activity in children with ADHD revealed by resting-state functional MRI. Brain Dev 29:83–91

Intraindividual Variability in ADHD and Its Implications for Research of Causal Links

Jonna Kuntsi and Christoph Klein

Contents

1 Introduction 69
2 Measurement of Variability 69
3 The Strength of the Association 72
    3.1 Phenotypic Association 72
    3.2 Genetic Association 75
4 Neural Basis 77
5 Theories 81
6 Nosological Specificity 84
7 Conclusions and Implications for Research and Intervention 85
References 86

Abstract Intraindividual variability (IIV) – reflecting short-term (within-session), within-person fluctuations in behavioral performance – and, specifically, reaction time (RT) variability, is strongly linked with attention-deficit hyperactivity disorder (ADHD) both at the phenotypic and genetic levels. Phenotypic case–control comparisons show a consistent and robust association between ADHD and RT variability across a broad range of cognitive tasks, samples, and age ranges (from childhood to adulthood). The association does not appear to be a nonspecific effect mediated by lower general cognitive ability. The finding from quantitative genetic

J. Kuntsi (✉)
King's College London, MRC Social, Genetic and Developmental Psychiatry Centre, Institute of Psychiatry, De Crespigny Park, London SE5 8AF, UK
e-mail: jonna.kuntsi@kcl.ac.uk

C. Klein
School of Psychology, Bangor University School of Psychology, Adeilad Brigantia, Penrallt Road, Gwynedd LL57 2AS, UK

Department of Child and Adolescent Psychiatry and Psychotherapy, University of Freiburg, Freiburg im Breisgau, Germany
e-mail: c.klein@bangor.ac.uk

studies of the shared genetic etiology between ADHD and RT variability is similarly robust, replicating across tasks, samples, and definitions of ADHD. Molecular genetic studies have produced intriguing initial findings: increasing sample sizes and replications across datasets remain priorities for future efforts. While the field has come a long way from considering increased RT variability in ADHD as the "noise" or "error" that we need to reduce in our data, the investigation of the causal pathways is only beginning. The neural basis of IIV is being investigated, with initial data pointing to a crucial role of fronto-striatal systems in controlling behavioral consistency. Several theories have been put forward to account for the observed IIV in ADHD, including accounts of arousal regulation, temporal processing and the "default-mode network." For the wider implications of the IIV phenomenon to be fully realized, we need to learn further about the underlying processes, their developmental context, and about shared and unique causal pathways across disorders where high RT variability is observed.

Keywords Attention-deficit hyperactivity disorder · Intraindividual variability · Reaction time variability

Abbreviations

BA	Brodman area
CPT	Continuous performance test
CV	Coefficient of variation
DAT	Dopamine transporter
DMN	Default-mode network
EEG	Electro-encephalography
FEF	Frontal eye fields
fMRI	Functional magnetic resonance imaging
GWAS	Genome wide association study
IIV	Intraindividual variability
MEG	Magneto-encephalography
MFFT	Matching familiar figures test
MPH	Methylphenidate
ODD	Oppositional Defiant Disorder
PET	Positron emission tomography
PFC	Prefrontal cortex: dl (dorsolateral), dm (dorsomedial), vl (ventrolateral)
RT	Reaction time
SMA	Supplementary motor area
SSRT	Stop signal reaction time
TBI	Traumatic brain injury
TMT	Trail making test
TOVA	Test of variables of attention
WAIS	Wechsler Adult Intelligence Scale

1 Introduction

Recent years have witnessed the transformation of the scientific investigation of intraindividual variability (IIV) [reflecting the short-term (within-session), within-person fluctuations in behavioral performance – and, specifically, reaction time (RT) variability in attention-deficit hyperactivity disorder (ADHD)] from a somewhat "oddball" topic to a hot one. Objectively, this rise in interest can be measured in the steep increase in published articles on this topic. Insightful early comments on the possible importance of RT variability in ADHD can be traced back to at least the 1970s (Firestone and Douglas 1975); yet, the topic was largely ignored by most investigators in the field until relatively recently. An influential review by Castellanos and Tannock, published in *Nature Reviews Neuroscience* in 2002, contributed to an emerging shift in the level of interest in the topic, by highlighting that *response variability is the one ubiquitous finding in ADHD research across a variety of speeded RT tasks, laboratories and cultures* (p. 624). The increased interest in the topic has opened up a discussion on whether a better understanding of IIV can help us learn more about the causal pathways in ADHD, as well as, potentially, the high rates of comorbidity between ADHD and several other disorders.

Here, we first review the evidence on the phenotypic and genetic association of RT variability with ADHD. We next review the evidence on its neural basis and consider theoretical approaches to the phenomenon and the nosological specificity issue: that is, whether increased RT variability can be observed in several disorders or is specific to ADHD.

2 Measurement of Variability

Increased IIV has been reported in ADHD for a wide range of behaviors, including classroom observational measures: e.g., attention and interference (Abikoff et al. 1977; Rapport et al. 2009), actigraph-based measures of activity (Wood et al. 2009), infrared motion analysis (Ohashi et al. 2010), motor timing (Rommelse et al. 2008), tapping responses (Ben-Pazi et al. 2006), and mood (reviewed in Skirrow et al. 2009). In the ADHD literature, IIV most commonly refers to RT variability. Reflecting this emphasis, this review focuses on variability in RTs. We comment on IIV in other aspects of performance only when relevant for the interpretation of RT variability.

In the vast majority of ADHD studies, intraindividual reaction time variability has been defined as the intraindividual standard deviation of RTs (referred to as "RT variability" in this review). While results obtained with this simple index are strong (see below), additional parameters have been derived to help identify the nature of the variability in ADHD. RT distributions are typically skewed and include an "ex-Gaussian" increased density of slow responses that could be explained by recurrent lapses of attention, for example. To cope with skewed RT distributions, an ex-Gaussian approach has been suggested (e.g., Heathcote et al. 2004; Leth-Steensen et al. 2000). This approach assumes that RT distributions represent the superposition (convolution) of a normal (Gaussian) and an exponential (ex-Gaussian) distribution and thus model the different parameters of these components separately. The output comprises three different parameters: μ and σ quantify the arithmetic mean and the standard deviation of the Gaussian part of the RT distribution (corresponding to the arithmetic mean and standard deviation of the Gaussian RT distribution, respectively), while τ models the mean of the exponential component. According to these models, the expectancy value of RT corresponds to $\mu + \tau$ and the variance of RT equals $\sigma^2 + \tau^2$ (Heathcote et al. 2004). While Gaussian and ex-Gaussian estimates of RT variability are conceptually different, statistically they may turn out to be highly redundant (Schmiedek et al. 2007). Also, results from the study of Leth-Steensen et al. (2000) suggest that, RT variability, or τ discriminate between patients with ADHD and controls equally well, and better than either μ or σ.

Another RT variability parameter is the coefficient of variation (CV = standard deviation of RT/mean RT), which controls for possible level (mean) differences. Alternatively, the mean can be taken into account when quantifying RT variability as well as differences between clinical groups in RTV using Analysis of Covariance (ANCOVA): By statistically leveling group differences in mean RT, only those group differences in RT variability are analyzed that cannot be explained already by group differences in mean RT. An empirical investigation of the links between mean differences and variability in RTs can be worthwhile. Nevertheless, we do not recommend an exclusive focus on RT variability parameters that control for mean RT in ADHD research because: (a) evidence suggests a shared etiology for increased mean RT and RT variability in ADHD; and (b) it has been argued that RTV is more fundamental than mean RT and the latter measure thus secondary to the former (Jensen 1992). Hence, indices such as CV may control, in part, for what one aims to study (Klein et al. 2006; Wood et al. 2010b; Kuntsi et al. 2010). Further measures of RT variability include the consecutive variance (Con; sqrt $(\Sigma(RT_i - RT_{i+1})^2/(n - 2))$; i = trial number, n = number of trials, sqrt = square root), quantifying the amount of moment-to-moment fluctuations (Klein et al. 2006), inter-quartile distances, and the range (although this last-mentioned parameter is not recommended because it is based on two data points only).

A conceptually different approach that offers a more precise characterization of the nature of RT variability focuses on the temporal structure of what can be considered a (reaction) time *series*. The Fourier analysis (Luck 2005) reconstructs the temporal signal as a superposition of sine waves of different frequencies and different strengths (power).

Using this technique, Gilden (2001) discovered a frequency characteristic of human RTs, which has a feature in common with a great number of diverse physical and biological systems: the so-called pink or 1/f noise (Fig. 1a). In 1/f noise, the

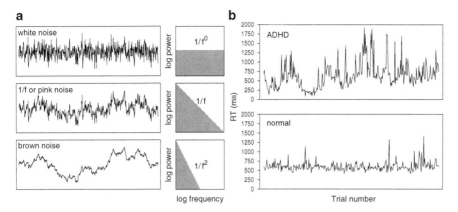

Fig. 1 1/f noise and a reaction time series of an individual with ADHD. (**a**) From Gilden (2001); (**b**) from Gilden and Hancock (2007)

power of the noise varies inversely with its frequency: relatively slow oscillations of high amplitude are convolved with relatively fast oscillations with low amplitude. Figure 1b displays the RT time series of an individual with ADHD. The similarity of this RT series with the pink noise seems obvious. In addition to the convolution of slow and fast frequencies, the RT series of the individual with ADHD seems to contain a linear trend suggesting that RT becomes longer with increasing trial number (for instance, due to time-on-task effects) and non-systematic RT spikes.

Di Martino et al. (2008) used another technique to investigate the temporal structure of the RT series of individuals with ADHD and controls: the "continuous wavelet transform" with Morlet wavelets. Wavelets are frequency "templates" that can be matched with a time series to extract the temporal course of the amplitude of specific frequency components. Di Martino et al. (2008) focused on slow oscillations because: (a) such oscillations have been found in basal ganglia neurons of awake, locally anesthetized rats and could be selectively modulated by dopaminergic medication; and (b) the brain's putative "default mode" network, proposed to underlie attention problems in ADHD (see Sect. 5 below), oscillates at such frequencies. Indeed, Di Martino and colleagues did find increased power in a slow-frequency band centered around 0.044 Hz that significantly separated participants with ADHD from typically developing children.

To summarize, RT series contain different types of IIV. RT variability is certainly sensitive to all of them, and it has proven useful to discover the extraordinary consistency of the phenomenon of increased RT variability in ADHD. Nevertheless, more specific parameters can help identify the nature of the variability in this disorder. With these methodological premises in mind, we next review the strength of the association of RT variability and ADHD.

3 The Strength of the Association

3.1 Phenotypic Association

Most cognitive tests that have been used in the ADHD literature require speeded responses and, as such, produce RT series that can be analyzed with respect to distribution statistics or temporal structure. With only a few exceptions, as yet, the analysis of the distribution statistics dominates the literature.

Among the most frequently used cognitive tests to study ADHD are the continuous performance test (CPT; Rosvold et al. 1956) and the go/no-go task (see: Aron and Poldrack 2005; Kerns et al. 2001; Schulz et al. 2004; Shue and Douglas 1992; Trommer et al. 1988). These two tasks "mirror" each other in that during the CPT, responding to "go" stimuli is unlikely and withholding responses to "no-go" stimuli is likely; in the go/no-go task, by contrast, responding to a go stimulus is likely and withholding responses to "no-go" stimuli is unlikely. The critical experimental variable in the CPT is therefore the proportion of go trials, since the smaller the likelihood of response inhibition, the greater the amount of response preparation and, hence, the more difficult is response inhibition. In the ADHD literature, the probability of go/no-go response inhibitions typically varies between about 25% and 50% of the trials, whereas the proportion of responses during the CPT is around 15%.

As reviewed in Klein et al. (2006), and reflecting the limited interest of earlier ADHD research in RT variability, less than 50% of the CPT studies derived RT variability as a measure from the CPT data at all. Of these, however, almost all reported increased variability in patients with ADHD. The two studies that did not report such an effect (Mahone et al. 2002; Mirsky et al. 1999) included an unusually dissimilar gender composition in the clinical groups. Importantly, several studies have shown that groups of participants with ADHD exhibited increased RT variability along with normal mean RT, suggesting that increased RT variability is not a consequence of, or secondary to, overall RT slowing. Similarly for the go/no-go task, RT variability has been reported in only some of the studies. Rubia et al. (2001), for instance, using a go/no-go task with 30% response inhibitions, reported increased RT variability in hyperactive children compared to typical controls, but not in psychiatric controls with disruptive behaviors. Increased RT variability was also reported by Mahone et al. (2002) for the TOVA-V, a go/no-go task with visual stimuli. [Additional questions that we have addressed in our research using go/no-go tasks are reviewed separately below (Klein et al. 2006; Kuntsi et al. 2009; Kuntsi et al. 2010; Uebel et al. 2010)].

Another popular task in ADHD research is the "stop task" (Logan and Cowan 1984) used in particular to test the behavioral disinhibition hypothesis of ADHD (Barkley 1997). RT variability was determined in most of the stop task studies we previously systematically reviewed (see Klein et al. 2006). Replicating findings obtained with the CPT, test statistics were in most cases larger for RT variability than mean RT. Only Daugherty et al. (1993) reported nonsignificant results for both mean RT and RT variability. Six studies reported slower and more variable responses in individuals with ADHD (Dimoska et al. 2003; Kuntsi et al. 2001; Oosterlaan and Sergeant 1995, 1998; Schachar et al. 1993; Scheres et al. 2001). Three studies administered two different tests (e.g., stop task and Conners' CPT; Purvis and Tannock 2000), and reported RT differences for only one of them, but differences in RT variability for both tasks (Purvis and Tannock 2000; Rubia et al. 1998; Schachar et al. 1993). Finally, eight studies reported no change in mean RT, but differences in RT variability (Jennings et al. 1997; McInerney and Kerns 2003; Nigg 1999; Pliszka et al. 2000; Rucklidge and Tannock 2002; Schachar and Logan 1990; Stevens et al. 2002; Willcutt et al. 2005). In two-thirds of the direct comparisons, test statistics for RT variability were again larger than those reported for stop signal reaction time (SSRT, the key inhibition index that measures the speed of the inhibition process); in the remaining comparisons, they were as often similar as they were smaller. This result is noteworthy, as Willcutt et al. (2005) had shown, in a large meta-analysis of 83 studies that had administered executive function tasks to a group with ADHD and another without ADHD (overall 6,700 participants tested), that the SSRT is among the best discriminators in ADHD-control comparisons. In a more recent study, Luman et al. (2009) measured RT variability in children with ADHD, children with ADHD and ODD, and typically developing controls. On the stop task, increased RT variability was found in the two clinical groups that, in terms of effect sizes, clearly surpassed the hypothesized slowing of SSRT by a factor of 3.2. Similarly, the group effect sizes in RT variability exceeded group effect sizes in median RT by a factor of 3.1.

Abnormally variable responding has been reported in ADHD across a variety of tasks, including selective attention (van der Meere and Sergeant 1988) and cancellation tasks (van der Meere et al. 1991), the Matching Familiar Figures Test (MFFT; Hopkins et al. 1979), pre-warned RT tasks (Cohen and Douglas 1972; Leth-Steensen et al. 2000), and the Stroop test (Leung and Connolly 1996; Hopkins et al. 1979). More recently, Oades and Christiansen (2008) used a Trail Making Test (TMT) and an RT task that involved cognitive switching between the identity or the number of figures, administering it to participants with ADHD, their siblings, and unrelated controls aged about 11 years. While participants with ADHD responded generally more variably than siblings or controls, the increase in RT variability during the switch, as compared to the no-switch condition, was greater in the patient group, even after controlling for level differences using the coefficient of variation.

Given the wide range of paradigms that report the high sensitivity of increased RT variability to ADHD, the questions arise whether IIV reflects a unitary phenomenon and whether other deficiencies may be secondary to the increased variability. These questions were addressed in our psychometric study that was based on data from 57 individuals with ADHD and 53 age- and gender-matched control participants who were assessed on the CPT, the stop task, the go/no-go task, as well as the 0-back and 1-back tasks (Klein et al. 2006). In support of the unitary construct hypothesis of increased RT variability in ADHD, a single principal component explained up to 67% of the interindividual differences in

IIV. Group comparisons revealed by far the strongest effect sizes for measures of dispersion (standard deviation, coefficient of variation, consecutive variance), followed by measures of central tendency and commission errors. Controlling statistically for RT variability, using ANCOVA, considerably reduced group differences in the other measures while, conversely, controlling groups for differences in mean RT or errors only slightly reduced the group differences in RT variability.

This pattern of differential associations with ADHD has recently been confirmed and extended in our analyses of data on the go/no-go task, a four-choice RT task (the "fast" task), and a choice-delay task from an international collaborative sample of 1,265 participants (ADHD probands, their siblings, and controls) aged from 6 to 18. The phenotypic correlation with ADHD was 0.39 for RT variability, 0.36 for mean RT, 0.22 for omission errors, 0.19 for commission errors, and −0.10 for a measure of choice impulsivity (Kuntsi et al. 2010). The findings on the etiological relationships between the variables, including analyses relevant to the "unitary construct" hypothesis, are reported in Sect. 3.2.

We have also reported an association with high RT variability for the continuum of ADHD symptoms in a population twin sample. In our analyses on go/no-go and fast data from 1,156 children, at a mean age of 8 years, we observed the association both for continuous ADHD symptom scores in the total sample and for in-group comparisons between children representing the top-scoring 5% on an ADHD composite score and the rest of the sample (Kuntsi et al. 2009).

An important topic in psychiatric research is the stability of a candidate marker variable in relation to the (potentially changing) clinical symptoms. Regarding IIV, this topic has been addressed by investigations of the early segment of childhood development. Increases in RT variability have been reported for children at risk for developing ADHD as young as 3–6 years of age, as shown by Berwid et al. (2005) for the parameter RT-SD derived from a combined CPT/go/no-go task. While the observed increases in mean RT in high-risk, as opposed to low-risk individuals, diminished with increasing age, the RT variability increases were independent of the participants' ages. This result suggests that increased IIV in ADHD is more robust to developmental changes than mean RT increases.

IIV has also been investigated in relation to the outcome of ADHD. Within this context and given the heterogeneity of adult ADHD outcomes, a clinically important topic is the distinction of "core" and persistent, as opposed to more "peripheral" and transient deficits that parallel the fluctuation and, possibly, recovery of symptoms. This leads to the question as to whether it is possible to predict persistence of ADHD symptoms through adolescence by deficits observed in childhood. Following this reasoning, Halperin et al. (2008) examined longitudinally the predictive validity of neuropsychological functioning in childhood for symptoms in late adolescence, using samples of 98 adolescents/young adults aged 18 years who had obtained an ADHD diagnosis as children and underwent neuropsychological examination at that time. The group was subdivided, based on the follow-up status, into those with persistent ($N = 44$), those with remitted ADHD symptoms ($N = 29$), and those who did not meet the criteria for persistence or remission. Analyses compared the whole clinical group and the two subgroups of remitted or persistent participants with ADHD with a sample of 85 typically developing controls (who were tested only as adolescents/young adults) on WAIS index scores, CPT, and Stroop parameters, as well as actigraph measures. Most interestingly, while the subgroup of "ADHD persisters" differed from controls in CPT hits, false alarms, d', and RT variability (but not mean RT), the subgroup of "ADHD remitters" differed from controls only in CPT RT variability (effect size d (ES): 0.57), Stroop word reading (ES: 0.55), and ankle actigraph activity (ES: 0.50). Based on these findings, Halperin et al. (2008) suggested that state regulatory mechanisms are likely to reflect a "core" underlying process in ADHD. The presence of increased IIV in ADHD remitters suggests that IIV is more trait-like than the symptoms of the disorder themselves.

Overall, the studies reviewed here indicate that ADHD shows a consistent and robust association with high RT variability across a broad range of cognitive tasks, samples, and age ranges (from childhood to adulthood). Future research should aim to establish further the extent to which this association varies with task demands and with the persistence versus remittance of symptoms.

Despite the strong evidence, as reviewed above, for the link between ADHD and increased RT variability, the association is significantly reduced or even disappears under certain conditions. In the population twin sample of 1,156 children, increased RT variability was no longer observed in the high ADHD symptom group in a fast task condition with rewards and a faster presentation rate of stimuli (1/s), and the association was also significantly reduced in a go/no-go task rewarded condition (Kuntsi et al. 2009). Similar comparisons within the fast task conditions and within the go/no-go task conditions in the large international collaborative sample indicated that RT variability decreased significantly more in the ADHD than control group from a baseline condition to a condition with rewards (with or without a faster event rate), although not quite to the level of the control group (Andreou et al. 2007; Uebel et al. 2010; see also Scheres et al. 2001). The joint effects of presentation rate and rewards suggest that "energetic" deficits are involved in increased IIV in ADHD.

3.2 Genetic Association

The strong phenotypic association between ADHD and RT variability raises the question of the extent to which genetic versus environmental factors account for this association. Given the high heritability for ADHD [recent estimates average at 76% (Faraone et al. 2005) and at 62% (Wood et al. 2010a)], much interest has focused on exploring possible shared genetic pathways. Quantitative genetic studies using twin and sibling designs can estimate the extent of shared genetic or familial influences on two (or more) phenotypes, such as ADHD and RT variability. Twin studies have the advantage over sibling designs in which "familial" influences can be separated into "genetic" and "shared environmental" (as well as "individual-specific environmental") influences. An advantage of sibling designs is that clinical samples of participants with ADHD diagnoses are easier to obtain, and, as such, the generalization of findings obtained in population-based twin samples to clinical ADHD samples can be investigated. It is the combined knowledge obtained across twin and sibling studies that can best inform us on the etiological pathways in ADHD.

The first suggestion of strong shared genetic influences on ADHD and RT variability came from our initial, small-scale twin study in 2001, where RT data were obtained from the stop task (Kuntsi and Stevenson 2001). A subsequent larger-scale family study of ADHD probands, siblings, and parents provided partial replication of this finding in indicating increased RT variability in both the ADHD probands and their mothers (Nigg et al. 2004). Focusing on dizygotic twin pairs, discordant for ADHD, and control twin pairs, Bidwell et al. (2007) reported increased RT variability on the stop task in the ADHD probands and their unaffected co-twins.

Further strong evidence for shared familial influences on ADHD and RT variability has emerged from the large collaborative study of a clinical sample of ADHD proband (combined subtype)-sibling pairs and control sibling pairs (Andreou et al. 2007; Uebel et al. 2010; Kuntsi et al. 2010). In model-fitting analyses of go/no-go and fast task data, we obtained a familial correlation with ADHD of 0.74 for RT variability, indicating that 74% of the familial influences on ADHD were shared with those on RT variability (Kuntsi et al. 2010). The familial correlation was also high (0.61) for mean RT, but noticeably lower for the other variables (0.48 for omission errors and 0.45 for commission errors). Further, a multivariate familial factor analysis across the different cognitive variables indicated a main RT familial factor, reflecting 85% of the familial variance of ADHD, which captured all familial influences on RT variability and 98% of those on mean RT (Kuntsi et al. 2010). RT variability and mean RT were indistinguishable at the familial level, with a familial correlation of 0.91. Initial analyses, where the two RT variability variables from the go/no-go and fast tasks were considered individually, had also indicated a high familial correlation (0.75) between them, suggesting that they measure largely the same underlying liability (Kuntsi et al. 2010). This supports the conclusion from previous phenotypic analyses that RT variability across several tasks (under baseline/unrewarded conditions) reflects a unitary construct (Klein et al. 2006). In further analyses, we also demonstrated how the association between ADHD and RT variability is largely independent (87%) of any contribution from etiological factors shared with IQ (Wood et al. 2011). This suggests that lower IQ does not account for the association between high RT variability and ADHD.

Key findings on the etiology of the ADHD-RT variability association from the clinical sibling-pair study closely match those obtained in the population twin sample of 1,314 children, using the same tasks. First, we obtained genetic correlations with ADHD symptoms scores of 0.68 for go/no-go task RT variability and 0.64 for fast task RT variability, indicating a similarly strong association (Wood et al. 2010b). Second, the genetic correlation across the two RT variability variables was also moderately high at 0.61. Third, we obtained a genetic correlation of 0.86 between mean RT and RT variability (Wood et al. 2010b), replicating the finding of shared etiology between mean RT and RT variability. Fourth, as in the clinical sample, the association between ADHD symptom scores and RT variability was largely independent (94%) of any contribution from etiological factors shared with IQ (Wood et al. 2010b). The comparability of the findings across the population twin sample and the clinical sibling-pair sample suggests that the familial covariance is largely genetic and supports the conceptualization of ADHD as the extreme of a continuously distributed trait.

Kebir et al. (2009) recently systematically reviewed the results of genetic studies investigating associations between putative susceptibility genes for ADHD and various neuropsychological traits relevant for the disorder, including RT variability. The genes that had attracted the largest number of studies were the dopamine-4 receptor gene (DRD4) and the dopamine transporter (DAT1). The CPT and derived tasks were the most commonly used cognitive measures. The most consistent result for DRD4 overall was, unexpectedly, the association between the *absence* of the 7-repeat risk allele and high RT variability. This finding was observed within ADHD samples only. For DAT1, the most replicated cognitive marker associated with the 10-repeat homozygosity was also high RT variability (conflicting results were obtained for commission and omission errors). Another approach is offered by genome-wide association studies (GWASs), which enable a hypothesis-free analysis of the entire genome, with power to detect genetic variants of small effect size. This approach is now being applied to RT variability data from an international ADHD collaboration.

Overall, the finding from quantitative genetic studies of the shared genetic etiology between ADHD and RT variability is robust, replicating across tasks, samples, and definitions of ADHD (diagnosis versus continuum of symptoms). Molecular genetic studies have produced intriguing initial findings; increasing sample sizes and replications across datasets remain priorities for future efforts.

4 Neural Basis

The empirical evidence that speaks to the question of the neural basis of IIV can be divided into three areas: anatomical, physiological, and pharmacological. Most of these findings have been obtained in studies on adults. Given the major developmental changes in brain structure and function that take place during childhood and adolescence (Klein and Feige 2005; Luciano 2010; Luna et al. 2008), the cited results are primarily relevant for an understanding of IIV as a (neuropsychological) construct. To what extent the reported findings also hold for typically and non-typically developing children and adolescents remains to be determined.

Regarding the *anatomical* evidence, among the clearest findings to point to the crucial role of frontal lobe regions for performance stability is the study of Stuss et al. (2003). Using an RT task that varied in complexity, the authors showed that lesions of the left or right dorsolateral PFC or the superior medial frontal lobes, but not inferior medial frontal or non-frontal lesions, increased RT variability. The auto-correlative structure of the RT series (i.e., the correlation of the RT in a given trial with the next, next but one, etc., RTs) and, here in particular, the correlation of the RT of subsequent trials (lag-1 auto-correlation) were not altered. The effects of traumatic brain injury (TBI) on different measures of IIV were investigated by Stuss et al. (1994) using RT tasks of different complexities and during two testing sessions 1 week apart. As increases in IIV could be the consequence of any kind of brain damage, the investigation of TBI patients with different etiologies is suited to provide conjecture or refutation of this assumption. TBI patients tended to respond generally more variably, particularly during the complex as compared to the simple tasks. This task difference was absent in controls. Patients and controls, however, did not differ in terms of linear RT trends or trial-to-trial RT auto-correlations. The overall weak effects of TBI on different aspects of IIV suggest that not all kinds of brain damage increase IIV. This result is hence not well suited to support the claim that IIV is "nonspecific."

When reviewing the *physiological* evidence, it is important to bear in mind that RT and its variation is measured with millisecond precision. Unlike anatomical evidence from lesion studies, which is able to identify brain regions that are *crucial* for a given function, physiological evidence is able to identify brain regions that are *involved* in a given function. Whether this involvement is crucial or of functional significance needs to be determined by examining brain–behavior relationships. Here, two approaches have been chosen: intraindividual and interindividual correlation analyses. Intraindividual correlations may establish that certain brain regions are, for instance, more (or less) activated preceding fast (or slow) RTs. Interindividual correlations may show that those who are, say, less variable in their responses are able to activate certain brain regions more (or less) than those who are more variable. Obviously, both approaches can be combined to investigate interindividual differences in intraindividual brain–behavior coupling.

Another important issue is the time scale. The precision of measuring RTs and their fluctuations has implications for the research design and data analysis: whether the temporal resolution of the physiological technique offers a corresponding temporal resolution or not. If not, the amplitude, but not the latency, of the brain response can be correlated with the latency of the behavioral reaction. Current brain-physiological techniques can be roughly grouped into those with excellent spatial and less convincing temporal resolution (such as functional magnetic resonance imaging (fMRI) or positron emission tomography (PET)) and those with excellent temporal and less favorable spatial resolution (such as electro-encephalography (EEG) or magneto-encephalography (MEG)) While the combinations of different techniques (e.g., MEG with fMRI and MRI; or EEG with MEG; or EEG with fMRI and MRI) to combine strengths and limit method-immanent weaknesses is becoming "state of the art," to our knowledge, no such method combinations have been used as yet to investigate the neural basis of IIV, whether in ADHD or in other populations.

The first group of techniques was chosen for the first direct examination of the neural correlates of IIV by Bellgrove et al. (2004). This study used an event-related fMRI design to investigate, in 42 healthy individuals, the brain-metabolic correlates of IIV, here assessed with intraindividual coefficient of variation, during a go/no-go task. During this task, the letters X and Y were presented in alternating order and subjects were required to respond to each of these letters. No-go stimuli were defined as an interruption of this alternation (e.g., X Y X X, with the fourth letter being the interruption). Greater behavioral variability (assessed with the individual coefficient of variation) was correlated behaviorally with poorer inhibitory performance and brain metabolically with increased brain activation during successful response inhibition in several brain regions, including left pre-central (BA 44/6, $r = 0.72$), right inferior frontal (BA 9, $r = 0.44$), bilateral middle frontal (BA 46, $0.42 \leq r \leq 0.52$), right inferior parietal (BA 40, $r = 0.35$), and thalamic ($r = 0.34$) regions. As these regions overlap substantially with those involved in response inhibition, Bellgrove and colleagues suggest that more variable individuals activate inhibitory networks more strongly because of a greater requirement for top-down executive control.

While the above studies focused on adult individuals, Simmonds et al. (2007) investigated the brain-metabolic correlates of RT variability in 30 healthy children aged 8–12 years using a go/no-go task. Go-trial activation in the anterior cerebellum and no-go-trial activation in the anterior supplementary motor area, the postcentral gyrus, the anterior cerebellum, and the inferior parietal lobule correlated with greater stability of RT, whereas higher RT variability correlated with greater prefrontal cortex and caudate activation. The latter correlation was interpreted as effortful information processing in tasks with minimal cognitive demands underlying increases in IIV. As the intraindividual coefficient of variation (ICV) was used as the measure of variability, the results can be considered as independent of the individual level of response speed.

Further insight into the neural underpinnings of IIV that are more specific to ADHD has been produced by Rubia et al. (2007). These authors administered an oddball task (e.g., Verleger 1988; Klein et al. 2000) to 17 medication-naïve participants with ADHD, aged 13 years, and 18 typically developing controls, aged 14 years (the two groups being comparable with respect to estimated IQ). While participants with ADHD and controls did not differ significantly in mean RT or the mean oddball effect (the increase in RT following the rare oddball stimuli), participants with ADHD responded more variably after standard stimuli and showed greater variability of the oddball effect on RT. During the rare oddball trials, as compared to the frequent standard trials, controls exhibited greater activation than participants with ADHD in several regions including the bilateral superior temporal gyrus and insula, as well as the left middle temporal gyrus, left inferior parietal lobe, tight posterior cingulate gyrus, caudate, putamen, and thalamus. For the two groups, different patterns of brain–behavior correlations emerged. In controls, the CV correlated negatively with activation increases during oddball, as compared to standard trials, in the left superior temporal lobe and insula, right inferior prefrontal and superior temporal lobe, and cerebellar vermis. By contrast, in participants with ADHD, CV correlated negatively with activation in the caudate, putamen, thalamus, left superior temporal lobe, and right cerebellum. These results suggest that different brain-metabolic correlates of IIV favor temporal-lobe attention allocation versus striato-thalamic output control in typically developing adolescents and adolescents with ADHD, respectively.

Another relatively recent proposal is that increased RT variability might arise from inadequate suppression during task performance of the "default-mode network" (DMN; Raichle and Snyder 2007; Sonuga-Barke and Castellanos 2007). The DMN concept arose from the observation that certain areas of the brain, which are not activated during resting states, decrease their activation in the transition from a resting state to a (variety of) task conditions (Raichle and Snyder 2007).

These networks have been found to be anti-correlated with networks that are activated during tasks, the "task-positive" networks (Fox et al. 2005). This means that the task-negative network is deactivated while the task-positive network is active, and vice versa. It has been postulated that these brain-intrinsic, resting state fluctuations continue during the execution of a task and produce IIV (Fox et al. 2006). This might explain why both RT series and spontaneous BOLD fluctuations exhibit the 1/f frequency distribution discussed in Sect. 2 (Fox et al. 2006). However, the failure to suppress default mode activity is associated with lapses of attention (Weissman et al. 2006), In a recent fMRI study, Kelly et al. (2008) investigated IIV from the angle of the "default mode" network theory in 26 healthy adults who were examined during rest and while performing an Eriksson Flanker Task. The authors identified a "default mode" network (including ventromedial PFC, posterior cingulate, precuneus, anterior temporal cortex, and lateral parietal cortex) and a "task-positive" network (including the dorsolateral prefrontal cortex (dlPFC), ventrolateral PFC (vlPFC), dorsomedial PFC (dmPFC), bilateral insula, premotor cortex, posterior and lateral parietal cortex, occipito-temporal cortex, and cerebellum). Individuals differed in the strength of the negative relationship between fluctuations in activation in these two networks, and, most importantly, the stronger this negative correlation, the less variable the RTs during the Flanker task. This result suggests that consistency of performance requires the stable suppression of the resting state "default mode" to activate the task-relevant functional systems of the brain.

Regarding the *pharmacological* evidence, some evidence points to a link between IIV and the dopaminergic system. MacDonald et al. (2009) studied D2 receptor binding with PET in 16 healthy adults. RT variability in two different tasks was inversely related to D2 receptor binding in the anterior cingulate ($-0.38 \leq r \leq -0.48$), the hippocampus ($-0.36 \leq r \leq -0.35$), and the orbitofrontal cortex ($-0.31 \leq r \leq -0.35$), but not the striatum ($r \leq -0.12$). These findings suggest the involvement of extrastriatal dopaminergic neurotransmission in the modulation of IIV in different cognitive tasks. The effects of methylphenidate (MPH), known to interact directly with the dopamine transporter (DAT; Krause et al., 2003), on IIV were investigated by DeVito et al. (2009) in 21 patients with ADHD, aged 7–13 years, using a double-blind, placebo-controlled crossover design of a single dose of placebo or methylphenidate. In keeping with the literature, RT variability for go-trials was greater in the ADHD than control group and discriminated the two groups better ($d = 1.53$) than did median go-RT ($d = 1.0$) or SSRT ($d = 1.28$). MPH reduced both RT variability ($d = 1.1$) and SSRT ($d = 1.54$). Reduced RT variability under methylphenidate has been reported in other studies (e.g., Heiser et al. 2004; Teicher et al. 2004), but was not replicated by Johnson et al. (2008b) when comparing participants with ADHD who were medication-naive at initial testing and after 6 weeks of treatment. However, methylphenidate did reduce the degree of fast moment-to-moment variability (faster than 0.0772 Hz) (Johnson et al. 2008b). A study by Lee et al. (2009) suggests that not only is RT variability reduced by MPH, but it may also predict ADHD patients' response (ADHD symptoms, and global clinical outcome) to methylphenidate because nonresponders were more variable at pre-methylphenidate baseline testing than responders. MacDonald et al. (2009), using PET in 16 healthy adults, found that RT variability in two different tasks was inversely related to D2 receptor binding in the anterior cingulate ($-0.38 \leq r \leq -0.48$), the hippocampus ($-0.36 \leq r \leq -0.35$), and the orbitofrontal cortex ($-0.31 \leq r \leq -0.35$), but not the striatum ($r \leq -0.12$). These findings suggest the involvement of extrastriatal dopaminergic neurotransmission in the modulation of IIV in different cognitive tasks.

A link between IIV and the GABA system was made by Pouget et al. (2009). These authors injected a $GABA_A$-receptor antagonist or a $GABA_A$-receptor agonist into the dorsolateral prefrontal cortex (dlPFC) of two monkeys performing a prosaccade task. While the agonist, muscimol, had no effect on RT variability and did not alter the auto-correlational structure of the RT series, the antagonist, bicuculline, resulted in the increased occurrence of slow or very slow responses and reduced RT auto-correlations up to lag 6.[1] These focal pharmacological perturbations occurred in a task with minimal cognitive load; further, they were specific to the dlPFC and could not be found after injection of $GABA_A$ agonists or antagonists into the pre-SMA or FEF.

The investigation of the neural underpinnings of IIV is undoubtedly in its infancy. Nevertheless, several exciting and promising research reports have been published that seem to converge in pointing to a crucial role of fronto-striatal systems in controlling behavioral consistency.

5 Theories

The "State Regulation" model, which is based on the Cognitive-Energetic Model of Sanders (Sanders 1983), proposes a key role for a nonoptimal energetic state in causing apparent cognitive impairments, such as high RT variability, in ADHD

---

[1]To analyze the temporal structure of reaction time series, reaction time series can be correlated with themselves (auto-correlations) by correlating a given RT with the next RT (lag 1), next-but-one RT (lag 2), and so on. A significant auto-correlation of lag 6 means that there is a relationship between the RT of a given trial and the RT six trials later.

(Sergeant 2005; van der Meere 2002). The model incorporates three energetic pools – activation, arousal, and effort – as well as a cortical "executive" system. The key proposal of the model is, as van der Meere (2002) writes, that: *the engine is intact (i.e. the basic information processing capacity is intact), but there is a problem with the petrol supply (i.e the utilization of the cognitive capacity depends on state factors such as incentives, event rate and presence/absence of the experimenter)* (p. 189). This model is in theoretical agreement with a recent model that explains inconsistent and inefficient performance, including RT variability, with deficiencies in the continuous energy supply (through glial cells) that is required, particularly during fact or complex information processing tasks (Russell et al. 2006).

According to the "Dynamic Developmental" theory of ADHD, altered motivational processes, in particular processes of reinforcement and extinction, leading to a failure to acquire complete and functional sequences of behavior, form the core of increased behavioral variability (Sagvolden et al. 2005). This model assumes that a hypo-functional dopaminergic neural system impairs the formation of stimulus–behavior response associations.

Other more recent models of ADHD incorporate aspects influenced by the State Regulation model, within a more complex two-process model. As reviewed in Sect. 3.1, Halperin and colleagues (Halperin and Schulz 2006; Halperin et al. 2008) investigated IIV in ADHD from a developmental perspective. The model proposes that RT variability reflects poor state regulation, perceptual sensitivity, and/or weak arousal mechanisms. The model makes a distinction between two neurocognitive processes: the proposed subcortical dysfunction, reflected in RT variability and linked with the etiology of ADHD, and prefrontally mediated executive control, linked with persistence or desistence of ADHD during adolescence. The developmental prediction is that the degree of recovery from ADHD symptoms is determined by the extent to which executive control, which develops throughout childhood and adolescence, can compensate for the more primary and enduring subcortical deficits. As reviewed in Sect. 3.1 above, a longitudinal investigation provided data consistent with the model: high RT variability was observed in both ADHD persisters and ADHD remitters, whereas compromised accuracy was observed in ADHD persisters, only (Halperin et al. 2008).

Another recent two-process model of ADHD that incorporates the RT variability findings is the Arousal-Attention model of Johnson, O'Connell, and colleagues (Johnson et al. 2007; O'Connell et al. 2008, 2009). This model, influenced by Posner (Posner and Petersen 1990), Paus (Paus et al. 1997), and Robertson (Robertson et al. 1998), suggests a distinction between bottom-up influences, from subcortical arousal structures, and top-down cortical control of the sustained attention system. Hence, the model distinguishes between two influences. The bottom-up vigilance decrement is reflected in continuous response control measures such as slow-frequency RT variability and linked to gradual decreases in arousal. The fluctuations in top-down control of attention over very brief time periods are reflected for example in commission errors on the sustained attention to response task (SART). Medication response and genotype effects were different for slow-versus fast-frequency RT variability, leading Johnson and colleagues (Johnson et al. 2008a, 2008b) to propose that while slow-frequency RT variability is likely to reflect arousal processes, fast-frequency RT variability may reflect top-down control.

The proposal in the Developmental (Halperin and Schulz 2006; Halperin et al. 2008) and Arousal-Attention (Johnson et al. 2007; O'Connell et al. 2008, 2009) models that the increased (slow) RT variability in ADHD relates to arousal regulation has obtained support from studies using electrophysiological (Loo and Smalley 2008) and skin conductance (O'Connell et al. 2008) measures. In the study from O'Connell et al. (2008), block-by-block increases in RT variability were accompanied by gradual decreases in arousal, suggesting a vigilance decrement. Further, the findings (discussed in Sect. 3a above) that RT variability in ADHD is not stable but shows greater than expected improvements under specific task manipulations, such as incentives or the presentation rate of stimuli, are also consistent with this proposal. The model of two familial factors that emerged from the multivariate factor analysis on the large collaborative sample (Kuntsi et al. 2010) – an "RT" factor, incorporating RT variability and accounting for 85% of the familial influences on ADHD, and an "error" factor, accounting for 12.5% of the familial influences on ADHD – is also potentially consistent with either the Developmental or the Arousal-Attention models. As such, we proposed that the first (RT) factor may represent arousal regulation and the second familial factor (errors) may represent brief reductions in the top-down control of sustained attention, leading to secondary inhibition deficits. The links from the model of two familial factors to the developmental predictions remain to be explicitly tested in future research.

The "Default Mode" account (see Sect. 4) has also been used to explain increased IIV in ADHD. Sonuga-Barke and Castellanos (2007) put forward the "default mode interference hypothesis," which states that spontaneous, low-frequency default mode activity persists into, or re-emerges, during states of active task processing and interferes with task performance. In this sense, ADHD can be hypothesized to reflect a "Default Mode Deficit Disorder." The previously cited results of Di Martino et al. (2008) are in accordance with this hypothesis. Also compatible is the hypothesis that increased IIV in ADHD reflects a unitary, cross-task generalizing construct (Klein et al. 2006; Wood et al. 2011; Kuntsi et al. 2010).

Finally, temporal processing deficits have been suggested to underlie increased IIV in ADHD (Castellanos and Tannock 2002). According to this scheme, deficits in the precise temporal representations might give rise to increased RT variability: for example, due to recurrent attentional lapses. The postulated link between temporal processing and RT variability seems intuitively plausible, and performance in time reproduction tasks and CPT RT variability have been reported to be correlated by $r = 0.59$ (Kerns et al. 2001). In Kerns and colleagues' study, however, group effect sizes for RT variability (1.10) were somewhat larger than those for time reproduction (0.97). Further, fluctuations in attentional performance could offer an alternative explanation for deficits in temporal processing tasks.

In summary, several models could account for the strong association between increased RT variability and ADHD. It is also possible that different models explain different elements of the same phenomena (e.g., the default-mode network versus arousal regulation models). Future progress in understanding the causal pathways will depend, in part, on the further development of models that incorporate explicit, a priori, testable hypotheses (Johnson et al. 2009). In this light, examples include the predictions from arousal regulation models on the effects of, for example, rewards on RT variability and on the association between fluctuations in arousal and simultaneous fluctuations in RT variability. We have come a long way from considering increased RT variability in ADHD as the "noise" or "error" that we need to reduce in our data – yet the investigation of the causal pathways is only beginning.

6 Nosological Specificity

The question of whether increased IIV is nosologically specific to ADHD can be answered only with reference to an (operational) definition of the variability. If we take RT variability as our measure of variability, the answer is clearly "no," because high RT variability has been reported for various conditions, including bipolar disorder (Bora et al. 2006; Brotman et al. 2009), schizophrenia (Kaiser et al. 2008; Schwartz et al. 1989; Schwartz et al. 1991; van den Bosch et al. 1996; Vinogradov et al. 1998), Parkinson's disease (Camicioli et al. 2008), and autism (Geurts et al. 2008).

Space restrictions do not allow a comprehensive review of the literature here, but we summarize a few selected studies that highlight some of the key issues. As ADHD frequently co-occurs with several of the other disorders, an initial question that needs to be tackled is whether the increased RT variability in another disorder is due to co-occurring ADHD symptoms. This was addressed in two studies focusing on ADHD and autism, but with conflicting results. Johnson et al. (2007) found increased RT variability in ADHD participants without autism but not in individuals with autism who did not have comorbid ADHD but Geurts et al. (2008) reported the opposite pattern. Geurts et al. (2008) also reported increased RT variability in an ADHD-autism comorbid group [such a group was not included in the Johnson et al. (2007) study]. It is not clear at this point what could explain the contrasting findings. The negative finding from Geurts et al. (2008) is inconsistent with several other studies too, as autism is frequently an exclusion criterion in ADHD studies (e.g., Kuntsi et al. 2010). ADHD often also co-occurs with bipolar disorder (reviewed in Skirrow et al. 2009). An initial study suggests that the high RT variability observed in bipolar disorder is not explained by co-occurring ADHD symptoms (Brotman et al. 2009), but evidence is as yet limited.

Where high RT variability is observed in another disorder, in the absence of co-occurring ADHD symptoms, the question arises as to whether they share the same underlying mechanisms. For co-occurring disorders or behavioral traits, genetic designs, such as twin studies, enable an initial investigation of the extent of shared genetic or environmental influences on the two (or more) disorders or traits. As such, behavior genetic studies allow investigation of both shared and independent etiological pathways between co-occurring disorders or traits. For example, twin studies suggest both shared and unique genetic influences on autism and ADHD behaviors (e.g., Ronald et al. 2008). To study the causes of the high RT variability at the level of cognitive or brain processes, tight experimental designs are required that enable one to test a specific hypothesis. We can address questions such as: do we observe a similar improvement in RT variability under rewarded conditions in another disorder, as we do in ADHD? Further, the detailed analysis methods that move beyond the consideration of RT variability (discussed in Sect. 2), as well as new methods for analyzing for example EEG variability data, offer promise for future cross-disorder investigations.

The nosological specificity issue is of course not unique to the observation of high RT variability. For example, various "executive function" impairments are observed in ADHD and many other disorders. Overall, the fascinating issues surrounding cross-disorder comparisons and investigations of both shared and unique causal pathways are undoubtedly one of the key directions for future research on IIV.

7 Conclusions and Implications for Research and Intervention

The strength of the phenotypic and genetic association of RT variability with ADHD, when ideally measured (under baseline conditions), is impressive. The association does not appear to be a nonspecific effect mediated by lower general cognitive ability. We are on a steep learning curve with regard to the neuroanatomical and neurophysiological correlates underlying this association, and further theoretical models are being refined and tested. While we are only at the beginning in trying to understanding the causal pathways, the initial data, models, and methods being developed lead us to predict considerable progress in the coming years. Recent models, such as the Developmental and Arousal-Attention models, which incorporate RT variability together with a second proposed underlying process, offer one solution to the conundrum of the "bigger picture": how to incorporate the findings on RT variability with the findings on other cognitive impairments in ADHD.

What will be the wider implications of research on IIV in ADHD? We wish to emphasize three issues here. First, indices of IIV have potential as biomarkers. This could lead to their use as objective measures within clinical settings, and improve predictions of clinical outcomes and appropriate targeting of clinical services. Second, cross-disorder comparisons on RT variability will lead to a better understanding of the underlying cognitive and brain mechanisms: this is essential for improved diagnosis and treatment. Third, research that investigates the association between RT variability and ADHD within a developmental context has potentially further implications for intervention. Our understanding of the processes that lead to remission versus persistence of ADHD in adolescence and adulthood is very limited. The model of Halperin and colleagues (Halperin and Schulz 2006; Halperin et al. 2008) provides one possible developmental framework. They propose a compensatory process, whereby it is the extent to which executive control can compensate for primary and enduring subcortical deficits (reflected in RT variability), which determines the degree of recovery from ADHD symptoms. The identification of such possible compensatory processes and interactions could lead to novel treatments that aim to simulate the development of the compensatory processes and thereby diminish clinical impairment.

References

Abikoff H, Gittelman-Klein R, Klein DF (1977) Validation of a classroom observe code for hyperactive children. J Consult Clin Psychol 45:772–783

Andreou P, Chen W, Christiansen H, Gabriels I, Heise A, Meidad S, Muller UC, Uebel H, Banaschewski T, Manor I, Neale B, Oades R, Roeyers H, Rothenberger A, Sham P, Steinhausen H-C, Asherson P, Kuntsi J (2007) Reaction time performance in ADHD: improvement under fast-incentive condition and familial effects. Psychol Med 37:1703–1716

Aron AR, Poldrack RA (2005) The cognitive neuroscience of response inhibition: relevance fir genetic research in attention-deficit/hyperactivity disorder. Biol Psychiatry 57:1285–1292

Barkley RA (1997) Behavioral inhibition, sustained attention, and executive functions: constructing a unifying theory of ADHD. Psychol Bull 121(1):65–94

Bellgrove MA, Hester R, Garavan H (2004) The functional neuroanatomical correlates of response variability: evidence from a response inhibition task. Neuropsychologia 42:1910–1916

Ben-Pazi H, Shalev RS, Gross-Tsur V, Bergman H (2006) Age and medication effect on rhythmic responses in ADHD: possible oscillatory mechanisms? Neuropsychologia 44:412–416

Berwid OG, Curko Kera EA, Marks DJ, Santra A, Bender HA, Halperin JA (2005) Sustained attention and response inhibition in young children at risk for attention deficit/ hyperactivity disorder. J Child Psychol Psychiatry 46:1219–1229

Bidwell LC, Willcutt EG, DeFries JC, Pennington BF (2007) Testing for neuropsychological endophenotypes in siblings discordant for attention-deficit/hyperactivity disorder. Biol Psychiatry 62:991–998

Bora E, Vahip S, Akdeniz F (2006) Sustained attention deficits in manic and euthymic patients with bipolar disorder. Prog Neuropsychopharmacol Biol Psychiatry 30:1097

Brotman MA, Rooney MH, Skup M, Pine DS, Leibenluft E (2009) Increased intrasubject variability in response time in youths with bipolar disorder and at-risk family members. J Am Acad Child Adolesc Psychiatry 48:628–635

Camicioli RM, Wieler M, de Frias CM, Martin WRW (2008) Early, untreated Parkinson's disease patients show reaction time variability. Neurosci Lett 441:77–80

Castellanos FX, Tannock R (2002) Neuroscience of attention-deficit/hyperactivity disorder: the search for endophenotypes. Neuroscience 3:617–628

Cohen NJ, Douglas VI (1972) Characteristics of the orienting response in hyperactive and normal children. Psychophysiology 9:238–245

Daugherty TK, Quay HC, Ramos L (1993) Response perseveration, inhibitory control, and central dopaminergic activity in childhood behavior disorders. J Genet Psychol 154:177–188

DeVito EE, Blackwell AD, Clark L, Kent L, Dezsery AM, Turner DC, Aitken MRF, Sahakian BJ (2009) Methylphenidate improves response inhibition but not reflection-impulsivity in children with attention-deficit hyperactivity disorder (ADHD). Psychopharmacology 201:531–539

Di Martino A, Ghaffari M, Curchack J, Reiss P, Hyde C, Vannucci M, Petkova E, Klein DF, Castellanos FX (2008) Decomposing intra-subject variability in children with attention-deficit/ hyperactivity disorder. Biol Psychiatry 64:607–614

Dimoska A, Johnstone SJ, Barry RJ, Clarke AR (2003) Inhibitory motor control in children with attention-deficit/hyperactivity disorder: event-related potentials in the stop-signal paradigm. Biol Psychiatry 54:1345–1354

Faraone SV, Perlis RH, Doyle AE, Smoller JW, Goralnick JJ, Holmgren MA, Sklar P (2005) Molecular genetics of attention-deficit/hyperactivity disorder. Biol Psychiatry 57:1313–1323

Firestone P, Douglas VI (1975) The effects of reward and punishment in reaction times and autonomic activity in hyperactive and normal children. J Abnorm Child Psychol 3:201–216

Fox MD, Snyder AZ, Vincent JL, Corbetta M, Van Essen DC, Raichle ME (2005) The human brain is intrinsically organized into dynamic, anticorrelated functional networks. Proc Natl Acad Sci USA 102:9673–9678

Fox MD, Snyder AZ, Zacks JM, Raichle ME (2006) Coherent spontaneous activity accounts for trial-to-trial variability in human evoked brain responses. Nat Neurosci 9:23–25

Geurts H, Grasman RPPP, Verte S, Oosterlaan J, Royers H, van Kammen SM, Sergeant JA (2008) Intra-individual variability in ADHD, autism spectrum disorders and Tourette's syndrome. Neuropsychologia 46:3030–3041

Gilden DL (2001) Cognitive emissions of 1/f noise. Psychol Rev 108:33–56

Gilden DL, Hancock H (2007) Response variability in attention-deficit disorders. Psychol Sci 18:796–802

Halperin JM, Schulz KP (2006) Revisiting the role of the prefrontal cortex in the pathophysiology of attention-deficit/hyperactivity disorder. Psychol Bull 132:560–581

Halperin JM, Trampush JW, Miller CJ, Marks DJ, Newcorn JH (2008) Neuropsychological outcome in adolescents/young adults with childhood ADHD: profiles of persisters, remitters and controls. J Child Psychol Psychiatry 64:958–966

Heathcote A, Brown S, Cousineau D (2004) QMPR: estimating lognormal, wald, weibull RT distributions with parameter-dependent lower bound. Behav Res Methods Instrum Comput 36:277–290

Heiser P, Frey J, Smidt J, Sommerlad C, Wehmeier PM, Hebebrand J, Remschmidt H (2004) Objective measurement of hyperactivity, impulsivity, and inattention in children with hyperkinetic disorders and after treatment with methylphenidate. Eur Child Adolesc Psychiatry 13:100–104

Hopkins J, Perlman T, Hechtman L, Weiss G (1979) Cognitive style in adults originally diagnosed as hyperactives. J Child Psychol Psychiatry 20:209–216

Jennings JR, van der Molen MW, Pelham W, Debski KB, Hoza B (1997) Inhibition in boys with attention deficit hyperactivity disorder as indexed by heart rate changes. Dev Psychol 33:308–318

Jensen AR (1992) The importance of intraindividual variation in reaction time. Personal Individ Differ 13:869–881

Johnson KA, Kelly SP, Bellgrove MA, Barry E, Cox M, Gill M, Robertson IH (2007) Response variability in attention deficit hyperactivity disorder: evidence for neuropsychological heterogeneity. Neuropsychologia 45:630–638

Johnson KA, Kelly SP, Robertson IH, Mulligan A, Daly M, Lambert D, McDonnell C, Connor TJ, Hawi Z, Gill M, Bellgrove MA (2008a) Absence of the 7-repeat variant of the DRD4 VTNR is associated with drifting sustained attention in children with ADHD but not in controls. Am J Med Genet B Neuropsychiatr Genet 147:927

Johnson KA, Barry E, Bellgrove MA, Cox M, Kelly SP, Daibhis A, Daly M, Keavey M, Watchorn A, Fitzgerald M, McNicholas F, Kirley A, Robertson IH, Gill M (2008b) Dissociation in response to methylphenidate on response variability in a group of medication naive children with ADHD. Neuropsychologia 46:1532–1541

Johnson KA, Wiersema JR, Kuntsi J (2009) What would Karl Popper say? Are current psychological theories of ADHD falsifiable? Behav Brain Funct 5:15

Kaiser S, Roth A, Rentrop M, Friederich HC, Bender S, Weisbrod M (2008) Intra-individual reaction time variability in schizophrenia, depression and borderline personality disorder. Brain Cogn 66:73–82

Kebir O, Tabbane K, Sengupta S, Joober R (2009) Candidate genes and neuropsychological phenotypes in children with ADHD: review of association studies. J Psychiatry Neurosci 34:88–101

Kelly AMC, Uddin LQ, Biswal BB, Castellanos FX, Milham MP (2008) Competition between functional brain networks mediates behavioral variability. NeuroImage 39:527–537

Kerns KA, McInerney RJ, Wilde N (2001) Investigation of time reproduction, working memory, and behavioral inhibition in children with ADHD. Child Neuropsychol 7:21–31

Klein C, Feige B (2005) An independent components analysis (ICA) approach to the study of developmental differences in the saccadic contingent negative variation. Biol Psychol 70:105–114

Klein C, Berg P, Rockstroh B, Andresen B (2000) Topography of the auditory P300 in schizotypal personality. Biol Psychiatry 45:1612–1621

Klein C, Wendling K, Huettner P, Ruder H, Peper M (2006) Intra-subject variability in attention-deficit hyperactivity disorder. Biol Psychiatry 60:1088–1097

Krause K-H, Dresel SH, Krause J, la Fougere C, Ackenheil M (2003) The dopamine transporter and neuroimaging in attention deficit hyperactivity disorder. Neurosci Biobehav Rev 27:605–613

Kuntsi J, Stevenson J (2001) Psychological mechanisms in hyperactivity: II. The role of genetic factors. J Child Psychol Psychiatry 42:211–219

Kuntsi J, Oosterlaan J, Stevenson J (2001) Psychological mechanisms in hyperactivity: I. Response inhibition deficit, working memory impairment, delay aversion, or something else? J Child Psychol Psychiatry 42:199–210

Kuntsi J, Wood AC, van der Meere J, Asherson P (2009) Why cognitive performance in ADHD may not reveal true potential: findings from a large population-based sample. J Int Neuropsychol Soc 15:570–579

Kuntsi J, Wood AC, Rijsdijk F, Johnson KA, Andreou P, Albrecht B, Arias-Vasquez A, Buitelaar JK, McLoughlin G, Rommelse NNJ, Sergeant JA, Sonuga-Barke EJS, Uebel H, van der Meere J, Banaschewski T, Gill M, Manor I, Miranda A, Mulas F, Oades R, Royers H, Rothenberger A, Steinhausen H-C, Faraone SV, Asherson P (2010) Separation of cognitive impairments in attention-deficit/hyperactivity disorder into 2 familial factors. Arch Gen Psychiatry 67:1159–1167

Lee SH, Song DH, Kim BN, Joung YS, Shin YJ, Yoo HJ, Shin DW (2009) Variability of response time as a predictor of methylphenidate treatment response in Korean children with attention deficit hyperactivity disorder. Yonsei Med J 50:650–655

Leth-Steensen C, Elbaz ZK, Douglas VI (2000) Mean response times, variability, and skew in the responding of ADHD children: a response time distributional approach. Acta Psychol 104:167–190

Leung PWL, Connolly KJ (1996) Distractibility in hyperactive and conduct-disordered children. J Child Psychol Psychiatry 37:305–312

Logan DG, Cowan WB (1984) On the ability to inhibit thought and action: a theory of an act of control. Psychol Rev 91:295–327

Loo SK, Smalley SL (2008) Preliminary report of familial clustering of EEG measures in ADHD. Am J Med Genet B Neuropsychiatry Genet 147:107–109

Luciano M (2010) Adolescent brain development: current themes and future directions (editorial). Brain Cogn 72:1–5

Luck SJ (2005) An introduction to the event-related potential technique. MIT Press, London

Luman M, van Noesel SJP, Papanikolau A, Van Oostenbruggen-Scheffer J, Veugelers D, Sergeant JA, Oosterlaan J (2009) Inhibition, reinforcement sensitivity and temporal information processing in ADHD and ADHD + ODD: evidence of a separate entity? J Abnorm Child Psychol 37:1123–1135

Luna B, Velanova K, Geier CF (2008) Development of eye movement control. Brain Cogn 68:293–308

MacDonald SWS, Cervenka S, Farde L, Nyberg L, Baeckman L (2009) Extrastriatal dopamine D2 receptor binding modulates intraindividual variability in episodic recognition and executive functioning. Neuropsychologia 47:2299–2304

Mahone EM, Hagelthorn KM, Cutting LE, Schuerholz LJ, Pelletier SF, Rawlins C, Singer HS, Denckla MB (2002) Effects of IQ on executive function measures in children with ADHD. Child Neuropsychol 8:52–65

McInerney RJ, Kerns KA (2003) Time reproduction in children with ADHD: motivation matters. Child Neuropsychol 9:91–108

Mirsky AF, Pascualvaca DM, Duncan CC, French LM (1999) A model of attention and its relation to ADHD. Ment Retard Dev Disabil Res Rev 5:169–176

Nigg JT (1999) The ADHD response-inhibition deficit as measured by the stop task: replication with DSM-IV combined type, extension, and qualification. J Abnorm Child Psychol 27:393–402

Nigg JT, Blaskey LG, Stawicki JA, Sachek J (2004) Evaluating the endophenotype model of ADHD neuropsychological deficit: results for parents and siblings of children with ADHD combined and inattentive subtypes. J Abnorm Psychol 113:614–625

Oades RD, Christiansen H (2008) Cognitive switching processes in young people with attentiondeficit/hyperactivity disorder. Arch Clin Neuropsychol 23:21–32

O'Connell RG, Bellgrove MA, Dockree PM, Lau A, Fitzgerald M, Robertson IH (2008) Self-alert training: volitional modulation of autonomic arousal improves sustained attention. Neuropsychologia 46:1379–1390

O'Connell RG, Dockree PM, Bellgrove MA, Turin A, Ward S, Foxe JJ, Robertson IH (2009) Two types of action error: electrophysiological evidence for separable inhibitory and sustained attention neural mechanisms producing error on go/no-go tasks. J Cogn Neurosci 21:1–12

Ohashi K, Vitalino G, Polcari A, Teicher MH (2010) Unraveling the nature of hyperactivity in children with attention-deficit hyperactivity disorder. Arch Gen Psychiatry 67:388–396

Oosterlaan J, Sergeant JA (1995) Response choice and inhibition in ADHD, anxious and aggressive children: the relationship between S-R-compatibility and stop signal task. In: Sergeant JA (ed) Eunethydis: European approaches to hyperkinetic disorder. Elsevier, Amsterdam, pp 225–240

Oosterlaan J, Sergeant JA (1998) Effects of reward and response cost on response inhibition in AD/HD, disruptive, anxious, and normal children. J Abnorm Child Psychol 26:161–174

Paus T, Zatorre RJ, Hofle N, Caramanos Z, Gotean J, Petrides M, Evans AC (1997) Time-related changes in neural systems underlying attention and arousal during the performance of an auditory vigilance task. J Cogn Neurosci 9:392–408

Pliszka SR, Liotti M, Woldorff MG (2000) Inhibitory control in children with attention-deficit hyperactivity disorder: event-related potentials identify the processing component and timing of an impaired right-frontal response-inhibition mechanism. Biol Psychiatry 48:238–246

Posner MI, Petersen SE (1990) The attention system of the human brain. Ann Rev Neurosci 13:25–42

Pouget P, Wattiez N, Rivaud-Pechout S, Gaymard B (2009) A fragile balance: perturbation of GABA mediated curcuit in prefrontal cortex generates high intraindividual performance variability. PLoS One 4:e5208. doi:10.1371/journal.pone.0005208

Purvis KL, Tannock R (2000) Phonological processing, not inhibitory control, differentiates ADHD and reading disability. J Am Acad Child Adol Psychiatry 39:485–494

Raichle ME, Snyder AZ (2007) A default mode of brain function: a brief history of an evolving idea. NeuroImage 37:1083–1090

Rapport MD, Kofler MJ, Alderson M, Timko TM, DuPaul GJ (2009) Variability of attention processes in ADHD: observations from the classroom. J Atten Disord 12:563–573

Robertson IH, Mattingley JB, Rorden C, Driver J (1998) Phasic alerting of neglect patients overcomes their spatial deficit in visual awareness. Nature 395:169–172

Rommelse NNJ, Altink ME, Oosterlaan J, Beem L, Buschgens CJM, Buitelaar JK, Sergeant JA (2008) Speed, variability, and timing of motor output in ADHD: which measures are useful for endophenotypic research? Behav Genet 38:121–132

Ronald A, Simonoff E, Kuntsi J, Asherson P, Plomin R (2008) Evidence for overlapping genetic influences on autistic and ADHD behaviours in a community twin sample. J Child Psychol Psychiatry 49:535–542

Rosvold H, Mirsky A, Sarason I (1956) A continuous performance test of brain damage. J Consult Psychol 20:343–350

Rubia K, Oosterlaan J, Sergeant JA, Brandeis D, Van Leeuwen TH (1998) Inhibitory dysfunction in hyperactive boys. Behav Brain Res 94:25–32

Rubia K, Taylor E, Smith AB, Oksannen H, Overmeyer S, Newman S (2001) Neuropsychological analyses of impulsiveness in childhood hyperactivity. Br J Psychiatry 179:138–143

Rubia K, Smith AB, Brammer MJ, Taylor E (2007) Temporal lobe dysfunction in medication-naïve boys with attention-deficit/hyperactivity disorder during attention allocation and its relation to response variability. Biol Psychiatry 62:999–1006

Rucklidge JJ, Tannock R (2002) Neuropsychological profiles of adolescents with ADHD: effects of reading difficulties and gender. J Child Psychol Psychiatry 43:988–1003

Russell VA, Oades RD, Tannock R, Killeen PR, Auerbach JG, Johansen EB, Sagvolden T (2006) Response variability in attention-deficit/hyperactivity disorder: a neuronal and glial energetics hypothesis. Behav Brain Funct 2:30

Sagvolden T, Johansen EB, Aase H, Russell VA (2005) A dynamic developmental theory of attention-deficit/hyperactivity disorder (ADHD) predominantly hyperactive/impulsive and combined subtypes. Behav Brain Sci 28:397–419

Sanders AF (1983) Towards a model of stress and performance. Acta Psychol 53:61–97

Schachar R, Logan GD (1990) Impulsivity and inhibitory control in normal development and childhood psychopathology. Dev Psychol 26:710–720

Schachar RJ, Tannock R, Logan G (1993) Inhibitory control, impulsiveness, and attention deficit hyperactivity disorder. Clin Psychol Rev 13:721–739

Scheres A, Oosterlaan J, Sergeant JA (2001) Response execution and inhibition in children with AD/HD and other disruptive disorders: the role of behavioural activation. J Child Psychol Psychiatry 42:347–357

Schmiedek F, Oberauer K, Wilhelm O, Suess HM, Wittmann WW (2007) Individual differences in components of reaction time distributions and their relations to working memory and intelligence. J Exp Psychol Gen 136:414–429

Schulz KP, Fan J, Tang CY, Newcorn JH, Buchsbaum MS, Cheung AM, Halperin JM (2004) Response inhibition in adolescents diagnosed with attention-deficit/hyperactivity disorder during childhood: an event-related fMRI study. Am J Psychiatry 161:1650–1657

Schwartz F, Carr AC, Munich RL, Glauber S, Lesser B, Murray J (1989) Reaction time impairment in schizophrenia and affective illness: the role of attention. Biol Psychiatry 25:540–548

Schwartz F, Munich RL, Carr A, Bartuch E, Lesser B, Rescigno D, Viegener B (1991) Negative symptoms and reaction time in schizophrenia. J Psychiatr Res 25:131–140

Sergeant JA (2005) Modeling attention-deficit/hyperactivity disorder: a critical appraisal of the cognitive-energetic model. Biol Psychiatry 11:1248–1255

Shue KL, Douglas VL (1992) Attention deficit hyperactivity disorder and the frontal lobe syndrome. Brain Cogn 20:104–124

Simmonds DJ, Fotedar SG, Suskauer SJ, Pekar JJ, Denckla MB, Mostofsky SH (2007) Functional brain correlates of response time variability in children. Neuropsychologia 45:2147–2157

Skirrow C, McLoughlin G, Kuntsi J, Asherson P (2009) Behavioural, neurocognitive and treatment overlap between attention deficit hyperactivity disorder and mood instability. Expert Rev Neurother 9:489–503

Sonuga-Barke EJS, Castellanos FX (2007) Spontaneous attentional fluctuations in impaired states and pathological conditions: a neurobiological hypothesis. Neurosci Biobehav Rev 31:977–986

Stevens J, Quittner AL, Zuckerman JB, Moore S (2002) Behavioural inhibition, self-regulation of motivation, and working memory in children with attention deficit hyperactivity disorder. Dev Neuropsychol 21:117–139

Stuss DT, Pogue J, Buckle L, Bondar J (1994) Characterization of stability of performance in patients with traumatic brain injury: variability and consistency on reaction time tests. Neuropsychology 8:316–324

Stuss DT, Murphy KJ, Binns MA, Alexander MP (2003) Staying on the job: the frontal lobes control individual performance variability. Brain 126:2363–2380

Teicher MH, Lowen SB, Polcari A, Foley M, McGreenery CE (2004) Novel strategy for the analysis of CPT data provides new insights into the effects of methylphenidate on attentional states in children with ADHD. J Child Adolesc Psychopharmacol 14:219–232

Trommer BL, Hoeppner JA, Lorber R, Armstrong KJ (1988) The Go-NoGo paradigm in attention deficit disorder. Ann Neurol 24:610–614

Uebel H, Albrecht B, Asherson P, Borger N, Butler L, Chen W, Christiansen H, Heise A, Kuntsi J, Schafer U, Andreou P, Manor I, Marco R, Meidad S, Miranda A, Mulligan A, Oades R, van der Meere J, Faraone SV, Rothenberger A, Banaschewski T (2010) Performance variability, impulsivity errors and the impact of incentives as gender-independent endophenotypes for ADHD. J Child Psychol Psychiatry 511:210–218

van den Bosch RJ, Rombouts RP, van Asma MJ (1996) What determines continuous performance task performance? Schiz Bull 22:643–651

van der Meere JJ (2002) The role of attention. In: Sandberg S (ed) Hyperactive disorders of childhood. Cambridge University Press, Cambridge, pp 162–213

van der Meere J, Sergeant J (1988) Acquisition of attention skill in pervasively hyperactive children. J Child Psychol Psychiatry 29:301–310

van der Meere J, Wekking E, Sergeant JA (1991) Sustained attention and pervasive hyperactivity. J Child Psychol Psychiatry 32:275–284

Verleger R (1988) Event-related potentials and cognition: a critique of the context updating hypothesis and an alternative interpretation of P3. Behav Brain Res 11:343–327

Vinogradov S, Poole JH, Willis-Shore J, Ober BA, Shenaut GK (1998) Slower and more variable reaction times in schizophrenia: what Do they signify? Schiz Res 32:183–190

Weissman DH, Roberts KC, Visscher KM, Woldorff MG (2006) The neural bases of momentary lapses in attention. Nat Neurosci 9:971–978

Willcutt EG, Doyle AE, Nigg JT, Faraone SV, Pennington BF (2005) Validity of the executive function theory of attention-deficit/hyperactivity disorder: a meta-analytic review. Biol Psychiatry 57:1336–1346

Wood AC, Asherson P, Rijsdijk F, Kuntsi J (2009) Is overactivity a core feature in ADHD? Familial and receiver operating characteristic curve analysis of mechanically assessed activity level. J Am Acad Child Adolesc Psychiatry 48:1023–1030

Wood AC, Buitelaar JK, Rijsdijk F, Asherson P, Kuntsi J (2010a) Rethinking shared environment as a source of variance underlying attention-deficit/ hyperactivity disorder symptoms: Comment on Burt (2009). Psychol Bull 36:331–340

Wood AC, Asherson P, van der Meere J, Kuntsi J (2010b) Separation of genetic influences on attention deficit hyperactivity disorder symptoms and reaction time performance from those on IQ. Psychol Med 40:1027–1037

Wood AC, Rijsdijk F, Johnson KA, Andreou P, Albrecht B, Arias-Vasquez A, Buitelaar JK, McLoughlin G, Rommelse NNJ, Sergeant JA, Sonuga-Barke EJS, Uebel H, van der Meere J, Banaschewski T, Gill M, Manor I, Miranda A, Mulas F, Oades RD, Royers H, Rothenberger A, Steinhausen H-C, Faraone SV, Asherson P, Kuntsi J (2011) The relationship between ADHD and key cognitive phenotypes is not mediated by shared familial effects with IQ. Psychol Med 41:861–871

Hypothalamic–Pituitary–Adrenocortical Axis Function in Attention-Deficit Hyperactivity Disorder

G. Fairchild

Contents

1 Introduction 94
2 The Hypothalamic–Pituitary–Adrenocortical Axis 95
3 What Is the Theoretical Basis for Investigating the
  Relationship Between ADHD and HPA Axis Abnormalities? 96
4 Basal Cortisol Secretion in ADHD and Related Disorders 97
5 Is ADHD Associated with Cortisol Hyporeactivity? 100
6 Recommendations for Future Research on HPA Activity
  in ADHD and Related Disorders 104
  6.1 Assess Cortisol in Relation to Waking Time 104
  6.2 Advantages of Assessing Cortisol Levels in Saliva 105
  6.3 Appropriate Conditions, Ethical Issues, and Sample Sizes
      in Psychoneuroendocrine Research 105
7 What Do We Need to Know About the Relationship
  Between ADHD and HPA Axis Activity? 106
8 Conclusions 108
References 108

Abstract The hypothalamic–pituitary–adrenocortical axis plays a critical role in mediating the physiological response to the imposition of stress. There are theoretical reasons to expect reduced basal cortisol secretion and cortisol hyporeactivity in hyperactive/impulsive or combined type attention-deficit hyperactivity disorder (ADHD). Early studies reported profound abnormalities in the diurnal rhythm of cortisol secretion or the cortisol response to stress in children with severe or persistent ADHD. However, subsequent work using larger samples or improved methods has not provided convincing evidence for changes in basal cortisol secretion in non-comorbid forms of ADHD. In contrast, children with ADHD and

G. Fairchild
Developmental Psychiatry Section, Department of Psychiatry, Cambridge University, Cambridge, UK
e-mail: gff22@cam.ac.uk

comorbid oppositional defiant disorder show lower basal cortisol concentrations and a blunted cortisol awakening response. With respect to cortisol reactivity to stress in ADHD, recent evidence has been mixed, with some studies reporting normal cortisol responses and others showing blunted cortisol responses in non-comorbid ADHD. Again, it appears important to consider whether comorbid disorders are present, because children with ADHD and comorbid disruptive behavior disorders exhibit blunted cortisol responses, whereas those with comorbid anxiety disorders show enhanced cortisol responses to stress. Longitudinal studies are required to investigate whether abnormalities in cortisol secretion play a causal role in the etiology of ADHD and related disruptive behavior disorders.

Keywords ADHD · Cortisol · Disruptive behavior disorder · Neuroendocrinology · Stress

Abbreviations

ACTH	Adrenocorticotropic hormone
AVP	Arginine vasopressin
BAS	Behavioral activation system
BIS	Behavioral inhibition system
CAR	Cortisol awakening response
CBG	Cortisol-binding globulin
CD	Conduct disorder
CRH	Corticotropin-releasing hormone
DBD	Disruptive behavior disorders
HPA	Hypothalamic–pituitary–adrenal (axis)
ODD	Oppositional defiant disorder
PVN	Paraventricular nucleus
SCID	Structured clinical interview for DSM-IV
TSST-C	Trier social stress test for children

1 Introduction

This chapter will consider the relationship between attention-deficit hyperactivity disorder (ADHD) and the hypothalamic–pituitary–adrenocortical (HPA) axis, a critical physiological system that mediates responses to stress. Stress can be defined as the imposition or perception of an environmental change or challenge, which could be positive or negative in valence but, in either case, requires an adaptive response by the organism (Herman and Cullinan 1997). The secretion of glucocorticoid hormones, principally cortisol in humans, is a key component of this adaptive response that alerts the organism to environmental or physiological changes and promotes the recovery of homeostasis. Dysregulation of HPA axis responses can have a detrimental effect on health and well being. For example, prolonged hypersecretion of glucocorticoids can impair psychological functioning and damage vulnerable brain structures, such as the hippocampus (Herbert et al. 2006). Although not a consistent observation, major depressive disorder is frequently associated with glucocorticoid hypersecretion, together with deficits in the negative feedback mechanisms that normally terminate HPA axis activity (Burke et al. 2005; Holsboer 2000).

Conversely, insufficient cortisol secretion may be harmful to health and psychological functioning (Heim et al. 2000a). Such a deficit has been linked with a compromised immune system and increased risk for physical illnesses (e.g., arthritis) and psychiatric disorders (e.g., chronic fatigue syndrome, post-traumatic stress disorder) (Cleare 2003; Yehuda et al. 2005; Yehuda et al. 1995). The main objective of this chapter is to review evidence linking ADHD with abnormalities in HPA axis activity and to explain what this can tell us about the pathophysiology of the disorder. Comorbid disorders will also be considered because they may moderate the relationship between ADHD and HPA axis dysfunction. The chapter will end with a discussion of gaps in our knowledge base and offer a number of recommendations for future research in this area.

2 The Hypothalamic–Pituitary–Adrenocortical Axis

Information about stressors in the environment is relayed from limbic brain regions, such as the amygdala and prefrontal cortex, to the paraventricular nucleus (PVN) of the hypothalamus. A population of neurons in this latter region responds to stress by releasing adrenocorticotropic hormone (ACTH) secretagogues, such as corticotropin-releasing hormone (CRH) and arginine vasopressin (AVP), into the pituitary portal circulation. CRH and AVP interact with receptors on corticotropic cells of the anterior pituitary to cause secretion of ACTH into the bloodstream. ACTH subsequently binds with receptors in the adrenal cortex to induce synthesis and secretion of cortisol.

Cortisol targets many organs and tissues and has effects on metabolism, such as promoting glycolysis. It also crosses the blood–brain barrier to act centrally and to regulate HPA axis activity by activating negative feedback mechanisms: effectively, cortisol inhibits its own production. Through its actions at glucocorticoid (GR) and mineralocorticoid (MR) receptors, particularly those expressed in limbic structures (such as the amygdala and hippocampus), cortisol is capable of modulating a range of psychological processes relating to learning and memory (de Quervain et al. 2009; Roozendaal 2000). CRH also plays a role in coordinating the adaptive response to stress through its effects on central structures such as the amygdala (Kalin et al. 1989; Kalin and Takahashi 1990; Swiergiel et al. 1993). CRH has anxiogenic properties, promoting vigilance and context-dependent motor responses, such as fight, flight, or freezing. All of these peptide- and steroid-based effects facilitate adaptive responses to stress in the short-term and help the organism to maintain or recover homeostasis.

As well as functioning as an alarm system at times of environmental challenge, the HPA axis exhibits a marked diurnal (or circadian) rhythm in humans. Cortisol secretion is highest in the morning, at the start of the activity cycle, and lowest immediately before or during sleep (Deuschle et al. 1997; Netherton et al. 2004; Rosmalen et al. 2005; Weber et al. 2000). Superimposed upon the early part of the rhythm is a characteristic increase in cortisol secretion within 1 h of waking (Clow et al. 2004; Pruessner et al. 1997): the cortisol awakening response (CAR). The suprachiasmatic nucleus of the hypothalamus is implicated in the initiation of the CAR (Clow et al. 2004, 2010), but different mechanisms may regulate the CAR relative to other components of the cortisol diurnal rhythm. In addition, genetic influences appear to be greater on cortisol concentrations measured in the morning, compared to the evening (Bartels et al. 2003; Wust et al. 2000). Of interest, the CAR may be less pronounced, or even absent, in childhood (Freitag et al. 2009; O'Connor et al. 2005), whereas many studies have observed an intact CAR in healthy adolescent populations (Fairchild et al. 2008; Oskis et al. 2009; Rosmalen et al. 2005).

3 What Is the Theoretical Basis for Investigating the Relationship Between ADHD and HPA Axis Abnormalities?

Gray (1982) proposed the existence of three interrelated neuropsychological systems subserving, respectively: fight or flight, reward sensitivity (mediated by the "behavioral activation system" or BAS), and punishment sensitivity (mediated by the "behavioral inhibition system" or BIS). According to this model, individuals differ in terms of the relative activity of these systems: some are particularly prone to seek out rewards (high BAS), while others are strongly motivated to avoid punishing stimuli (high BIS). Individuals with psychopathological conditions may represent the extreme end of the spectrum on either of these continua (e.g., anxiety disorders may be the result of excessive BIS activity) or a functional imbalance between these systems. Gray argued that the BIS responds to conditioned stimuli (signals) for punishment and nonreward so as to induce passive avoidance and extinction of behavior. BIS output causes the cessation of ongoing behaviors, focuses the organism's attention on environmental cues, and increases nonspecific physiological arousal (the most relevant feature for the present chapter).

Gray originally argued that the functions of the BIS were mediated by the septohippocampal system and its connections to the prefrontal cortex. Following early suggestions by Quay (1997) and Barkley (1997), a great deal of work in the field has been directed at trying to understand ADHD as a consequence of functional impairment of the BIS. Such impairment is argued to give rise to the hyperactive/impulsive symptoms of ADHD as well as neurocognitive impairments in visual and verbal working memory, emotion regulation, and motor fluency (Barkley 1997). As implied by its hypothetical role in increasing arousal levels, an underactive BIS may lead to reduced cortisol levels both at baseline, and particularly under conditions of psychological challenge, in ADHD. It is *critical* to distinguish between cortisol levels measured under resting conditions (referred to as 'basal' cortisol) and cortisol measures under psychological or physical stress (referred to as 'cortisol reactivity' or 'cortisol responses to stress'). Given that impaired BIS activity is suggested to be relatively specific to the combined or predominantly hyperactive/impulsive types of ADHD (Barkley 1997; Quay 1997), one prediction arising from this theory is that abnormalities in HPA axis activity should be observed in these two subtypes but not in the predominantly inattentive type of ADHD.

4 Basal Cortisol Secretion in ADHD and Related Disorders

This review of empirical findings will start by considering studies that have investigated basal or resting cortisol levels in ADHD.

An early study measured the diurnal rhythm of cortisol secretion in a group of 30 children with ADHD, comparing their results with those of a control group of adults and a psychiatric control group (21 children with autism). Saliva samples were collected at 2-h intervals in the morning. Further samples were obtained in the afternoon and evening to characterize the diurnal profile. Absolute cortisol levels were not reported and so it is not possible to tell whether ADHD was associated with abnormally low or high levels relative to the other groups. Nevertheless, the authors found that a majority (57%) of the children with ADHD failed to show a normal diurnal rhythm, whereas only 20% of the autistic subjects and 10% of the adult controls exhibited abnormalities in diurnal rhythm (Kaneko et al. 1993). When the results for the ADHD group were subdivided in terms of severity of hyperactivity, dysregulation of the cortisol rhythm was more common in those with moderate-to-severe hyperactivity, relative to those with mild hyperactivity.

The authors also used the Dexamethasone Suppression Test, which involves giving subjects a synthetic glucocorticoid in the evening and then measuring plasma cortisol concentrations the following morning. Due to the negative feedback effects of glucocorticoid administration on HPA axis activity, most subjects show suppression of cortisol secretion in the morning after dexamethasone administration. Kaneko et al. (1993) found that cortisol nonsuppression was significantly more common in the children with ADHD, relative to adult controls, implying that HPA axis negative feedback mechanisms function less effectively in ADHD. Again, HPA axis abnormalities were more common in those with severe hyperactivity: 78% of subjects in this subgroup showed cortisol nonsuppression, compared with only 20% of the mild hyperactivity ADHD subgroup.

Although these results are certainly interesting, the authors' failure to report absolute cortisol concentrations makes this study difficult to interpret. It would have been preferable to include a group of healthy children, rather than using adult controls. There were also differences in terms of procedure between the adults and the children, with adults taking a higher dose of dexamethasone closer in time to the morning cortisol assessment, and sex differences between the clinical and control groups. It is unclear whether ADHD participants were assessed for comorbid disorders that could have impacted upon HPA axis activity, such as disruptive behavior disorders (DBDs), anxiety, or depression. Finally, the absence of a normal cortisol diurnal rhythm has not been replicated in subsequent studies (see below), and it seems unlikely that such marked disturbances in the pattern of cortisol secretion are characteristic of most individuals with ADHD.

A study that compared patients with ADHD and comorbid oppositional defiant disorder (ODD; $n = 32$) and healthy control subjects ($n = 25$) found lower basal cortisol levels in the ADHD plus ODD group (Kariyawasam et al. 2002). A post hoc examination of the effects of psychostimulant administration showed that reductions in cortisol concentration were specific to the ADHD plus ODD subjects who were not taking methylphenidate or amphetamine: this finding implies that psychostimulants increase cortisol secretion. It would have been helpful to have included a non-comorbid ADHD group to determine whether it was ADHD or ODD that was driving these results and to provide further data on the effects of psychostimulants. Another limitation of this study is that cortisol concentration was measured at a single time-point only, which may have occurred at a different point in the diurnal rhythm of the two groups (although all samples were collected in the afternoon). An observation of consistently lower levels across the diurnal cycle would have been more convincing.

Methodological improvements in cortisol assessment were implemented in a small-scale study of 18 children with ADHD and 71 healthy controls that found normal waking cortisol levels and an intact CAR in the ADHD group, overall (Blomqvist et al. 2007). However, a post hoc comparison of 13 participants with high levels of hyperactivity/impulsivity ADHD symptoms and the healthy control group showed that the hyperactive ADHD subgroup (who met criteria for combined type ADHD) did not show a rise in cortisol levels in the 30 min after waking. This finding is broadly consistent with the results described above showing that HPA axis function is more aberrant in subjects with moderate-to-severe forms of hyperactivity. Limitations of this study included the use of a small ADHD group, the apparent lack of assessment for comorbid psychiatric illnesses, and lack of matching for gender.

A recent study that also used up-to-date saliva collection methodology found a normal CAR in children with ADHD alone or ADHD plus conduct disorder (CD), but a reduced CAR in those with ADHD and comorbid ODD (Freitag et al. 2009). This finding appeared to be driven by a general reduction in cortisol secretion in the latter group, rather than a specific effect on the awakening response, since they showed lower cortisol levels both at waking and +30 min after waking. This observation of a blunted CAR, specifically in those with ADHD plus comorbid

ODD, illustrates the importance of collecting information about comorbid disorders, a point which will be discussed in more detail below. The study also showed that increases in cortisol concentrations, from waking to +30 min postawakening, were smaller in participants who had experienced higher levels of psychosocial adversity.

A large ($n = 1,768$) general population study assessing the CAR and evening cortisol levels in preadolescent children found only a weak (positive) association between self-reported ADHD symptoms and evening cortisol (Sondeijker et al. 2007). There was no relationship between cortisol concentrations and parent-reported ADHD symptoms, arguably a more meaningful and reliable measure of psychopathology than self-reported symptoms. Moreover, the magnitude of the CAR was not related to symptoms of ODD or CD. One possibility is that these null findings reflect the limited variance in externalizing symptomatology that occurs in general population samples: the situation may have differed if a clinical sample had been used. However, given the statistical power provided by such a large sample, the most parsimonious interpretation of these results is that ADHD symptoms are not associated with clinically significant changes in *basal* cortisol secretion, except perhaps in extreme samples.

This conclusion is further reinforced by a recent study that characterized the diurnal profile of cortisol secretion in 28 adults with ADHD (mainly the combined type) and 28 healthy control subjects. Saliva samples were collected directly after waking and +30 min postawakening to assess the CAR. Further samples were obtained at 1700 and 2300 h to provide information about the diurnal rhythm. The results showed almost identical cortisol diurnal profiles in the ADHD and control groups (Hirvikoski et al. 2009), with both groups exhibiting a pronounced cortisol rhythm and a normal CAR. However, this study was limited by a high incidence of comorbid disorders in the ADHD group, with many participants meeting current or past criteria for major depressive disorder, generalized anxiety disorder, or borderline personality disorder.

In addition to research focusing specifically on aspects of basal cortisol secretion, such as the diurnal cortisol rhythm or the CAR, a large number of studies have reported data relevant to the issue of altered basal cortisol secretion in ADHD. These experiments assessed prestress or baseline cortisol levels before stress induction (and therefore under relatively controlled conditions). The majority of these earlier experiments found no differences between ADHD participants and healthy controls in baseline cortisol levels (Blomqvist et al. 2007; Hirvikoski et al. 2009; Lackschewitz et al. 2008; Luby et al. 2003; Randazzo et al. 2008; Snoek et al. 2004). An exception was a study that observed slightly lower prestress cortisol levels in children with combined type ADHD (van West et al. 2009).

To summarize these results: although a number of early studies provided evidence for reduced or dysregulated basal cortisol in ADHD (particularly in its persistent form or when it involves severe hyperactivity), the methodologically strongest or largest studies have not demonstrated abnormalities in basal cortisol secretion. Furthermore, studies that collected data on cortisol secretion before stress induction largely failed to show alterations in prestress cortisol levels in ADHD, regardless of subtype. While the literature does provide tentative evidence that combined type ADHD, or particularly ADHD with comorbid ODD, may be associated with slightly lower basal cortisol levels, or a blunted CAR, it is unclear whether these reductions are of clinical or prognostic significance. In addition, some of the studies purporting to have measured "basal" cortisol may have inadvertently assessed stress-induced cortisol levels, particularly if they only obtained one saliva sample in a setting that was novel to the child (e.g., a laboratory or a psychiatric clinic). It is partly for this reason that most current studies of basal cortisol secretion typically involve collecting multiple samples across the day while the participants are going about their normal routines (i.e., in naturalistic conditions). In the next section, studies that assessed cortisol reactivity during stress in ADHD will be reviewed and evaluated.

5   Is ADHD Associated with Cortisol Hyporeactivity?

An early study reported that children with persistent ADHD were characterized by reduced cortisol reactivity during performance of a neuropsychological test battery, relative to those who showed a remission of their ADHD (King et al. 1998). Baseline cortisol levels before neurocognitive testing were also slightly lower in the group with persistent ADHD compared to those who would subsequently remit. Limitations of this study included a small sample size, the lack of a matched control group, and incomplete characterization of participants. In addition, only one poststress saliva sample was obtained, leaving open the possibility that the groups differed in the latency of their stress response. From what we now know about factors determining whether a cortisol response occurs (see below), it is surprising that such large increases in cortisol were seen in the nonpersistent ADHD group when performing a battery of neurocognitive tasks. In fact, the children with nonpersistent ADHD may have been the group showing an abnormal pattern of cortisol reactivity, rather than the group with persistent ADHD.

Another study used separation from the caregiver and a play task, involving induction of mild frustration, as separate psychosocial stressors, in a cohort of preschool children. Although the focus of the study was major depressive disorder, it is relevant to the current review because a psychiatric control group of children with ADHD or ODD was included, together with a healthy control group. No differences were found between subjects with ADHD or ODD and healthy controls in salivary cortisol responses to caregiver separation (cortisol went down in both groups). Cortisol increased in both groups following performance of the frustrating task (Luby et al. 2003). In contrast, the depressed group showed a weak increase in cortisol during caregiver separation and a further increase during the frustration task.

Cortisol levels were also assessed on three consecutive evenings: analysis of these data revealed no group differences in evening basal cortisol. As will be discussed in a later section, it is not ideal to collapse across the diagnoses of ADHD and ODD and, as such, the results of this study are difficult to interpret. However, they do not provide evidence for disrupted HPA axis activity in either of these externalizing disorders because their pattern of reactivity was similar to that of healthy controls. A further limitation of the study was that participants were given a snack by the experimenters up to 30 min before baseline cortisol assessment. This may have increased cortisol levels and made it difficult to demonstrate effects of caregiver separation, since "prestress" levels were already relatively high. Nevertheless, work of this nature is important because little is known about associations between cortisol secretion and ADHD or ODD in younger age groups.

More recently, researchers have used standardized psychosocial stressors, such as the Trier Social Stress Test (TSST), to assess cortisol reactivity in clinical groups. This test involves giving a public speech in front of audience, and typically also a video camera, to induce the feeling of being socially evaluated (Kirschbaum et al. 1993). The advantage of using standardized tests of this kind is that the results can be compared to those generated by other research groups, and the stressor is more likely to be effective than those generated on an ad hoc basis. Relevant to this point, many everyday experiences that might be considered stressful do not appear to elicit cortisol responses in the majority of children or adolescents (Gunnar et al. 2009). For example, discussing conflictual topics with parents did not elicit increased cortisol secretion in a majority of healthy adolescents (Klimes-Dougan et al. 2001). In addition, even in children who report high levels of dental anxiety, cortisol levels typically did not increase during dental examinations (Blomqvist et al. 2007).

A recent meta-analytic review of 208 studies of stress reactivity in adults found that, to elicit cortisol or ACTH increases, stressors must threaten the individual's goals (Dickerson and Kemeny 2004). These might include not only the goal of physical self-preservation but also the goal of preserving the "social self." Thus, tasks that threaten the social self, such as being socially evaluated while giving a speech in front of a panel of judges, reliably increase HPA axis activity. Tasks that require intense mental effort but which do not involve threat to the social self, such as challenging neurocognitive tasks or mental arithmetic, were far less effective in provoking cortisol responses. The meta-analysis also found that uncontrollability and unpredictability were important components of effective stressors and, that these factors acted synergistically to enhance the effects of threatening the individual's central goals (Dickerson and Kemeny 2004). The TSST encompasses many of these elements, including social evaluation, achievement stress, and loss of control, and thus represents an effective psychological stressor in its original format or in the version modified for children (Foley and Kirschbaum 2010).

The question of whether patterns of cortisol reactivity differ between ADHD subtypes was addressed in a study that compared children with predominantly inattentive versus combined types of ADHD (van West et al. 2009). They used the modified Trier Social Stress Test for Children (TSST-C) to induce psychological stress in their participants. Children with combined type ADHD showed blunted cortisol responses during the test, relative to control subjects and those with predominantly inattentive type ADHD, whereas the latter two groups did not differ. Baseline cortisol levels were also slightly lower in children with combined type ADHD compared to healthy controls. These results therefore support the distinction made in the DSM-IV between predominantly inattentive and combined types of ADHD, and the suggestions that these forms of ADHD constitute qualitatively distinct and unrelated disorders (Barkley 1997; Diamond 2005). They are also in line with theoretical proposals described in an earlier section, which hold that attenuated cortisol responses are a consequence of an underactive BIS. As noted above, this underactivity is suggested to be specific to predominantly hyperactive/impulsive or combined types of ADHD.

One point of note is that although ADHD subjects with comorbid disorders were excluded from the study, the combined type ADHD group had significantly higher scores on the delinquency and aggression subscales of the Teacher Report Form and Child Behavior Checklist compared to the other two groups: this subthreshold externalizing comorbidity could have influenced the results. In addition, it seems likely that many of the subjects are at high risk for developing CD/ODD in the future, even though they did not meet full DSM-IV criteria at the time of assessment at age 6–12 years. Although the authors were not able to recruit a predominantly hyperactive/impulsive type ADHD group, which would have provided stronger evidence for a link between cortisol hyporeactivity and the hyperactive/impulsive cluster of ADHD symptoms, this study had several strengths, including the use of a standardized stressor; a comprehensive psychiatric assessment; a reasonably large sample size; and group matching for age, gender, socioeconomic status, and IQ.

Another study that investigated cortisol reactivity to the TSST-C, specifically in children with predominantly inattentive type ADHD, reported blunted cortisol responses to stress. This pattern was observed in subjects meeting diagnostic criteria for ADHD, relative to both healthy controls and those with subthreshold levels of predominantly inattentive type ADHD symptoms (Randazzo et al. 2008). These findings are clearly at variance with those reported above by van West et al. (2009), showing normal stress reactivity to the TSST-C in children with predominantly inattentive type ADHD. Of possible relevance in this respect is that the sample size used in the Randazzo et al. (2008) study was small: only seven participants met full criteria for predominantly inattentive type ADHD.

A recent study examined cortisol and cardiovascular responses to psychological stress in adults with ADHD compared to control subjects (Lackschewitz et al. 2008). Although cardiovascular responses were blunted during stress in the ADHD group, there were no group differences in cortisol reactivity. There was, however, a trend toward lower cortisol levels at baseline and throughout the procedure in the ADHD group. Interestingly, the ADHD participants reported experiencing greater subjective stress than control subjects, suggesting a discrepancy between emotional and physiological components of the stress response. A potential limitation of this study was that one-third of the ADHD subjects had comorbid depression/anxiety disorders, which may enhance cortisol responses. In addition, although all ADHD subjects were assessed for current comorbid psychiatric disorders, using the structured clinical interview for DSM-IV (SCID), it is unclear whether they were screened for, or met, lifetime criteria for DBDs such as ODD or CD. This could be problematic because cortisol hyporeactivity may be a trait abnormality even in subjects with remission of DBD symptoms.

Snoek et al. (2004) assessed cortisol reactivity in children with ADHD alone, ADHD plus comorbid ODD, or ODD alone to examine whether reduced cortisol reactivity was specific to DBD, such as ODD or CD, or whether this pattern extended to externalizing disorders in general. Compared with healthy controls, children with ODD alone, or ADHD plus ODD, showed blunted cortisol responses, whereas those with ADHD alone exhibited a normal pattern of cortisol reactivity. Group differences were evident only under stressful conditions: there were no differences in baseline cortisol in the early afternoon. These data indicate that cortisol reactivity is normal in non-comorbid ADHD, but reactivity may be blunted in individuals with a comorbid DBD.

This study had a number of limitations, including relatively high rates of comorbid anxiety disorder in all three clinical groups. Also, many of the subjects with ADHD or ADHD plus ODD were taking methylphenidate on the day of testing. However, it is worth noting that chronic use of psychostimulant medication is reported to have no detectable effects on serum cortisol levels (Weizman et al. 1987). Furthermore, the presence of comorbid anxiety disorders should have affected all three clinical groups to an equal extent. Thus, neither of these factors appears to be a plausible explanation for differences between the "ADHD alone" and the ODD groups.

Further evidence that comorbid disorders may play an important role in influencing cortisol reactivity comes from a large study of clinic-referred ADHD patients who underwent venipuncture to provide blood samples for genotyping (Hastings et al. 2009). This relatively naturalistic stressor appeared effective in eliciting cortisol responses, and the researchers subsequently examined whether comorbid anxiety or DBD influenced stress reactivity. They hypothesized that these conditions would have opposing effects on cortisol secretion. They also investigated whether there were differences in cortisol reactivity between ADHD subtypes. As predicted, children with ADHD and a comorbid anxiety disorder displayed larger cortisol responses, whereas those with ADHD and comorbid DBD showed blunted cortisol responses to stress. Children with ADHD alone and those with ADHD and comorbid anxiety plus DBD formed intermediate groups between those with either internalizing or externalizing comorbidity, only. Unfortunately, this study did not include a healthy control group; hence, it is unclear whether the children with pure ADHD exhibited a normal or a blunted cortisol response relative to those without any psychiatric disorder. The results of the analyses comparing ADHD subtypes did not reveal differences in baseline or stress-induced cortisol. However, the effect of DBD comorbidity on cortisol reactivity appeared to differ according to ADHD subtype: there were no differences between combined type ADHD without DBD and combined type ADHD with DBD, whereas comorbid DBD led to lower cortisol reactivity in those with predominantly inattentive or predominantly hyperactive–impulsive types of ADHD.

In summary, findings on cortisol reactivity in ADHD have been mixed, with several studies providing convincing evidence for blunted cortisol responses to psychosocial stress in ADHD (particularly combined type ADHD), but a similar number of experiments revealing normal patterns of cortisol reactivity in ADHD. In fact, studies using almost identical methodology and clinical samples with the same presenting diagnosis (e.g., predominantly inattentive type ADHD) have reported entirely conflicting results (cf., Randazzo et al. 2008; van West et al. 2009).

One way of reconciling some of these discrepancies may be to consider the prevalence of comorbid DBD symptoms or diagnoses in the ADHD participants, since ODD and CD have repeatedly been shown to be associated with cortisol hyporeactivity during psychosocial stress (Fairchild et al. 2008; Popma et al. 2006; van Goozen et al. 1998, 2000). The importance of assessing for comorbid DBDs in patients with ADHD is shown most clearly in studies that have explicitly compared subjects with ADHD alone, versus those with ADHD and comorbid ODD or ODD alone. Cortisol responses were normal in those with pure ADHD but blunted in subjects with ODD (with or without ADHD) (Snoek et al. 2004). This position is broadly supported by a study explicitly investigating the impact of comorbid internalizing and externalizing disorders on cortisol reactivity in ADHD (Hastings et al. 2009), which found that subjects with ADHD and comorbid DBD showed weaker cortisol responses to stress than those with pure ADHD. The authors also found that subjects with ADHD and comorbid anxiety disorders displayed larger cortisol responses during stress than those with ADHD alone. Thus, in general, cortisol reactivity is increased in individuals with internalizing disorders and reduced in those with externalizing disorders such as ODD and CD. When these forms of comorbidity occur together, they appear to cancel each other out in terms of their effects on cortisol reactivity.

6 Recommendations for Future Research on HPA Activity in ADHD and Related Disorders

6.1 Assess Cortisol in Relation to Waking Time

A key point of this review has been that the HPA axis is a dynamic system that not only responds to psychological and physical stress but also exhibits a marked diurnal rhythm and a CAR close to the start of the activity cycle. As a consequence, in studies of basal cortisol secretion in psychiatric disorders, it is critical to assess cortisol levels in relation to the waking time of the individual being assessed. This is particularly important when morning cortisol levels are being measured because subtle group differences in waking time could create the erroneous impression that there are group differences in cortisol secretion (when, in fact, the CAR is simply occurring at a different time in the clinical group). Thus, measuring cortisol levels at waking and +30 min after waking and recording the time of awakening are strongly advised. As well as increasing the validity and reliability of group comparisons of morning cortisol secretion, assessment of the CAR means that researchers obtain a sensitive measure of physiological reactivity. Additional measurements at +45 and +60 min are desirable since they permit enhanced characterization of the latency and profile of the CAR and may provide information about efficacy of negative feedback mechanisms. However, it is also important to use protocols that most participants can implement in their everyday lives and to consider that it may not be practical to collect these extra morning samples when studying young children. Furthermore, because afternoon and evening cortisol levels may be more readily influenced by the effects of psychopathology, subjective mood states, and environmental factors, it is advisable to collect additional samples during these periods to enable characterization of the diurnal cortisol profile. Finally, wherever feasible, cortisol levels should be assessed across two or more days to assess reliability and stability of group differences in cortisol secretion or dysregulation of the cortisol diurnal profile.

6.2 Advantages of Assessing Cortisol Levels in Saliva

Undergoing venipuncture can be stressful (Hastings et al. 2009), and so measuring cortisol levels in saliva rather than serum is usually preferable because it avoids this confound. In addition, salivary cortisol is a better measure of biologically available (so-called free) cortisol levels: much of the cortisol measured in serum is bound to cortisol-binding globulin (CBG) and unable to act at corticosteroid receptors. A further advantage is that most participants regard provision of saliva samples as more acceptable than giving blood samples. Thus, researchers maximize the number of individuals willing to take part in their studies by adopting the former method. Clearly, if cortisol is being measured under naturalistic conditions (i.e., in the participants' homes), then obtaining saliva is the only practical method of doing this. Electronic devices can be used to monitor subject compliance with saliva collection protocols (Broderick et al. 2004), but this may not be economically feasible in large population-level studies. An alternative approach is to ask participants to record their saliva collection times in a "diary"; they can also use this to make a note of potential confounds that may influence cortisol levels, such as taking exercise, using caffeinated drinks, or smoking cigarettes before saliva collection.

6.3 Appropriate Conditions, Ethical Issues, and Sample Sizes in Psychoneuroendocrine Research

Several studies have shown that the incremental increase in cortisol levels during stress is broadly similar whether the stressor is applied in the morning or in the afternoon (Kudielka et al. 2004). Nevertheless, it is recommended that researchers restrict their testing to a set period in the day (e.g., mid- to late-afternoon). This should improve detection of increases in cortisol levels following stress and provide other advantages, such as making group differences in baseline levels more interpretable. As noted above, the use of a standardized stressor and acclimatization to the laboratory setting are also highly desirable in terms of achieving increases in cortisol levels in most control participants. Many studies of stress reactivity in children have failed to meet these criteria, largely due to the use of weak psychological stressors (Gunnar et al. 2009). Of course, this issue has an ethical dimension because researchers should not seek to induce intensely negative emotional states and they have a duty to protect their participants – particularly when working with children. However, to elicit measurable and robust cortisol responses, it is necessary to temporarily threaten the individual's "social self" by making them feel as though they are being negatively socially evaluated and that they have lost control of the situation in some way (Dickerson and Kemeny 2004).

In addition, the tests need to be adequately powered, and so sample sizes need to be relatively large in stress reactivity research because there is considerable interindividual variability in cortisol responses to psychological stress. Even in typically developing adolescents, when using a standardized stressor, cortisol levels may drop in some individuals and rise by 400–500% in others. As a consequence, it is difficult to demonstrate significant group differences if sample sizes are small.

7 What Do We Need to Know About the Relationship Between ADHD and HPA Axis Activity?

Several important issues remain unresolved. First, does cortisol hyporeactivity play a causal role in the etiology of either ADHD or common comorbid disorders such as ODD, or are any changes in cortisol secretion a consequence of these disorders? Related to this is another question: does cortisol hyporeactivity reflect a state-like effect or does it represent a trait-like vulnerability factor that increases risk for developing ADHD or ODD/CD in a probabilistic fashion? Furthermore, is such a deficit present before development of the full ADHD syndrome or even after remission of symptoms? Longitudinal studies are required to answer these questions since, to date, almost all research in this area has been cross-sectional in basis (i.e., the studies have made comparisons between cases and control subjects at a specific point in time). Second, what neurobiological mechanisms are responsible for the pattern of cortisol hyporeactivity observed in ADHD and related externalizing disorders, such as ODD? For example, are there fundamental deficits in the hypothalamic–pituitary–adrenal axis (such as reduced pituitary size or adrenal insensitivity to ACTH) in individuals with ADHD/ODD? Or, as seems more likely, is there some impairment of "processive" aspects of the stress response, mediated by dysfunction in limbic circuits that convey information about stressors in the environment to the hypothalamus? Accordingly, it may be informative to relate volumes in key brain structures that influence HPA axis activity (such as amygdala and hippocampus) and patterns of cortisol secretion in disorders such as ADHD.

Third, if individuals with ADHD, or related externalizing disorders, retain a normal pattern of cortisol reactivity, does this act as a protective factor and is it predictive of better outcomes? Preliminary findings in children with ODD suggest that cortisol hyporeactivity during stress is a potential biomarker for a poor response to psychological treatment (van de Wiel et al. 2004). This could be a worthwhile issue to investigate in relation to psychological (or possibly pharmacological) treatments for ADHD.

Fourth, does cortisol hyporeactivity reflect a general deficit in physiological arousal, or an adaptation to prenatal or early life stress, in those with ADHD? Studies in animals and humans have demonstrated that prenatal stress can exert a programming effect on the HPA axis (Glover et al. 2010; Lupien et al. 2009). For example, fetal exposure to elevated glucocorticoid concentrations at certain periods during gestation permanently alters HPA axis activity in rodents (Kapoor et al. 2006). Although the evidence for similar programming effects in humans is sparse, recent studies have reported elevated basal cortisol in children whose mothers experienced prenatal anxiety late in pregnancy (Huizink et al. 2008; O'Connor et al. 2005), although another study found the largest effects if prenatal anxiety occurred early in gestation (Van den Bergh et al. 2008). In contrast to these results, a study investigating the impact of adverse life events during pregnancy upon cortisol secretion in offspring (tested in adulthood) found normal basal cortisol secretion and exaggerated responses to the TSST in those exposed to prenatal stress (Entringer et al. 2009). These findings, which provide suggestive evidence for fetal programming of the HPA axis in humans, may be relevant, given that prenatal anxiety also appears to act as a risk factor for the development of ADHD symptoms and externalizing symptoms (O'Connor et al. 2003; Van den Bergh and Marcoen 2004).

Since it is well established that environmental adversity is linked with ADHD (Biederman et al. 2002), it may also be instructive to consider the consequences of early life or chronic psychosocial adversity. Findings on the long-term impact of early life stress and maltreatment on cortisol reactivity are mixed: some studies show heightened cortisol responses in adult survivors of sexual abuse (Heim et al. 2000b), but others show blunted cortisol responses in victims of physical abuse and maltreatment (Carpenter et al. 2007, 2009; Elzinga et al. 2008). In relation to the consequences of chronic adversity, a number of studies have revealed hypocortisolism in healthy individuals living under conditions of ongoing stress, or patients living with chronic pain or physical illness (Heim et al. 2000a). These studies illustrate the potential importance of studying early experiences that may have a programming effect on the HPA axis and measuring current exposure to everyday stressors. This is because both factors may impact upon basal cortisol secretion and cortisol reactivity. In addition, measuring these factors may enhance our understanding of the mechanisms underlying alterations in HPA axis activity in ADHD and related externalizing disorders such as ODD and CD.

8 Conclusions

The HPA axis is a key physiological system, which mediates responses to stress and also exhibits a diurnal rhythm and a response to awakening. There are theoretical reasons to expect an association between hyperactive/impulsive forms of ADHD and reduced cortisol secretion, particularly under conditions of stress. Although early studies reported provocative results on cortisol secretion in ADHD, suggesting marked dysregulation of the HPA axis, more recent work has provided little convincing evidence for alterations in basal (resting) cortisol secretion in non-comorbid forms of ADHD. Where ADHD occurs together with ODD, there may be a moderate reduction in basal cortisol secretion or the magnitude of the CAR. The results of studies assessing cortisol reactivity during stress in ADHD do not show a consistent pattern, although blunted cortisol responses are more commonly reported in combined type relative to predominantly inattentive type ADHD. Again, the presence or absence of comorbid illnesses may be important, with cortisol hyporeactivity typically observed in those with ADHD plus ODD or CD and exaggerated cortisol responses seen in those with comorbid internalizing disorders such as generalized anxiety disorder.

The vast majority of studies on basal cortisol secretion or cortisol reactivity in ADHD have been cross-sectional in design and, as a consequence, little is known about the prognostic value of measuring HPA axis activity. It is also not known whether abnormalities in cortisol secretion play a causal role in the etiology of ADHD, or are merely a consequence of living with ADHD [which may be accompanied by higher levels of perceived stress in everyday life (Hirvikoski et al. 2009)]. Evidently, longitudinal research in this area is merited and may be highly informative.

Acknowledgment The author was supported by a Project Grant from the Wellcome Trust (#083140) during the writing of this chapter.

References

Barkley RA (1997) Behavioral inhibition, sustained attention, and executive functions: constructing a unifying theory of ADHD. Psychol Bull 121:65–94

Bartels M, Van den Berg M, Sluyter F, Boomsma DI, de Geus EJ (2003) Heritability of cortisol levels: review and simultaneous analysis of twin studies. Psychoneuroendocrinology 28:121–137

Biederman J, Faraone SV, Monuteaux MC (2002) Differential effect of environmental adversity by gender: Rutter's index of adversity in a group of boys and girls with and without ADHD. Am J Psychiatry 159:1556–1562

Blomqvist M, Holmberg K, Lindblad F, Fernell E, Ek U, Dahllof G (2007) Salivary cortisol levels and dental anxiety in children with attention deficit hyperactivity disorder. Eur J Oral Sci 115:1–6

Broderick JE, Arnold D, Kudielka BM, Kirschbaum C (2004) Salivary cortisol sampling compliance: comparison of patients and healthy volunteers. Psychoneuroendocrinology 29:636–650

Burke HM, Davis MC, Otte C, Mohr DC (2005) Depression and cortisol responses to psychological stress: a meta-analysis. Psychoneuroendocrinology 30:846–856

Carpenter LL, Carvalho JP, Tyrka AR, Wier LM, Mello AF, Mello MF et al (2007) Decreased adrenocorticotropic hormone and cortisol responses to stress in healthy adults reporting significant childhood maltreatment. Biol Psychiatry 62:1080–1087

Carpenter LL, Tyrka AR, Ross NS, Khoury L, Anderson GM, Price LH (2009) Effect of childhood emotional abuse and age on cortisol responsivity in adulthood. Biol Psychiatry 66:69–75

Cleare AJ (2003) The neuroendocrinology of chronic fatigue syndrome. Endocr Rev 24:236–252

Clow A, Thorn L, Evans P, Hucklebridge F (2004) The awakening cortisol response: methodological issues and significance. Stress 7:29–37

Clow A, Hucklebridge F, Stalder T, Evans P, Thorn L (2010) The cortisol awakening response: more than a measure of HPA axis function. Neurosci Biobehav Rev 35:97–103

de Quervain DJ, Aerni A, Schelling G, Roozendaal B (2009) Glucocorticoids and the regulation of memory in health and disease. Front Neuroendocrinol 30:358–370

Deuschle M, Schweiger U, Weber B, Gotthardt U, Korner A, Schmider J et al (1997) Diurnal activity and pulsatility of the hypothalamus-pituitary-adrenal system in male depressed patients and healthy controls. J Clin Endocrinol Metab 82:234–238

Diamond A (2005) Attention-deficit disorder (attention-deficit/hyperactivity disorder without hyperactivity): a neurobiologically and behaviorally distinct disorder from attention-deficit/hyperactivity disorder (with hyperactivity). Dev Psychopathol 17:807–825

Dickerson SS, Kemeny ME (2004) Acute stressors and cortisol responses: a theoretical integration and synthesis of laboratory research. Psychol Bull 130:355–391

Elzinga BM, Roelofs K, Tollenaar MS, Bakvis P, van Pelt J, Spinhoven P (2008) Diminished cortisol responses to psychosocial stress associated with lifetime adverse events a study among healthy young subjects. Psychoneuroendocrinology 33:227–237

Entringer S, Kumsta R, Hellhammer DH, Wadhwa PD, Wust S (2009) Prenatal exposure to maternal psychosocial stress and HPA axis regulation in young adults. Horm Behav 55:292–298

Fairchild G, van Goozen SH, Stollery SJ, Brown J, Gardiner J, Herbert J et al (2008) Cortisol diurnal rhythm and stress reactivity in male adolescents with early-onset or adolescence-onset conduct disorder. Biol Psychiatry 64:599–606

Foley P, Kirschbaum C (2010) Human hypothalamus-pituitary-adrenal axis responses to acute psychosocial stress in laboratory settings. Neurosci Biobehav Rev 35(1):91–96

Freitag CM, Hanig S, Palmason H, Meyer J, Wust S, Seitz C (2009) Cortisol awakening response in healthy children and children with ADHD: impact of comorbid disorders and psychosocial risk factors. Psychoneuroendocrinology 34:1019–1028

Glover V, O'Connor TG, O'Donnell K (2010) Prenatal stress and the programming of the HPA axis. Neurosci Biobehav Rev 35:17–22

Gray JA (1982) The neuropsychology of anxiety: an enquiry into the functions of the septo-hippocampal system. Oxford University Press, Oxford

Gunnar MR, Talge NM, Herrera A (2009) Stressor paradigms in developmental studies: what does and does not work to produce mean increases in salivary cortisol. Psychoneuroendocrinology 34:953–967

Hastings PD, Fortier I, Utendale WT, Simard LR, Robaey P (2009) Adrenocortical functioning in boys with attention-deficit/hyperactivity disorder: examining subtypes of ADHD and associated comorbid conditions. J Abnorm Child Psychol 37:565–578

Heim C, Ehlert U, Hellhammer DH (2000a) The potential role of hypocortisolism in the pathophysiology of stress-related bodily disorders. Psychoneuroendocrinology 25:1–35

Heim C, Newport DJ, Heit S, Graham YP, Wilcox M, Bonsall R et al (2000b) Pituitary-adrenal and autonomic responses to stress in women after sexual and physical abuse in childhood. J Am Med Assoc 284:592–597

Herbert J, Goodyer IM, Grossman AB, Hastings MH, de Kloet ER, Lightman SL et al (2006) Do corticosteroids damage the brain? J Neuroendocrinol 18:393–411

Herman JP, Cullinan WE (1997) Neurocircuitry of stress: central control of the hypothalamo-pituitary-adrenocortical axis. Trends Neurosci 20:78–84

Hirvikoski T, Lindholm T, Nordenstrom A, Nordstrom AL, Lajic S (2009) High self-perceived stress and many stressors, but normal diurnal cortisol rhythm, in adults with ADHD (attention-deficit/hyperactivity disorder). Horm Behav 55:418–424

Holsboer F (2000) The corticosteroid receptor hypothesis of depression. Neuropsychopharmacology 23:477–501

Huizink AC, Bartels M, Rose RJ, Pulkkinen L, Eriksson CJ, Kaprio J (2008) Chernobyl exposure as stressor during pregnancy and hormone levels in adolescent offspring. J Epidemiol Community Health 62:e5

Kalin NH, Takahashi LK (1990) Fear-motivated behavior induced by prior shock experience is mediated by corticotropin-releasing hormone systems. Brain Res 509:80–84

Kalin NH, Shelton SE, Barksdale CM (1989) Behavioral and physiologic effects of CRH administered to infant primates undergoing maternal separation. Neuropsychopharmacology 2:97–104

Kaneko M, Hoshino Y, Hashimoto S, Okano T, Kumashiro H (1993) Hypothalamic-pituitary-adrenal axis function in children with attention-deficit hyperactivity disorder. J Autism Dev Disord 23:59–65

Kapoor A, Dunn E, Kostaki A, Andrews MH, Matthews SG (2006) Fetal programming of hypothalamo-pituitary-adrenal function: prenatal stress and glucocorticoids. J Physiol 572:31–44

Kariyawasam SH, Zaw F, Handley SL (2002) Reduced salivary cortisol in children with comorbid attention deficit hyperactivity disorder and oppositional defiant disorder. Neuroendocrinol Lett 23:45–48

King JA, Barkley RA, Barrett S (1998) Attention-deficit hyperactivity disorder and the stress response. Biol Psychiatry 44:72–74

Kirschbaum C, Pirke KM, Hellhammer DH (1993) The 'Trier Social Stress Test' – a tool for investigating psychobiological stress responses in a laboratory setting. Neuropsychobiology 28:76–81

Klimes-Dougan B, Hastings PD, Granger DA, Usher BA, Zahn-Waxler C (2001) Adrenocortical activity in at-risk and normally developing adolescents: individual differences in salivary cortisol basal levels, diurnal variation, and responses to social challenges. Dev Psychopathol 13:695–719

Kudielka BM, Schommer NC, Hellhammer DH, Kirschbaum C (2004) Acute HPA axis responses, heart rate, and mood changes to psychosocial stress (TSST) in humans at different times of day. Psychoneuroendocrinology 29:983–992

Lackschewitz H, Huther G, Kroner-Herwig B (2008) Physiological and psychological stress responses in adults with attention-deficit/hyperactivity disorder (ADHD). Psychoneuroendocrinology 33:612–624

Luby JL, Heffelfinger A, Mrakotsky C, Brown K, Hessler M, Spitznagel E (2003) Alterations in stress cortisol reactivity in depressed preschoolers relative to psychiatric and no-disorder comparison groups. Arch Gen Psychiatry 60:1248–1255

Lupien SJ, McEwen BS, Gunnar MR, Heim C (2009) Effects of stress throughout the lifespan on the brain, behaviour and cognition. Nat Rev Neurosci 10:434–445

Netherton C, Goodyer I, Tamplin A, Herbert J (2004) Salivary cortisol and dehydroepiandrosterone in relation to puberty and gender. Psychoneuroendocrinology 29:125–140

O'Connor TG, Heron J, Golding J, Glover V (2003) Maternal antenatal anxiety and behavioural/emotional problems in children: a test of a programming hypothesis. J Child Psychol Psychiatry 44:1025–1036

O'Connor TG, Ben-Shlomo Y, Heron J, Golding J, Adams D, Glover V (2005) Prenatal anxiety predicts individual differences in cortisol in pre-adolescent children. Biol Psychiatry 58:211–217

Oskis A, Loveday C, Hucklebridge F, Thorn L, Clow A (2009) Diurnal patterns of salivary cortisol across the adolescent period in healthy females. Psychoneuroendocrinology 34:307–316

Popma A, Jansen LM, Vermeiren R, Steiner H, Raine A, Van Goozen SH et al (2006) Hypothalamus pituitary adrenal axis and autonomic activity during stress in delinquent male adolescents and controls. Psychoneuroendocrinology 31:948–957

Pruessner JC, Wolf OT, Hellhammer DH, Buske-Kirschbaum A, von Auer K, Jobst S et al (1997) Free cortisol levels after awakening: a reliable biological marker for the assessment of adrenocortical activity. Life Sci 61:2539–2549

Quay HC (1997) Inhibition and attention deficit hyperactivity disorder. J Abnorm Child Psychol 25:7–13

Randazzo WT, Dockray S, Susman EJ (2008) The stress response in adolescents with inattentive type ADHD symptoms. Child Psychiatry Hum Dev 39:27–38

Roozendaal B (2000) 1999 Curt P. Richter award. Glucocorticoids and the regulation of memory consolidation. Psychoneuroendocrinology 25:213–238

Rosmalen JG, Oldehinkel AJ, Ormel J, de Winter AF, Buitelaar JK, Verhulst FC (2005) Determinants of salivary cortisol levels in 10-12 year old children; a population-based study of individual differences. Psychoneuroendocrinology 30:483–495

Snoek H, Van Goozen SH, Matthys W, Buitelaar JK, van Engeland H (2004) Stress responsivity in children with externalizing behavior disorders. Dev Psychopathol 16:389–406

Sondeijker FE, Ferdinand RF, Oldehinkel AJ, Veenstra R, Tiemeier H, Ormel J et al (2007) Disruptive behaviors and HPA-axis activity in young adolescent boys and girls from the general population. J Psychiatr Res 41:570–578

Swiergiel AH, Takahashi LK, Kalin NH (1993) Attenuation of stress-induced behavior by antagonism of corticotropin-releasing factor receptors in the central amygdala in the rat. Brain Res 623:229–234

van de Wiel NM, van Goozen SH, Matthys W, Snoek H, van Engeland H (2004) Cortisol and treatment effect in children with disruptive behavior disorders: a preliminary study. J Am Acad Child Adolesc Psychiatry 43:1011–1018

Van den Bergh BR, Marcoen A (2004) High antenatal maternal anxiety is related to ADHD symptoms, externalizing problems, and anxiety in 8- and 9-year-olds. Child Dev 75:1085–1097

Van den Bergh BR, Van Calster B, Smits T, Van Huffel S, Lagae L (2008) Antenatal maternal anxiety is related to HPA-axis dysregulation and self-reported depressive symptoms in adolescence: a prospective study on the fetal origins of depressed mood. Neuropsychopharmacology 33:536–545

van Goozen SH, Matthys W, Cohen-Kettenis PT, Gispen-de Wied C, Wiegant VM, van Engeland H (1998) Salivary cortisol and cardiovascular activity during stress in oppositional-defiant disorder boys and normal controls. Biol Psychiatry 43:531–539

van Goozen SH, Matthys W, Cohen-Kettenis PT, Buitelaar JK, van Engeland H (2000) Hypothalamic-pituitary-adrenal axis and autonomic nervous system activity in disruptive children and matched controls. J Am Acad Child Adolesc Psychiatry 39:1438–1445

van West D, Claes S, Deboutte D (2009) Differences in hypothalamic-pituitary-adrenal axis functioning among children with ADHD predominantly inattentive and combined types. Eur Child Adolesc Psychiatry 18:543–553

Weber B, Lewicka S, Deuschle M, Colla M, Vecsei P, Heuser I (2000) Increased diurnal plasma concentrations of cortisone in depressed patients. J Clin Endocrinol Metab 85:1133–1136

Weizman R, Dick J, Gil-Ad I, Weitz R, Tyano S, Laron Z (1987) Effects of acute and chronic methylphenidate administration on beta-endorphin, growth hormone, prolactin and cortisol in children with attention deficit disorder and hyperactivity. Life Sci 40:2247–2252

Wust S, Federenko I, Hellhammer DH, Kirschbaum C (2000) Genetic factors, perceived chronic stress, and the free cortisol response to awakening. Psychoneuroendocrinology 25:707–720

Yehuda R, Kahana B, Binder-Brynes K, Southwick SM, Mason JW, Giller EL (1995) Low urinary cortisol excretion in Holocaust survivors with posttraumatic stress disorder. Am J Psychiatry 152:982–986

Yehuda R, Golier JA, Kaufman S (2005) Circadian rhythm of salivary cortisol in Holocaust survivors with and without PTSD. Am J Psychiatry 162:998–1000

Brain Processes in Discounting: Consequences of Adolescent Methylphenidate Exposure

Walter Adriani, Francesca Zoratto, and Giovanni Laviola

Contents

1 Introduction 115
2 Behavioral Features of ADHD and Adolescence 116
    2.1 ADHD and the Use of MPH 116
    2.2 Adolescence and Its Modeling 117
3 Dopaminergic Peculiarities in Rodent Models of Adolescence 119
    3.1 Understanding Adolescence from a Neurobiological Basis 119
    3.2 Pharmacological Imaging of MPH in the Adolescent Forebrain 120
4 Discounting Processes in ADHD Modeling 122
    4.1 Self-Control Behavior Following Adolescent MPH 123
    4.2 Methodological Remarks on Discounting Tasks 125
5 Consequences of Adolescent MPH Exposure 128
    5.1 Persistent Brain Metabolic Outcomes of Adolescent MPH 128
    5.2 Consequences on Striatal Gene Expression of Adolescent MPH Exposure 130
6 Concluding Remarks and Future Perspectives 131
References 136

Abstract Traits of inattention, impulsivity, and motor hyperactivity characterize children diagnosed with attention-deficit/hyperactivity disorder (ADHD), whose inhibitory control is reduced. In animal models, crucial developmental phases or experimental transgenic conditions account for peculiarities, such as sensation-seeking and risk-taking behaviors, and reproduce the beneficial effects of psychostimulants. An "impulsive" behavioral profile appears to emerge more extremely in rats when

W. Adriani, F. Zoratto, and G. Laviola
Section of Behavioural Neuroscience, Department of Cell Biology & Neurosciences, Istituto Superiore di Sanitá, Viale Regina Elena 299, I-00161, Rome, Italy
e-mail: walter.adriani@iss.it; francesca.zoratto@iss.it; giovanni.laviola@iss.it

forebrain dopamine (DA) systems undergo remodeling, as in adolescence, or with experimental manipulation tapping onto the dopamine transporter (DAT). Ritalin® (methylphenidate, MPH), a DAT-blocking drug, is prescribed for ADHD therapy but is also widely abused by human adolescents. Administration of MPH during rats' adolescence causes a long-term modulation of their self-control, in terms of reduced intolerance to delay and diminished proneness for risk when reward is uncertain. Exactly the opposite profile emerges when exogenous alteration of DAT levels is achieved via lentiviral transfection. Both adolescent MPH exposure and DAT-targeting transfection lead to enduring hyperfunction of dorsal striatum and hypofunction of ventral striatum. Together with upregulation of prefronto-cortical phospho-creatine, striatal upregulation of selected genes (like serotonin 7 receptor gene) suggests that enhanced inhibitory control is generated by adolescent MPH exposure. Operant tasks, which assess the balance between motivational drives and inhibitory self-control, are thus useful for investigating reward-discounting processes and their modulation by DAT-targeting tools. In summary, due to the complexity of human studies, preclinical investigations of rodent models are necessary to understand better both the neurobiology of ADHD-like symptoms' etiology and the long-term therapeutic safety of adolescent MPH exposure.

Keywords Delay-discounting · Long-term effects · Magnetic resonance · Preclinical models · Psychostimulants · Rats

Abbreviations

5-HT	5-Hydroxytryptamine (serotonin)
5-HT7	5-Hydroxytryptamine (serotonin) receptor 7 (protein)
ADHD	Attention-deficit/hyperactivity disorder
BOLD	Blood-oxygen level-dependent
CBF	Cerebral blood flow
CBV	Cerebral blood volume
DA	Dopamine
DAT	Dopamine transporter
DBH	Dopamine-β-hydroxylase
dStr	Dorsal striatum
fMRI	Functional magnetic resonance imaging
H-MRS	Proton magnetic resonance spectroscopy
Htr7	5-Hydroxytryptamine (serotonin) receptor 7 (gene)
ID	Intolerance-to-Delay (task)
mITI	Mean inter-trial interval
MPH	Methylphenidate
NAcc	Nucleus accumbens
NET1	Norepinephrine transporter (for uptake 1 (neuronal))
PD	Probabilistic-Delivery (task)

PFC	Prefrontal cortex
phMRI	Pharmacological magnetic resonance imaging
pnd	Postnatal day
RT	Response time
RTT	Response-time task
SERT	Serotonin transporter
SNAP-25	Synaptosomal-associated peptide-25
SSRI	Selective serotonin reuptake inhibitor
TO	Timeout

1 Introduction

The traits of inattention, impulsivity, and motor hyperactivity are core characteristics of attention-deficit/hyperactivity disorder (ADHD). Increased motor and exploratory activity, as well as a propensity for sensation-seeking, risk-taking, and other "extreme" behaviors, appear to be even more pronounced in ADHD subjects when they become adolescents: namely, at a developmental stage when inhibitory control is further reduced (Chambers and Potenza 2003). As such, "risky business" is an additional, characteristic complication of adolescents with ADHD (Fareri et al. 2008). Here, we focus on the features of ADHD in adolescence, with a special emphasis on a possible "meaning" for such a peculiar profile of symptoms or traits (see also the dedicated box at the end of the chapter). Specifically, we sought to model (1) "origins" or etiology, as studied by modern techniques such as lentiviral transfection in vivo, and (2) sequelae or "consequences," possibly linked with neurobehavioral adaptive responses elicited by psychostimulant drugs.

In the first part of this chapter, we summarize literature that explores both the behavioral features of human adolescence and the neurobiological peculiarities of rodent models for adolescence (see Sects 2 and 3). In the second part of this chapter, we review some preclinical work that highlights the need for epidemiological and clinical investigations to help answer a crucial question: does psychostimulant exposure affect the developmental trajectories in humans and, if so, to what extent? Indeed, we report preclinical evidence of unease about the use of methylphenidate (MPH), which is commonly prescribed (as Ritalin®) to treat ADHD (Scheffler et al. 2007). With the aim of analyzing the consequences of MPH consumption among adolescents, we focus on the long-term consequences of juvenile exposure to MPH in rodent models (see Sects 4 and 5).

Animal models have added value from a translational perspective because it is possible to use approaches which are virtually impossible with humans. Thus, in addition to the use of MPH, the same protein and gene which are targeted by this drug can be reached via direct transgenetic approaches. Permanent interference with gene and protein expression and function is possible nowadays using lentiviral tools (Park 2007; Adriani et al. 2009, 2010a). In particular, these have been used to investigate whether a lentiviral-mediated direct intervention of the dopamine transporter (DAT), the MPH target and site of action, leads to similar, persistent neurobiological adaptations, when compared to adolescent MPH exposure. In both cases, results included enduring changes in gene expression, together with hypofunction of the nucleus accumbens (NAcc) and hyperfunction of dorsal striatum (dStr). The profile of discounting behavior was, however, symmetrical if not opposite. Here, we discuss brain processes which may explain the similarities and discrepancies between these two models.

At the end of this chapter, we conclude with an in-depth presentation of evolutionary perspectives on ADHD. Indeed, the "origins" of ADHD-like symptoms or traits have also been studied from the point of view of evolutionary biology, which considers questions such as: could these traits have, or have had, a potentially adaptive value? Could they have been advantageous at any time or in any environment? In this special focus box, we provide a short overview of the contrasting evolutionary explanations that have been proposed. We consider the adaptive value (if any) of having (or maintaining) a proportion of ADHD individuals in the adolescent population, as assessed in the context of evolutionary (or Darwinian) psychiatry.

2 Behavioral Features of ADHD and Adolescence

2.1 ADHD and the Use of MPH

ADHD is a chronic neurobehavioral disease, classically considered to be an executive dysfunction characterized by "poor" decision making. Selective impairment of those cognitive processes, which evaluate pros and cons in terms of costs and benefits of alternative possibilities, is thought to be especially relevant. ADHD can also be viewed as a motivational dysfunction, arising from altered processing of reward value by fronto-striatal circuits, and characterized by attempts to escape or avoid any situation that requires waiting (or "wasting") of time, such as: slow gathering of information, withholding of impulses, or prolonged focusing of attention (Sagvolden et al. 1993; Puumala et al. 1996; Sadile et al. 1996). From this perspective, ADHD may be a consequence of a psychological inability to give a correct account of, and to represent mentally, reward that is not immediately accessible. As such, reward that is delivered in the distant future, and requires "patience" to be attained, is "discounted." By contrast, reward that is (or gives an impression of being) immediate, or available in the near future (by some "quick action"), generates the motivation for its attainment. In both cases, the value of an anticipated reward is often interpreted incorrectly (i.e., "discounted") by ADHD patients because of a greater impact of adverse factors, such as: effort needed to obtain the reinforcer; time needed to complete such effort; waiting time required before delivery of the reinforcer; as well as diminished concern about the potential risk of unforeseen interference and/or negative unexpected outcomes.

In the past two decades, the prescription and production rates of MPH have increased dramatically. Notwithstanding this trend, there is concern that the rise in therapeutic use of MPH might coincide with a rise in its nontherapeutic, nonmedical (mis)use. Intranasal MPH produces a euphoric sensation or "high" in humans (Volkow and Swanson 2003), and rats readily self-administer this drug intravenously (Collins et al. 1984; Fletcher et al. 2007; Botly et al. 2008). College students and adolescents might be attracted to MPH for its attention-focusing, weight-loss, or euphoric effects. Indeed, the majority of college students in the USA have reported that the primary reason for its use was to improve academic performance (Bogle and Smith 2009). In spite of the fact that MPH has little abuse potential, especially if taken orally (as prescribed) (Volkow et al. 1999, 2002; Svetlov et al. 2007), the reinforcing effects of this compound depend upon the route of administration and should not be neglected. The annual prevalence of illicit MPH use as a recreational, performance-improving drug in the USA is 4% (predominantly intranasal) and rising (Klein-Schwartz and McGrath 2003; McCabe et al. 2004). Despite this evidence, the long-term effects of nonmedical, recreational MPH use and abuse during adolescence are still unknown.

So far, there have been few systematic studies on the sequelae of such widespread MPH misuse in human adolescents; few published investigations have assessed why MPH is so widely consumed, for recreational purpose and not just as a therapy for ADHD (Arria and Wish 2006; Wilens et al. 2008; Bogle and Smith 2009), and whether MPH may serve as a gateway to adult substance misuse (Barkley et al. 2003; Hechtman and Greenfield 2003; Ptacek et al. 2009).

2.2 Adolescence and Its Modeling

Pharmacological pathways (and the psychotropic effects) of MPH are similar, although not identical, to those targeted (and elicited) by amphetamine and cocaine. All these compounds act mainly through the dopaminergic system, by blocking the dopamine (DA) transporter (DAT), thereby increasing the extracellular concentration of DA in the terminal field of mesocorticolimbic neurons (Volkow et al. 2001, 2002). In line with this, a simple reason for psychostimulant use in human adolescents is the age-related need of (and search for) any DA-releasing stimuli (Laviola et al. 1999). The unique developmental profile of the brain dopaminergic system, during the transition from childhood to adult life (Chambers and Potenza 2003; Brenhouse and Andersen 2011), may contribute to a higher sensitivity for motivating stimuli and to the psychostimulant effects of abused drugs. In other words, functional differences in the adolescent versus adult forebrain dopaminergic systems (Teicher et al. 1995; Marco et al. 2011) appear to underlie a deviant overexpression of behaviors that are particularly rewarding at this age (such as the search for novel environments and/or social playmates; Andersen et al. 2000; Kalsbeek et al. 1988; Laviola et al. 1999). Specific natural events, such as exploration of novel environments (Rebec et al. 1997a,b), have been hypothesized to result in a greater DA release and hence to be particularly rewarding at this age, when compared to adulthood (Laviola et al. 2003; Tirelli et al. 2003). Consistently, lentiviral manipulation of the dopaminergic tone in adult animals does alter novelty seeking and social approach, together with a reduction in anxiety and increased propensity to risk (Adriani et al. 2009, 2010a). Interestingly, however, an unbalanced and exaggerated version of the adolescent behavioral repertoire is found in ADHD individuals, where affiliative and playful behaviors are relatively less prevalent and there is a notable (perhaps excessive) drive for sensation-seeking and risk-taking behavior (Nemoda et al. 2011).

During adolescence, dramatic hormonal and physical changes occur within the brain, thus providing grounds for interactions with ADHD-like symptoms. Because of a lower basal function within the mesolimbic dopaminergic system, adolescents might experience reduced motivation (see also Spear 2000; Doremus-Fitzwater et al. 2010). Such an age-dependent alteration could underlie motivational abnormality such as "boredom" and "dissatisfaction," which are often seen in human adolescents. Hence, the search for salient stimuli is perhaps a strategy to satisfy a strong inner drive to overcome such a "poorly stimulating" physiological condition (Nightingale and Fischhoff 2002). Interestingly, in humans with ADHD or otherwise vulnerable personality, a general under-arousal is closely associated with emotional deficiency, thus leading to exaggerated sensation-seeking behavior (Herpertz and Sass 2000).

Human adolescents are typically prone to taking risks, a reckless behavior, and have, in general, a disinhibited conduct (Arnett 1992). According to Zuckerman (1984), the sensation-seeking drive during adolescence is characterized by "the continuing necessity to experiment with various, novel and complex sensations." This could justify the "curiosity" (and first approaches) toward psychoactive agents (Wills et al. 1994). Indeed, hazardous behaviors include highly rewarding and highly challenging activities, such as the search for and use of psychoactive substances (Wills et al. 1999; Adriani and Laviola 2004; Schepis et al. 2008; Doremus-Fitzwater et al. 2010). A prominent sensation-seeking drive at adolescence may have adaptive benefits in the development of independence and survival without parental protection. Actually, an adaptive role of an adolescent-like phase during life has been proposed for hominid evolution (Lockwood et al. 2007). In this framework, increments in human lifespan were accompanied by an extension of the adolescent phase that enabled a longer and protected period for the exploration of new environments and social dynamics, thus allowing the emergence of innovative behaviors.

The developmental stage of "adolescence" has been modeled in laboratory rodents, for which this term covers the broad postnatal period from weaning (usually, at postnatal day ("pnd") 21) to young adulthood (set conventionally at pnd 60). According to the literature, rodent adolescence can be divided into 3 phases: from pnd 28 to pnd 35, from pnd 35 to pnd 42, from pnd 42 to pnd 49. These 3 phases have been proposed to model for 9–12, 12–15, 15–18 year-old humans, respectively (Kellogg et al. 1998; Spear 2000; Laviola et al. 2003). As far as the development of neurobiological pathways is concerned, adolescent rodents display both enhanced reward-related and reduced anxiety-related profiles in response to novelty, leading to a novelty-prone behavioral profile. Furthermore, adolescent rats and mice show increased social interactions, especially in the form of playful and associative behaviors (Meaney and Stewart 1981; Panksepp 1981; Terranova et al. 1993, 1998). In summary, some features of human adolescence can be reproduced, at least to some extent, in animal models. These can be considered as adequate and reliable, since both "face" and "construct" validity can be demonstrated (see Spear 2000).

3 Dopaminergic Peculiarities in Rodent Models of Adolescence

3.1 Understanding Adolescence from a Neurobiological Basis

Developmental discontinuities in the brain have been hypothesized to explain the neural bases of adolescent behavior. The human brain continues to develop during adolescence (Paus et al. 1999; Sowell et al. 2003), and enhanced metabolism, as measured by elevated blood and oxygen flow, is found in adolescent brains (Chugani et al. 1987). Clinical studies with magnetic resonance imaging (MRI) have shown robust decreases in grey matter density in several areas of brain growth during the transitions between childhood, adolescence, and adulthood (De Bellis et al. 2001; Sowell et al. 2001). More recently, MRI studies in humans (for a short review specifically about ADHD, see Altabella et al. 2011) have indicated continued white matter development throughout adolescence, predominantly in the frontal part of the brain. These changes are likely to be the result of increases in axonal caliber and/or increased myelination (Lebel et al. 2008). There are also reports of continued microstructural changes in white matter during late adolescence. Refinement of projection and association fibers, which lasts until early adulthood, parallels an improvement in cognitive performance (Bava et al. 2010). In summary, a gradual loss of synapses, together with a strengthening of surviving synaptic connections, occurs during adolescence (Blakemore 2008; Casey et al. 2008; Giedd 2008; Schmithorst and Yuan 2010).

Remarkable remodeling of cortical and limbic circuits has been identified as a biological hallmark of adolescence across mammalian species. A number of circuits in the adolescent brain are established and refined through axonal overgrowth and plasticity. In teenagers, this is followed by a phase of dramatic pruning (Chechik et al. 1999). Autoradiographic studies of adolescent, nonhuman primates found overproduction and subsequent pruning of GABAergic, adrenergic, cholinergic, serotonergic, and dopaminergic receptors in several forebrain areas (Lidow et al. 1991). Peculiarities in the maturation of both DA and serotonin (5-HT) neurotransmitter systems have also been reported among adolescent rodents, possibly leading to discontinuities in mesolimbic and prefronto-cortical function (Chambers and Potenza 2003).

Acute effects, as well as long-term consequences, of psychostimulant drugs during development have been evaluated in rats and mice. The impact of pharmacological manipulation on DA systems, which has been extensively studied using adult animals, may be even greater during adolescence. This is because dopaminergic systems undergo key maturational and remodeling processes during this developmental period, which include proliferation and maturation of axon terminals and synapses (Stamford 1989; Teicher et al. 1995). A phase of massive overproduction before pruning may characterize adolescent rats around 5–6 weeks of life, leading to redundant structures and circuits that are functional only intermittently (Spear and Brake 1983; Spear 2000; Brenhouse and Andersen 2011). Brain dopaminergic receptors are similarly over-expressed before puberty, although pruning decreases this receptor density to adult levels thereafter (for a review, see Teicher et al. 1995; Brenhouse and Andersen 2011). Expression of D1 and D2 receptors within the striatum peaks in adolescent rats between pnd 28 and 40 (Tarazi et al. 1998, 1999). As far as the nigrostriatal system is concerned, a reduction in the basal rate of DA release and a reduced pool of readily releasable dopamine have been reported in adolescent rats (Stamford 1989). In the same study, adolescent rats also showed higher uptake and lower release of dopamine when compared to adults. As a consequence, at this age, there is a lower basal concentration of extracellular dopamine in the striatum (Andersen and Gazzara 1993). These neurochemical features of dopaminergic systems have implications for adolescent behavior because of the well-known link between dopamine and motivation (Ikemoto and Panksepp 1999; Salamone and Correa 2002; Berridge 2007). Indeed, several studies have reported that the effects of MPH and other dopaminergic DA indirect agonists, such as cocaine and amphetamine, differ in adolescent and adult rats and mice (Spear and Brake 1983; Laviola et al. 1995; Bolanos et al. 1998; White and Kalivas 1998).

3.2 Pharmacological Imaging of MPH in the Adolescent Forebrain

Studies of mechanisms of MPH action have focused extensively on the limbic cortico-striatal circuit, which processes emotional information and goal-directed behavior (Alexander et al. 1990). In particular, diverse key nodes within this circuit have been considered, including: the prefrontal cortex (PFC), the dorsal (dStr), and the ventral (i.e., nucleus accumbens, NAcc) striatum as well as hippocampus and thalamus. The NAcc receives information from PFC, hippocampus, and also amygdala, and projects to motor output structures such as the globus pallidus and also (indirectly) the medio-dorsal thalamus (Alexander et al. 1990). The PFC is involved in resolution of conflicting decisions through planning, feedback regulation, and inhibition of behavior (Dalley et al. 2004). The PFC is underactive when poor decision making is expressed (Eshel et al. 2007) and is still immature at adolescence: therefore, it is likely to contribute for the proneness of adolescents to risk-taking behaviors (Chambers et al. 2003).

We recently used pharmacological magnetic resonance imaging (phMRI) to investigate the neurobiological substrates that could account for the different effects of MPH in adult versus adolescent rats (for a short review on the characterization of ADHD performed with magnetic resonance techniques, see Altabella et al. 2011). Acute administration of MPH in rats evoked changes in intensity of blood-oxygen level-dependent (BOLD) signal in specific brain regions. The extent, direction (positive or negative), and temporal profile of BOLD responses to MPH, detected in a number of selected regions, were markedly different in adolescent and adult rats (Canese et al. 2009). As expected for adult rats, acute MPH increased the BOLD signal specifically in NAcc and PFC. Similarly, a recent study (Hewitt et al. 2005) reported a BOLD effect in the PFC and NAcc of adult male rats following MPH administration. Previous literature on phMRI mainly describes a positive BOLD signal in response to diverse psychostimulants, such as cocaine (Marota et al. 2000; Febo et al. 2005) and amphetamine (Chen et al. 1997; Dixon et al. 2005), although there are some discrepant findings (Preece et al. 2007). In contrast, we observed a generalized decrease in BOLD signal intensity in the adolescent group after treatment with MPH. Once again, the temporal profile of BOLD changes was similar for NAcc and PFC, with changes taking place soon after MPH administration.

Reasons for a negative BOLD response and its implications regarding neural activity are still poorly understood and controversial. In physiological terms, an increase in oxygen consumption within the brain should be accompanied by a compensatory increase in cerebral blood flow (CBF) and/or volume (CBV) to maintain a regular oxygen supply. Some authors have associated negative BOLD responses with local decreases in neuronal activity (Shmuel et al. 2006). These results agree with those of Easton and coworkers who also found widespread negative BOLD in the striatum of young adult rats (Easton et al. 2006, 2007) treated with guanfacine or atomoxetine, two alternative drugs for ADHD therapy. As with guanfacine and atomoxetine, MPH appears to have the ability to "turn down" striatal activity in the immature brain of adolescent and young adult rats, which may account for its beneficial modulatory role in the treatment of ADHD (Levy and Swanson 2001).

On the other hand, a negative BOLD signal has also been reported as a consequence of oxygen consumption without a feedback hemodynamic response (Rother et al. 2002). In fact, time-resolved fMRI studies have shown an early decrease in BOLD signal intensity followed by the well-known BOLD signal increase. The latter is due to increased CBF and/or CBV as a consequence of an overcompensating hemodynamic response (Frahm et al. 1996). Indeed, Harel et al. (2002) demonstrated that brain areas with positive BOLD signals experience an increase in CBV, while regions exhibiting a prolonged negative BOLD signal undergo a decrease in CBV (Harel et al. 2002). A human study in a patient with severely disturbed cerebral autoregulation (Rother et al. 2002) reported a negative BOLD effect, which can be interpreted as local oxygen consumption in the absence of a feedback hemodynamic response. Similarly, the negative BOLD signal that we have observed in adolescent rats (immediately after MPH administration) could be interpreted as an increase in oxygen consumption that was not followed by a prompt hemodynamic response. It could therefore be speculated that mechanisms for the feedback reallocation of CBF, following neural activation, are still immature during adolescence (Sicard and Duong 2005).

Alternatively, redistribution of CBF and/or CBV might not be sufficient to overcome oxygen consumption if this activity-induced demand is excessive. Negative BOLD responses have been reported in human infants and may indicate excessive oxygen consumption due to a developmental increase in synaptic density (Muramoto et al. 2002). Similarly, one reason to justify an excessive neuronal activity in MPH-treated adolescent rats may be found in maturational processes occurring at this age. Indeed, during adolescence, brain maturation and refinement include overproduction and selective elimination (pruning) of synapses, fibers and cells (see Sect. 3.1). Waves of overproduction of neural structures, which can justify a negative BOLD, might well occur in adolescent rodents and may parallel, in terms of magnitude, the same events happening to human brains during infancy (Sowell et al. 2003; Lenroot and Giedd 2006; Giedd 2008).

4 Discounting Processes in ADHD Modeling

Together with inattention and hyperactivity, impulsivity is a key symptom of ADHD. The impulsive behavior can be defined in several ways. One of the most widely adopted paradigms, the Intolerance-to-Delay (ID) task, assumes that impulsive subjects are intolerant to situations when reward is delayed: i.e., smaller, immediate reinforcers are preferred to larger rewards that are available only after a delay (Evenden and Ryan 1996; Evenden 1999). Alternatively, to probe individual (in)tolerance to frustration of anticipated reward, another proposed way is to exploit reward uncertainty, instead of delay. Current investigations are exploring the validity of a Probabilistic-Delivery (PD) task (Mobini et al. 2000; Adriani and Laviola 2006), which involves a choice between a smaller and certain reinforcer versus a larger but probabilistic one. Specifically, in this task, the large prize is occasionally and randomly withheld by the feeding device, so that the experimental subject faces a "loss." The progressive accumulating "losses" over time clearly have consequences for long-term payoff. Such a task also provides information reflecting (in)ability to cope with nonregularly delivered reinforcement.

These two operant-behavior tasks share many characteristics. In terms of experimental apparatus, both involve two alternative devices (e.g., levers or nose-poking holes, where the animal can express its choice), and computer-controlled delivery of reinforcers (e.g., food or liquids) that differ in size, time to delivery (delay) and probability of actual delivery (uncertainty). Other important features of these tasks are inherent to the trial/session schedule. For instance, the total number of choice opportunities given to the subject may be fixed (i.e., the session ends after the last trial) and independent of total time needed to complete the task. Alternatively, the total duration of the experimental session may be fixed (minutes, hours) and thus independent of the total number of trials actually completed. In addition, a timeout (TO) period, during which the devices are inactive, usually follows each reward delivery.

Finally, the temporal schedule of trials may be "strict," in which animals are given just a short, signaled window to express their choice, and failure to do so results in "omissions" (i.e., uncompleted trials). Alternatively, they are left to express their next choice at any time (in this case, time before next response may vary considerably among individuals). Importantly, both the ID and the PD are "free-choice" tasks in which animals are not required to learn, or to pay attention to, any signal that is intended to impose (or guide) their next choice. Conversely, as in the case of response-time tasks (RTT), there would be "correct" trials versus "errors" (and the task evaluates performance, not preference). As such, these two ID and PD tasks measure individuals' spontaneous preference for either of the options, which is the main parameter denoting the subjective performance.

4.1 Self-Control Behavior Following Adolescent MPH

The behavior of adult rats exposed to MPH during adolescence is opposite to that of adult rats whose DAT levels were manipulated via a lentiviral transfection. The former (MPH rats) and the latter (DAT rats) appear to be characterized by reduced (MPH rats) or increased (DAT rats) basal impulsivity in the ID task, respectively, compared to control animals (Adriani et al. 2007). For MPH-pretreated rats, the value assigned to the delayed reward was discounted much less, and hence great enough to allow subjects to select this option more often, despite a progressive increase in the waiting intervals. On the contrary, for DAT-transfected rats, the value assigned to the delayed reward was discounted much more, and hence low enough to allow subjects to select this option less often, because of a progressive increase in the waiting intervals. Such interpretations agree with the well-known therapeutic action of this drug in ADHD patients and with the putative role of DAT alterations in ADHD etiology. However, other explanations are possible (from now on, the two distinct models, MPH rats and DAT rats, are described together and compared, the former out of brackets and the latter within brackets). First, MPH-pretreatment (versus DAT transfection) might bias rats' perception of magnitude, thus over (versus under)-estimating the amount of a large reward, or vice versa under (versus over)-estimating the value of the small reward. Since it is well known that perception of reward-amount relies on the mesolimbic branch of the dopaminergic system (Salamone and Correa 2002; Berridge 2007), such a perceptive-bias phenomenon would fit with under-activation of the ventral striatum (see Sect. 5.1). Second, MPH-pretreated (but not DAT-transfected) rats may be biased in terms of flexibility: i.e., there may be a rigid preference for one of the two devices. This biased choice may be then attributed to a strong acquired preference for the large-reward device. This strategy (which develops before the introduction of delays) may turn out to be hard to change, thereby imposing a state of relative behavioral inflexibility. Such a predominant rigidity of newly developed strategies could fit with overactivation of dorsal-striatal dopaminergic systems (see Sect. 5.1). A final alternative is a bias in perception of time: i.e., the MPH-pretreated (versus DAT-transfected) rats may be under (versus over)-estimating the time that has elapsed during delay intervals. Obviously, such a bias would reduce (versus enhance) the subjective impact of the longest delays and diminish (versus exacerbate) the intolerance for waiting before reward delivery.

A different range of explanations shall be proposed to discuss findings from a PD task, exploiting a reward-rarefaction paradigm. Note that a severe reward-uncertainty challenge would probe willingness for risk-taking, as we proposed recently (Adriani and Laviola 2006). First of all, under these probabilistic conditions, control rats continued to choose the large reward (which was fivefold in size), despite the vast majority of their nose-poking choices not triggering any food-pellet delivery (i.e., with more than 80% of omitted deliveries): in other words, receiving a larger reward eventually and instantly (a "rare-but binge" event) was consistently preferred. By contrast, receiving a much smaller reward, but certainly and at a regular pace (i.e., one nosepoke for one pellet), is clearly not attractive. This observation is consistent with preference for "binge" reinforcement typically found in naive, food-restricted rats (Hastjarjo et al 1990; Kaminski and Ator 2001). Notably, this choice strategy is maintained in spite of adverse consequences on total foraging, i.e., despite a decreasing payoff in the long-term (see Sect. 4.2).

Interestingly, rats with adolescent exposure to MPH displayed a marked reaction when the frequency of large-reinforcement events was decreased, in which they displayed an increase of nose-poking responses for the small, certain option. Conversely, DAT-transfected rats continued to choose the large reward, even under conditions of extreme rarefaction, when control rats eventually were forced to shift toward the safer option. These enduring effects of adolescent MPH (versus DAT transfection) can be explained by: (1) increased (versus decreased) "economical" efficiency; and/or (2) enhanced (versus impaired) behavioral "flexibility" (Romani et al. 2010). In the first case, MPH-exposed (versus DAT-transfected) rats might be able (versus unable) to perceive as accumulating the omitted rewards (i.e., to "sum up" the "losses" across time) and/or estimate their adverse effects on the total payoff. An implication of this hypothesis would be that rats acquire and process this information, trial after trial across the session, and they behave accordingly. A corollary is that control animals and, to an even larger extent, DAT-transfected ones are attracted (and their choices merely driven) by the "unitary" size of bulk wins (that are eventual, but rare). In the second case, performance exhibited by MPH-pretreated (versus DAT-transfected) rats would imply enhanced (versus reduced) flexibility of choice, namely a lower (versus greater) likelihood to stick to behavioral strategies once they are established. Notably, however, neither adolescent MPH exposure nor DAT-directed transgenesis can lead to opposite consequences (i.e., inflexible patterns in one task and more flexible ones in the other). Results obtained in these two models with either task may help to discard all explanations based solely on flexibility levels. Reduced (versus increased) temporal discounting and decreased (versus enhanced) attraction for "binge" reward thus emerge as the most solid interpretations for MPH versus DAT rats, respectively.

Specifically, the MPH rats showed a twofold increase in choice level for a delayed reinforcer and for a low-risk certain-payoff one, whereas the opposite was true for DAT rats. From ID task data, MPH-pretreated (versus DAT-transfected) animals may have discovered (versus ignored) the adaptive value of waiting or, at least, to be able (versus unable) to inhibit the aversive reaction that waiting instinctively generates. In other words, these rats seem to display more (versus less) "patience" when facing the ever-increasing delay intervals. Similarly, from PD task data, MPH (versus DAT) rats may present a more "smooth" (versus "sharp") strategy, with a clear (in)capacity to move away from the tempting attractiveness of the risky option. Outcomes of such a strategy include: for MPH rats, a useful dilution of the uncertainly rewarded periods; for DAT rats, a reduction in the overall payoff in terms of increased loss. Interestingly, in both ID and PD protocols, a "better" decision (leading to an "optimal" outcome) occurred in rats exposed to MPH during adolescence (Adriani et al. 2006, 2007), while the opposite was true for rats following DAT transfection (Adriani et al. 2009, 2010a). MPH exposure during adolescence was able to generate more self-controlled adult rats in either task (Fig. 1); lentiviral manipulation of DAT levels resulted in exacerbation of both "impulsive aversion" and "temptation by risk." In summary, the profiles produced in rats by adolescent MPH administration versus DAT transgenesis can be termed "more (versus less) efficient," respectively.

4.2 *Methodological Remarks on Discounting Tasks*

We emphasize that these two tasks are tailored so that a choice shift toward small reward versus a sustained preference for a large reward have two opposite meanings. While the former denotes "impulsive aversion" in the ID task and the latter denotes "temptation by risk" in the PD task, both appear as suboptimal performance profiles, a notion that is particularly evident in the case of DAT rats. It is worth noting once again that the profiles of preference, exhibited by MPH-exposed rats, consistently shifted toward an optimal performance. However, in terms of benefit, these profiles, although conceptually similar, took opposite forms: an increase in large-reward preference in the ID task (Adriani et al. 2007) versus a decrease in large-reward preference in the PD task (Adriani et al. 2006).

We recently formulated a theoretical framework to interpret the performance of laboratory rats in these symmetrically tailored ID and PD tasks (Adriani and Laviola 2006). A landmark in both protocols is the "indifferent" point: i.e., the specific level (of delay or uncertainty) at which the animals can choose either option freely with no effect on the overall economic convenience.

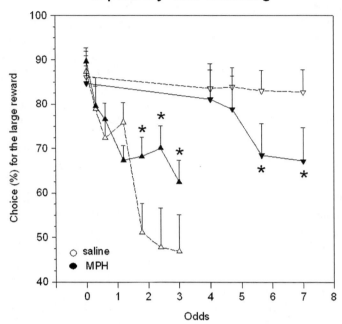

Fig. 1 Comparison of Intolerance-to-Delay (ID) and Probabilistic-Delivery (PD) tasks in adult rats following adolescent MPH administration. Rats were exposed during adolescence (pnd 30–46) with MPH (2 mg/kg once daily) or saline, and were then tested at adulthood (pnd > 90). Animals were placed in an operant cage with nose-poking as operandum, providing a choice between immediate (ID task) and certain (PD task) delivery of a small reward versus delayed (ID task) or uncertain (PD task) delivery of a large reward. Delay and probability are represented by their odds-equivalent values (formulae: $Odds = Delay/mITI$; $Odds = 1/p - 1$, respectively) on X-axis. The indifferent point is set at Odds = 4 (i.e., at $P = 20\%$ and at delay = 100 s, respectively). *Impulsivity (upward triangles) from the ID task* – mean choice (%) for the large but delayed reward (Adriani et al. 2007) on Y-axis. The adolescent MPH exposure reduced the shift toward the small and immediate reward, observed for delay > 45 s (i.e., at Odds > 1.8). Such increase of preference for the large but delayed reward suggests a reduced intolerance to delay. *$P < 0.05$ when comparing MPH with saline pretreatment. *Risk proneness (downward triangles) from the PD task* – mean choice (%) for the large but uncertain reward (Adriani et al. 2006) on Y-axis. The adolescent MPH exposure produced a flexible response for $P < 15\%$ (i.e., at Odds > 5.6), consisting of a more marked shift toward the small and certain reward, that contrasted reinforcement rarefaction. Reprinted (modified) from Adriani and Laviola (2006) (Open Access Source)

In the PD task, if the ratio between the large and small rewards is fivefold then the indifferent point is at "P" $= 20\%$. In the ID task, two consecutive reinforcers will always be spaced by a mean inter-trial interval (mITI), composed by the timeout (TO), set by the experimenter, plus a further period of spontaneous waiting by the subjects (termed response time, RT). Hence, when delays are introduced, the temporal outline is an alternation of delays and mITIs. Any delay interval (activated by large-reward choices) impedes the experimental animal from performing further choices. As an alternative, the subject may prefer a series of small-reward choices that could be spaced by the mITI alone. In a sense, many consecutive small-reward choices are equivalent to a single long delay leading to large reward delivery. If the large reward is fivefold compared to the small one, then the indifference point is when the delay interval is fourfold greater than the mITI (Adriani et al. 2010b). In our laboratory, the mITI is usually 25 s and the indifference point is therefore at delay = 100 s. Once the indifference point is established, the range of values providing worthy information is easily recognized. In brief, to run a protocol providing useful information, any "inner drive" of interest (e.g., delay-aversion, or risk-proneness) shall push animals into a choice that necessarily leads to a suboptimal outcome. Self-control is then defined as the ability to effect an optimal response (Stephens and Anderson 2001) by directing choices onto the opposite operandum (nose-poking hole or lever to press). The protocol must never load both instances (i.e., the inner drive and the optimal payoff) on the same operandum because it would be impossible to discriminate whether any preference for that operandum is due to payoff-detecting processes ("economical efficiency") or to the "inner drive" itself.

As illustrated in Fig. 1, the conflict between opposite reward options (large versus small) differs in the two tasks. For the ID task, at least before the indifference point (i.e., for delay < 100 s), the economical benefit unequivocally loads onto the large reward option. In other words, to maximize the payoff, subjects should retain a large-reward preference: delay aversion is here the "inner drive" that rules against this benefit by generating a subjective temporal intolerance that discounts large/late rewards. Conversely, for the PD task at "P" values beyond the indifference point (i.e., $20\% > P > 0\%$), economical benefit loads unequivocally onto the small reward option. Thus, to maximize the payoff, subjects should be flexible enough to abandon their large-reward preference: this requires a self-control effort to overcome the "innate drive" that justifies its attractiveness.

Many factors act together to push animals toward a large reward, even though this is delivered quite rarely. One factor is insensitivity to risk, whereby the subjects perceive (as they should) neither uncertainty of the outcome (usually, a source of aversion before choice) nor the punishment of "losses" (represented by randomly and frequently omitted delivery of reward). Another factor is habit-induced rigidity, under which the subject seems to behave according to a consolidated choice strategy. Such form of inflexibility may be due to a failure of negative reinforcement, namely a lack of feedback-reaction to the aversion and/or to the punishment just described. A third factor is temptation to gamble, whereby the motivational impact of the magnitude of the reward seems to monopolize the subject's attention over any other reward feature. It is possible that risk of punishment under uncertainty becomes attractive as a secondary conditioned feature, and this is because the "binge" reward may be generating a peak of positive reinforcement when eventually released by the feeding device and received by the subject. Whatever factor is prevalent in PD tasks, the suboptimal preference for large, rarefied reward (namely, the innate attraction for a "rare-but binge" event) is an index of "risk-proneness."

5 Consequences of Adolescent MPH Exposure

Over recent years, an increasing amount of data from animal models has provided evidence for enduring neurobehavioral effects of adolescent MPH, both with imposed- and self-administration (Burton et al. 2010). These persistent and long-lasting consequences, which are still evident at adulthood, can be beneficial for ADHD therapy, as is the case for the long-term modulation of self-control abilities, described above (see Sect. 4.1).

However, there are also examples of detrimental MPH-induced changes. Exposure to MPH in adolescence modifies emotional and motivational responses later in life, whereby MPH decreases sensitivity to rewarding stimuli (Brandon et al. 2001; Andersen et al. 2002) and enhances emotionality as well as depressive-like symptoms (Bolanos et al. 2003; Carlezon et al. 2003). First, exposure to MPH (at therapeutic dosages, usually ranging 1–3 mg/kg/day during adolescence) rendered animals less responsive to natural rewards (such as sucrose, novelty-induced activity and sex) when compared with control animals (Bolanos et al. 2003; Ferguson and Boctor 2010). In contrast, MPH-treated animals were more stress-sensitive and displayed an increase in fear- and anxiety-like responses (Britton et al. 2007; Vendruscolo et al. 2008) as well as depressive-like effects in the forced-swim test (Carlezon et al. 2003).

Such neurobiological rearrangements, verified preclinically, again raise some concern about the safety of MPH, although not in the case of prescription of therapeutic doses to ADHD children. Rather, there should be more attention to the extensive misuse by adolescents, arising from illicit (or off-label) access to this psychostimulant (Marco et al. 2011).

5.1 Persistent Brain Metabolic Outcomes of Adolescent MPH

Based on the arguments for increased versus decreased self-control abilities (see Sect. 4.1), a functional rearrangement within dopaminergic pathways would be expected in rats, following either MPH exposure during adolescence versus transgenic alteration of DAT levels obtained through lentiviral-mediated transfection. Such functional changes were investigated in key forebrain areas (Adriani et al. 2007). Among the magnetic resonance techniques (see Altabella et al. 2011 for a short review), we have used proton magnetic resonance spectroscopy (^{1}H-MRS) to study in vivo levels of total creatine, a metabolite involved in bioenergetics. The concentration of this key metabolite was proposed to be a specific index for the functional activity of key forebrain areas (Adriani et al. 2007; Marco et al. 2011).

The total creatine signal comprises both creatine and phospho-creatine (an energy reservoir). Specifically, a great deal of energy is used for brain signaling processes, particularly Na^+ transport to ensure maximal activity. The margin of safety between generated and required energy is small (Ames 2000). Since phospho-creatine plays an important role in maintaining high ATP levels, especially during acute energetic challenges (Gadian 1995), increased levels appear highly relevant for a good behavioral performance. Thus, the supply of energy may impose an upper limit on the activity of neurons in specific brain regions, thereby influencing the extent to which that region will resist fatigue in challenging conditions. The bioenergetic capacity of brain areas may therefore predict their contribution to processing of neural information and hence to the final behavioral output.

In our hands, hyperfunction within the dStr and, conversely, hypofunction of the NAcc were suggested in adult rats both exposed to adolescent MPH and transfected with DAT-directed lentiviruses. The striatum has a critical role in habit formation and behavioral flexibility (Ragozzino 2003; Yin et al. 2004; Balleine et al. 2007). Enhanced function of the dorsal striatum (dStr) may enable both habit-driven performance and the development of more flexible and innovative behavioral strategies in response to an environmental challenge (i.e., delay of reward or its uncertainty). Therefore, adult rats following a DAT-directed transgenesis may be expressing a habit-prone phenotype (Adriani et al. 2009, 2010a). Conversely, rats exposed to MPH during adolescence may ultimately be characterized by an increased coping capacity (i.e., more accuracy and competence in challenging situations; Marco et al. 2011).

On the other hand, the ventral striatum (i.e., NAcc) accounts for the affective evaluation of outcomes and favors a feedback to redirect future choice. Intact prefronto-cortical projections to the NAcc are a crucial part of the motivational system, involved in "wanting" to reach a given goal. They are also pivotal for behavioral flexibility, helping to overcome any obstacles toward completion of required work (Cardinal et al. 2004; Christakou et al. 2004). More generally, mesolimbic pathways projecting to the NAcc are a major determinant of the maximal effort that is subjectively affordable (Salamone et al. 2005). In the framework of ID and PD tasks, the physiological role of NAcc should be: (1) to subserve attraction for reward according to its magnitude (primary value); (2) to further sustain such attraction in spite of "losses" due to delivery omission (innate drive for a "binge-although rare" reward); and (3) eventually to support delay-induced aversion (innate drive for immediacy), at least when waiting times become excessive (Sonuga-Barke 2005). In this light, a NAcc functional downregulation was observed in adult rats if they were exposed to MPH during adolescence or following DAT-directed transgenesis. In response to long delay intervals and to strong uncertainty, depending on whether we consider MPH or DAT rats respectively, a decreased activity of ventro-striatal dopaminergic pathways may reduce or produce intolerant aversion to delay, and may facilitate or impede the subject to direct less effort toward uncertain payoff. What is then a major determinant of the behavior actually observed?

An upregulation of the phospho-creatine/creatine ratio has been reported in the PFC following adolescent MPH administration but not in the case of lentiviral DAT manipulation. This change (reflecting enhanced energy metabolism within this brain region) may explain the opposite profiles observed in rats after adolescent MPH versus DAT transfection, although a causal link between these factors cannot be demonstrated. Increased PFC function may be responsible for tuning the functional activity within the NAcc and redirecting the role played by the dStr. This notion seems in line with experimental evidence for separate functional roles of PFC projections to these two target regions (Christakou et al. 2001, 2004; Rogers et al. 2001). Specifically, the PFC-to-dorsal striatum (dStr) circuit subserves attention devoted to executive planning and to the selection of behavioral actions, while the pathway from PFC-to-ventral striatum (NAcc) integrates and elaborates information about the motivational consequences of each action's outcome. These findings may therefore suggest how prefronto-cortical areas buffer and redirect subcortical activity, playing an apparently pivotal role to determine the final behavioral profile in ID and PD tasks. Following an MPH-dependent rearrangement, but not in the case of direct DAT manipulation (by transfection with lentiviral tools), enhanced function of the PFC may: (1) favor a less steep devaluation of incentive values (via a cognitive buffering of downregulated NAcc function) and (2) permit the elaboration of flexible behavioral strategies (via an upregulated dStr function), allowing better overall coping with uncertainty- and delay-induced adverse contingencies. Conversely, the lack of such a role may explain detrimental behavior: in rats whose DAT levels were altered due to lentiviral transfection, a reduced NAcc function may (1) support a steep temporal discounting, and (2) favor a more rigid, habit-based strategy thus explaining a suboptimal attraction for "binge" reinforcement.

5.2 Consequences on Striatal Gene Expression of Adolescent MPH Exposure

Repeated psychostimulant exposure causes stable changes in gene expression and persistent alterations in dendritic morphology within the mesocorticolimbic DA system, the main pathway of the brain reward circuitry (Self 2004; Wolf et al. 2004; Canales 2005). Similar plastic changes also occur with MPH exposure. Consistently, molecular evidence suggests that this is especially true if exposure occurs during adolescence (Yano and Steiner 2007). The effects of acute and repeated MPH treatment on molecules of neuronal signaling and neuroplasticity (including transcription factors, neuropeptides, and components of second messenger cascades) have been recently evaluated. Such molecular effects are comparable with those produced by cocaine and amphetamine (Brandon et al. 2003; Brandon and Steiner 2003; Shen and Choong 2006; Banerjee et al. 2009). Some differences observed between MPH and cocaine/amphetamine treatment support the notion that MPH produces fewer neuro-adaptations, and might provide a molecular basis for the reduced addiction liability of MPH compared with cocaine and amphetamine and other psychostimulants (Yano and Steiner 2005, 2007).

By genome-wide expression profiling, some 700 genes show changes in their striatal expression at the end of a 2-week adolescent MPH treatment. Among these, five genes encode subtypes or subunits of neurotransmitter receptors (*Grik2*,

Gabrg1, Gabrb3, Htr7, Adra1b). However, of these five genes, only transcripts for the glutamate kainate-2 (*Grik2* gene, the protein also known as GluR6 or KA2) and the serotonin 7 receptor (*Htr7* gene, the protein also known as 5-HT7) were still found upregulated as an enduring sequela in adulthood (Adriani et al. 2006; Marco et al. 2011). Hence, while most of the neurotransmission within basal ganglia is transiently affected during the adolescent treatment, only some of these alterations are enduring.

It is well established that the serotonergic 5-HT7 pathways play a role in timing behavior and circadian rhythm (Morrissey et al. 1993; Wogar et al. 1993). In addition, blockade of this receptor enhances synaptic serotonin release and transmission, an effect similar to selective serotonin reuptake inhibitors (SSRIs) (Guscott et al. 2005; Bonaventure et al. 2007). Thus, the notion that persistent upregulation of *Htr7* gene expression (as seen after chronic adolescent MPH) parallels reduced behavioral impulsivity and a profile of reward insensitivity is particularly intriguing. A key role of *Htr7* gene/5-HT7 protein in behavioral flexibility and/or reward-devaluation processes has been proposed recently (Leo et al. 2009). Further experiments are needed to establish a causal role of *Htr7*/5-HT7 expression and/or function in the behavioral phenotype, as observed in tests for self-control and decision-making abilities.

6 Concluding Remarks and Future Perspectives

Despite having a neuropharmacological profile similar to that of amphetamine and cocaine (Kallman and Isaac 1975; Patrick and Markowitz 1997; Challman and Lipsky 2000), MPH (Ritalin®) is one of the most frequently prescribed drugs for the treatment of ADHD (Levin and Kleber 1995; Accardo and Blondis 2001). MPH therapy (in children and adolescents) is effective in the management of ADHD symptoms, including cognitive impulsivity (Seeman and Madras 1998). However, beyond medical administration in ADHD therapy, illicit use and possibly misuse of Ritalin® by adolescents raise concerns for public health, because of its potential long-term consequences. Preclinical data reviewed here also open new topics or fields of investigation. For instance, given the identification of persistent overexpression of the striatal *Htr7* gene, and since this neural pathway is a functional determinant of impulsive behavior (Leo et al. 2009), this gene and related brain systems should be further investigated, possibly by means of selective pharmacological agonists, which are unfortunately not available yet. There is ground for formulating working hypotheses about the role of this gene in ADHD etiology and, alternatively, its use as a possible new target for innovative approaches in ADHD therapy.

In conclusion, preclinical investigation is useful for providing insights into phenomena (and mechanisms) that cannot easily be addressed in humans. The use of adolescent rodents is worthy because of the opportunity to study the impact of drugs on the immature brain, thus casting light into their potential for unwanted

Special Focus Box: Evolutionary Perspectives on ADHD

ADHD: The Supposed Adaptive Advantages

Nowadays ADHD is recognized to be common in children of many different races and societies worldwide. Differences in prevalence observed among countries (e.g., 4–7% in the USA versus 1–3% in the EU) are due to cultural reasons, with the impact of this disease being not always recognized by the society as a whole, and/or to medical management, with diverse diagnostic guidelines set by each nation's medical community (Faraone et al. 2003).

We provide here a brief glance on evolutionary psychiatry and genetic studies in relation to ADHD, trying to answer the following question: "Could susceptibility to ADHD be the result of evolutionary forces that shaped human behavior?" In the available scientific literature, conflicting answers are reported, since different schools of thought propose contrasting evolutionary explanations.

A popular hypothesis about the origins of ADHD, proposed by Hartmann in 1993, suggests that it could be a form of adaptive behavior. This hypothesis, known as the "Hunter–Farmer theory," proposes ADHD as an evolved trait that was advantageous to "hunters" in past environments, but then became maladaptive within societies based on "farming" (Hartmann 2003). Interestingly, Hartmann speculated that people with ADHD retained some of the former "hunter" characteristics, but he himself stated: "It's not hard science, and was never intended to be." Nonetheless, many researchers have used the "Hunter–Farmer theory" as a working hypothesis for the origin of ADHD and have proposed it to be an anachronistic behavioral trait (e.g., Jensen et al. 1997; Austin 1998; Arcos-Burgos and Acosta 2007). Specifically, Arcos-Burgos and Acosta (2007) noted that ADHD is a common variant, rather than a rare one, among behavioral phenotypes, and that the genetic variants conferring susceptibility to ADHD are even totally fixed in some populations. During millions of years of human evolution, more fitness (i.e., reproductive success and survival) might have been provided to those people possessing ADHD behavioral traits through several selective advantages: faster response to predators, best hunting performance, more effective territorial defense and improvement of mobility and settling capacity. The sudden change of human society in recent centuries has brought new social needs, such as cognitive planning skills, design and manufacturing of goods and attention to multiple stimuli. As a consequence, most frequently the ADHD phenotype, with its inattentiveness and hyperactivity, is highly maladaptive for contemporary social settings, for example in a school classroom where hypoactivity and hyperattention are required (Arcos-Burgos and Acosta 2007).

Genetic evidence gives a fundamental contribution to verification of evolutionary explanations for ADHD. Work from several groups has led to

(continued)

the identification of candidate genes and pathways strongly implicated in the etiology of ADHD. These are primarily related to dopamine and serotonin receptor subpopulations and transporters (Grady et al. 2003; Acosta et al. 2004; Bobb et al. 2005a; Faraone et al. 2005). Several genes, such as DAT, DBH, DRD1, DRD2, DRD4, DRD5, SERT, HTR1B, NET1, and SNAP-25 (Bobb et al. 2005a,b; Brookes et al. 2006; Khan and Faraone 2006), have allelic variants conferring susceptibility to ADHD. In particular, the DRD4 gene (that encodes the D4 receptor for dopamine) is likely to be involved in impulsivity, reward anticipation, and addictive behavior. One variant of the DRD4 gene, the seven repeat allele (DRD4/7R allele), is relatively recent (it originated about 40,000 years ago during upper Paleolithic era), and has been linked with greater food and drug-craving, novelty-seeking, and ADHD symptoms. In particular, Ding et al. (2002) demonstrated that this genetic variant has been subject to positive selection under selective pressure, meaning that carriers of this allele displayed some selective advantage.

It is important to emphasize that the effects of different genetic variants conferring susceptibility to ADHD have been studied almost exclusively in industrialized countries. Little research has been carried out in nonindustrial, subsistence environments, despite the fact that such environments may be more similar to those where much of human genetic evolution has taken place. In this perspective, a recent study of particular interest (Eisenberg et al. 2008) was conducted on men of the Ariaal (a community in northern Kenya, chronically undernourished) who have two genetic polymorphisms of the DRD4 gene: the 48 base pair (bp) repeat polymorphism (DRD4/48bp) and the seven repeat minor allele (DRD4/7R allele). While men with the DRD4/7R allele were better nourished in the nomadic population, they were less well nourished in the population living in settled communities. Thus, the 7R allele seemed to confer additional adaptive benefits in the nomadic, but was maladaptive in sedentary conditions. Such an allele would favor: increased impulsivity, novelty-seeking, aggression, violence, and/or hyperactivity (Eisenberg et al. 2008). In ancient societies, individuals with those ADHD-like traits might have been more proficient in tasks involving risk or competition (i.e., war, hunting, or mating rituals), so that their hyperactively phenotype may be seen as evolutionarily beneficial (Williams and Taylor 2006). However, the same behavioral tendencies of men with the DRD4/7R allele might well be less suited for farming or selling goods at market among humans in a settled community.

In a diametrically opposite perspective, others reported that ADHD has probably no value for adaptation, and that no hunting conditions have existed in which ADHD could have provided more fitness (i.e., survival or reproductive advantage). Goldstein and Barkley (1998) suggested that those who do not suffer from ADHD clearly have an adaptive advantage across multiple generations. In their view, ADHD symptoms reflect a weakness that impedes the development of efficient self-regulation and self-control, and individuals

(continued)

with these traits fall into the bottom tail of a normal curve (Goldstein and Barkley 1998). The ability to wait, plan, and cooperate are primary elements for an effective hunter. Greater hunting success and more opportunities to reproduce could be the result of small enhancements in executive function, so that hunting itself acted as an evolutionary force for the emergence of executive function, rather than the other way round (Brody 2001). In an evolutionary scale, ADHD phenotype should reduce, rather than increase, the survival and reproductive advantage conveyed by executive functions and self-regulation (Barkley 1985). Deficits in these domains should reflect impairment in any environment and in any period (Brody 2001). This interpretation is supported by observations of nonhuman primates, such as the chimpanzee (Wrangham and Peterson 1996; Ghiglieri 1999) and current primitive human cultures (Sagan 1977). During their patrols, chimpanzees maintain prolonged silence and behave as a coordinated group. The most likely targets of these patrols are individuals of other groups who are inattentive, impulsive/aggressive, and therefore isolated from their group. In the wild, impulsive and hyperactive behavior would probably have lethal consequences.

References

Acosta MT, Arcos-Burgos M, Muenke M (2004) Attention deficit/hyperactivity disorder (ADHD): complex phenotype, simple genotype? Genet Med 6:1–15

Arcos-Burgos M, Acosta MT (2007) Tuning major gene variants conditioning human behavior: the anachronism of ADHD. Curr Opin Genet Dev 17:234–238

Austin L (1998) Selective advantages of attention deficit hyperactivity. Across Species Comparisons and Psychopathology 11:6

Barkley R (1985) Social interactions of hyperactive children: developmental changes, drug effects, and situational variation. In: McMahon R, Peters R (eds) Childhood disorders: behavioral-developmental approaches. Brunner/Mazel, New York, pp 218–243

Bobb AJ, Addington AM, Sidransky E, Gornick MC, Lerch JP, Greenstein DK, Clasen LS, Sharp WS, Inoff-Germain G, Wavrant-De Vrieze F, Arcos-Burgos M, Straub RE, Hardy JA, Castellanos FX, Rapoport JL (2005a) Support for association between ADHD and two candidate genes: NET1 and DRD1. Am J Med Genet Part B 134:67–72

Bobb AJ, Castellanos FX, Addington AM, Rapoport JL (2005b) Molecular genetic studies of ADHD: 1991 to 2004. Am J Med Genet Part B 132:109–125

Brody JF (2001) Evolutionary recasting: ADHD, mania and its variants. J Affect Dis 65:197–215

Brookes K, Xu X, Chen W, Zhou K, Neale B, Lowe N, Anney R, Franke B, Gill M, Ebstein R, Buitelaar J, Sham P, Campbell D, Knight J, Andreou

(continued)

P, Altink M, Arnold R, Boer F, Buschgens C, Butler L, Christiansen H, Feldman L, Fleischman K, Fliers E, Howe-Forbes R, Goldfarb A, Heise A, Gabriëls I, Korn-Lubetzki I, Johansson L, Marco R, Medad S, Minderaa R, Mulas F, Müller U, Mulligan A, Rabin K, Rommelse N, Sethna V, Sorohan J, Uebel H, Psychogiou L, Weeks A, Barrett R, Craig I, Banaschewski T, Sonuga-Barke E, Eisenberg J, Kuntsi J, Manor I, McGuffin P, Miranda A, Oades RD, Plomin R, Roeyers H, Rothenberger A, Sergeant J, Steinhausen HC, Taylor E, Thompson M, Faraone SV, Asherson P (2006) The analysis of 51 genes in DSM-IV combined type attention deficit hyperactivity disorder: association signals in DRD4, DAT1 and 16 other genes. Mol Psychiatry 11:934–953

Ding YC, Chi HC, Grady DL, Morishima A, Kidd JR, Kidd KK, Flodman P, Spence MA, Schuck S, Swanson JM, Zhang YP, Moyzis RK (2002) Evidence of positive selection acting at the human dopamine receptor D4 gene locus. Proc Natl Acad Sci U S A 99:309–314

Eisenberg DTA, Campbell B, Gray PB, Sorenson MD (2008) Dopamine receptor genetic polymorphisms and body composition in undernourished pastoralists: an exploration of nutrition indices among nomadic and recently settled Ariaal men of northern Kenya. BMC Evol Biol 8:173

Faraone SV, Sergeant J, Gillberg C, Biederman J (2003) The worldwide prevalence of ADHD: is it an American condition? World Psychiatry 2:104–113

Faraone SV, Perlis RH, Doyle AE, Smoller JW, Goralnick JJ, Holmgren MA, Sklar P (2005) Molecular genetics of attention deficit/hyperactivity disorder. Biol Psychiatry 57:1313–1323

Ghiglieri M (1999) The dark side of man: tracing the origins of male violence. Perseus, Reading, MA, USA

Goldstein S, Barkley RA (1998) ADHD, hunting and evolution: "just so" stories. ADHD Report 6:1–4

Grady DL, Chi HC, Ding YC, Smith M, Wang E, Schuck S, Flodman P, Spence MA, Swanson JM, Moyzis RK (2003) High prevalence of rare dopamine receptor D4 alleles in children diagnosed with attention-deficit hyperactivity disorder. Mol Psychiatry 8:536–545

Hartmann T (1993) Attention deficit disorder: different perception. Underwood, New York

Hartmann T (2003) The Edison gene: ADHD and the gift of the hunter child. Park Street Press, Rochester, VT, USA

Jensen PS, Mrazek D, Knapp PK, Steinberg L, Pfeffer C, Schowalter J, Shapiro T (1997) Evolution and revolution in child psychiatry: ADHD as a disorder of adaptation. J Am Acad Child Adol Psychiatry 36:1672–1681

Khan SA, Faraone SV (2006) The genetics of ADHD: a literature review of 2005. Curr Psychiatry Rep 8:393–397

Sagan C (1977) Dragons of Eden: speculations on the evolution of human intelligence. Random House, NY, USA

(continued)

Williams J, Taylor E (2006) The evolution of hyperactivity, impulsivity and cognitive diversity. J R Soc Interface 3:399–413
Wrangham R, Peterson D (1996) Demonic males: apes and the origins of human violence. Houghton Mifflin, NY, USA

enduring consequences. These types of studies may help to promote public awareness about possible risks deriving from psychostimulants when they are illicitly abused by adolescents. This is a delicate issue for which evidence-based information is missing and too often ideology prevails.

Acknowledgments This study was supported by the "ADHD-sythe" young-investigator grant and by the "NeuroGenMRI" ERAnet project, "Prio-Med-Child" call (to WA), Italian Ministry of Health; ERARE-EuroRETT Network ERAR/6 (to GL). The authors wish to thank the European Mind and Metabolism Association (EMMA, www.emmaweb.org), L.T. Bonsignore and N. Francia for technical support.

References

Accardo P, Blondis TA (2001) What's all the fuss about Ritalin? J Pediatr 138:6–9
Adriani W, Laviola G (2004) Windows of vulnerability to psychopathology and therapeutic strategy in the adolescent rodent model. Behav Pharmacol 15:341–352
Adriani W, Laviola G (2006) Delay aversion but preference for large and rare rewards in two choice tasks: implications for the measurement of self-control parameters. BMC Neurosci 7:52
Adriani W, Leo D, Greco D, Rea M, di Porzio U, Laviola G, Perrone-Capano C (2006) Methylphenidate administration to adolescent rats determines plastic changes on reward-related behavior and striatal gene expression. Neuropsychopharmacology 31:1946–1956
Adriani W, Canese R, Podo F, Laviola G (2007) 1H MRS-detectable metabolic brain changes and reduced impulsive behavior in adult rats exposed to methylphenidate during adolescence. Neurotoxicol Teratol 29:116–125
Adriani W, Boyer F, Gioiosa L, Macrì S, Dreyer JL, Laviola G (2009) Increased impulsive behavior and risk proneness following lentivirus-mediated dopamine transporter overexpression in rats' nucleus accumbens. Neuroscience 159:47–58
Adriani W, Boyer F, Leo D, Canese R, Podo F, Perrone-Capano C, Dreyer JL, Laviola G (2010a) Social withdrawal and gambling-like profile after lentiviral manipulation of DAT expression in the rat accumbens. Int J Neuropsychopharmacol 13:1329–42
Adriani W, Zoratto F, Romano E, Laviola G (2010b) Cognitive impulsivity in animal models: role of response time and reinforcing rate in delay intolerance with two-choice operant tasks. Neuropharmacology 58:694–701
Alexander GE, Crutcher MD, DeLong MR (1990) Basal ganglia-thalamocortical circuits: parallel substrates for motor, oculomotor, "prefrontal" and "limbic" functions. Prog Brain Res 85:119–146
Altabella L, Strolin S, Villani N, Zoratto F, Canese R (2011) Magnetic resonance imaging and spectroscopy in the study in ADHD syndrome: a short review. Biophys Bioeng Lett 4:23–34
Ames A III (2000) CNS energy metabolism as related to function. Brain Res Rev 34:42–68

Andersen SL, Gazzara RA (1993) The ontogeny of apomorphine-induced alterations of neostriatal dopamine release: effects on spontaneous release. J Neurochem 61:2247–2255

Andersen SL, Thompson AT, Rutstein M, Hostetter JC, Teicher MH (2000) Dopamine receptor pruning in prefrontal cortex during the periadolescent period in rats. Synapse 37:167–169

Andersen SL, Arvanitogiannis A, Pliakas AM, LeBlanc C, Carlezon WA Jr (2002) Altered responsiveness to cocaine in rats exposed to methylphenidate during development. Nat Neurosci 5:13–14

Arnett J (1992) Reckless behavior in adolescence: a developmental perspective. Dev Rev 12:339–373

Arria AM, Wish ED (2006) Nonmedical use of prescription stimulant drugs among students. Pediatr Ann 35:565–571

Balleine BW, Delgado MR, Hikosaka O (2007) The role of the dorsal striatum in reward and decision-making. J Neurosci 27:8161–8165

Banerjee PS, Aston J, Khundakar AA, Zetterström TS (2009) Differential regulation of psychostimulant-induced gene expression of brain derived neurotrophic factor and the immediate-early gene Arc in the juvenile and adult brain. Eur J Neurosci 29:465–476

Barkley RA, Fischer M, Smallish L, Fletcher K (2003) Does the treatment of attention-deficit/hyperactivity disorder with stimulants contribute to drug use/abuse? A 13-year prospective study. Pediatrics 111:97–109

Bava S, Jacobus J, Mahmood O, Yang TT, Tapert SF (2010) Neurocognitive correlates of white matter quality in adolescent substance users. Brain Cogn 72:347–354

Berridge KC (2007) The debate over dopamine's role in reward: the case for incentive salience. Psychopharmacology (Berl) 191:391–431

Blakemore SJ (2008) The social brain in adolescence. Nat Rev 9:267–277

Bogle KE, Smith BH (2009) Illicit methylphenidate use: a review of prevalence, availability, pharmacology, and consequences. Curr Drug Abuse Rev 2:157–176

Bolanos CA, Glatt SJ, Jackson D (1998) Subsensitivity to dopaminergic drugs in periadolescent rats: a behavioral and neurochemical analysis. Brain Res 111:25–33

Bolanos CA, Barrot M, Berton O, Wallace-Black D, Nestler EJ (2003) Methylphenidate treatment during pre- and periadolescence alters behavioral responses to emotional stimuli at adulthood. Biol Psychiatry 54:1317–1329

Bonaventure P, Kelly L, Aluisio L, Shelton J, Lord B, Galici R, Miller K, Atack J, Lovenberg TW, Dugovic C (2007) Selective blockade of 5-hydroxytryptamine (5-HT) 7 receptors enhances 5-HT transmission, antidepressant-like behavior, and rapid eye movement sleep suppression induced by citalopram in rodents. J Pharmacol Exp Ther 321:690–698

Botly LC, Burton CL, Rizos Z, Fletcher PJ (2008) Characterization of methylphenidate self-administration and reinstatement in the rat. Psychopharmacology (Berl) 199:55–66

Brandon CL, Steiner H (2003) Repeated methylphenidate treatment in adolescent rats alters gene regulation in the striatum. Eur J Neurosci 18:1584–1592

Brandon CL, Marinelli M, Baker LK, White FJ (2001) Enhanced reactivity and vulnerability to cocaine following methylphenidate treatment in adolescent rats. Neuropsychopharmacology 25:651–661

Brandon CL, Marinelli M, White FJ (2003) Adolescent exposure to methylphenidate alters the activity of rat midbrain dopamine neurons. Biol Psychiatry 54:1338–1344

Brenhouse HC, Andersen SL (2011) Developmental trajectories during adolescence in males and females: a cross-species understanding of underlying brain changes. Neurosci Biobehav Rev 35:1687–703

Britton GB, Segan AT, Sejour J, Mancebo SE (2007) Early exposure to methylphenidate increases fear responses in an aversive context in adult rats. Dev Psychobiol 49:265–75

Burton CL, Nobrega JN, Fletcher PJ (2010) The effects of adolescent methylphenidate self-administration on responding for a conditioned reward, amphetamine-induced locomotor activity, and neuronal activation. Psychopharmacology (Berl) 208:455–68

Canales JJ (2005) Stimulant-induced adaptations in neostriatal matrix and striosome systems: transiting from instrumental responding to habitual behavior in drug addiction. Neurobiol Learn Mem 83:93–103

Canese R, Adriani W, Marco EM, De Pasquale F, Lorenzini P, De Luca N, Fabi F, Podo F, Laviola G (2009) Peculiar response to methylphenidate in adolescent compared to adult rats: a phMRI study. Psychopharmacology (Berl) 203:143–53

Cardinal RN, Winstanley CA, Robbins TW, Everitt BJ (2004) Limbic corticostriatal systems and delayed reinforcement. Ann N Y Acad Sci 1021:33–50

Carlezon WA Jr, Mague SD, Andersen SL (2003) Enduring behavioral effects of early exposure to methylphenidate in rats. Biol Psychiatry 54:1330–1337

Casey BJ, Jones RM, Hare TA (2008) The adolescent brain. Ann N Y Acad Sci 1124:111–126

Challman TD, Lipsky JJ (2000) Methylphenidate: its pharmacology and uses. Mayo Clin Proc 75:711–721

Chambers RA, Potenza MN (2003) Neurodevelopment, impulsivity, and adolescent gambling. J Gambl Stud 19:53–84

Chambers RA, Taylor JR, Potenza MN (2003) Developmental neurocircuitry of motivation in adolescence: a critical period of addiction vulnerability. Am J Psychiatry 160:1041–1052

Chechik G, Meilijson I, Ruppin E (1999) Neuronal regulation: a mechanism for synaptic pruning during brain maturation. Neural Comput 11:2061–2080

Chen YC, Galpern WR, Brownell AL, Matthews RT, Bogdanov M, Isacson O, Keltner JR, Beal MF, Rosen BR, Jenkins BG (1997) Detection of dopaminergic neurotransmitter activity using pharmacologic MRI: correlation with PET, microdialysis, and behavioral data. Magn Reson Med 38:389–398

Christakou A, Robbins TW, Everitt BJ (2001) Functional disconnection of a prefrontal cortical-dorsal striatal system disrupts choice reaction time performance: implications for attentional function. Behav Neurosci 115:812–825

Christakou A, Robbins TW, Everitt BJ (2004) Prefrontal cortical-ventral striatal interactions involved in affective modulation of attentional performance: implications for corticostriatal circuit function. J Neurosci 24:773–780

Chugani HT, Phelps ME, Mazziotta JC (1987) Positron emission tomography study of human brain functional development. Ann Neurol 22:487–497

Collins GB, Pippenger CE, Janesz JW (1984) Links in the chain: an approach to the treatment of drug abuse on a professional football team. Cleve Clin Q 51:485–92

Dalley JW, Cardinal RN, Robbins TW (2004) Prefrontal executive and cognitive functions in rodents: neural and neurochemical substrates. Neurosci Biobehav Rev 28:771–784

De Bellis MD, Keshavan MS, Beers SR, Hall J, Frustaci K, Masalehdan A, Noll J, Boring AM (2001) Sex differences in brain maturation during childhood and adolescence. Cereb Cortex 11:552–557

Dixon AL, Prior M, Morris PM, Shah YB, Joseph MH, Young AM (2005) Dopamine antagonist modulation of amphetamine response as detected using pharmacological MRI. Neuropharmacology 48:236–245

Doremus-Fitzwater TL, Varlinskaya EI, Spear LP (2010) Motivational systems in adolescence: Possible implications for age differences in substance abuse and other risk-taking behaviors. Brain Cogn 72:114–123

Easton N, Shah YB, Marshall FH, Fone KC, Marsden CA (2006) Guanfacine produces differential effects in frontal cortex compared with striatum: assessed by phMRI BOLD contrast. Psychopharmacology (Berl) 189:369–385

Easton N, Marshall F, Fone K, Marsden C (2007) Atomoxetine produces changes in cortico-basal thalamic loop circuits: assessed by phMRI BOLD contrast. Neuropharmacology 52:812–826

Eshel N, Nelson EE, Blair RJ, Pine DS, Ernst M (2007) Neural substrates of choice selection in adults and adolescents: development of the ventrolateral prefrontal and anterior cingulate cortices. Neuropsychologia 45:1270–1279

Evenden JL (1999) Varieties of impulsivity. Psychopharmacology (Berl) 146:348–361

Evenden JL, Ryan CN (1996) The pharmacology of impulsive behaviour in rats: the effects of drugs on response choice with varying delays of reinforcement. Psychopharmacology (Berl) 128:161–170

Fareri DS, Martin LN, Delgado MR (2008) Reward-related processing in the human brain: developmental considerations. Dev Psychopathol 20:1191–211

Febo M, Segarra AC, Nair G, Schmidt K, Duong TQ, Ferris CF (2005) The neural consequences of repeated cocaine exposure revealed by functional MRI in awake rats. Neuropsychopharmacology 30:936–943

Ferguson SA, Boctor SY (2010) Cocaine responsiveness or anhedonia in rats treated with methylphenidate during adolescence. Neurotoxicol Teratol 32:432–442

Fletcher PJ, Tenn CC, Sinyard J, Rizos Z, Kapur S (2007) A sensitizing regimen of amphetamine impairs visual attention in the 5-choice serial reaction time test: reversal by a D1 receptor agonist injected into the medial prefrontal cortex. Neuropsychopharmacology 32:1122–1132

Frahm J, Kruger G, Merboldt KD, Kleinschmidt A (1996) Dynamic uncoupling and recoupling of perfusion and oxidative metabolism during focal brain activation in man. Magn Reson Med 35:143–148

Gadian DG (1995) NMR and its applications to living systems. Oxford University Press, Oxford

Giedd JN (2008) The teen brain: insights from neuroimaging. J Adolesc Health 42:335–343

Guscott M, Bristow LJ, Hadingham K, Rosahl TW, Beer MS, Stanton JA, Bromidge F, Owens AP, Huscroft I, Myers J, Rupniak NM, Patel S, Whiting PJ, Hutson PH, Fone KC, Biello SM, Kulagowski JJ, McAllister G (2005) Genetic knockout and pharmacological blockade studies of the 5-HT7 receptor suggest therapeutic potential in depression. Neuropharmacology 48:492–502

Harel N, Lee SP, Nagaoka T, Kim DS, Kim SG (2002) Origin of negative blood oxygenation level-dependent fMRI signals. J Cereb Blood Flow Metab 22:908–917

Hastjarjo T, Silberberg A, Hursh SR (1990) Risky choice as a function of amount and variance in food supply. J Exp Anal Behav 53:155–61

Hechtman L, Greenfield B (2003) Long-term use of stimulants in children with attention deficit hyperactivity disorder: safety, efficacy, and long-term outcome. Paediatr Drugs 5:787–794

Herpertz SC, Sass H (2000) Emotional deficiency and psychopathy. Behav Sci Law 18:567–580

Hewitt KN, Shah YB, Prior MJ, Morris PG, Hollis CP, Fone KC, Marsden CA (2005) Behavioural and pharmacological magnetic resonance imaging assessment of the effects of methylphenidate in a potential new rat model of attention deficit hyperactivity disorder. Psychopharmacology (Berl) 180:716–723

Ikemoto S, Panksepp J (1999) The role of nucleus accumbens dopamine in motivated behavior: a unifying interpretation with special reference to reward-seeking. Brain Res Rev 31:6–41

Kallman WM, Isaac W (1975) The effects of age and illumination on the dose-response curves for three stimulants. Psychopharmacologia 40:313–318

Kalsbeek A, Voorn P, Buijs RM, Pool CW, Uylings HB (1988) Development of the dopaminergic innervation in the prefrontal cortex of the rat. J Comp Neurol 269:58–72

Kaminski BJ, Ator NA (2001) Behavioral and pharmacological variables affecting risky choice in rats. J Exp Anal Behav 75:275–97

Kellogg CK, Awatramani GB, Piekut DT (1998) Adolescent development alters stressor-induced Fos immunoreactivity in rat brain. Neuroscience 83:681–689

Klein-Schwartz W, McGrath J (2003) Poison centers' experience with methylphenidate abuse in pre-teens and adolescents. J Am Acad Child Adolesc Psychiatry 42:288–94

Laviola G, Wood RD, Kuhn C, Francis R, Spear LP (1995) Cocaine sensitization in periadolescent and adult rats. J Pharmacol Exp Ther 275:345–357

Laviola G, Adriani W, Terranova ML, Gerra G (1999) Psychobiological risk factors for vulnerability to psychostimulants in human adolescents and animal models. Neurosci Biobehav Rev 23:993–1010

Laviola G, Macri S, Morley-Fletcher S, Adriani W (2003) Risk-taking behavior in adolescent mice: psychobiological determinants and early epigenetic influence. Neurosci Biobehav Rev 27:19–31

Lebel C, Walker L, Leemans A, Phillips L, Beaulieu C (2008) Microstructural maturation of the human brain from childhood to adulthood. NeuroImage 40:1044–1055

Lenroot RK, Giedd JN (2006) Brain development in children and adolescents: insights from anatomical magnetic resonance imaging. Neurosci Biobehav Rev 30:718–729

Leo D, Adriani W, Cavaliere C, Cirillo G, Marco EM, Romano E, di Porzio U, Papa M, Perrone-Capano C, Laviola G (2009) Methylphenidate to adolescent rats drives enduring changes of accumbal Htr7 expression: implications for impulsive behavior and neuronal morphology. Genes Brain Behav 8:356–68

Levin FR, Kleber HD (1995) Attention-deficit hyperactivity disorder and substance abuse: relationships and implications for treatment. Harvard Rev Psychiatry 2:246–258

Levy F, Swanson JM (2001) Timing, space and ADHD: the dopamine theory revisited. Aust N Z J Psychiatry 35:504–511

Lidow MS, Goldman-Rakic PS, Rakic P (1991) Synchronized overproduction of neurotransmitter receptors in diverse regions of the primate cerebral cortex. Proc Natl Acad Sci U S A 88:10218–10221

Lockwood CA, Menter CG, Moggi-Cecchi J, Keyser AW (2007) Extended male growth in a fossil hominin species. Science 318:1443–1446

Marco EM, Adriani W, Ruocco LA, Canese R, Sadile AG, Laviola G (2011) Neurobehavioral adaptations to methylphenidate: the issue of early adolescent exposure. Neurosci Biobehav Rev 35:1722–39

Marota JJ, Mandeville JB, Weisskoff RM, Moskowitz MA, Rosen BR, Kosofsky BE (2000) Cocaine activation discriminates dopaminergic projections by temporal response: an fMRI study in Rat. NeuroImage 11:13–23

McCabe SE, Teter CJ, Boyd CJ, Guthrie SK (2004) Prevalence and correlates of illicit methylphenidate use among 8th, 10th, and 12th grade students in the United States, 2001. J Adolesc Health 35:501–504

Meaney MJ, Stewart J (1981) Neonatal-androgens influence the social play of prepubescent rats. Horm Behav 15:197–213

Mobini S, Chiang TJ, Ho MY, Bradshaw CM, Szabadi E (2000) Effects of central 5-hydroxytryptamine depletion on sensitivity to delayed and probabilistic reinforcement. Psychopharmacology (Berl) 152:390–397

Morrissey G, Wogar MA, Bradshaw CM, Szabadi E (1993) Effect of lesions of the ascending 5-hydroxytryptaminergic pathways on timing behaviour investigated with an interval bisection task. Psychopharmacology (Berl) 112:80–85

Muramoto S, Yamada H, Sadato N, Kimura H, Konishi Y, Kimura K, Tanaka M, Kochiyama T, Yonekura Y, Ito H (2002) Age-dependent change in metabolic response to photic stimulation of the primary visual cortex in infants: functional magnetic resonance imaging study. J Comput Assist Tomogr 26:894–901

Nemoda Z, Szekely A, Sasvari-Szekely M (2011) Psychopathological aspects of dopaminergic gene polymorphisms in adolescence and young adulthood. Neurosci Biobehav Rev 35:1665–86

Nightingale EO, Fischhoff B (2002) Adolescent risk and vulnerability: overview. J Adolesc Health 31:3–9

Panksepp J (1981) The ontogeny of play in rats. Dev Psychobiol 14:327–332

Park F (2007) Lentiviral vectors: are they the future of animal transgenesis? Physiol Genomics 31:159–173

Patrick KS, Markowitz JS (1997) Pharmacology of methylphenidate, amphetamine enantiomers, and penoline in attention deficit/hyperactivity disorder. Hum Psychopharmacol 12:527–546

Paus T, Zijdenbos A, Worsley K, Collins DL, Blumenthal J, Giedd JN, Rapoport JL, Evans AC (1999) Structural maturation of neural pathways in children and adolescents: in vivo study. Science 283:1908–1911

Preece MA, Sibson NR, Raley JM, Blamire A, Styles P, Sharp T (2007) Region-specific effects of a tyrosine-free amino acid mixture on amphetamine-induced changes in BOLD fMRI signal in the rat brain. Synapse 61:925–932

Ptacek R, Kuzelova H, Paclt I, Zukov I, Fischer S (2009) ADHD and growth: anthropometric changes in medicated and non-medicated ADHD boys. Med Sci Monit 15:CR595–599

Puumala T, Ruotsalainen S, Jakala P, Koivisto E, Riekkinen P Jr, Sirvio J (1996) Behavioral and pharmacological studies on the validation of a new animal model for attention deficit hyperactivity disorder. Neurobiol Learn Mem 66:198–211

Ragozzino ME (2003) Acetylcholine actions in the dorsomedial striatum support the flexible shifting of response patterns. Neurobiol Learn Mem 80:257–267

Rebec GV, Christiansen JRC, Guerra C, Bardo MT (1997a) Regional and temporal differences in realtime dopamine efflux in the nucleus accumbens during free-choice novelty. Brain Res 776:61–67

Rebec GV, Grabner CP, Johnson M, Pierce RC, Bardo MT (1997b) Transient increases in cathecolaminergic activity in medial prefrontal cortex and nucleus accumbens shell during novelty. Neuroscience 76:707–714

Rogers RD, Baunez C, Everitt BJ, Robbins TW (2001) Lesions of the medial and lateral striatum in the rat produce differential deficits in attentional performance. Behav Neurosci 115:799–811

Romani C, Adriani W, Manciocco A, Vitale A, Laviola G (2010) Evaluation of impulsive behaviour in Callithrix jacchus: a pilot study. In: Abstract Book, V European Conference on Behavioural Biology (ECBB), Ferrara, Italy, July 15–18, p 126

Rother J, Knab R, Hamzei F, Fiehler J, Reichenbach JR, Buchel C, Weiller C (2002) Negative dip in BOLD fMRI is caused by blood flow-oxygen consumption uncoupling in humans. NeuroImage 15:98–102

Sadile AG, Pellicano MP, Sagvolden T, Sergeant JA (1996) NMDA and non-NMDA sensitive [L-3H]glutamate receptor binding in the brain of the Naples high- and low-excitability rats: an autoradiographic study. Behav Brain Res 78:163–174

Sagvolden T, Pettersen MB, Larsen MC (1993) Spontaneously hypertensive rats (SHR) as a putative animal model of childhood hyperkinesis: SHR behavior compared to four other rat strains. Physiol Behav 54:1047–1055

Salamone JD, Correa M (2002) Motivational views of reinforcement: implications for understanding the behavioral functions of nucleus accumbens dopamine. Behav Brain Res 137:3–25

Salamone JD, Correa M, Mingote SM, Weber SM (2005) Beyond the reward hypothesis: alternative functions of nucleus accumbens dopamine. Curr Opin Pharmacol 5:34–41

Scheffler RM, Hinshaw SP, Modrek S, Levine P (2007) The global market for ADHD medications. Health Aff (Millwood) 26:450–457

Schepis TS, Adinoff B, Rao U (2008) Neurobiological processes in adolescent addictive disorders. Am J Addict 17:6–23

Schmithorst VJ, Yuan W (2010) White matter development during adolescence as shown by diffusion MRI. Brain Cogn 72:16–25

Seeman P, Madras BK (1998) Anti-hyperactivity medication: methylphenidate & amphetamine. Mol Psychiatry 3:386–396

Self DW (2004) Regulation of drug-taking and -seeking behaviors by neuroadaptations in the mesolimbic dopamine system. Neuropharmacology 47:242–255

Shen RY, Choong KC (2006) Different adaptation in ventral tegmental dopamine neurons in control vs ethanol exposed rats after methylphenidate treatment. Biol Psychiatry 59:635–642

Shmuel A, Augath M, Oeltermann A, Logothetis NK (2006) Negative functional MRI response correlates with decreases in neuronal activity in monkey visual area V1. Nat Neurosci 9:569–577

Sicard KM, Duong TQ (2005) Effects of hypoxia, hyperoxia, and hypercapnia on baseline and stimulus-evoked BOLD, CBF, and CMRO2 in spontaneously breathing animals. NeuroImage 25:850–858

Sonuga-Barke EJ (2005) Causal models of attention-deficit/hyperactivity disorder: from common simple deficits to multiple developmental pathways. Biol Psychiatry 57:1231–1238

Sowell ER, Delis D, Stiles J, Jernigan TL (2001) Improved memory functioning and frontal lobe maturation between childhood and adolescence: a structural MRI study. J Int Neuropsychol Soc 7:312–322

Sowell ER, Peterson BS, Thompson PM, Welcome SE, Henkenius AL, Toga AW (2003) Mapping cortical change across the human life span. Nat Neurosci 6:309–315

Spear LP (2000) The adolescent brain and age-related behavioral manifestations. Neurosci Biobehav Rev 24:417–463

Spear LP, Brake SC (1983) Periadolescence: age-dependent behavior and psycho-pharmacological responsivity in rats. Dev Psychobiol 16:83–109

Stamford JA (1989) Development and ageing of the rat nigrostriatal dopamine system studied with fast cyclic voltammetry. J Neurochem 52:1582–1589

Stephens DW, Anderson D (2001) The adaptive value of preference for immediacy: when shortsighted rules have farsighted consequences. Behav Ecol 12:330–339

Svetlov SI, Kobeissy FH, Gold MS (2007) Performance enhancing, non-prescription use of Ritalin: a comparison with amphetamines and cocaine. J Addict Dis 26:1–6

Tarazi FI, Tomasini EC, Baldessarini RJ (1998) Postnatal development of dopamine D4-like receptors in rat forebrain regions: comparison with D2-like receptors. Brain Res 110:227–233

Tarazi FI, Tomasini EC, Baldessarini RJ (1999) Postnatal development of dopamine D1-like receptors in rat cortical and striatolimbic brain regions: An autoradiographic study. Dev Neurosci 21:43–49

Teicher MH, Andersen SL, Hostetter JC Jr (1995) Evidence for dopamine receptor pruning between adolescence and adulthood in striatum but not nucleus accumbens. Brain Res 89:167–172

Terranova ML, Laviola G, Alleva E (1993) Ontogeny of amicable social behavior in the mouse: gender differences and ongoing isolation outcomes. Dev Psychobiol 26:467–481

Terranova ML, Laviola G, de Acetis L, Alleva E (1998) A description of the ontogeny of mouse agonistic behavior. J Comp Psychol 112:3–12

Tirelli E, Laviola G, Adriani W (2003) Ontogenesis of behavioral sensitization and conditioned place preference induced by psychostimulants in laboratory rodents. Neurosci Biobehav Rev 27:163–178

Vendruscolo LF, Izídio GS, Takahashi RN, Ramos A (2008) Chronic methylphenidate treatment during adolescence increases anxiety-related behaviors and ethanol drinking in adult spontaneously hypertensive rats. Behav Pharmacol 19:21–27

Volkow ND, Swanson JM (2003) Variables that affect the clinical use and abuse of methylphenidate in the treatment of ADHD. Am J Psychiatry 160:1909–18

Volkow ND, Wang GJ, Fowler JS, Gatley SJ, Logan J, Ding YS, Dewey SL, Hitzemann R, Gifford AN, Pappas NR (1999) Blockade of striatal dopamine transporters by intravenous methylphenidate is not sufficient to induce self-reports of "high". J Pharmacol Exp Ther 288:14–20

Volkow ND, Wang G, Fowler JS, Logan J, Gerasimov M, Maynard L, Ding Y, Gatley SJ, Gifford A, Franceschi D (2001) Therapeutic doses of oral methylphenidate significantly increase extracellular dopamine in the human brain. J Neurosci 21:RC121

Volkow ND, Wang GJ, Fowler JS, Logan J, Franceschi D, Maynard L, Ding YS, Gatley SJ, Gifford A, Zhu W, Swanson JM (2002) Relationship between blockade of dopamine transporters by oral methylphenidate and the increases in extracellular dopamine: therapeutic implications. Synapse 43:181–187

White FJ, Kalivas PW (1998) Neuroadaptations involved in amphetamine and cocaine addiction. Drug Alcohol Depend 51:141–153

Wilens TE, Adler LA, Adams J, Sgambati S, Rotrosen J, Sawtelle R, Utzinger L, Fusillo S (2008) Misuse and diversion of stimulants prescribed for ADHD: a systematic review of the literature. J Am Acad Child Adolesc Psychiatry 47:21–31

Wills TA, Vaccaro D, McNamara G (1994) Novelty seeking, risk taking, and related constructs as predictors of adolescent substance use: an application of Cloninger's theory. J Subst Abuse 6:1–20

Wills TA, Sandy JM, Shinar O (1999) Cloninger's constructs related to substance use level and problems in late adolescence: a mediational model based on self-control and coping motives. Exp Clin Psychopharmacol 7:122–134

Wogar MA, Bradshaw CM, Szabadi E (1993) Does the effect of central 5-hydroxytryptamine depletion on timing depend on motivational change? Psychopharmacology (Berl) 112:86–92

Wolf ME, Sun X, Mangiavacchi S, Chao SZ (2004) Psychomotor stimulants and neuronal plasticity. Neuropharmacology 47:61–79

Yano M, Steiner H (2005) Methylphenidate (Ritalin) induces Homer1a and zif268 expression in specific corticostriatal circuits. Neuroscience 132:855–865

Yano M, Steiner H (2007) Methylphenidate and cocaine: the same effects on gene regulation? Trends Pharmacol Sci 28:588–596

Yin HH, Knowlton BJ, Balleine BW (2004) Lesions of dorsolateral striatum preserve outcome expectancy but disrupt habit formation in instrumental learning. Eur J Neurosci 19:181–189

Zuckerman M (1984) Sensation seeking: a comparative approach to a human trait. Behav Brain Sci 7:413–471

Attention Deficit Hyperactivity Disorder and Substance Use Disorders

Oscar G. Bukstein

Contents

1 ADHD and SUD.. 146
   1.1 Terminology.. 146
   1.2 Prevalence of ADHD-SUD Comorbidity..................................... 147
2 ADHD and the Risk for the Development of SUD..................................... 148
3 Smoking and ADHD 151
4 Dopamine and the SUD–ADHD Relationship... 151
   4.1 Neuropsychological Functioning 151
   4.2 Heritability 153
   4.3 Genetics 154
5 Does Treatment for ADHD Contribute to the Development of SUDs?.................. 156
   5.1 Potential Mechanisms 156
   5.2 Studies of Clinical Populations... 158
6 Misuse of Psychostimulants and ADHD... 159
7 Effects of Pharmacological Treatment on ADHD-SUD................................ 160
8 Guidelines for Management of Patients with Comorbid ADHD-SUD 162
9 Summary 164
References.. 165

Abstract During the past two decades, there has been an increased recognition that Attention Deficit Hyperactivity Disorder (ADHD) is overrepresented in treatment and community populations of both adolescents and adults with substance use disorders (SUDs). This chapter explores this relationship, including a review of the prevalence of this comorbidity, ADHD and the risk for the development of SUDs. Possible neurobiological underpinnings of the relationship are also discussed. Because of the salience of the association between smoking (tobacco) and ADHD, this topic is included in the discussion of substance use and SUDs.

O.G. Bukstein
University of Texas Health Science Center at Houston, Houston, Texas
e-mail: Oscar.G.Bukstein@uth.tmc.edu

Keywords ADHD · Adolescents · Drug abuse · Substance use disorders

Abbreviations

APA	American Psychiatric Association
AUD	Alcohol use disorder
CBT	Cognitive behavioral therapy
CD	Conduct disorder
CPT	Continuous performance test
DA	Dopamine
DAT	Dopamine transporter
DEA	Drug Enforcement Agency
FDA	Food and Drug Administration
MAO-A	Monoamine Oxidase (A isoform)
MDA	Multi-modal treatment study of ADHD (MTA)
MI/CBT	Motivational interviewing/cognitive behavioral therapy
MPH	Methamphetamine
OR	Odds ratio
PET	Positron emission tomography
SODAS	Spheroidal oral drug absorption system
SPECT	Single photon emission computerized tomography
SUD	Substance abuse disorder
VNTR	Variable-number tandem repeat

1 ADHD and SUD

1.1 Terminology

Problems pertaining to the use of alcohol or other psychoactive substances are often combined under the generic term of "abuse" when there are three distinct concepts to consider: abuse, misuse, and diversion. Abuse is a diagnostic term described by the Diagnostic and Statistical Manual, Fourth Edition (DSM-IV), indicating recurrent use with problems relating to hazardous situations, legal difficulties, fulfillment of major obligations, social or interpersonal problems (APA 2000). Abuse and dependence are maladaptive patterns of drug use that produce clinically significant impairment. Single use *does not* equal abuse. Misuse can be defined as use for a purpose not consistent with medical guidelines (e.g., use of a medication not prescribed for the individual using the medication, modifying the dose, use to achieve euphoria and/or using with other nonprescribed psychoactive substances (WHO 2010; Wilens et al. 2006). Diversion is the transfer of medication from the individual for whom it was prescribed to one for whom it is not prescribed

(Wilens et al. 2006). While abuse (or dependence) connotes pathology related to substance use, diversion, and misuse do not. Individuals who meet DSM-IV abuse or dependence criteria may divert or misuse.

1.2 Prevalence of ADHD-SUD Comorbidity

Past research has established that attention deficit hyperactivity disorder (ADHD) is more prevalent in clinical and community populations of individuals with substance use disorder (SUD: Wilens 2007). Adolescents and adults with ADHD are more likely than those without ADHD to be diagnosed with a SUD (Biederman et al. 1995; Flory et al. 2003; Katusic et al. 2005; McGough et al. 2005). In a study of adolescents, those with ADHD were 6.2 times more likely to have SUD (21.9%) than were the matched controls (4.4%; Katusic et al. 2005). In adults with persistent ADHD, the 47.0% prevalence of a comorbid SUD is significantly higher than the 38.0% rate in the matched controls (McGough et al. 2005).

Perhaps owing to referral bias, SUD treatment populations of adolescents and adults are more likely to be diagnosed with ADHD than would be expected in community samples (Carroll and Rounsaville 1993; Gordon et al. 2004). For instance in a community sample, ADHD was reported in more than 1/3 of the sample of those also diagnosed with a SUD, which is substantially higher than the 4.4% incidence rate of ADHD reported in the adult US population (Kessler et al. 2006). Also, in a report from 4,930 adolescents and 1,956 adults who were admitted for substance abuse treatment in multisite studies, two thirds of this sample had a co-occurring mental health problem in the year prior to treatment admission (Chan et al. 2008). Approximately half of the adolescents under 15 years of age had ADHD and about one third of adults met criteria for ADHD (Chan et al. 2008). Moreover, in a general population study of subjects aged 18–44 years in ten countries in the Americas, Europe, and the Middle East ($n = 11,422$), 12.5% of those diagnosed with ADHD reported a SUD and among those with a SUD, 11.1% met ADHD screening criteria (Fayyad et al. 2007).

A considerable body of literature describes the comorbidity of ADHD with alcohol use disorders (AUD), nicotine use and dependence, and other SUDs, with the latter being accounted for mostly by cannabis-use disorders. For instance, ADHD is associated with both earlier and more frequent alcohol relapses (Ercan et al. 2003) and lower likelihood of cannabis-treatment completion in adolescents (White et al. 2004). Also, individuals with ADHD show tobacco smoking rates that are higher than in the general population and/or nondiagnosed controls among both adults (41–42% versus 26% for ADHD and non-ADHD, respectively; Lambert and Hartsough 1998; Pomerleau et al. 1995) and adolescents (19.0–46% versus 10–24% for ADHD and non-ADHD, respectively; Lambert and Hartsough 1998; Milberger et al. 1997a, b; Molina and Pelham 2003; Rohde et al. 2004). Kollins et al. (2005) reported that ADHD symptoms, even at levels below the threshold required to make a diagnosis, carry a risk for / increase the risk of smoking. These SUDs involve the most prevalent SUDs. Due to low prevalence of other SUDs, especially in community studies, establishing a relationship between ADHD and these less-frequent outcomes is difficult. Thus, this chapter deals specifically with alcohol, nicotine, and other nonalcohol SUDs.

2 ADHD and the Risk for the Development of SUD

There is an increased risk of development of SUDs among those who have ADHD in childhood, although in most studies the risk is partially, if not entirely, mediated by the presence of conduct disorder (CD).

Several studies have reported ADHD as an independent risk factor for SUD, even in the absence of other psychiatric comorbidity (Biederman et al. 1995; Milberger et al. 1997b). Other factors may further increase the risk of SUD in patients with ADHD. These factors include the presence of comorbid disruptive behavior disorders (DBDs) (e.g., oppositional defiant disorder and CD; August et al. 2006; Barkley et al. 1990; Biederman et al. 1997; Katusic et al. 2005; Mannuzza et al. 1991; McGough et al. 2005) and comorbid bipolar disorder (Biederman et al. 1997; Wilens et al. 1999). Among these characteristics, the most consistently reported predictive factor has been comorbid CD, which in most studies completely mediated the increased prevalence of SUD in individuals with ADHD (Barkley et al. 1990; McGough et al. 2005). Some studies (Flory et al. 2003) have reported how comorbid ADHD and CD uniquely contribute to a more severe form of SUD, which is characterized by an increased risk for dependence on illicit drugs.

In a meta-analysis quantifying risk for development of SUDs for young children diagnosed with ADHD, Charach et al. (2011) reported that children with ADHD are at risk for developing AUD (OR = 1.35) by early adulthood and for self-reported nicotine use by middle adolescence (OR = 2.36). Children with ADHD are also at risk for nonalcohol drug use disorders as young adults (OR = 3.48). These figures represent unadjusted odds ratios as neither the presence of childhood conduct problems nor the subsequent development of CD in adolescence is taken into account. Most studies have reported risks for one or more of the three most common specific substances: alcohol, tobacco, and cannabis. However, estimates of SUDs in adulthood clearly identify that many individuals use more than one substance at a time and may develop disorders to multiple substances without specificity. Unfortunately, few studies provide estimates of combined psychoactive SUD or the nonalcohol SUD. Although some of the published literature indicates that childhood ADHD increases the risk of nonalcohol, nonnicotine psychoactive SUDs, the available evidence does not yet confirm a strong relationship.

In another recent meta-analysis, Lee et al. (2011) examined longitudinal studies that prospectively followed children with and without ADHD into adolescence or adulthood. Children with ADHD were significantly more likely to have ever used nicotine and other substances, but not alcohol. Children with ADHD were also more likely to develop SUD (i.e., abuse/dependence) for nicotine, alcohol, marijuana, cocaine, and other substances (i.e., unspecified). Children with ADHD were at least 1.5 times more likely to develop SUD across diverse forms of substances, including nearly 3 times higher for nicotine dependence. Furthermore, empirical tests of potential moderators for outcomes with heterogeneity in effect-size estimates, consisting of demographic or methodological features that varied across studies, were not significant. Few studies addressed ADHD and comorbid DBDs, thus preventing a formal meta-analytic review. A summary of the results of these studies indicates that comorbid DBD complicates inferences about the specificity of ADHD effects on substance use outcomes. Overall, findings from previous studies of ADHD and comorbid ODD/CD predicting substance use outcomes strongly suggest that the relation between ADHD and substance outcomes in the literature, and potentially in this meta-analysis, may be partially or fully accounted for by the comorbidity between ADHD and ODD/CD, which is strongly related to substance outcomes.

Many of the studies involve clinical samples of ADHD youth, which might lead to a referral bias toward increased severity (i.e., the presence of comorbidity such as SUDs). In the follow up of the Multi-Modal Treatment Study of ADHD (MTA), MTA children had significantly higher rates of delinquency (27.1% versus 7.4%) and substance use (17.4% versus 7.8%) relative to local normative comparison group at 36 months (Molina et al. 2007a). By 24 and 36 months, more days of prescribed medication were associated with more serious delinquency but not substance use.

As part of the Pittsburgh ADHD Longitudinal Study (Molina et al. 2007b), 364 children diagnosed with ADHD were interviewed an average of 8 years after diagnosis and index treatment, either as adolescents (11–17 years of age) or as young adults (18–28 years of age). When compared with demographically similar age-matched participants without ADHD, recruited as adolescents ($n = 120$), episodic heavy drinking (measured as >5 drinks per occasion), drunkenness, DSM-IV AUD symptoms and DSM-IV AUD were elevated among 15- to 17-year-old probands, but not among younger adolescents. Among young adults, drinking quantity and AUD were elevated among probands with antisocial personality disorder (ASPD).

Two large prospective studies reported discrepant results in examining the risk for the development of SUDs in those having been diagnosed with ADHD as children. A 25-year longitudinal study of a birth cohort of 1,265 New Zealand-born children (Fergusson et al. 2007) examined the relationship between measures of conduct and attentional problems obtained in middle childhood (7–9 years) and adolescence (14–16 years) and measures of substance use, abuse, and dependence from 18–25 years. Conduct problems in childhood and adolescence were generally related to later substance use, abuse, and dependence even after controlling for attentional problems and confounders. However, attentional problems were largely unrelated to later substance use, abuse, and dependence. Any association between early attentional problems and later substance use abuse and dependence was largely mediated via the association between conduct and attentional problems.

In another large prospective study of 11-year-old twins (760 females and 752 males), dimensional and categorical measures of ADHD and CD were associated with subsequent initiation of tobacco, alcohol, and illicit drug use by 14 years of age, as well as with onset of actual SUDs or pathologic use by 18 years of age (Elkins et al. 2007). For boys and girls, hyperactivity/impulsivity predicted initiation of all types of substance use, nicotine dependence and cannabis abuse/dependence, even when controlling for CD at two time points. However, relationships between inattention and substance outcomes disappeared when hyperactivity/impulsivity and CD were controlled for, with the possible exception of nicotine dependence. A categorical diagnosis of ADHD-predicted tobacco and illicit drug use only. A diagnosis of CD between 11 and 14 years of age was a powerful predictor of substance disorders by 18 years of age. The investigators concluded that even a single symptom of ADHD or CD is associated with increased risk.

Using a community sample, another prospective study of 428 children, aged 12 years and free from any SUD at grade 7, were assessed over three consecutive years, using a standardized psychiatric interview (Elkins et al. 2007). With the outcome being the age of onset of a SUD, the most significant predictive factors for the development of adolescent SUD included male gender, attention-deficit hyperactivity disorder, CD, and sibling use of tobacco.

Failure in previous research to consistently observe relationships between ADHD and substance use and abuse outcomes could be due to reliance on less-sensitive categorical diagnoses rather than dimensional measures of ADHD symptoms (e.g., impulsivity or inattention). Similarly, in a study of Brazilian youth, 968 male adolescents (15–20 years of age) were screened for SUD in their households (Szobot et al. 2007). Of the subjects who were screened as positive, 61 cases with *a SUD* were identified and compared with controls without a nonalcohol or alcohol SUD, matched by age and proximity with the case's household. Adolescents with ADHD presented a higher risk for illicit SUD than youths without ADHD, even after adjusting for potential confounders (e.g., CD, ethnicity, religion, and estimated IQ). In the French GAZEL prospective study, 916 subjects aged 7–18 were recruited from the general population and surveyed in 1991 and 1999 (Galer et al. 2008). In males, after controlling for CD, neither hyperactivity-inattention nor inattention symptoms independently increased the liability to later substance use.

Age of onset and number of ADHD criteria may hold a clue to the level of risk for the development of an SUD. Faraone et al. (2007a) compared four groups of adults: (1) ADHD subjects who met all DSM-IV criteria for childhood onset ADHD, (2) late-onset ADHD subjects who met all criteria except the age at onset criterion, (3) subthreshold ADHD subjects who did not meet full symptom criteria, and (4) non-ADHD subjects who did not meet any of the above criteria. Cigarette and marijuana use was greater in all ADHD groups compared to non-ADHD controls. The late onset and full ADHD groups were more likely to have endorsed ever having a problem due to use of cigarettes, alcohol, or marijuana and reported more problems resisting use of drugs or alcohol. The full ADHD group was more likely than the other groups to have reported "getting high" as their reason for using their preferred drug. Not surprisingly, adults with ADHD have elevated rates of substance use and related impairment compared to those without the diagnosis.

3 Smoking and ADHD

Unlike the equivocal literature on the status of ADHD as an independent risk factor for SUD, investigators have shown that ADHD is a specific, independent risk factor for tobacco use in clinical and high-risk samples, after controlling for comorbid CD (Molina and Pelham 2003; Milberger et al. 1997a, b). Individuals with ADHD report earlier initiation of smoking than their peers without ADHD and are more likely to progress from initiation to regular smoking (Milberger et al. 1997a; Molina and Pelham 2003; Rohde et al. 2004). In addition to higher rates of smoking, individuals with ADHD seem to have more difficulty quitting. One study reported that the percentage of ever-smokers who quit smoking was lower among adults with ADHD (29%) compared to the general population (48.5%), suggesting that individuals with ADHD have greater difficulty quitting (Pomerleau et al. 1995). Another study reported that a history of childhood ADHD was predictive of worse smoking cessation outcomes, even after controlling for demographic, baseline smoking variables and depression symptoms. Two studies of population-based samples have shown that levels of ADHD symptoms predict nicotine use and dependence. Among current regular smokers, self-reported numbers of both hyperactive–impulsive and inattentive ADHD symptoms predicted the number of cigarettes smoked per day (Kollins et al. 2005).

In a population-based sample of over 15,000 young adults, Kollins et al. (2005) identified a linear relationship between the number of retrospectively self-reported ADHD symptoms and the lifetime risk of regular smoking and a negative association between the number of ADHD symptoms and the age of onset of smoking. The investigators also noted a positive association between the number of ADHD symptoms and number of cigarettes smoked per day among current smokers.

McClernon et al. (2008a, b) assessed smoking withdrawal in a clinical sample of adults diagnosed with ADHD, observing Group × Smoking Condition interactions on several parameters of CPT performances. Finally, main effects or trends for group differences were observed on several measures including self-reported confusion and hostility, continuous performance test (CPT) reaction time variability, CPT commission errors, and accuracy. Consistent with other studies, these results indicate that smoking abstinence impaired both response inhibition [e.g., errors of commission and measures of attentional control (Bekker et al. 2005; Hatsukami et al. 1989; Powell et al. 2002; Zack et al. 2001)].

4 Dopamine and the SUD–ADHD Relationship

4.1 Neuropsychological Functioning

Evidence of impaired brain dopamine activity in individuals with ADHD (Volkow and Swanson 2008) could explain why individuals with ADHD are at greater risk for the abuse of drugs or use of tobacco/nicotine. Such psychoactive drugs acutely, but transiently, raise the concentration of extracellular dopamine in the brain and could temporarily improve ADHD symptoms. In contrast, chronic drug abuse decreases dopaminergic transmission in the brain (Volkow et al. 2007). Inasmuch as dopamine modulates the activity of brain regions (including the prefrontal cortex, striatum, hippocampus, and amygdala) implicated in ADHD symptoms (i.e., attention, executive function, and impulsivity) chronic drug exposure in ADHD individuals could exacerbate the symptoms of the disorder (Volkow and Swanson 2003).

Individuals with ADHD and SUD commonly share poor insight as to the magnitude of their deficits or resulting impairments (Barkley et al. 2002; Goldstein et al. 2007, 2009a). Goldstein et al. (2009a) suggested that the discordance between self-reported motivation and goal-driven behavior in individuals with cocaine dependence is mirrored by brain-behavior dissociations in tasks of reward processing, behavioral monitoring, and emotional suppression (Goldstein et al. 2009b). Correlations with neuropsychological performance support the notion that neurocognitive dysfunction underlies such compromised self-awareness, which is frequently mislabeled as "denial" (which assumes *a priori* knowledge and intent to negate, or minimize, the severity of symptoms) (Rinn et al. 2002). Many of these deficits are similar to those in ADHD. Reduced activity in the anterior cingulate cortex, mostly affecting selective attention and inhibitory control tasks, is a common observation in users of cocaine, heroin, alcohol, cannabis, and other drugs (Garavan and Stout 2005).

The "switch" from voluntary drug use to habitual and progressively compulsive drug use represents a transition at the neural level from prefrontal cortical to striatal control over drug-seeking and drug-taking behaviors. There is also a progression from ventral to more dorsal domains of the striatum, mediated at least in part by its stratified dopaminergic innervation (Everitt et al. 2008). The dorsal anterior cingulate might be central to on-line cognitive control and consequential for the decision making of drug users through adjustments in response options (including risky options). The intensity of activation of the anterior cingulate cortex is predictive of successful treatment outcome in alcoholics and methamphetamine users (Grüsser et al. 2004; Paulus et al. 2005) and can be improved by acute cocaine administration (Garavan et al. 2008). This restorative effect is likely to be dose specific (Fillmore et al. 2006) in cocaine users as illustrated in fMRI studies in cocaine users (Garavan and Stout 2005).

Drug-seeking behavior is paralleled by an increase in dopamine release in the dorsolateral striatum. Local infusion of dopamine receptor antagonists into this brain region reduces such behavior. Miller and Cohen (2001) suggest that the lateral parts of the dorsal striatum dominate the stimulus response instrumental process in which drug-seeking behavior becomes a response habit, which is triggered and maintained by drug-associated stimuli. The presentation of drug cues to cocaine-addicted individuals both induces drug craving and activates the dorsal striatum. Further, there is a reduction in dopamine D2 receptor density in abstinent alcoholics, cocaine, heroin, and methamphetamine-addicted individuals (Volkow et al. 2003). Drug addiction or compulsive drug seeking may be viewed as a maladaptive stimulus-response habit in which the ultimate goal of the behavior has been devalued, perhaps through tolerance to the rewarding effects of the drug (Everitt et al. 2008). Limited or compromised insight into the presence and/or severity of addiction in drug-addicted individuals could partly be driven by this switch to an automatic and habitual system, which might operate outside awareness: automatic processes require less attention, effortful control, and conscious awareness (Miller and Cohen 2001). The limited awareness may also enhance the influence of automatized action schemata (Tiffany 1990) leading to uncontrollable drug-seeking behaviors (especially during high-risk situations) (Rohsenow et al. 1994). Similar deficits in metacognition are postulated for ADHD in adults (Barkley 2010). Individuals with ADHD also often lack insight into their own behavior and, consequently, underestimate their symptomatology as well as their competence in social and academic functioning (positive attributional bias) (Barkley et al. 2002; Hoza et al. 2004; Owens et al. 2007). It is possible that the "denial" noted in SUDs and the poor insight noted in ADHD represent similar deficits.

4.2 Heritability

As previously mentioned, smoking (tobacco) provides a good model for the relationship between tobacco and other drug use and ADHD (see: Kollins et al. 2005). Likely the result of decreased dopaminergic transmission in prefrontal cortex and anterior cingulate after abstinence from nicotine (Powell et al. 2002), evidence supports a central role for dopamine in modulating both response inhibition and attentional control. For example, the therapeutic effects of the stimulant, methylphenidate (MPH), which exerts its action by blocking the dopamine transporter (DAT), particularly in striatal areas (Volkow et al. 1995, 2009a), then increases DA signaling in corticostriatal circuits and improves attentional and inhibitory functioning (Castellanos et al. 1996; Solanto 1998; Volkow et al. 2007, 2009a). The role of dopamine in response inhibition is also supported by evidence that genes associated with dopamine functioning (e.g., DRD4 and DAT) are associated with differences in inhibitory control, even in healthy, non-ADHD individuals (Congdon et al. 2007). Given evidence that nicotine withdrawal results in downregulation of dopamine receptors (Hildebrand et al. 1998; Rahman et al. 2004), it stands to reason that the mechanism through which response inhibition and attentional control were disrupted in the Kollins et al.'s (2009) study was associated with abstinence-induced changes in dopamine transmission. However, these results suggest that nicotine withdrawal may result in even more pronounced disruptions in dopamine functioning and subsequent behavioral performance for some tasks. This relative difference may help explain why individuals with ADHD report greater difficulty in quitting smoking compared to those without ADHD (Pomerleau et al. 1995).

ADHD and smoking are both highly heritable: genetic factors account for 60–80% and 56% of the two phenotypes, respectively (Faraone et al. 2005; Li et al. 2004). Candidate gene studies have identified a number of similar genetic markers associated with both ADHD and smoking phenotypes, suggesting that several common neurobiological mechanisms may give rise to this comorbidity (Li et al. 2004; Maher et al. 2002; Munafo et al. 2004; Todd et al. 2005). The genetic substrates of ADHD and smoking behavior appear to overlap considerably, with a number of candidate genes, most notably DRD4 and DAT, exhibiting associations with both ADHD and smoking phenotypes (Faraone et al. 2005; Munafo et al. 2004). In addition, several studies have examined the relationship among genes, smoking, and ADHD. One study examined interactions between gene and ADHD symptoms and found effects for both DRD2 and, among females, MAO-A (McClernon et al. 2008a, b). This study also reported that carriers of the DRD2/ANKK1 *Taq*1 A2/A2 allele, with six or more hyperactivity–impulsivity symptoms, were almost twice as likely to have a history of smoking as individuals carrying the A1 allele.

Genes regulating nicotinic receptor function are another group of potential targets that point to potential genetic overlaps between ADHD and smoking (McClernon and Kollins 2008; McClernon et al. 2008a, b). Results from studies examining relationships between the CHRNA4 gene and ADHD have been mixed. One study found an association between variation in this gene and a quantitative phenotype of ADHD (Todd et al. 2003), three studies reported small associations between the CHRNA4 gene and ADHD phenotypes (Brookes et al. 2006; Guan et al. 2009; Lee et al. 2008), while another failed to identify any association between the gene and ADHD (Ken et al. 2001).

Despite the considerable amount of descriptive work that has characterized associations between ADHD and smoking, relatively little research has been conducted. Kollins et al. (2009) proposed a model linking smoking and ADHD, and speculated on the mechanisms underlying this common comorbidity. From a neuropharmacological perspective, ADHD is hypothesized to result from aberrant striatal dopaminergic systems that result in disrupted dopaminergic transmission in corticostriatal circuits, which can give rise to the characteristic deficits in executive functioning observed in ADHD patients. Individuals with ADHD are hypothesized to have lower DA tone, which may amplify the phasic DA response stimulated by nicotine. This, in turn, may enhance the reward salience of smoking in this population. Deficits in attentional and inhibitory control functions are reduced by nicotine, which negatively reinforces its continued use. Upon quitting, or attempting to quit, smoking, individuals with ADHD experience greater withdrawal symptom severity and greater disruption of inhibitory control, which increase the likelihood of relapse. Finally, higher baseline levels of impulsivity and greater sensitivity to salient reward-related cues may confound efforts to maintain smoking abstinence.

4.3 Genetics

Genetic studies have identified polymorphisms of several genes that govern expression of dopamine receptors (e.g., *DRD4)* or the dopamine transporter *(DAT1)* and are associated with ADHD. Brain imaging studies have reported disrupted transmission in dopamine systems, which are involved in the core ADHD symptoms of impulsivity and inattention. There is also increasing evidence for both reward and motivation deficits, as related to dopamine system functioning, in patients with ADHD. Using PET imaging, Volkow et al. (2009a) reported lower DRD2/DRD3 receptor and DAT availability in the nucleus accumbens and midbrain in a group of adults with ADHD when compared to a control group. This finding supports the hypothesis that the dopamine reward pathway is dysfunctional in ADHD, particularly when DRD2/DRD3 receptor measures (e.g., binding density) in the nucleus accumbens are correlated with attentional measures. As a low density of DRD2DR/D3 receptors in the nucleus accumbens has been associated with the risk for SUDs, this may underlie the observed increase in prevalence of SUD among those with ADHD.

Both *DAT1* and *DRD4* genes have been associated with SUDs. Guindalini et al. (2006) demonstrated an association between cocaine dependence and a variable-number tandem repeat (VNTR) allele in the gene encoding *DAT1* (SLC6A3) (the DAT). *DRD4* VNTR polymorphism may contribute to cue-elicited craving in heroin dependence (Shao et al. 2006). However, little is known about how genetics affect brain response to MPH in subjects with ADHD/SUDs, despite the dopaminergic actions of drugs of abuse (Volkow et al. 2004) and despite findings from epigenetic studies that drug exposure might influence gene expression (Renthal and Nestler 2008). Szobot et al. (2008a) conducted a study using [Tc^{99m}]TRODAT-1 and computed tomography (SPECT) before, and after three weeks of the stimulant medication MPH-SODAS (Spheroidal Oral Drug Absorption System), documenting that MPH reduced DAT availability in adolescents with ADHD/SUDs (Szobot et al. 2008a).

Szobot et al. (2011) then evaluated whether ADHD risk alleles at *DRD4* and *DAT1* genes could predict the change in striatal DAT occupancy after treatment with MPH in adolescents with ADHD/SUDs. Seventeen adolescents with ADHD/SUDs underwent a single photon emission SPECT scan with [Tc^{99m}]TRODAT-1 at baseline and after three weeks on MPH. Caudate and putamen DAT binding were measured and DAT changes were compared according to subjects' genotype. The investigators reported that the combination of both *DRD4* 7-repeat allele (7R) and homozygosity for the *DAT1* 10-repeat allele (10/10) was associated with a reduced DAT change. This was evident even after adjusting the results for potential confounders. Thus, in patients with ADHD/SUDs, combined *DRD4* 7R and *DAT1* 10/10 could index MPH-reduced DAT occupancy.

The DAT, most densely expressed in the basal ganglia, is under the influence of several adaptive factors. Psychoactive drugs can modify striatal DA transmission and cause more prolonged changes in DAT regulation (Daws et al. 2002). Although presence of *DRD4*-7 allele was associated with poorer clinical response to stimulants in individuals with ADHD (Hamarman et al. 2004), little is known about the associated brain imaging patterns. It was documented that subjects without the *DRD4*-7 allele who smoked nicotine during the PET scan had a greater smoking-induced DA release in the ventral caudate and nucleus accumbens (Brody et al. 2006). Thus, although *DRD4* locates mostly in the frontal cortex, this receptor seems to play an important role in striatal DA transmission, which can also be measured by brain-imaging studies.

5 Does Treatment for ADHD Contribute to the Development of SUDs?

5.1 Potential Mechanisms

ADHD is believed to be the result, in part, of aberrant dopamine function. In line with this theory, psychostimulants may be effective in treating ADHD because they increase extracellular dopamine concentration (Schiffer et al. 2006). MPH binds with the DAT, whereas amphetamines trigger presynaptic dopamine release. These increases in extracellular dopamine, following administration of a pyschostimulant, contribute to the alerting effects in individuals with or without ADHD (Rapoport et al. 1980; Volkow et al. 2002). Stimulants also increase central activation of mesolimbic dopamine "reward" pathways in the nucleus accumbens, which is an important common substrate of abuse potential in humans (Di Chiara and Imperato 1988; Koob and Nestler 1997). As a result, products containing MPH or α-methylphenethylamine (amphetamine) are classified as Schedule II controlled substances by the Drug Enforcement Administration. Although the foregoing review highlights the effects of acute psychostimulant treatment on dopamine function, there may be important differences in how these drugs affect dopamine transmission and other brain functions following chronic administration. A recent review provides evidence that chronic amphetamine exposure, even at clinically relevant doses, may result in considerable changes in brain morphology and function (Advokat 2007).

Kollins et al. (2009) compared the reinforcing and subjective effects of oral MPH in 33 adults ($n = 16$ with ADHD; $n = 17$ free from psychiatric diagnoses). The subjects completed four pairs of experimental sessions, each of which included a sampling session and a self-administration session. During sampling sessions, subjects received in randomized order 0 (placebo), 20, 40, or 60 mg MPH. Both groups showed robust effects of MPH on subjective endpoints. Main effects of group were noted on subjective effects involving concentration and arousal. Thus, compared to placebo, MPH produced reinforcing effects for the ADHD group and not for the control group. Increases in stimulant-related subjective effects in non-ADHD subjects were not associated with drug reinforcement. The reinforcing effects of MPH, as measured by the experimental task, were higher in adults diagnosed with ADHD compared to a group of nondiagnosed adults. Non-ADHD subjects showed differences only for the subject-rated effects measures on items generally assessing cognitive status (concentration) or affect/arousal. These findings suggest that the reinforcing effects of MPH do not necessarily coincide with the subjective effects of the drug. Since MPH increases DA neurotransmission at therapeutic doses, through blockade of the DAT, the drug may reinforce behavior more strongly in patients with ADHD because it likely restores DA activity to more normal levels. The lack of reinforcing effects of MPH in the non-ADHD individuals is surprising given the well-established abuse potential of MPH (Kollins et al. 2001). However, several previous studies have failed to show that MPH functions as a reinforcer under routine laboratory conditions (Chait 1994; Roehrs et al. 1999; Stoops et al. 2005). In studies that have shown MPH to function as a reinforcer, it often occurs under specific environmental conditions, such as sleep deprivation or prior to a high-demand task (Roehrs et al. 1999; Stoops et al. 2005).

A number of neuropharmacologic mechanisms contribute to the observed differences in abuse rates between oral stimulants that are used in the treatment of ADHD and drugs of abuse. A key factor in determining abuse potential is the ability of a compound to yield rapid absorption and increases in central drug concentrations, enabling rapid increases in central synaptic dopamine concentrations, followed by rapid clearance (Resnick et al. 1977, Volkow and Swanson 2003). Whether or not a given psychostimulant yields a rapid on–off effect is modulated by several factors, including route of administration, pharmacokinetic profile, and dose. Rapid effects are most likely achieved via intravenous, intranasal, and inhaled routes of administration (Balster and Schuster 1973; Resnick et al. 1977; Volkow et al. 1995). Route of administration is perhaps the most important factor in modulating the timing and rate of increase in central dopaminergic transmission; both intravenous and intranasal administrations produce more rapid increases in central dopamine concentrations than does oral administration. Volkow et al. (1995) observed that oral and intravenous MPH produced comparable changes in striatal dopamine concentration (i.e., "a high"): intravenous but not oral MPH induced a euphoric response. This response appears directly related to differences in the *rate* of dopamine changes, which were much faster with intravenous administration of MPH (which peak after 6–10 min) compared with oral administration (which peak after 60–90 min; Volkow and Swanson 2003). These studies further show that the absolute concentration of extracellular dopamine is not relevant to producing a high, although it may affect its intensity (Volkow et al. 1995; Volkow and Swanson 2003). Even when administered orally, both MPH and amphetamine produce subjective and reinforcing effects suggestive of abuse potential (Kollins et al. 2001; Rush et al. 1998). The rate and timing of peak brain concentration following administration of a drug also depends on its dose; a higher dose will deliver more drug centrally per unit of time than will a lower dose, thus producing the euphoria sought by abusers (Coetzee et al. 2002; Volkow and Swanson 2003).

The drug formulation may also have an effect on abuse potential. To reduce the need for multiple daily doses, a number of extended-release formulations of MPH and amphetamine have been developed, characterized by a slower rise in plasma concentrations over a longer duration. These should, theoretically, have a lower potential for abuse (Volkow and Swanson 2003). Two studies evaluated this hypothesis: both found lower subjective ratings of "likeability" or other subjective ratings of "like drug," "drug effects," "high," "good effects" in sustained-release stimulants when compared with immediate-release (Spencer et al. 2006; Kollins et al. 1998). In a comparison of the subjective effects of immediate-release (40 mg) MPH versus osmotic-release (90 mg) MPH in healthy volunteers, Spencer et al. (2006) found that average peak drug concentrations in plasma and DAT blockade were comparable for the two formulations, but that both were reached several hours earlier with the immediate-release formulation. These studies support the notion that both the overall dose and the rapidity of central dopamine changes are critical in determining the abuse potential of a given drug. In two separate studies of adults and children diagnosed with ADHD, orally administered MPH failed to produce the characteristic increases in subjective ratings of abuse potential (e.g., "feel like talking," "energetic," "heart beating fast"), despite patients' preferences for oral MPH over placebo (Fredericks and Kollins 2004; MacDonald and Kollins 2005). These studies suggest that patient preferences for MPH may be related more to ADHD symptom relief than to euphoria or other markers of abuse potential.

Based on the literature, the conditions under which stimulants are currently used clinically (i.e., given to patients with ADHD in small oral doses or extended-release preparations) make them potentially less liable to abuse than cocaine or methamphetamine, which typically are administered via intravenous or intranasal routes and result in more rapid effects (Volkow 2006; Volkow and Swanson 2003).

5.2 Studies of Clinical Populations

Prospective observational studies have concluded that the long-established clinical practice of using stimulant medication to treat young children with ADHD does not affect the risk for substance abuse in adulthood (Volkow et al. 2008).

One study compared subgroups of children with ADHD who received treatment with stimulants in childhood or adolescence with those who did not (Biederman et al. 2008a, b; Wilens et al. 2008a, b, c). Another study evaluated the age when stimulant treatment was initiated in childhood and its relationship to drug abuse in adulthood (Volkow et al. 2008). This study also compared the prevalence of substance abuse in persons with ADHD with a comparison group. Both studies document high rates (up to 45%) of SUDs in their adult cohorts. These two studies evaluated clinically referred samples, but medication status was not randomized, making the findings vulnerable to referral bias. The samples were also small, especially for the subgroup not treated with stimulants. Treatment with stimulants was initiated (as usual) at an average age between 8 and 9 years, with most children discontinuing treatment after an average of 2–6 years. Thus, most individuals were probably exposed to stimulants for only a short time during childhood, and only a few patients were exposed to stimulants during adolescence.

Preclinical studies have revealed that exposure to stimulant drugs during adolescence, but not in childhood, increased sensitivity to the rewarding effects of cocaine. As a consequence, prospective studies of larger samples of adolescents treated with stimulant medications are necessary in order to evaluate more carefully the consequences of stimulant exposure during this developmental period (Volkow et al. 2008). In the Mannuzza et al. study (2008), subjects with late initiation of stimulant medication (ages 8–12) had greater substance abuse that was mediated by an increase in ASPD in adulthood. The subgroup with early treatment (before the age of 8) did not differ from comparison subjects in lifetime rates of nonalcohol substance use (27% versus 29%, respectively). In a retrospective study from Europe, two groups ($n = 17$ and $n = 74$) were defined on the basis of data from archives at the Expert team for ADHD for Middle and Northern Norway. Treatment with stimulants in childhood/youth contributed to alcohol abuse, substance abuse, and criminality (Goksoyr and Nottestad 2008). In a retrospective study of 206 ADHD adults ($n = 79$ late-onset ADHD, $n = 127$ full ADHD) grouped by lifetime history of ADHD treatment (no treatment, past treatment, current and past treatment), there were no differences in the prevalence of cigarette smoking, alcohol, or drug abuse or dependence, as well as no differences in 1-month prevalence of any use or use more than 20 times (Faraone et al. 2007b). There were also no differences in complications of drug or alcohol use across groups.

The presence of ADHD as a risk factor for the development of SUD and the high prevalence of ADHD-SUD comorbidity prompts clinicians to consider treatment of ADHD as an important element of intervention. Although long-term positive outcomes, particularly as a result of medication, have not been established for these interventions in these populations, the existing literature points to no increased risk in attempting to do so.

6 Misuse of Psychostimulants and ADHD

Diversion and misuse are widespread, especially in high school and college students. In a survey of junior/senior high school students in Canada (total survey sample = 13,549; *5.3% of whom were prescribed stimulants*), 15% reported having given away some of their medication, 7% having sold some of their medication, 4% having experienced theft of their medication, and 3% having been forced to give up some of their medication in the 12 months before the survey (Poulin 2001). In a series of publications based on school-based surveys, McCabe et al. 2004, 2006) reported that about 8% of non-ADHD students in middle school, high school, and college engaged in nonmedical use of stimulants. A primary source for misuse is apparently from prescriptions for medical use diverted by sale or other means. Increased diversion may be related to the use of stimulant medications as "cognitive enhancers" for the general population (i.e., those not diagnosed with ADHD: Sahakian and Morein-Zamir 2007). In a study of adults (18–49 years of age) and college students, a majority of the participants endorsed desire for increased productivity (i.e., for cognitive enhancement) as the primary motivation for misuse. Swanson and Volkow (2009) suggest that when stimulants are prescribed for adolescents and adults who are seeking treatment for themselves, there may be a higher rate of diversion for nonmedical use than for children whose parents are seeking treatment for them. Although performance enhancement appears to be the primary reason for stimulant misuse and diversion among college students, over 40% of these students also reported using stimulants to get "high." In another survey of high school students, lifetime misuse rates ranged from 2.7% in 8th graders to 5% in 12th graders and were decreased for students making higher grades (McCabe et al. 2004).

7 Effects of Pharmacological Treatment on ADHD-SUD

Currently, there is modest empirical support for the use of pharmacotherapy with only several randomized double-blind clinical trials in this population. These trials in adults with cooccurring SUD-ADHD include trials of atomoxetine, MPH, and bupropion. The three trials that included patients concurrently enrolled in outpatient cognitive behavioral therapy (CBT) for SUD showed no differences between medication (MPH or bupropion) and placebo for ADHD or SUD (Schubiner et al. 2002; Levin et al. 2006, 2007). In two published, controlled stimulant medication trials for ADHD in adolescents with SUD, both found medication – pemoline (Riggs et al. 2004) and MPH-SODAS (Szobot et al. 2008b) to be more efficacious than placebo for ADHD in adolescents not in SUD treatment. Neither study observed an impact on substance use. In a randomized double-blind placebo trial of pemoline in adolescents with comorbid ADHD and SUD and of MPH in adult cocaine abusers with ADHD, stimulant medication improved ADHD symptoms but had little, if any, effect on substance use (Riggs et al. 2004) In a 12-week randomized, control trial comparing bupropion, MPH, and placebo in 98 methadone-maintained adults with ADHD and cocaine dependence or abuse, Levin et al. (2006) reported reduced ADHD symptoms in all three groups with no differences in outcome between the groups. No reports of misuse or worsening of cocaine use were noted among those randomized to cocaine.

In a recent randomized, double-blind trial of atomoxetine and placebo treatment of 70 adolescents with ADHD and SUD, with both groups receiving motivational interviewing/cognitive behavioral therapy (MI/CBT) for SUD, Thurstone et al. (2010) reported no difference between the atomoxetine + MI/CBT and placebo + MI/CBT groups in ADHD or SUD. The investigators propose that the MI/CBT and/or a placebo effect may have contributed to a large treatment response in the placebo group and obscured any differences between the atomoxetine and placebo groups. In a recently completed double-blind placebo controlled study of osmotic release oral system (OROS) MPH in over 300 adolescents, aged 14–18 with ADHD and a nonopiate SUD, Riggs (2009) administered CBT treatment for SUD to all subjects. Both groups improved on ADHD and SUD measures, although there were no differences in adolescent report of ADHD symptoms or substance use (number of days used in past month). However, the OROS MPH group had lower ADHD scores than the placebo group on parent report. There were few significant adverse events.

Several nonstimulants offer additional options although the efficacy of these agents has yet to be tested in controlled trials in adults. Atomoxetine is a noradrenergic reuptake inhibitor shown to be superior to placebo in the treatment of ADHD in children, adolescents, and adults (Michelson et al. 2003). Due to the absence of evidence regarding abuse potential (atomoxetine does not produce subjective effects similar to MPH), atomoxetine is not listed as a scheduled drug by the Drug Enforcement Agency (DEA) (Heil et al. 2002). In a recent study of adults with DSM-IV diagnoses of ADHD and alcohol abuse and/or dependence, and who were abstinent from alcohol at least 4 days (maximum 30 days) before study randomization, subjects received atomoxetine (25–100 mg daily) or placebo for 12 weeks (Wilens et al. 2008a, b, c). ADHD symptoms were improved in the atomoxetine cohort compared to placebo. Although no differences between treatment groups occurred in time-to-relapse of heavy drinking, the cumulative heavy drinking days were reduced 26% in atomoxetine-treated subjects versus placebo.

Bupropion is a noradrenergic and dopamine reuptake blocker with stimulant-like effects that is in current use as an antidepressant (Davidson and Connor 1998). Bupropion appears to have a low abuse potential on physiological measures compared with dextroamphetamine (Griffith et al. 1983). The approval by the Food and Drug Administration (FDA) for the use of bupropion for smoking cessation and its efficacy in controlled clinical trials (Hurt et al. 1997) suggests the potential value of this agent for addictive disorders, including patients with comorbid ADHD, comorbid depression, and/or smoking behavior.

Modafinal is marketed for the treatment of daytime sleepiness in narcolepsy and is listed as a schedule IV agent by the DEA. Modafinil has stimulant effects on specific areas of the brain, as well as an alerting effect that may ameliorate some of the vegetative symptoms of cocaine withdrawal (Dackis et al. 2003). Dose-related effects on inhibition in normal volunteers suggest a positive effect of modafinil in reducing impulsive responding (Turner et al. 2004), an effect noted in a follow-up study in patients with ADHD (Dackis et al. 2005). A double-blind, placebo-controlled trial in 62 cocaine-dependent men and women reported increased cocaine abstinence in the modafinil group (Rugino and Copley 2001). Trials in children and adolescents with ADHD show improvement in ADHD symptoms for modafinil over placebo while demonstrating its safety and tolerability in doses up to 425 mg/day (Biederman et al. 2005). However, modafinil appears to block DATs and increase extracellular dopamine in the human brain, including the nucleus accumbens, as do other drugs with a high potential for dependence (Volkow et al. 2009a, b). Modafinil also can function as a reinforcer and these reinforcing effects are influenced by behavioral demands following drug administration, similar to those of other stimulant drugs (Stoops et al. 2005). These results suggest a heightened awareness for potential abuse of, and dependence on, modafinil in vulnerable populations.

Two recent studies examined the effect of therapeutic stimulant treatment for smoking (tobacco) behavior in adults with ADHD. Gehricke et al. (2011) asked 15 smokers with ADHD to complete a continuous performance task (CPT) and a smoking withdrawal questionnaire during the following four conditions: (1) ADHD medication + cigarette smoking, (2) ADHD medication + overnight abstinence, (3) placebo + cigarette smoking, and (4) placebo + overnight abstinence. Medication consisted of the medication that the subjects had been taking prior to the study and included one on OROS MPH, ten on amphetamines, and four on atomoxetine. During the field monitoring phase, participants were asked to provide salivary cotinine samples and complete electronic diaries about smoking, smoking urge, ADHD symptoms, and stress in everyday life for two days on ADHD medication and for two days on placebo. Results of the experimental phase showed that ADHD medication improved task performance on the CPT and reduced withdrawal during overnight abstinence. During the field monitoring phase, ADHD medication reduced salivary cotinine levels compared to placebo. In addition, the electronic diary revealed that ADHD medication improved difficulty with concentrating during no smoking events and stress. These findings suggest that ADHD medication may be used to aid smoking withdrawal and cessation in smokers with ADHD.

Winhusen et al. (2010) conducted a randomized, double-blind, placebo-controlled, 11-week trial with a 1-month follow-up in adults (aged 18–55 years) meeting DSM-IV criteria for ADHD and interested in quitting smoking. The subjects were randomly assigned to OROS-MPH titrated to 72 mg/d ($n = 127$) or placebo ($n = 128$). All participants received brief weekly individual smoking cessation counseling for 11 weeks and 21 mg/d nicotine patches starting on the smoking quit day (day 27) through study week 11. Eighty percent of the subjects completed the trial. Prolonged abstinence rates, 43.3% and 42.2%, for the OROS-MPH and placebo groups, respectively, did not differ significantly. Relative to placebo, OROS-MPH evidenced a greater reduction in DSM-IV ADHD-RS score ($P < 0.0001$) and in cigarettes per day during the postquit phase ($P = 0.016$). Medication discontinuation did not differ significantly between treatments.

In summary, the literature is mixed, at best, and currently does not support specific, aggressive treatment of ADHD in all SUD treatment populations (Wilens et al. 2005). Clearly, there are significant limitations to the literature including use in chronic populations, choice of agent and formulation, high levels of participant dropout, and possibly inadequate duration of treatment. In adult populations, the common existence of other psychiatric comorbidities such as mood and anxiety disorders with both ADHD or SUDs may limit effectiveness of medications. The relative absence of abuse or worsening of the SUD is reassuring. However, clinicians can rarely show the same degree of monitoring and vigilance as researchers.

8 Guidelines for Management of Patients with Comorbid ADHD-SUD

Among the challenges of managing patients with comorbid ADHD and SUD are: screening/assessment for ADHD in SUD patients: making decisions (which to treat first), compliance issues, and preventing abuse/diversion of prescribed agents.

Among the factors leading to underdiagnosis are that clinicians do not ask about ADHD and inability to recall symptoms prior to age 7. The presence of other psychiatric disorders (e.g., hypomania, depression) may make the diagnosis of ADHD more difficult, depending on the clinician to recall symptoms when carrying out an assessment, and not recognizing that symptoms may be fewer, less obvious, or be compensated for in adolescents or adults.

Given the high prevalence of ADHD-SUD comorbidity, it is critical to consider this comorbidity in the context of an assessment for either ADHD or SUD. For optimal assessment, the clinician should collect detailed substance use history and be particularly vigilant with high-risk groups (adolescents, family SUD history, comorbid CD, or ASPD). Even if criteria for SUD are met, the clinician must be able to differentiate independent symptoms/impairment resulting from ADHD symptoms rather than those arising from the pharmacologic properties of illicit drugs or alcohol, SUD-related problems, or other comorbid psychiatric disorders, such as depression or anxiety. For a more definitive diagnosis, the clinician should collect detailed information about ADHD in childhood, supported whenever possible by objective data sources, and evaluate ADHD symptoms independently of drug effects.

Regarding the decision of which problem to treat first – ADHD or the SUD(s) – the general rule is to administer the treatment that will have the most immediate effect on outcome(s). This is almost always treatment for SUDs, although concurrent treatment is always possible. Once some stabilization in substance use is achieved, additional attention to the diagnosis and treatment of ADHD can proceed.

In terms of the treatment options for comorbid ADHD and SUD, physicians in the United States can select from stimulants and nonstimulants (stimulants include short-, intermediate-, and long-acting MPH: e.g., OROS, bead, or patch technology), d-MPH (in both short- and long-acting formulations), short- and long-acting amphetamine and mixed salts amphetamine. Among the nonstimulants are atomoxetine, guanfacine XR, modafinil, bupropion, venlafaxine, tricyclic antidepressants (e.g., imipramine, nortriptyline) and α_2-adrenoceptor agonists (clonidine, guanfacine). Only the stimulants, atomoxetine and guanfacine XR, are approved by the FDA.

In selecting an appropriate agent for patients with comorbid ADHD and SUD, there are a number of considerations. If stimulants are being considered, the physician must determine if the patient is reliable, if there are family members or close (nonsubstance-abusing) friends involved in the treatment plan, and if patient/family has been adequately informed of potential risks involved in using stimulants. Nonstimulant options should be strongly considered and are preferred in patients with cocaine or stimulant-use disorders. The substance-use history of the patient is also a critical consideration. While past use with an established period of abstinence and/or ongoing SUD treatment prompts less concern, more recent use and/or problems related to substance use, a history of stimulant or amphetamine abuse, may preclude stimulant treatment.

As medication treatment alone is seldom sufficient to treat SUD in the patient with comorbid ADHD and SUD, the clinician should utilize one or more of several evidence-based psychosocial interventions for SUD. These include: CBT, motivational interviewing (MI), contingency management, family therapy, and 12-step (Minnesota Model) approaches (Galanter and Kleber 2008).

A staged approach to pharmacotherapy in patients with comorbid ADHD and SUD, largely based on an assessment of the risk or severity of SUDs, may be useful in making clinical decisions about pharmacotherapy for adolescents or adults with ADHD and SUD (Wilens et al. 2005). Risk is determined by the history and current level of alcohol or other illicit drug use. With increasing risk, the level of substance use intervention also increases as the physician increases the monitoring of drug use (including toxicology) and possible abuse or diversion of the therapeutic agent.

For patients having SUD(s), there are a number of safety issues requiring physician oversight and monitoring, including: overdose potential, the use of prescribed medications with other stimulants, interaction with other nonstimulant drugs, psychoeducation (to make users aware of contraindications and precautions), and, in view of recent changes in ADHD medication package inserts, consideration and screening for cardiovascular risk. The physician or clinician can monitor for red flags, indicating a high suspicion for diversion or misuse, including evidence of: continued abuse or dependence, medication-related emergencies, demands for short or immediate-release stimulant compounds, repeated discordant pill counts, lost prescriptions, continuously escalating doses, infrequent prescription use, cardiovascular symptoms (e.g., palpitations, syncope, shortness of breath), and/or symptoms of psychosis or mania. To manage or prevent problems related to diversion or misuse when using medications with abuse potential, the physician should carefully monitor compliance and apply necessary psychoeducation whenever possible.

9 Summary

ADHD and SUD are a common psychiatric comorbidity among both adults and adolescents. While research has indentified ADHD as a risk factor for the development of SUD and a potential important target for intervention in clinical populations with SUD, future research needs to focus on elucidating the elements of this risk, including the direct role of ADHD in increasing SUD risk, the molecular basis and neuropsychological processes underlying risk, and the possible differential response of individuals with ADHD to various substances of abuse. Finally, despite equivocal results of ADHD-SUD intervention trials, future treatment studies are needed across different ADHD intervention modalities (medications and psychosocial), including different SUD clinical populations, and in combination with specific SUD treatment modalities. Such treatment studies should expand outcomes to include the effect of ADHD intervention on engagement and completion of SUD treatment and to prevent SUD relapse.

References

Advokat C (2007) Update on amphetamine neurotoxicity and its relevance to the treatment of ADHD. J Atten Disord 11:8–16

American Psychiatric Association (2000) Diagnostic and Statistical Manual, Fourth Edition Text Revision (DSM-IV TR). American Psychiatric Publishing, Arlington, VA

August GJ, Winters KC, Realmuto GM, Fahnhorst T, Botzet A, Lee S (2006) Prospective study of adolescent drug use among community samples of ADHD and non-ADHD participants. J Am Acad Child Adolesc Psychiatry 45:824–832

Balstey RL, Schuster CR (1973) Fixed-interval schedule of cocaine reinforcement: effect of dose and infusion duration. J Exp Anal Behav 20:119–129

Barkley RA (2010) Attention-deficit/hyperactivity disorder, executive functioning, and self-regulation. In: Baumeister RF, Vohs KD (eds) Handbook of self-regulation: research, theory, and applications, 2nd edn. Guilford, New York, NY, pp 551–564

Barkley RA, Fischer M, Edelbrock CS, Smallish L (1990) The adolescent outcome of hyperactive children diagnosed by research criteria: I. An 8-year prospective follow-up study. J Am Acad Child Adolesc Psychiatry 29:546–557

Barkley R, Fischer M, Smallish L, Fletcher K (2002) The persistence of attention deficit hyperactivity disorder into young adulthood as a function of reporting source and definition of disorder. J Abnorm Psychol 111:279–289

Bekker EM, Bocker KB, Van Hunsel F, van den Berg MC, Kenemans JL (2005) Acute effects of nicotine on attention and response inhibition. Pharmacol Biochem Behav 82:539–548

Biederman J, Wilens T, Mick E, Milberger S, Spencer TJ, Faraone SV (1995) Psychoactive substance use disorders in adults with attention deficit hyperactivity disorder (ADHD): effects of ADHD and psychiatric comorbidity. Am J Psychiatry 152:1652–1658

Biederman J, Wilens T, Mick E, Faraone SV, Weber W, Curtis S et al (1997) Is ADHD a risk factor for psychoactive substance use disorders? Findings from a four-year prospective follow-up study. J Am Acad Child Adolesc Psychiatry 36:21–29

Biederman J, Swanson JM, Wigal SB, Kratochvil CJ, Boellner SW, Earl CQ et al (2005) Efficacy and safety of modafinil film-coated tablets in children and adolescents with attention-deficit/hyperactivity disorder: results of a randomized, double-blind, placebo-controlled, flexible-dose study. Pediatrics 116:e777–e784

Biederman J, Monuteaux MC, Spencer T, Wilens TE, Macpherson HA, Faraone SV (2008a) Stimulant therapy and risk for subsequent substance use disorders in male adults with ADHD: a naturalistic controlled 10-year follow-up study. Am J Psychiatry 165:597–603

Biederman J, Petty CR, Wilens TE, Fraire MG, Purcell CA, Mick E et al (2008b) Familial risk analyses of attention deficit hyperactivity disorder and substance use disorders. Am J Psychiatry 165:107–115

Brody AL, Mandelkern MA, Olmstead RE, Scheibal D, Hahn E, Shiraga S et al (2006) Gene variants of brain dopamine pathways and smoking-induced dopamine release in the ventral caudate/nucleus accumbens. Arch Gen Psychiatry 63:808–816

Brookes K, Xu X, Chen W, Zhou K, Neale N, Lowe N et al (2006) The analysis of 51 genes in DSM-IV combined type attention deficit hyperactivity disorder: association signals in DRD4, DAT1 and 16 other genes. Mol Psychiatry 1:934–953

Carroll KM, Rounsaville BJ (1993) History and significance of childhood attention deficit disorder in treatment-seeking cocaine abusers. Compr Psychiatry 34:75–82

Castellanos FX, Giedd JN, Marsh WL, Hamburger SD, Vaituzis AC, Dickstein DP et al (1996) Quantitative brain magnetic resonance imaging in attention-deficit hyperactivity disorder. Arch Gen Psychiatry 53:607–616

Chait LD (1994) Reinforcing and subjective effects of methylphenidate in humans. Behav Pharmacol 5:281–288

Chan YF, Dennis ML, Funk RR (2008) Prevalence and comorbidity of major internalizing and externalizing problems among adolescents and adults presenting to substance abuse treatment. J Subst Abuse Treat 34:14–24

Charach A, Yeung E, Climans T, Lillie E (2011) Childhood attention deficit/hyperactivity disorder and future substance use disorders: comparative meta-analyses. J Am Acad Child Adolesc Psychiatry 50:9–21

Coetzee M, Kaminer Y, Morales A (2002) Megadose intranasal methylphenidate (ritalin) abuse in adult attention deficit hyperactivity disorder. Subst Abus 23:165–169

Congdon E, Lesch KP, Canli T (2007) Analysis of DRD4 and DAT polymorphisms and behavioral inhibition in healthy adults: implications for impulsivity. Am J Med Genet B Neuropsychiatr Genet 147B:27–32

Dackis CA, Lynch KG, Yu E, Samaha FF, Kampman KM, Cornish JW et al (2003) Modafinil and cocaine: a double-blind, placebo-controlled drug interaction study. Drug Alcohol Depend 70:29–37

Dackis CA, Kampman KM, Lynch KG, Pettinati HM, O'Brien CP (2005) A double-blind, placebo-controlled trial of modafinil for cocaine dependence. Neuropsychopharmacology 30:205–211

Davidson JT, Connor KM (1998) Bupropion sustained release: a therapeutic overview. J Clin Psychiatry 58(suppl 4):25–31

Daws LC, Callaghan PD, Moron JA, Kahlig KM, Shippenberg TS, Javitch JA et al (2002) Cocaine increases dopamine uptake and cell surface expression of dopamine transporters. Biochem Biophys Res Commun 290:1545–1550

Di Chiara G, Imperato A (1988) Drugs abused by humans preferentially increase synaptic dopamine concentrations in the mesolimbic system of freely moving rats. Proc Natl Acad Sci U S A 85:5274–5278

Elkins IJ, McGue M, Iacono WG (2007) Prospective effects of attention-deficit/hyperactivity disorder, conduct disorder, and sex on adolescent substance use and abuse. Arch Gen Psychiatry 64:1145–1152

Ercan ES, Coskunol H, Varan A, Toksoz K (2003) Childhood attention deficit/hyperactivity disorder and alcohol dependence: a 1-year follow-up. Alcohol 38:352–356

Everitt BJ, Belin D, Economidou D, Robbins TW (2008) Neural mechanisms underlying the vulnerability to develop compulsive drug-seeking habits and addiction. Phil Trans R Soc Lond B Biol Sci 363:3125–3135

Faraone SV, Perlis RH, Doyle AE, Smoller JW, Goralnick JJ, Holmgren MA et al (2005) Molecular genetics of attention-deficit/hyperactivity disorder. Biol Psychiatry 57:1313–1323

Faraone SV, Wilens TE, Petty C, Natshel K, Spencer T, Biederman J (2007a) Substance use among ADHD adults: implications of late onset and subthreshold diagnoses. Am J Addict 16 (S1):24–34

Faraone SV, Biederman J, Wilens TE, Adamsen JJ (2007b) A naturalistic study of the effects of pharmacotherapy on substance use disorders among ADHD adults. Psychol Med 37:1743–1752

Fayyad J, De Graaf R, Kessler R, Slonso J, Angermeyer M, Demyttenaere K et al (2007) Cross-national prevalence and correlates of adult attention-deficit hyperactivity disorder. Br J Psychiatry 190:402–409

Fergusson DM, Horwood LJ, Ridder EM (2007) Conduct and attentional problems in childhood and adolescence and later substance use, abuse and dependence: results of a 25-year longitudinal study. Drug Alcohol Depend 88(Suppl 1):S14–S26

Fillmore MT, Rush CR, Hays L (2006) Acute effects of cocaine in two models of inhibitory control: implications of non-linear dose effects. Addiction 101:1323–1332

Flory K, Milich R, Lynam DR, Leukefeld C, Clayton R (2003) Relation between childhood disruptive behavior disordersand substance use and dependence symptoms in young adulthood: individuals with symptoms of attention-deficit/hyperactivity disorder and conduct disorder are uniquely at risk. Psychol Addict Behav 17:151–158

Fredericks EM, Kollins SH (2004) Assessing methylphenidate preference in ADHD patients using a choice procedure. Psychopharmacology 175:391–398

Galanter M, Kleber HD (eds) (2008) Textbook of substance abuse treatment, 4th edn. American Psychiatric Publishing, Arlington, VA

Galer C, Bouvard MP, Messiah A, Fombonne E (2008) Hyperactivity-inattention symptoms in childhood and substance use in adolescence: the youth gazel cohort. Drug Alcohol Depend 94:30–37

Garavan H, Stout JC (2005) Neurocognitive insights into substance abuse. Trends Cogn Sci 9:195–201

Garavan H, Kaufman JN, Hester R (2008) Acute effects of cocaine on the neurobiology of cognitive control. Phil Trans R Soc Lond B Biol Sci 363:3267–3276

Gehricke JG, Hong N, Wigal TL, Chan V, Doan A (2011) ADHD medication reduces cotinine levels and withdrawal in smokers with ADHD. Pharmacol Biochem Behav 98:485–491

Goksoyr PK, Nottestad JA (2008) The burden of untreated ADHD among adults: the role of stimulant medication. Addict Behav 33:342–346

Goldstein RZ, Tomasi D, Alia-Klein N, Tomasi D, Zhang L, Cottone LA et al (2007) Is decreased prefrontal cortical sensitivity to monetary reward associated with impaired motivation and self-control in cocaine addiction? Am J Psychiatry 164:43–51

Goldstein RZ, Craig AD, Bechara A, Garavan H, Childress AR, Paulus MP et al (2009a) The neurocircuitry of impaired insight in drug addiction. Trends Cogn Sci 13:372–380

Goldstein RZ, Alia-Klein N, Tomasi D et al (2009b) Anterior cingulated cortex hypoactivations to an emotionally salient task in cocaine addiction. Proc NAtl Acad Sci U S A 106:9453

Gordon SM, Tulak F, Troncale J (2004) Prevalence and characteristics of adolescent patients with co-occurring ADHD and substance dependence. J Addict Dis 23:31–40

Griffith JD, Carranza J, Griffith C, Miller LI (1983) Bupropion: clinical essay for amphetamine-like potential. J Clin Psychiatry 44:206–208

Grüsser SM, Wrase J, Klein S, Hermann D, Smolka MN, Ruf M et al (2004) Cue-induced activation of the striatum and medial prefrontal cortex is associated with subsequent relapse in abstinent alcoholics. Psychopharmacology (Berl) 175:296–302

Guan L, Wang B, Chen Y, Yang L, Li J, Qian Q et al (2009) A high-density single-nucleotide polymorphism screen of 23 candidate genes in attention deficit hyperactivity disorder: suggesting multiple susceptibility genes among Chinese Han population. Mol Psychiatry 14:546–554

Guindalini C, Howard M, Haddley K, Lanaranjeira R, Collier D, Ammar N et al (2006) A dopamine transporter gene functional variant associated with cocaine abuse in a Brazilian sample. Proc Natl Acad Sci U S A 103:4552–4557

Hamarman S, Fossella J, Ulger C (2004) Dopamine receptor 4 (DRD4) 7-repeat allele predicts methylphenidate dose response in children with attention deficit hyperactivity disorder: a pharmacogenetic study. J Child Adolesc Psychopharmacol 14:564–574

Hatsukami D, Fletcher L, Morgan S, Keenan R, Amble P (1989) The effects of varying cigarette deprivation duration on cognitive and performance tasks. J Subst Abuse 1:407–416

Heil SH, Holmes HW, Bickel WK, Higgins ST, Badger GJ, Laws HF et al (2002) Comparison of the subjective, physiological, and psychomotor effects of atomoxetine and methylphenidate in light drug users. Drug Alcohol Depend 67:149–156

Hildebrand BE, Nomikos GG, Hertel P, Schilstrom B, Svensson TH (1998) Reduced dopamine output in the nucleus accumbens but not in the medial prefrontal cortex in rats displaying a mecamylamine-precipitated nicotine withdrawal syndrome. Brain Res 779:214–225

Hoza B, Gerdes AC, Hinshaw SP et al (2004) Self-perceptions of competence in children with ADHD and comparison children. J Consult Clin Psychol 72:382–391

Hurt RD, Sachs PL, Glover ED, Offord KP, Johnston JA, Dale LC et al (1997) A comparison of sustained-release bupropion and placebo for smoking cessation. N Engl J Med 337:1195–1202

Katusic SK, Barbaresi WJ, Colligan RC, Weaver AL, Leibson CL, Jacobsen SJ (2005) Psychostimulant treatment and risk for substance abuse among young adults with a history of attention-deficit/hyperactivity disorder: a population-based, birth cohort study. J Child Adolesc Psychopharmacol 15:764–776

Ken L, Greene E, Holmes J, Thapar A, Gill M, Hawi Z et al (2001) No association between CHRNA7 microsatellite markers and attention-deficit hyperactivity disorder. Am J Med Genet 105:686–689

Kessler RC, Adler L, Barkley R, Biederman J, Conners CK, Demler O et al (2006) The prevalence and correlates of adult ADHD in the United States: results from the National Comorbidity Survey Replication. Am J Psychiatry 163:716–723

Kollins SH, Rush CR, Pazzaglia PJ, Ali JA (1998) Comparison of acute behavioral effects of sustained-release and immediate-release methylphenidate. Exp Clin Psychopharmacol 6:367–374

Kollins SH, MacDonald EK, Rush CR (2001) Assessing the abuse potential of methylphenidate in nonhuman and human subjects: a review. Pharmacol Biochem Behav 68:611–627

Kollins SH, McClernon FJ, Fuemmeler BF (2005) Association between smoking and attention-deficit/hyperactivity disorder symptoms in a population-based sample of young adults. Arch Gen Psychiatry 62:1142–1147

Kollins SH, English J, Robinson R, Hallyburton M, Chrisman AK (2009) Reinforcing and subjective effects of methylphenidate in adults with and without attention deficit hyperactivity disorder (ADHD). Psychopharmacology 204:73–83

Koob GF, Nestler EJ (1997) The neurobiology of drug addiction. J Neuropsychiatry Clin Neurosci 9:482–497

Lambert NM, Hartsough CS (1998) Prospective study of tobacco smoking and substance dependencies among samples of ADHD and non-ADHD participants. J Learn Disabil 31:533–544

Lee J, Laurin N, Crosbie J, Ickowicz A, Pathare T, Malone M et al (2008) Association study of the nicotinic acetylcholine receptor alpha4 subunit gene, CHRNA4, in attention-deficit hyperactivity disorder. Genes Brain Behav 7:53–60

Lee SS, Humphreys KL, Flory K, Liu R, Glass K (2011) Prospective association of childhood attention-deficit/hyperactivity disorder (ADHD) and substance use and abuse/dependence: a meta-analytic review. Clin Psychol Rev 31:328–341

Levin FR, Evans SM, Brooks DJ, Kalbag AP, Garawi F, Nunes EV (2006) Treatment of methadone-maintained patients with adult ADHD: double-blind comparison of methylphenidate, bupropion, and placebo. Drug Alcohol Depend 81:137–148

Levin FR, Evans SM, Brooks DJ, Garawi F (2007) Treatment of cocaine dependent treatment seekers with adult ADHD: double-blind comparison of methylphenidate and placebo. Drug Alcohol Depend 87:20–29

Li MD, Ma JZ, Beuten J (2004) Progress in searching for susceptibility loci and genes for smoking-related behaviour. Clin Genet 66:382–392

MacDonald FE, Kollins SH (2005) A pilot study of methylphenidate preference assessment in children diagnosed with attention-deficit/hyperactivity disorder. J Child Adolesc Psychopharmacol 15:729–741

Maher BS, Marazita ML, Ferrell RE et al (2002) Dopamine system genes and attention deficit hyperactivity disorder: a meta-analysis. Psychiatr Genet 12:207–215

Mannuzza S, Klein RG, Bonagura N, Malloy P, Giampino TL, Addalli KA (1991) Hyperactive boys almost grown up. V. Replication of psychiatric status. Arch Gen Psychiatry 48:77–83

Mannuzza S, Klein RG, Truong NL, Moulton JL, Roizen ER, Howell KH et al (2008) Age of methylphenidate treatment initiation in children with ADHD and later substance abuse: prospective follow-up into adulthood. Am J Psychiatry 165:604–609

McCabe SE, Teter CJ, Boyd CJ, Guthrie SK (2004) Prevalence and correlates of illicit methylphenidate use among 8th, 10th, and 12th grade students in the United States, 2001. J Adolesc Health 35:501–504

McCabe SE, Teter CJ, Boyd CJ (2006) Medical use, illicit use, and diversion of prescription stimulant medication. J Psych Drugs 38:43–56

McClernon FJ, Kollins SH (2008) ADHD and smoking: from genes to brain to behavior. Ann NY Acad Sci 1141:131–147

McClernon FJ, Fuemmeler BF, Kollins SH, Kail ME, Ashley-Koch AE (2008a) Interactions between genotype and retrospective ADHD symptoms predict lifetime smoking risk in a sample of young adults. Nicotine Tob Res 10:117–127

McClernon FJ, Kollins SH, Lutz AM, Fitzgerald DP, Murray DW, Redman C et al (2008b) Effects of smoking abstinence on adult smokers with and without attention deficit hyperactivity disorder: results of a preliminary study. Psychopharmacology 197:95–105

McGough JJ, Smalley SL, McCracken JT, Yang M, Del'Homme M, Lynn DE et al (2005) Psychiatric comorbidity in adult attention deficit hyperactivity disorder: findings from multiplex families. Am J Psychiatry 162:1621–1627

Michelson D, Adler L, Spencer T, Reimherr FW, West SA, Allen AJ et al (2003) Atomoxetine in adults with ADHD: two randomized, placebo-controlled studies. Biol Psychiatry 53:112–120

Milberger S, Biederman J, Faraone SV, Chen L, Jones J (1997a) ADHD is associated with early initiation of cigarette smoking in children and adolescents. J Am Acad Child Adolesc Psychiatry 36:37–44

Milberger S, Biederman J, Faraone SV, Wilens T, Chu MP (1997b) Associations between ADHD and psychoactive substance use disorders. Findings from a longitudinal study of high-risk siblings of ADHD children. Am J Addict 6:318–329

Miller EK, Cohen JD (2001) An integrative theory of prefrontal cortex function. Annu Rev Neurosci 24:167–202

Molina BS, Pelham WE (2003) Childhood predictors of adolescent substance use in a longitudinal study of children with ADHD. J Abnorm Psychol 112:497–507

Molina BS, Flory K, Hinshaw SP, Greiner AR, Arnold LE, Swanson JM et al (2007a) Delinquent behavior and emerging substance use in the MTA at 36 months: prevalence, course, and treatment effects. J Am Acad Child Adolesc Psychiatry 46:1028–1040

Molina BS, Pelham WE, Gnagy EM (2007b) Attention-deficit/hyperactivity disorder risk for heavy drinking and alcohol use disorder is age specific. Alcohol Clin Exp Res 31:643–654

Munafo M, Clark T, Johnstone E, Murphy M, Walton R (2004) The genetic basis for smoking behavior: a systematic review and meta-analysis. Nicotine Tob Res 6:583–597

Novak SP, Kroutil LA, Williams RL, Van Brunt DL (2007) The nonmedical use of prescription ADHD medications: results from a national Internet panel. Subst Abuse Treat Prev Policy 2:32

Owens J, Goldfine M, Evangelista N, Hoza B, Kaiser NM (2007) A critical review of self-perceptions and the positive illusory bias in children with ADHD. Clin Child Fam Psychol Rev 10:335–351

Paulus MP, Tapert SF, Schuckit MA (2005) Neural activation patterns of methamphetamine-dependent subjects during decision making predict relapse. Arch Gen Psychiatry 62:761–768

Pomerleau OF, Downey KK, Stelson FW, Pomerleau CS (1995) Cigarette smoking in adult patients diagnosed with attention deficit hyperactivity disorder. J Subst Abuse 7:373–378

Poulin C (2001) Medical and nonmedical stimulant use among adolescents: from sanctioned to unsanctioned use. Can Med Assoc J 165:1039–1044

Powell J, Dawkins L, Davis RE (2002) Smoking, reward responsiveness, and response inhibition: tests of an incentive motivational model. Biol Psychiatry 51:151–163

Rahman S, Zhang J, Engleman EA, Corrigall WA (2004) Neuroadaptive changes in the mesoaccumbens dopamine system after chronic nicotine self-administration: a microdialysis study. Neuroscience 129:415–424

Rapoport JL, Buchsbaum MS, Weingartner H, Zahn TP, Ludlow C, Mikkelsen EJ (1980) Dextroamphetamine. Its cognitive and behavioral effects in normal and hyperactive boys and normal men. Arch Gen Psychiatry 37:933–943

Renthal W, Nestler EJ (2008) Epigenetic mechanisms in drug addiction. Trends Mol Med 14:341–350

Resnick RB, Kestenbaum RS, Schwartz LK (1977) Acute systemic effects of cocaine in man: a controlled study by intranasal and intravenous routes. Science 195:696–698

Riggs P (2009) Multi-site study of OROS-MPH for ADHD in substance-abusing adolescents. In: Scientific Proceedings of the 56th Annual Meeting of the American Academy of Child and Adolescent Psychiatry. Honolulu, HI, pp 94–95

Riggs PD, Hall SK, Mikulich-Gilbertson SK, Lohman M, Kayser A (2004) A randomized controlled trial of pemoline for attention-deficit/hyperactivity disorder in substance-abusing adolescents. J Am Acad Child Adolesc Psychiatry 43:420–429

Rinn W, Desai N, Rosenblatt H, Gastfriend DR (2002) Addiction denial and cognitive dysfunction: a preliminary investigation. J Neuropsychiatry Clin Neurosci 14:52–57

Roehrs T, Papineau K, Rosenthal L, Roth T (1999) Sleepiness and the reinforcing and subjective effects of methylphenidate. Exp Clin Psychopharmacol 7:145–150

Rohde P, Kahler CW, Lewinsohn PM, Brown RA (2004) Psychiatric disorders, familial factors, and cigarette smoking: II. Associations with progression to daily smoking. Nicotine Tob Res 6:119–132

Rohsenow DJ, Monti PM, Rubonis AV, Sirota AD, Niaura RS, Colby SM et al (1994) Cue reactivity as a predictor of drinking among male alcoholics. J Consult Clin Psychol 62:620–626

Rugino TA, Copley TC (2001) Effects of modafinil in children with attention-deficit/hyperactivity disorder: an open-label study. J Am Acad Child Adolesc Psychiatry 40:230–235

Rush CR, Kollins SH, Pazzaglia PJ (1998) Discriminative stimulus and participant-rated effects of methylphenidate, bupropion, and triazolam in d-amphetamine-trained humans. Exp Clin Psychopharmacol 6:32–44

Sahakian B, Morein-Zamir SC (2007) Cognitive enhancement: Professor's little helper. Nature 450:1157–1159

Schiffer WK, Volkow ND, Fowler JS, Alexoff DL, Logan J, Dewey SL (2006) Therapeutic doses of amphetamine or methylphenidate differentially increase synaptic and extracellular dopamine. Synapse 59:243–251

Schubiner H, Saules KK, Arfken CL, Johanson CE, Schuster CR, Lockhart N et al (2002) Double-blind placebo-controlled trial of methylphenidate in the treatment of ADHD patients with comorbid cocaine dependence. Exp Clin Psychopharmacol 10:286–294

Shao C, Li Y, Jiang K, Zhang D, Xu Y, Lin L et al (2006) Dopamine D4 receptor polymorphism modulates cue-elicited heroin craving in Chinese. Psychopharmacology 186:185–190

Solanto MV (1998) Neuropsychopharmacological mechanisms of stimulant drug action in attention-deficit hyperactivity disorder: a review and integration. Behav Brain Res 94:127–152

Spencer TJ, Biederman J, Ciccone PE, Madras BK, Dougherty DD, Bonab AA et al (2006) PET study examining pharmacokinetics, detection and likeability, and dopamine transporter receptor occupancy of short- and long-acting oral methylphenidate. Am J Psychiatry 163:387–395

Stoops WW, Lile JA, Fillmore MT, Glaser PE, Rush CR (2005) Reinforcing effects of modafinil: influence of dose and behavioral demands following drug administration. Psychopharmacology 182:186–193

Swanson JM, Volkow ND (2009) Psychopharmacology: concepts and opinions about the use of stimulant medications. J Child Psych Psychiatry 50:180–193

Szobot CM, Rohde LA, Bukstein O, Molina BSG, Martins C, Ruaro P et al (2007) Is attention-deficit/hyperactivity disorder associated with illicit substance use disorders in male adolescents? A community-based case-control study. Addiction 102:1122–1130

Szobot CM, Shih MC, Schaefer T, Júnior N, Hoexter MQ, Kai Fuet Y et al (2008a) Methylphenidate DAT binding in adolescents with attention-deficit/hyperactivity disorder comorbid with Substance Use Disorder–a single photon emission computed tomography with [Tc$^{(99m)}$] TRODAT-1 study. NeuroImage 40:1195–1201

Szobot CM, Rohde LA, Katz B et al (2008b) A randomized crossover clinical study showing that methylphenidate-SODAS improves attention-deficit/hyperactivity disorder symptoms in adolescents with substance use disorder. Braz J Med Biol Res 41:250–257

Szobot CM, Roman T, Hutz JP, Shih MC, Hoexter MQ, Junior N et al (2011) Molecular imaging genetics of methylphenidate response in ADHD and substance use comorbidity. Synapse 65:154–159

Thurstone C, Riggs PD, Salomonsen-Sautel S, Mikulich-Gilbertson SK (2010) Randomized, controlled trial of atomoxetine for attention-deficit/hyperactivity disorder in adolescents with substance use disorder. J Am Acad Child Adolesc Psychiatry 49:573–582

Tiffany ST (1990) A cognitive model of drug urges and drug-use behavior: role of automatic and nonautomatic processes. Psychol Res 97:147–168

Todd RD, Lobos EA, Sun LW, Neuman RJ (2003) Mutational analysis of the nicotinic acetylcholine receptor alpha 4 subunit gene in attention deficit/hyperactivity disorder: evidence for association of an intronic polymorphism with attention problems. Mol Psychiatry 8:103–108

Todd RD, Huang H, Smalley SL, Nelson SF, Willcutt BF, Pennington BF et al (2005) Collaborative analysis of DRD4 and DAT genotypes in population-defined ADHD subtypes. J Child Psychol 46:1067–1073

Turner DC, Clark L, Dowson J, Robbins TW, Sahakian BJ (2004) Modafinil improves cognition and response inhibition in adult attention deficit/hyperactivity disorder. Biol Psychiatry 55:1031–1040

Volkow ND (2006) Stimulant medications: how to minimize their reinforcing effects? Am J Psychiatry 163:359–361

Volkow ND, Swanson JM (2003) Variables that affect the clinical use and abuse of methylphenidate in the treatment of ADHD. Am J Psychiatry 160:1909–1918

Volkow ND, Swanson JM (2008) Does childhood treatment of ADHD with stimulant medication affect substance abuse in adulthood? Am J Psychiatry 165:553–555

Volkow ND, Ding YS, Fowler JS, Wang GJ, Logan J, Gatley JS et al (1995) Is methylphenidate like cocaine? Studies on their pharmacokinetics and distribution in the human brain. Arch Gen Psychiatry 52:456–463

Volkow ND, Fowler JS, Wang GJ, Ding YS, Gatley SJ (2002) Role of dopamine in the therapeutic and reinforcing effects of methylphenidate in humans: results from imaging studies. Eur Neuropsychopharmacol 12:557–566

Volkow ND, Fowler JS, Wang G-J (2003) The addicted human brain: insights from imaging studies. J Clin Invest 111:1444–1451

Volkow ND, Fowler JS, Wang G-J (2004) The addicted human brain viewed in the light of imaging studies: brain circuits and treatment strategies. Neuropharmacology 47(Suppl 1):3–13

Volkow ND, Wang G-J, Newcorn J, Telang F, Solanto MV, Fowler JS et al (2007) Depressed dopamine activity in caudate and preliminary evidence of limbic involvement in adults with attention-deficit/hyperactivity disorder. Arch Gen Psychiatry 64:932–940

Volkow ND, Wang G-J, Kollins S, Wigal TM, Newcorn JH, Telang F et al (2009a) Evaluating dopamine reward pathway in ADHD: clinical implications. JAMA 302:1084–1091

Volkow ND, Fowler JS, Logan J, Alexoff D, Zhu W, Telang F et al (2009b) Effects of modafinil on dopamine and dopamine transporters in the male human brain: clinical implications. JAMA 301:1148–1154

White AM, Jordan JD, Schroeder KM, Acheson SK, Georgi BD, Sauls G et al (2004) Predictors of relapse during treatment and treatment completion among marijuana-dependent adolescents in an intensive outpatient substance abuse program. Subst Abus 25:53–59

WHO (2010) www.who.int/substnce_abuse/terminology/who_lexicon/en/index.html

Wilens TE (2007) The nature of the relationship between attention-deficit/hyperactivity disorder and substance use. J Clin Psychiatry 68(Suppl 1):4–8

Wilens TE, Biederman J, Millstein RB, Wozniak J, Hahesy AL, Spencer TJ (1999) Risk for substance use disorders in youths with child- and adolescent-onset bipolar disorder. J Am Acad Child Adolesc Psychiatry 38:680–685

Wilens TE, Monuteaux MC, Snyder LE, Moore H (2005) The clinical dilemma of using medications in substance-abusing adolescents and adults with attention-deficit/hyperactivity disorder: what does the literature tell us? J Child Adolesc Psychopharmacol 15:787–798

Wilens TE, Gignac M, Swezey A, Monuteaux M, Biederman J (2006) Characteristics of adolescents and young adults with ADHD who divert or misuse their prescribed medications. J Am Acad Child Adolesc Psychiatry 45:408–414

Wilens TE, Adler LA, Adams J et al (2008a) Misuse and diversion of stimulants prescribed for ADHD: a systematic review of the literature. J Am Acad Child Adolesc Psychiatry 47:21–31

Wilens TE, Adamson J, Monuteaux MC, Faraone SV, Schillinger M, Westerberg D et al (2008b) Effect of prior stimulant treatment for attention-deficit/hyperactivity disorder on subsequent risk for cigarette smoking and alcohol and drug use disorders in adolescents. Arch Pediatr Adolesc Med 162:916–921

Wilens TE, Adler LA, Weiss MD, Michelson D, Ramsey JL, Moore RJ et al (2008c) Atomoxetine treatment of adults with ADHD and comorbid alcohol use disorders. Drug Alcohol Depend 96:145–154

Winhusen TM, Somoza EC, Brigham GS, Liu DS, Green CA, Covey LS et al (2010) Impact of attention-deficit/hyperactivity disorder (ADHD) treatment on smoking cessation intervention in ADHD smokers: a randomized, double-blind, placebo-controlled trial. Negat Results Biomed 71:1680–1688

Zack M, Belsito L, Scher R, Eissenberg T, Corrigall WA (2001) Effects of abstinence and smoking on information processing in adolescent smokers. Psychopharmacology (Berl) 153:249–257

Linking ADHD, Impulsivity, and Drug Abuse: A Neuropsychological Perspective

Gonzalo P. Urcelay and Jeffrey W. Dalley

Contents

1 Linking ADHD, Impulsivity, and Drug Abuse: A Neuropsychological Perspective 174
2 Evidence for Executive Dysfunction in Drug Addiction 175
3 Dimensions of Impulsivity 176
4 Executive Dysfunction and ADHD Heterogeneity: Beyond a Single "Core" Deficit 178
5 Neural Substrates of Impulsivity 180
6 Modeling Vulnerability to Drug Addiction 182
7 Impulsivity and Choice Behavior: A Psychological Perspective 183
8 The Hippocampus, Environmental Stimuli, and Delayed Rewards 185
9 Linking Impulsivity, Executive Dysfunction, and Propensity to Drug Addiction 190
10 Conclusions 191
References 192

Abstract In this chapter, we consider the relevance of impulsivity as both a psychological construct and endophenotype underlying attention-deficit/hyperactivity disorder (ADHD) and drug addiction. The case for executive dysfunction in ADHD and drug addiction is critically reviewed in the context of dissociable cognitive control processes mediated by the dorsolateral prefrontal cortex (DLPFC), the orbital and ventral medial prefrontal cortex (VMPFC). We argue that such neuroanatomical divisions within the prefrontal cortex are likely to account for the multidimensional basis of impulsivity conceptually categorized in terms of "motoric" and "choice" impulsivity. The relevance of this distinction for the etiology of ADHD and drug addiction is integrated within a novel theoretical framework. This scheme embraces animal learning theory to help explain the heterogeneity of impulse control disorders, which are exemplified by ADHD as a vulnerability disorder for drug addiction.

G.P. Urcelay
Behavioural and Clinical Neuroscience Institute, Department of Experimental Psychology, University of Cambridge, Downing St., Cambridge CB2 3EB, UK

J.W. Dalley (✉)
Behavioural and Clinical Neuroscience Institute, Department of Experimental Psychology & Department of Psychiatry, University of Cambridge, Downing St., Cambridge CB2 3EB, UK
e-mail: jwd20@cam.ac.uk

Keywords Addiction · ADHD · Basal ganglia · Discounting · Dopamine · Hot and cold cognition · Impulsivity · Learning · Limbic system · Motor inhibition · Prefrontal cortex

Abbreviations

5-CSRTT	5-Choice serial reaction time task
5-HT	5-Hydroxytryptamine (serotonin)
ADHD	Attention deficit hyperactivity disorder
CR	Conditioned reinforcer
DA	Dopamine
DAT	Dopamine transporter
DLPFC	Dorsolateral prefrontal cortex
DRL	Differential response to low rates of reinforcement
fMRI	Functional magnetic resonance imaging
IFG	Inferior frontal gyrus
IL	Infralimbic (cortex)
mPFC	Medial prefrontal cortex
NAcb	Nucleus accumbens
O	Outcome
OFC	Orbitofrontal cortex
PET	Positron emission tomography
PIT	Pavlovian to instrumental transfer
R	Response
R_D	Response followed by a delayed outcome
R_I	Response followed by an immediate outcome
S	Stimulus
SN	Substantia nigra
SSRT	Stop-signal reaction time
SST	Stop-signal task
TMS	Transcranial magnetic stimulation
VMPFC	Ventromedial prefrontal cortex
VTA	Ventral tegmental area

1 Linking ADHD, Impulsivity, and Drug Abuse: A Neuropsychological Perspective

Attention-deficit/hyperactivity disorder (ADHD) is a prevalent neurodevelopmental brain disorder that manifests early in life and is characterized by inattentiveness, hyperactivity, and impulsivity (American Psychiatric Association 2000; Solanto et al. 2001; Castellanos and Tannock 2002). Despite increasing research interest in ADHD, the neurobiological etiology of this brain disorder remains largely unknown. Based on the strong clinical parallels between ADHD and frontal lobe patients, ADHD has been conceptualized as a disorder of executive dysfunction (Barkley 1997), specifically mediated by deficient behavioral inhibition. In contrast, others have argued for alternative accounts, including an altered sensitivity to reinforcement (Sonuga-Barke 2003). However, it remains unclear whether impaired executive function in ADHD is a consequence of altered reinforcer sensitivity or whether these seemingly divergent, causal mechanisms co-exist and thereby develop simultaneously.

The psychological construct of impulsivity, which features prominently in ADHD, can take many forms including intolerance to delayed gratification or an inability to stop an already initiated motor response. Impulsivity has long been associated with addiction and substance use disorders (reviewed by Verdejo-García et al. 2008), but in ways that are poorly understood. The origin of impulsivity in addicted individuals (e.g., see Kirby and Petry 2004) is hypothesized to arise from two, potentially interacting, mechanisms: either as a consequence of chronic drug abuse, or as a preexisting behavioral trait that predisposes individuals to harmful drug use (Perry and Carroll 2008). Indeed, interactions between predispositions and drug exposure are believed to make an important contribution to deteriorating behavioral outcomes (de Wit 2009). In this chapter, we review the available evidence for "top-down" executive dysfunction in ADHD and drug addicts, alike, and discuss the putative role of impulsivity phenotypes in the etiology of these disorders. Finally, we highlight the challenge of explaining how environmental variables can interact with predispositions to influence the emergence of complex syndromes associated with impaired behavioral self-control.

2 Evidence for Executive Dysfunction in Drug Addiction

Executive functioning is an umbrella term used to define several components of cognitive control that help optimize performance in complex task settings. It includes distinct cognitive processes, such as: attentional selection, monitoring, behavioral inhibition, task switching, planning, and decision making (Dalley et al. 2004; Fuster 2000; Miller 2000). Although there is no consensus on the taxonomy of executive functions (Roberts et al. 1998), there is general agreement that the concept of executive functions refers to several dissociable processes that can be fractionated according to neuroanatomical divisions of the frontal cortex (Robbins 1996).

Chronic drug users display deficits in multiple tasks used to assess cognitive and motor control. These deficits may increase the likelihood of seeking drugs and impede the progression of rehabilitation programs (Rogers and Robbins 2001). Neuropsychological tests suggest that chronic stimulant users show profound deficits in tasks of visuomotor performance, attention, and verbal memory (reviewed in Bolla et al. 1998; Rogers and Robbins 2001). For example, stimulant abusers show a pattern of impairment, which is similar to that of patients with focal lesions of the orbitofrontal cortex (OFC) and is characterized by longer decision times and risky choices (choices for the least probable outcome). Moreover, in stimulant-dependent addicts, risky choice was correlated with years of abuse (Rogers et al. 1999; but see Verdejo-García et al. 2008). This pattern of choice was mimicked by central serotonin (5-HT) depletion, suggesting that monoaminergic dysfunction may underlie cognitive impairment in drug addicts (Rogers et al. 1999; also see Verdejo-García et al. 2008). A follow-up study found further evidence for frontal-executive dysfunction in chronic stimulant and opiate abusers. This assessment suggested a dissociation between amphetamine and opiate chronic users, with the former showing an impairment in extradimensional attentional shifts and the latter showing an impairment in intradimensional shifts, paralleled by learning deficits and impairments in visuospatial tasks (Ornstein et al. 2000).

Deficits in executive functions in addicts are hypothesized to be caused by alterations in a network that includes the prefrontal cortex (PFC), limbic system (including the hippocampus and amygdala), basal ganglia (ventral and dorsal striatum), and the thalamus (Everitt and Robbins 2005). This neural circuit can be segregated into two clear interacting, apparently competitive systems: a "bottom-up" system, reflecting the influence of environmental stimuli, which trigger cognitive (i.e., cravings) and behavioral (i.e., motoric) responses, and a competing "top-down" control system, which maintains online representations of long-term goals. Deficits in "executive functions" in drug addicts could thus reflect heightened responsiveness to environmental variables, or a lack of control exerted by the PFC (Belin et al. 2009; Jentsch and Taylor 1999).

3 Dimensions of Impulsivity

Impulsivity is a complex, multidimensional psychological construct (Bari et al. 2011; Evenden 1999) involving a maladaptive tendency for fast, unplanned behavior with little foresight on the consequences of such behavior (Moeller et al. 2001). Clearly, impulsivity can sometimes be advantageous but is more likely to result in detrimental outcomes. As defined above, therefore, "impulsivity" is distinct from "compulsivity", which refers to a persistent tendency to behave in a repetitive, habitual way despite the adverse consequences of such behavior (Fineberg et al. 2010). This distinction is relevant to the emergence of drug addiction, specifically as the endpoint of a series of transitions from impulsivity to compulsivity. At this point, subjects no longer exert volitional control over drug intake but progressively come to seek and take drugs compulsively, despite mounting adverse consequences (Everitt and Robbins 2005).

Increasingly, impulsivity is categorized into two broad domains: *impulsive action* and *impulsive choice* (Winstanley et al. 2006). The first has a clear motoric component and is defined as an inability to withhold from making a response that has already been initiated. A sensitive test of impulsive action is the stop-signal task (SST) developed by Logan and collaborators (Logan 1994). In the SST, subjects are asked to respond as quickly as they can to a "go" signal, but to inhibit their response when presented with a "stop" signal. The intervals between the onsets of the go and stop signals are varied to allow an estimation of the "stop-signal reaction time" (SSRT). SSRT is calculated (in milliseconds) by subtracting the mean Stop-signal delay (the interval between the presentation of the Go signal and the Stop signal) from the mean reaction time to the Go signal. In a rodent version of this task, rats or mice are required to make a rapid lever press (R_1) followed by a second response on an alternative lever (R_2) to obtain a reward (i.e., a sucrose pellet). In a small subset of trials, a signal (the stop signal) is presented after the R_1 has been made and subjects are required to refrain from responding to receive a reward. In these so-called stop trials, the signal is presented when R_2 has already been initiated and the rodents must withhold from responding (Eagle and Robbins 2003). In fact, this task is a refined version of the conventional Go/no-Go task in which a stop signal is presented after subjects have initiated a response to a "go" signal (Hester and Garavan 2004).

The second category of impulsive behavior involves making a choice between two responses presented simultaneously (R_1 and R_2). One of these choices is followed by an immediate, small reward and the other by a delayed but larger reward. In other words, subjects are presented with the opportunity to choose between a small immediate reward and a larger delayed reward. Self-controlled subjects choose the larger delayed reward and maximize total gains, whereas impulsive subjects choose the small, immediate reward (Ainslie 1975).

The distinction between *impulsive choice* and *impulsive action* is important in the context of executive functioning in ADHD and drug addiction because it provides an opportunity to elucidate the underlying etiology and behavioral mechanisms of such disorders. Implicit in this assumption is the notion that intermediate markers (endophenotypes) can help identify distinct components of clinical syndromes that are assumed to be polygenic in origin (Gottesman and Gould 2003). For example, the distinction between diverse forms of impulsivity and their behavioral characterisation can help to differentiate whether common manifestations of impulsivity are observed in addiction, ADHD, and other disorders within the general category of impulse control disorders (American Psychiatric Association 2000).

Previously, it has been shown that childhood ADHD is accompanied by steepened reward discounting functions (Sonuga-Barke 1994; Sonuga-Barke et al. 1992). However, this early research also showed that the ability to detect differences between ADHD and control participants depended on the choice of parameters used in the task. In particular, when the task was configured in such a way that choosing the delayed reward would increase the total time to complete the task, ADHD individuals opted for the immediate reward and thus showed steepened delayed discounting functions. However, if the task was adjusted so that choosing the immediate reward was followed by a longer waiting time until the next trial was initiated, these differences disappeared (Sonuga-Barke et al. 1992). As a consequence, the difference between ADHD individuals and controls normalized when the choice of immediate reward did not decrease the total time taken to complete the task. This finding led Sonuga-Barke (1994) to propose the Delay

Aversion Hypothesis: namely, that ADHD individuals tend to opt for immediate rewards because their perception of passage of time is longer and so is more aversive than for control subjects. Williams (2010) recently extended this idea by suggesting that such deficits reflect failures in the adjustment of the perception of passage of time in ADHD individuals, which are hypothesized to reflect constant changes in minor internal states (e.g., sleepiness, frustration).

Not only is impulsivity a defining characteristic of ADHD, but it is also prevalent in drug addiction (Verdejo-García et al. 2008). Choices made by addicted individuals in real life resemble those made by impulsive subjects in choice impulsivity tasks. Drug addicts display a tendency to choose immediate rewards (i.e., the pleasure produced by a drug "high") over the long-term benefits of a healthy life. These choices eventually conflict with the subject's everyday activities and have long-term detrimental effects on the subject's well-being. Studies using animal models, together with functional brain imaging techniques, reveal several key regions that are compromised in drug addiction, including the PFC, OFC, and striatum (Everitt et al. 2008; Kalivas and O'Brien 2008; Koob and Volkow 2010; Nestler 2001). Dysfunction in such regions is likely to underlie poor decision-making and persistent drug-seeking behavior in human addicts (Verdejo-García et al. 2008).

4 Executive Dysfunction and ADHD Heterogeneity: Beyond a Single "Core" Deficit

Executive dysfunction has been documented extensively in children with ADHD (Barkley 1997; Castellanos et al. 2006; Lijffijt et al. 2005) and in drug addicts (Bolla et al. 1998; Rogers and Robbins 2001). Findings in ADHD subjects led Barkley (1997) to propose that deficient behavioral inhibition (i.e., motor impulsivity) is a core deficit in individuals with ADHD. However, as mentioned above, ADHD has also been associated with increased choice impulsivity, or an inability to wait for delayed rewards, which has been linked with suboptimal reward processing (Luman et al. 2010). This suggests that the neuropsychological profile of ADHD may be more heterogeneous than previously thought, which challenges the assumption of a single core deficit in ADHD (Castellanos et al. 2005, 2006; Sonuga-Barke 2003, 2005).

A major determinant for this change in direction was a comprehensive study in which ADHD individuals and controls were tested in both choice and motor impulsivity tasks (Solanto et al. 2001; see also Lambek et al. 2010). This study found no association between motor and choice impulsivity in children with ADHD, suggesting that the contribution of each deficit to ADHD diagnosis was somewhat independent. This suggested that, consistent with diagnostic criteria, clinical heterogeneity can obscure identification of putative markers that are key to treating ADHD. This study thus reinforced the notion that dissociable domains of ADHD may help us to understand underlying neuropsychological processes that are based on functionally specialized subregions of the frontal cortex. In fact, studies have replicated these findings (Lambek et al. 2010) and also revealed three dissociable processes that account for ADHD: two are the forms of impulsivity described above, and the third that is captured by timing deficits (Sonuga-Barke et al. 2010).

Castellanos et al. (2006) proposed that each of the main factors contributing to ADHD symptomatology (i.e., inattention/hyperactivity and impulsivity) represent a failure in executive functions related to sensory/motoric (i.e., "cold") and emotional (i.e., "hot") processing, respectively. Thus, cortical regions implicated in abstract and sensory cognitive processing (e.g., DLPFC) are considered "cold" regions, whereas regions involved in emotional processing (orbital and medial prefrontal cortex VMPFC) are deemed "hot" regions (Roiser et al. 2009). This taxonomy is hypothesized to depend on cognitive sensory/motoric and limbic regions that project differentially to the DLPFC and VMFC (see Fig. 1). Because ADHD individuals display alterations in reward-learning related to impulsive-choice (which reflects ventral striatum-VMPFC dysfunction) and also in motor impulsivity

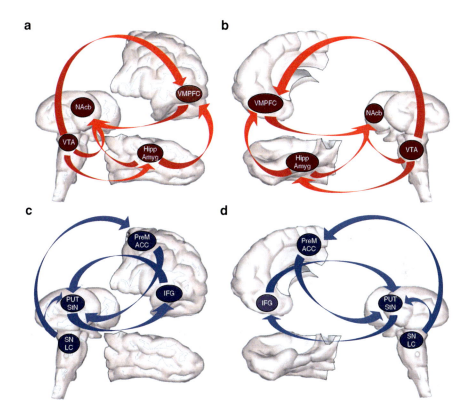

Fig. 1 Schematic depiction of hot (**a**, **b**) and cold (**c**, **d**) neural circuitry mediating emotional and sensoriomotor processing, respectively. *VTA* ventral tegmental area, *NAcb* nucleus accumbens, *VMPFC* ventromedial prefrontal cortex, *Hipp* hippocampus; *Amyg* amygdala, *SN* substantia nigra, *LC* locus ceruleous, *PUT* putamen (dorsal striatum), *StN* subthalamic nucleus, *IFG* inferior frontal gyrus, *PreM* premotor area, *ACC* anterior cingulate cortex

(as assessed by the stop-signal task), it is likely that neurally dissociable regions of the PFC are causally involved in distinct profiles of behavioral impairment in ADHD. Performance on the stop signal task and delay-discounting tasks accurately predicts 90% of ADHD cases but performance on each task alone predicts a much lower proportion of clinical cases (Solanto et al. 2001; also see Lambek et al. 2010 for similar findings).

5 Neural Substrates of Impulsivity

The neural substrates of impulsivity involve a network of prefrontal, limbic, and striatal regions. The striatum is crucial for both forms of impulsivity as it receives dense projections from a number of limbic regions, such as the hippocampus, amygdala, and PFC, which collectively have been implicated in impulsive behavior (Cheung and Cardinal 2005; Dalley et al. 2008; Winstanley et al. 2004). The PFC contains several distinct regions that are critically involved in: planning future actions; deciding in situations of conflict; use of abstract rules, and the online representation of relationships between stimuli beyond those simply defined by the physical attributes of a particular stimulus (Dalley et al. 2004). It is therefore not surprising that the PFC is implicated in choice and motor impulsivity, as measured by delay-discounting and SSRTT, respectively. For example, increased activity of prefrontal regions during delay-discounting is typically observed, which distinguishes between the VMPFC and the DLPFC. Thus, the VMPFC shows selective activation when subjects choose immediate rewards, whereas the DLPFC is activated when subjects make a choice that is independent of the delay (McClure et al. 2004).

The functional imaging studies reviewed above suggest that PFC activity is associated with impulsive behavior, but do not provide direct causal evidence that this is the case. Causal evidence for a role of the PFC in choice impulsivity was recently demonstrated with the use of transcranial magnetic stimulation (TMS; Figner et al. 2010). In this experiment, subjects who received TMS of the DLPFC (which has effects analogous to a temporary lesion) showed increased preference for immediate rewards. This effect was not observed after the TMS had ceased or following "sham" stimulation. Similarly, stopping capacity on the stop signal task was impaired following TMS of the inferior frontal gyrus (IFG; Chambers et al. 2006), which is consistent with similar impairments observed in individuals with lesions in the IFG (Aron et al. 2003).

There are many functional and anatomical homologies between ventral and dorsal mPFC in the rat and the VMPFC and the DLPFC (respectively) in the monkey (Vertes 2006). These provide an important foundation for integrating findings from human and nonhuman animals concerning the relation between neural substrates of executive dysfunctions and different forms of impulsive behavior. In human imaging studies, choice for large, delayed rewards sometimes correlates with activation of the VMPFC and sometimes with activation of the DLPFC

(Daniel and Pollmann 2010). These two frontal regions are extreme endpoints of a continuum linking hot and cold processing in the normal brain. Indeed, McClure et al. (2004) found that these two systems control choice in a delayed-discounting task. One of these was involved in choices for immediate rewards (the beta regions, comprising the mOFC, medial prefrontal cortex, posterior cingulate cortex, and ventral striatum). The second was involved in higher-order cognitive functions responsible for choices that are independent of delay (the delta regions, comprising lateral prefrontal regions, supplementary, and pre-motor regions). Of note, these areas show opposite patterns of activation during choices for immediate and delayed rewards, suggesting that these parallel circuits may be partly in competition during choice.

The VMPFC also appears to have a prominent involvement in situations in which there is a clear conflict between choices (i.e., large preference for delayed reward or for the immediate reward) whereas dorsolateral regions of the PFC control situations in which there is no clear preference for one of the alternatives: for example, when a subject's choices are close to indifference (Marco-Pallarés et al. 2010). Thus, convergent evidence suggests that the DLPFC and VMPFC are psychologically dissociable in humans in relation to impulse decision-making.

The general idea that a system related to emotional processing encodes choice for imminent small rewards, while a second system for cognitive control allows rational decisions about larger rewards, is supported by clinical evidence. Regions strongly implicated in successful inhibition of motor impulsivity such as the IFG are located in close proximity with delta (i.e., cognitive) regions (Eagle and Baunez 2010), whereas regions implicated in impulsive choice, such as the OFC, comprise mainly beta regions (i.e., emotional processing). Both distinctions can predict ADHD diagnosis and together they can account for much of the large variance in the clinical population (Castellanos et al. 2006; Solanto et al. 2001). In studies using rodents, lesions of the OFC impair successful performance in the stop signal task, but lesions of the infralimbic (IL) cortex (which is situated more medial and posterior to the OFC) do not (Eagle et al. 2008). In the 5-CSRTT, which measures a form of impulsivity highly related to choice impulsivity (see below), lesions of the IL cortex increase premature responses (impulsive behavior, whereas lesions of the OFC increase perseverative responses, presumably related to compulsive lack of control (Chudasama et al. 2003). Taken together, this evidence suggests that fractionable components of the PFC control distinct processes related to choice and motor impulsivity.

Additional support for the idea that a gradient of impulsivity can be delineated at behavioral and neural levels stems from recent evidence in animal models of drug addiction using controlled schedules of drug self-administration. Such studies reveal a progressive gradient of neural dysfunction. Initially, this involves hot (emotional) brain regions and gradually extends to include cold (sensory/motoric) brain regions (DLPFC) after extended drug self-administration (Beveridge et al. 2008). In a recent review of the neural effects of chronic drug exposure, Beveridge and colleagues concluded that "most consistent deficits in functional activity are within ventromedial orbital and anterior cingulate cortices, regardless of task"

(Beveridge et al. 2008, p 3260). In support of this view, monkeys exposed to 5 days, 3.3 months, or 1.2 years of cocaine self-administration showed a progressive reduction in metabolic activity starting in the ventromedial orbital region and eventually extending to include the DLPFC (Letchworth et al. 2001). Thus, initial cocaine self-administration affects mainly the so-called "hot" cognitive systems, whereas longer exposure to cocaine produces more pervasive effects on PFC function, especially the DLPFC (Shackman et al. 2009). Consistent with this notion are recent findings in rats showing that prolonged access to cocaine self-administration impairs PFC-dependent cognitive functions such as working memory (Dalley et al. 2005), and findings in monkeys showing deficits in motor inhibition (measured in the stop signal task) after chronic cocaine self-administration (Liu et al. 2009).

6 Modeling Vulnerability to Drug Addiction

Impulsive behavior and novelty-(sensation) seeking have long been associated with drug addiction in humans, whether as a causal mechanism or as a consequence of repeated episodes of drug taking (Verdejo-García et al. 2008; Wills et al. 1994; Zuckerman 1990). Increasing evidence suggests that certain personality traits, including the seeking out of intense forms of sensation, novelty, and impulsivity may predispose to drug abuse and addiction (Adams et al. 2003; Sher et al. 2000; Verdejo-García et al. 2008). Indeed recent work on prospective studies in adolescents indicates that impulsivity may pre-date drug use and contribute to the development of addiction (Nigg et al. 2006; Wong et al. 2006).

Consistent with a causal role of impulsivity in drug addiction, we have recently shown that a naturally occurring form of impulsivity in rats predicts both the escalation of intravenous cocaine self-administration (Dalley et al. 2007) and the subsequent emergence of compulsive cocaine taking (Belin et al. 2008). Rats were screened for impulsivity on the 5-CSRTT, which was defined by the number of anticipatory responses made before the onset of a food-predictive, brief light stimulus (Robbins 2002). In subsequent work, we have shown that the high impulsivity phenotype is best described by a deficit in waiting for delayed rewards (Robinson et al. 2009). Indeed, rats exhibiting high levels of impulsive choice for food reinforcement also escalate their cocaine self-administration more readily than do low impulsive rats (Anker et al. 2009; Perry et al. 2005).

One of the best studied animal models of novelty-(sensation) seeking is the high responder rat, which displays high levels of locomotor activity in a novel environment compared with low responder rats. The high responder rat shows an increased propensity to acquire stimulant self-administration, in particular when low doses are infused (Piazza et al. 1989). However, it does not predict features of addiction that resemble the behavior of addicts, such as compulsive cocaine-seeking (Belin et al. 2008). In addition, high responder rats not only show lower choice-impulsivity than low responder rats, but they also show increased motor impulsivity under a schedule looking for a differential response to low rates of reinforcement (DRL; Flagel et al. 2010). We suggest that this is equivalent to "cold" impulsivity. Rats bred for high levels of responsiveness to a novel environment (hereafter high responder) also show an increased approach response to stimuli that predict food rewards (sign tracking), rather than to the location of delivery of reward (goal tracking). That is, rats are exposed to stimulus-food pairings in which the insertion of a lever inside the chamber is the stimulus, and delivery of a food pellet the reward. Although presses on the lever have no effect on the delivery of rewards, so-called "sign trackers" increase their number of responses to the lever at the expense of waiting in the food hopper for reward delivery (i.e., "goal-trackers"; Flagel et al. 2010). Sign-trackers have also been proposed as a putative model for addiction. However, these rats do not acquire cocaine self-administration faster than rats that tend to approach the location of the food (i.e., goal-trackers) (Robinson and Flagel 2009; Flagel et al. 2011).

In summary, subjects with preexisting high levels impulsivity as determined by premature responses in the 5-CSRTT (Dalley et al. 2007; Diergaarde et al. 2008) or choice impulsivity (Perry et al. 2005) seem to be particularly prone to escalate cocaine or nicotine intake. In addition, high impulsive animals in the 5-CSRTT (Dalley et al. 2007), which also behave impulsively in temporal discounting (the critical measure of choice impulsivity; Robinson et al. 2009), are prone to develop addiction-like patterns of behavior modeled on the criteria used in the clinical diagnosis of substance abuse disorders (Belin et al. 2008). High responder rats selected for their reactivity to novel environments and rats selected for their tendencies to display Pavlovian approach responses (i.e., sign trackers) do not seem to escalate faster self-administration of stimulants, but rather display increased reinstatement of responding after presentations of the drug-predictive stimulus (Saunders and Robinson 2010). This latter phenotypic characteristic is not unique to sign trackers because high impulsive animals selected on the basis of their premature responses, also show increased resumption of cocaine seeking after punishment-induced abstinence (Economidou et al. 2009).

7 Impulsivity and Choice Behavior: A Psychological Perspective

As discussed above, there are at least two dissociable psychological constructs of impulsivity (choice and motor impulsivity) and each is governed by distinct neuroanatomical networks (see: Fig. 1). Increased choice impulsivity for food rewards predicts escalation of cocaine self-administration, as does impulsive responding on the 5-CSRTT (Dalley et al. 2007; Diergaarde et al. 2008; Perry and Carroll 2008). Given that drug self-administration initially affects ventral regions of mPFC (Beveridge et al. 2008; Letchworth et al. 2001) and progressively shifts toward more dorsal regions of the striatum and PFC with increasing drug exposure, it has been hypothesized that drug addiction is a brain disorder that progresses from ventral to dorsal frontostriatal brain regions (Porrino et al. 2007). Thus, choice (or waiting) impulsivity may be an early marker or determinant of vulnerability to drug addiction, whereas motor impulsivity deficits observed in addicts may be the end point in which the detrimental effects of chronic drug exposure are better revealed.

Individuals with ADHD show profound differences in choice impulsivity relative to normal healthy controls and, under most circumstances, they opt for small, immediate rewards. This may be a consequence of a failure to encode the magnitude of the larger reward (Sonuga-Barke et al. 1992). Reward encoding is mediated by midbrain dopaminergic neurons that originate in the ventral tegmental area and substantia nigra (VTA/SN) and project to the striatum, in particular to the nucleus accumbens (NAcb). In humans, the involvement of the NAcb is reliably revealed when real rewards (as opposed to abstract rewards) are employed. For example, a recent study compared NAcb activation during performance of two feedback-based tasks: one task used monetary rewards, the other used relevant feedback. NAcb activation was higher during the anticipation of a monetary reward than task-related (or cognitive) feedback (Daniel and Pollmann 2010). This suggests that ventral–striatal activation (i.e., hot processing) is greater when the task involves biologically significant rewards (also see McClure et al. 2007). Moreover, these regions may be disrupted in individuals with ADHD. Thus, a recent positron emission tomography (PET) imaging study in 53 nonmedicated adults with ADHD found that dopamine D2/3 receptors and dopamine transporter (DAT) availability are reduced in the left midbrain and NAcb (Volkow et al. 2009). In addition, an fMRI study showed decreased activation in the ventral striatum of adult ADHD individuals (relative to controls) during performance on a discounting task with monetary rewards (Plichta et al. 2009). Taken together, these studies provide strong support for the notion that altered reward sensitivity is a core deficit in ADHD individuals and that altered dopamine transmission in the VTA-NAcb pathway contributes to the altered responsiveness of ADHD subjects to biologically relevant rewards. Altered reward sensitivity has also been proposed as a candidate mechanism in drug addiction, specifically because drugs of abuse usurp dopaminergic (and other) systems implicated in reward prediction error (Dalley and Everitt 2009; Stephens et al. 2010). Deregulated dopaminergic inputs may contribute to an addict's failure to encode delayed rewards in favor of immediate pleasure. In addition, some drugs of abuse (e.g., psychostimulants) increase dopaminergic transmission, in part by compensating for the general reward deficits and sense of anhedonia often experienced by addicts.

This failure to encode delayed rewards may be due to impaired encoding of response-outcome associations (R_D-O) during delayed reward, in which subjects assess R-O associations relative to alternative stimuli (S) or responses (R_I). Critical to the present discussion is the functional role of these stimuli including the training context, which have their own associations with the outcome (S-O; or R_I-O). For example, relative to immediate rewards, delayed rewards result in stronger context-outcome (S-O) associations, but it is not entirely clear how this additional strength of the context affects instrumental choice and performance (Cardinal et al. 2002). That is, in free-operant procedures with only one response alternative (R), subjects assess the causal relation of their responses and the outcome relative to all other stimuli (e.g., the context) that are present in the session. In a choice situation with two response alternatives each followed by an immediate or delayed reward (R_I and R_D, respectively), the causal relation of R_D and the delayed O will be assessed in relation to that of R_I and the immediate O: context becomes a signal for availability rather than a competing stimulus. For example, studies using instrumental conditioning have found opposite effects of contextual extinction during/after instrumental free-operant training. Whereas context extinction decreases instrumental performance when the outcome (O) is presented immediately after the response (Pearce and Hall 1979), it increases responding when outcomes are delayed (Dickinson et al. 1992; Reed and Reilly 1990). Self-administration of cocaine or heroin also varies as a function of environmental cues in both addicts and rodents (Caprioli et al. 2009): this supports the idea that different drugs of abuse interact with contextual stimuli in different and sometimes opposite ways.

8 The Hippocampus, Environmental Stimuli, and Delayed Rewards

The hippocampus projects extensively to the ventral striatum, in particular to the shell subregion of the NAcb (Groenewegen et al. 1999), and is implicated in ADHD (Plessen et al. 2006; Volkow et al. 2007). It is generally agreed that the NAcb shell processes contextual information, whereas the NAcb core processes Pavlovian cues, consistent with segregated neuroanatomical inputs from the hippocampus and amygdala, respectively (Cardinal et al. 2002). Instrumental learning and choice can be modulated by several factors, including the presence of Pavlovian stimuli during training. However, in the absence of choice it is now becoming clear that training contexts can both compete with, and facilitate, instrumental performance depending on the training parameters (Reed and Reilly 1990; Pearce and Hall 1979). With immediate reinforcement, extinction of the context after instrumental training reduces instrumental performance in a subsequent test, revealing that the context facilitated instrumental action (Pearce and Hall 1979). With similar training parameters, but delayed reinforcement, the opposite effect is found; context extinction after instrumental training increases instrumental performance (Reed and Reilly 1990).

Lesions of the hippocampus produce a similar pattern of results. For example, after hippocampal lesions, subjects learn more slowly about immediate reinforcement but more quickly when reinforcement is delayed (Cheung and Cardinal 2005). Such effects contrast with lesions of NAcb, which increase instrumental learning when there is no delay between responding and reward, but retard learning when delays are imposed (Cardinal and Cheung 2005). Thus, under immediate *versus* delayed reward conditions, the context seems to exert opposite effects on instrumental performance. Contextual competition and facilitation also depend on other parameters, such as the number of levers (response alternatives) present in the training environment. In delayed, free-operant training with only one response alternative, subjects assess the R-O representation in relation to that between other stimuli, such as the context (Fig. 2, right). In a choice situation with two response alternatives each followed by an immediate or delayed reward (R_I and R_D, respectively), R_D is compared with a second R_I rather than the training context (Fig. 2, left). The fact that in the absence of choice the context seems to compete under delayed reward conditions does not necessarily suggest that this is the only role that the context exerts. In delayed reinforcement with choice between two

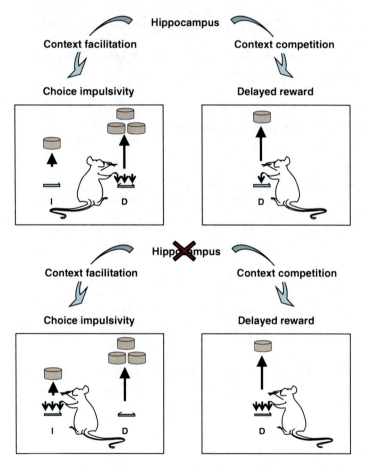

Fig. 2 Hypothetical functions of the hippocampus and the NAcb in coding the effects of context on choice impulsivity and delayed reward (*top*). The context can either compete (*right*) or facilitate (*left*) instrumental performance with delayed rewards. The NAcb mediates performance in both situations, but the critical contribution of context can give rise to two opposing outcomes. When the context bridges the response-outcome relation (*left*), subjects opt for the delayed reward and thus display reduced impulsive choice. Hippocampal lesions (*bottom*), which reduce context processing, induce impulsive choice (*left*). Conversely, in a single-lever situation involving delayed rewards (*top right*), the context competes with R for control over behavior. In this situation, absence of context processing increases responding for delayed rewards (*bottom right*)

response alternatives, the context acts as a bridge between any given R_D and the delayed O, consistent with several reports suggesting that the hippocampus is necessary for delayed (i.e., trace) conditioning (Shors 2004). This explains why, in hippocampus-lesioned subjects, failing to bridge these two representations impairs encoding for the delayed outcome and results in a shift in choice toward the immediate reward (Cheung and Cardinal 2005). These results show that the context can compete with, or facilitate, instrumental performance depending on the number of response alternatives and on whether rewarding outcomes are presented immediately or after a delay (Fig. 2).

Similarly, an issue that awaits clarification concerns the function of discrete signals presented during the delay between the R and the O. This is relevant because contextual stimuli are present during these intervals and influence choice, but the exact function that allows contexts to bridge the delay between the R and the O is obscured by long periods during the intervals between trials on which O is not presented. These putative roles can be clarified with the use of discrete stimuli. For example, due to associations with the outcome, contextual stimuli or discrete signals present between the response and outcome may support instrumental performance through a process called conditioned reinforcement, which depends on dopamine function in the NAcb (Taylor and Robbins 1986). However, the relationship between conditioned reinforcement processes and impulsivity, which both depend on dopaminergic mechanisms (Dalley et al. 2007; van Gaalen et al. 2006), is unclear.

Because the symptoms of ADHD are relieved by drugs that augment dopamine function (Solanto 2002), it is widely hypothesized that ADHD and impulsivity are mediated by an insufficiency of dopamine function (Castellanos and Tannock 2002; Solanto 2002; Volkow et al. 2007). Decreased DA transmission can explain steepened delayed-discounting in ADHD subjects through two mechanisms. Low-tonic dopaminergic transmission (Volkow et al. 2007) increases signal-to-noise ratios in reward neural circuitry and makes external (distractor) stimuli more salient and attention-grabbing (Sikström and Söderlund 2007), whereas increased attention to nontarget stimuli is presumably mediated by increased phasic dopamine responses. In a situation that imposes a choice between immediate and delayed rewards, this explanation predicts steep delay-discounting functions due to a heightened phasic dopamine response to the immediate reward, which competes with (i.e., overshadows) reinforcement produced by the delayed reward. Alternatively, ADHD individuals may display impulsive discounting because interfering stimuli, during the interval between the choice and the presentation of the reward, impair learning about stimuli that guide response selection (Johansen et al. 2009). Both these explanations predict that lesions of the NAcb should increase impulsive behavior, either due to decreased signal-to-noise ratios or increased learning about stimuli that guide response selection toward the immediate reward (Cardinal et al. 2001).

Although these two explanations seem compatible in principle, the second posits that learning about distracting stimuli leads to ineffective learning about discriminative stimuli (such as contextual cues), which are critical for choice behavior. This account predicts that hippocampal lesions, which decrease processing of contextual stimuli (Holland and Bouton 1999), should increase impulsivity. This is because contextual stimuli determine response selection and, in the absence of context processing, preference for the immediate choice should be increased (Fig. 2). One explanation for this lack of agreement may arise from the assumptions made regarding the role of contextual information. For example, Johansen et al. (2009) assume that the context and other stimuli influence choice in the same way as discriminative stimuli: that is by signaling an R-O relation (Colwill and Rescorla 1990). This process is similar to Pavlovian-to-instrumental transfer (PIT), in which stimuli that have previously been associated with the outcome increase instrumental behavior. However, evidence suggests that differences between ADHD and control participants tend to be observed when sessions are adjusted, depending on the choices made; this culminates in relatively massed trials when subjects always choose the immediate reward (Sonuga-Barke et al. 1992). Massed training increases the strength of associations between contextual (i.e., environmental) stimuli and the outcome because subjects experience the presentation of the outcome simultaneously with the context. Therefore, the context is likely to compete with (i.e., overshadow) responses when waiting intervals are relatively short. The question then is how to determine the specific influence of the context in a choice situation, as each choice (immediate *versus* delayed) has a different temporal relationship with the outcome.

Overall, the evidence summarized above suggests that environmental stimuli can change instrumental conditioning with delayed reward and choice behavior through at least two mechanisms; first by acting as discrete cues (which makes the context akin to a conditioned reinforcer (CR), and, second, by signaling an R-O association through its discriminative properties (a process analogous to Pavlovian-to-instrumental transfer (PIT)). Unfortunately, little is known about the conditions that favor the increase in instrumental performance observed in CR and PIT procedures. In a recent study, Crombag et al. (2008) observed that brief stimuli favor the development of CR but not PIT, whereas long stimuli favor robust PIT but not CR. Although the mechanisms underlying this dissociation are not entirely clear, it suggests that briefly exposed stimuli (i.e., comparable to a schedule with massed trials) are more likely to become conditioned reinforcers and therefore potentiate instrumental responding for delayed outcomes. This potentiation is likely to occur when the outcome is delayed rather than presented immediately.

Using cocaine as an outcome, Di Ciano and Everitt (2003) reported that relatively short stimuli that have been paired with self-administered drug can support the acquisition of a new response (suggesting strong CR properties) but these stimuli fail to invigorate instrumental behavior when presented noncontingently. On the contrary, noncontingent presentation of discriminative stimuli of long duration, which previously predicted drug availability, did invigorate responding for cocaine under similar training conditions, suggesting that the ability of environmental signals to support cocaine seeking is largely dependent on their duration and predictive relationship with the primary reinforcer (O). Together, these data indicate that stimulus duration (and trial spacing, in the case of contexts) determines the reinforcing and invigorating properties of CSs associated with primary reinforcers. Why this is the case is not entirely clear and is discussed next.

One possibility is that CR and PIT engage different psychological processes, each favored by the different training conditions. This would be consistent with multiple neural dissociations underpinning CR and PIT (Corbit and Balleine 2005; Corbit et al. 2007). For example, CR depends on the basolateral amygdala, whereas some forms of PIT depend on the central amygdala (Cardinal et al. 2002). Regardless of this question, it is clear that both processes contribute to instrumental reinforcement with delayed rewards, and that these differences may stem from alternative roles of environmental information present during the interval between R and O. Thus, it could be argued that conditioned reinforcement by contextual stimuli may support responding for delayed rewards, and that ADHD individuals fail to encode these aspects of contextual stimuli. This proposal is consistent with differences in choice behavior between ADHD and control individuals (discussed above). Specifically, when the intertrial interval was adjusted, based on the choice outcome of each trial, control subjects showed preferential choice for the immediate outcome, just like ADHD individuals (Sonuga-Barke et al. 1992).

In summary, our analysis suggests that contextual and discrete stimuli can influence instrumental behavior and choice in at least two ways. First, by virtue of their strong associations with primary reinforcers (O), brief stimuli can support instrumental action by acting as a CR. Second, they can signal the availability of delayed outcomes through its discriminative properties, a process which is favored by long-lasting stimuli. A remaining issue, therefore, is how psychostimulant drugs, such as d-amphetamine and methylphenidate, which are clinically efficacious in ADHD (Solanto 2002), interact with these mechanisms. This was tested explicitly in a study by Cardinal et al. (2000) in groups of rats trained on a delayed-discounting task, with or without a bridging signal between the choice phase and the delivery of the food outcome. One group of rats received training in which the house-light of the chamber was illuminated before choice, but was turned off after the choice response (R) and remained off until the presentation of the delayed reward (R-O). For a second group of animals, the house-light was illuminated during the entire trial (including the interval between response and delayed reward). A third group of animals received a light stimulus after a response had been made until the outcome was delivered. The experimental design thus tests how different stimuli presented during the interval between choice and outcome presentation (continuous house-light, a discrete light stimulus, or no stimulus) influence delayed discounting. It was observed that d-amphetamine increased impulsivity in the absence of any stimulus between the response and outcome, but decreased impulsivity when there was a cue presented between the response and outcome (i.e., discrete light stimulus). These results strongly suggest that d-amphetamine potentiates the conditioned reinforcing effects of the light cue thereby increasing choice for the larger, delayed reward (Cardinal et al. 2000).

A similar dissociation was observed recently in rats selected for high *versus* low baseline levels of delayed discounting (i.e., choice) impulsivity (Zeeb et al. 2010). In high-impulsive rodents, presentation of a signal increased impulsive choice, but the opposite result was observed in low-impulsive rodents. Moreover, inactivation of the OFC increased impulsive choice in high impulsive rodents when no cue was presented, but decreased impulsive choice in low-impulsive rats when a cue was presented between the response and outcome. In short, the signal facilitated responding for the delayed reward in low-impulsive rats but decreased it in high-impulsive rats. The OFC is postulated to play a role in processing reward-predictive, Pavlovian cues (Schoenbaum et al. 2009). However, this does not explain why OFC inactivation produced opposite effects, depending on baseline levels of impulsivity. It is plausible to speculate that the role of the OFC depends on basal levels of DA transmission, but the opposite interaction is not readily anticipated by current views on the OFC function. These results challenge simple views on behavior and brain function and suggest a need for an approach that considers sensitivity to basal levels of stimulation from the environment. This is relevant for ADHD because adequate stimulation by the environment (as is the case when the environment is novel) decreases differences between ADHD individuals and controls (Abikoff et al. 1996).

9 Linking Impulsivity, Executive Dysfunction, and Propensity to Drug Addiction

In this chapter, we review evidence suggesting that environmental factors, such as discrete signals and contexts, can have opposite effects on instrumental performance and choice. Impulsive subjects either fail to use these signals, or use them in a way that biases their responses toward immediate, smaller rewards. ADHD and early stages of drug addiction are characterized by a lack of impulse control: this depends on dysfunctional dopamine neurotransmission, which translates into reward deficits at a psychological level. One question is how does this particular form of impulsivity or endophenotype predispose subjects to drug addiction?

Addiction can be conceptualized as a series of transitions from recreational, occasional drug exposure to habitual and, ultimately, compulsive drug use that is immune to negative consequences (Everitt and Robbins 2005). Increased impulsivity may decrease the age at first exposure, while increasing the incidence and amounts of drug consumed. These two factors, in concert, may accelerate higher rates of drug use in impulsive subjects (Rogers and Robbins 2001; de Wit 2009). However, according to our present theoretical analysis, there is a second mechanism through which impulsive endophenotypes may predispose to drug addiction. By failing to encode delayed outcomes, and so preferentially choosing immediate, small magnitude rewards, individuals with ADHD may be especially susceptible to the development of *habitual* behavior. Habitual behavior is an automatic behavior that is independent of the value of the goal (Dickinson 1985).

Recent unpublished experiments have begun to shed light on the potential link between delayed reward and the transition to habits (Urcelay et al., unpublished observation). In rodents trained to lever-press for food rewards, we have observed that when rewards are delayed (20 s) instrumental behavior is less dependent on the value of the outcome and so is more habitual. Moreover, long exposure to the training context (extinction) restores goal-dependent behavior in animals trained with a 20 s delay between the R and the O. These findings suggest that the failure to encode delayed goals may accelerate the transition to habitual behavior. It could be argued that failure to encode reward value or magnitude predisposes to habitual behavior and that this explains why endophenotypes, characterized by high impulsive choice are not only more prone to escalate during cocaine self-administration, but also continue drug-seeking despite negative consequences (Belin et al. 2008; Dalley et al. 2007; Diergaarde et al. 2008).

10 Conclusions

Although the present analysis of dual PFC control over striatal-based behaviors may be an oversimplification, it is consistent with a dual-locus explanation of ADHD (Castellanos et al. 2006) and also with deficits observed in drug addicts (Verdejo-García et al. 2008). The rodent literature we have reviewed also shows parallels with the notion of "cool" and "hot" processing, suggesting continuity in an analysis involving preclinical animal models and observations in clinical subgroups (Solanto et al. 2001). Perhaps, the strength of this approach is that it ultimately relies on detailed neuroanatomical studies that suggest gradients of projections from different prefrontal and limbic regions to the basal ganglia (Haber 2003; Voorn et al. 2004). Within this framework, "hot" and "cold" frontostriatal networks provide a useful working model to explain the pattern of dysfunction in cocaine addicts. Moreover, studies assessing the impact of vulnerable neural and behavioral endophenotypes suggest that the regions first affected by cocaine exposure (ventral network; hot) are also affected in impulsive subjects with no prior exposure to drugs or ADHD medication. Thus, individuals exhibiting high levels of choice impulsivity may be especially prone to develop drug addiction. In contrast, high levels of motor impulsivity (dorsal network; cold) may instead relate to the later stages of drug addiction as a manifestation of the direct neurotoxic effects of chronic drug exposure.

A major challenge for future research will be to elucidate the genetic basis of choice impulsivity and determine how environmental influences (context, drug exposure) interact and modify the transition from initial goal-directed drug use to habitual and ultimately compulsive drug abuse. A second challenge is to refine current behavioral protocols used in the laboratory to increase the accuracy of clinical diagnosis (i.e., DSM5). The evidence summarized here highlights the prevalence of different forms of impulsivity in both ADHD and addiction, thus suggesting that impulsivity is critical in the identification of the diverse behavioral patterns observed in these conditions.

Acknowledgments This research was supported by a Wellcome Trust Programme Grant (089589/Z/09/Z) awarded to T.W. Robbins, B. J. Everitt, A. C. Roberts, B.J. Sahakian, and JWD, MRC grants to B.J. Everitt (G0600196) and JWD (G0401068, G0701500) and by the

European Community's Sixth Framework Programme ("Imagen" LSNM-CT-2007-037286). GPU was supported by a Marie Curie Intra-European Fellowship (PIEF-GA-2009-237608) awarded by the European Commission. This review reflects only the authors' views and the European Community is not liable for any use that may be made of the information contained therein. The authors thank Andrea Bari for insightful discussions. Figure 1 was adapted from best-lemming, brain, Google SketchUp 3D warehouse, http://sketchup.google.com/3dwarehouse/details?mid=bdde6508945af6e2a4dd6527f4a3f142 (accessed 8 July, 2010).

References

Abikoff H, Courtney ME, Szeibel PJ, Koplewicz HS (1996) The effects of auditory stimulation on the arithmetic performance of children with ADHD and nondisabled children. J Learn Disabil 29:238–246

Adams JB, Heath AJ, Young SE, Hewitt JK, Corley RP, Stallings MC (2003) Relationships between personality and preferred substance and motivations for use among adolescent substance abusers. Am J Drug Alcohol Abuse 29:691–712

Ainslie G (1975) Specious reward: a behavioral theory of impulsiveness and impulse control. Psychol Bull 82:463–496

American Psychiatric Association (2000) Diagnostic and statistical manual of mental disorders: DSM-IV-TR. American Psychiatric Association, Washington, DC

Anker JJ, Perry JL, Gliddon LA, Carroll ME (2009) Impulsivity predicts the escalation of cocaine self-administration in rats. Pharmacol Biochem Behav 93:343–348

Aron AR, Fletcher PC, Bullmore ET, Sahakian BJ, Robbins TW (2003) Stop-signal inhibition disrupted by damage to right inferior frontal gyrus in humans. Nat Neurosci 6:115–116

Bari A, Robbins TW, Dalley JW (2011) Impulsivity. In: Olmstead M (ed) Animal models of drug addiction, Neuromethods (in press) DOI 10.1007/978-1-60761-934-5_14. Springer Science + Business Media, LLC 2011

Barkley RA (1997) Behavioral inhibition, sustained attention, and executive functions: constructing a unifying theory of ADHD. Psychol Bull 121:65–94

Belin D, Mar AC, Dalley JW, Robbins TW, Everitt BJ (2008) High impulsivity predicts the switch to compulsive cocaine-taking. Science 320:1352–1355

Belin D, Jonkman S, Dickinson A, Robbins TW, Everitt BJ (2009) Parallel and interactive learning processes within the basal ganglia: relevance for the understanding of addiction. Behav Brain Res 199:89–102

Beveridge TJR, Gill KE, Hanlon CA, Porrino LJ (2008) Review. Parallel studies of cocaine-related neural and cognitive impairment in humans and monkeys. Philos Trans R Soc Lond B Biol Sci 363:3257–3266

Bolla KI, Cadet JL, London ED (1998) The neuropsychiatry of chronic cocaine abuse. J Neuropsychiatry Clin Neurosci 10:280–289

Caprioli D, Celentano M, Dubla A, Lucantonio F, Nencini P, Badiani A (2009) Ambience and drug choice: cocaine- and heroin-taking as a function of environmental context in humans and rats. Biol Psychiatry 65:893–899

Cardinal RN, Cheung THC (2005) Nucleus accumbens core lesions retard instrumental learning and performance with delayed reinforcement in the rat. BMC Neurosci 6:9

Cardinal RN, Robbins TW, Everitt BJ (2000) The effects of d-amphetamine, chlordiazepoxide, alpha-flupenthixol and behavioural manipulations on choice of signalled and unsignalled delayed reinforcement in rats. Psychopharmacology (Berl) 152:362–375

Cardinal RN, Pennicott DR, Sugathapala CL, Robbins TW, Everitt BJ (2001) Impulsive choice induced in rats by lesions of the nucleus accumbens core. Science 292:2499–2501

Cardinal RN, Parkinson JA, Hall J, Everitt BJ (2002) Emotion and motivation: the role of the amygdala, ventral striatum, and prefrontal cortex. Neurosci Biobehav Rev 26:321–352

Castellanos FX, Tannock R (2002) Neuroscience of attention-deficit/hyperactivity disorder: the search for endophenotypes. Nat Rev Neurosci 3:617–628

Castellanos FX, Sonuga-Barke EJS, Scheres A, Di Martino A, Hyde C, Walters JR (2005) Varieties of attention-deficit/hyperactivity disorder-related intra-individual variability. Biol Psychiatry 57:1416–1423

Castellanos FX, Sonuga-Barke EJS, Milham MP, Tannock R (2006) Characterizing cognition in ADHD: beyond executive dysfunction. Trends Cogn Sci (Regul Ed) 10:117–123

Chambers CD, Bellgrove MA, Stokes MG, Henderson TR, Garavan H, Robertson IH, Morris AP, Mattingley JB (2006) Executive 'brake failure' following deactivation of human frontal lobe. J Cogn Neurosci 18:444–455

Cheung THC, Cardinal RN (2005) Hippocampal lesions facilitate instrumental learning with delayed reinforcement but induce impulsive choice in rats. BMC Neurosci 6:36

Chudasama Y, Passetti F, Rhodes SEV, Lopian D, Desai A, Robbins TW (2003) Dissociable aspects of performance on the 5-choice serial reaction time task following lesions of the dorsal anterior cingulate, infralimbic and orbitofrontal cortex in the rat: differential effects on selectivity, impulsivity and compulsivity. Behav Brain Res 146:105–119

Colwill RM, Rescorla RA (1990) Evidence for the hierarchical structure of instrumental learning. Anim Learn Behav 18:71–82

Corbit LH, Balleine BW (2005) Double dissociation of basolateral and central amygdala lesions on the general and outcome-specific forms of pavlovian-instrumental transfer. J Neurosci 25:962–970

Corbit LH, Janak PH, Balleine BW (2007) General and outcome-specific forms of Pavlovian-instrumental transfer: the effect of shifts in motivational state and inactivation of the ventral tegmental area. Eur J Neurosci 26:3141–3149

Crombag HS, Galarce EM, Holland PC (2008) Pavlovian influences on goal-directed behavior in mice: the role of cue-reinforcer relations. Learn Mem 15:299–303

Dalley JW, Everitt BJ (2009) Dopamine receptors in the learning, memory and drug reward circuitry. Semin Cell Dev Biol 20:403–410

Dalley JW, Cardinal RN, Robbins TW (2004) Prefrontal executive and cognitive functions in rodents: neural and neurochemical substrates. Neurosci Biobehav Rev 28:771–784

Dalley JW, Theobald DEH, Berry D, Milstein JA, Lääne K, Everitt BJ, Robbins TW (2005) Cognitive sequelae of intravenous amphetamine self-administration in rats: evidence for selective effects on attentional performance. Neuropsychopharmacology 30:525–537

Dalley JW, Fryer TD, Brichard L, Robinson ESJ, Theobald DEH, Lääne K, Peña Y, Murphy ER, Shah Y, Probst K, Abakumova I, Aigbirhio FI, Richards HK, Hong Y, Baron J, Everitt BJ, Robbins TW (2007) Nucleus accumbens D2/3 receptors predict trait impulsivity and cocaine reinforcement. Science 315:1267–1270

Dalley JW, Mar AC, Economidou D, Robbins TW (2008) Neurobehavioral mechanisms of impulsivity: Fronto-striatal systems and functional neurochemistry. Pharmacol Biochem Behav 90:250–260

Daniel R, Pollmann S (2010) Comparing the neural basis of monetary reward and cognitive feedback during information-integration category learning. J Neurosci 30:47–55

de Wit H (2009) Impulsivity as a determinant and consequence of drug use: a review of underlying processes. Addict Biol 14:22–31

Di Ciano P, Everitt BJ (2003) Differential control over drug-seeking behavior by drug-associated conditioned reinforcers and discriminative stimuli predictive of drug availability. Behav Neurosci 117:952–960

Dickinson A (1985) Actions and habits: the development of behavioural autonomy. Philos Trans R Soc Lond B Biol Sci 308:67–78

Dickinson A, Watt A, Griffiths WJH (1992) Free-operant acquisition with delayed reinforcement. Q J Exp Psychol B 45:241–258

Diergaarde L, Pattij T, Poortvliet I, Hogenboom F, de Vries W, Schoffelmeer ANM, De Vries TJ (2008) Impulsive choice and impulsive action predict vulnerability to distinct stages of nicotine seeking in rats. Biol Psychiatry 63:301–308

Eagle DM, Baunez C (2010) Is there an inhibitory-response-control system in the rat? Evidence from anatomical and pharmacological studies of behavioral inhibition. Neurosci Biobehav Rev 34:50–72

Eagle DM, Robbins TW (2003) Inhibitory control in rats performing a stop-signal reaction-time task: effects of lesions of the medial striatum and d-amphetamine. Behav Neurosci 117:1302–1317

Eagle DM, Baunez C, Hutcheson DM, Lehmann O, Shah AP, Robbins TW (2008) Stop-signal reaction-time task performance: role of prefrontal cortex and subthalamic nucleus. Cereb Cortex 18:178–188

Economidou D, Pelloux Y, Robbins TW, Dalley JW, Everitt BJ (2009) High impulsivity predicts relapse to cocaine-seeking after punishment-induced abstinence. Biol Psychiatry 65:851–856

Evenden JL (1999) Varieties of impulsivity. Psychopharmacology (Berl) 146:348–361

Everitt BJ, Robbins TW (2005) Neural systems of reinforcement for drug addiction: from actions to habits to compulsion. Nat Neurosci 8:1481–1489

Everitt BJ, Belin D, Economidou D, Pelloux Y, Dalley JW, Robbins TW (2008) Review. Neural mechanisms underlying the vulnerability to develop compulsive drug-seeking habits and addiction. Philos Trans R Soc Lond B Biol Sci 363:3125–3135

Figner B, Knoch D, Johnson EJ, Krosch AR, Lisanby SH, Fehr E, Weber EU (2010) Lateral prefrontal cortex and self-control in intertemporal choice. Nat Neurosci 13:538–539

Fineberg NA, Potenza MN, Chamberlain SR, Berlin HA, Menzies L, Bechara A, Sahakian BJ, Robbins TW, Bullmore ET, Hollander E (2010) Probing compulsive and impulsive behaviors, from animal models to endophenotypes: a narrative review. Neuropsychopharmacology 35:591–604

Flagel SB, Robinson TE, Clark JJ, Clinton SM, Watson SJ, Seeman P, Phillips PEM, Akil H (2010) An animal model of genetic vulnerability to behavioral disinhibition and responsiveness to reward-related cues: implications for addiction. Neuropsychopharmacology 35:388–400

Flagel SB, Clark JJ, Robinson TE, Mayo L, Czuj A, Willuhn I, Akers CA, Clinton SM, Phillips PEM, Akil H (2011) A selective role for dopamine in stimulus-reward learning. Nature 469:53–57

Fuster JM (2000) Prefrontal neurons in networks of executive memory. Brain Res Bull 52:331–336

Gottesman II, Gould TD (2003) The endophenotype concept in psychiatry: etymology and strategic intentions. Am J Psychiatry 160:636–645

Groenewegen HJ, Mulder AB, Beijer AVJ, Wright CI, Lopes da Silva FH, Pennartz CMA (1999) Hippocampal and amygdaloid interactions in the nucleus accumbens. Psychobiology 27:149–164

Haber SN (2003) The primate basal ganglia: parallel and integrative networks. J Chem Neuroanat 26:317–330

Hester R, Garavan H (2004) Executive dysfunction in cocaine addiction: evidence for discordant frontal, cingulate, and cerebellar activity. J Neurosci 24:11017–11022

Holland PC, Bouton ME (1999) Hippocampus and context in classical conditioning. Curr Opin Neurobiol 9:195–202

Jentsch JD, Taylor JR (1999) Impulsivity resulting from frontostriatal dysfunction in drug abuse: implications for the control of behavior by reward-related stimuli. Psychopharmacology (Berl) 146:373–390

Johansen EB, Killeen PR, Russell VA, Tripp G, Wickens JR, Tannock R, Williams J, Sagvolden T (2009) Origins of altered reinforcement effects in ADHD. Behav Brain Funct 5:7

Kalivas PW, O'Brien C (2008) Drug addiction as a pathology of staged neuroplasticity. Neuropsychopharmacology 33:166–180

Kirby KN, Petry NM (2004) Heroin and cocaine abusers have higher discount rates for delayed rewards than alcoholics or non-drug-using controls. Addiction 99:461–471

Koob GF, Volkow ND (2010) Neurocircuitry of addiction. Neuropsychopharmacology 35:217–238

Lambek R, Tannock R, Dalsgaard S, Trillingsgaard A, Damm D, Thomsen PH (2010) Validating neuropsychological subtypes of ADHD: how do children with and without an executive function deficit differ? J Child Psychol Psychiatry 51:895–904

Letchworth SR, Nader MA, Smith HR, Friedman DP, Porrino LJ (2001) Progression of changes in dopamine transporter binding site density as a result of cocaine self-administration in rhesus monkeys. J Neurosci 21:2799–2807

Lijffijt M, Kenemans JL, Verbaten MN, van Engeland H (2005) A meta-analytic review of stopping performance in attention-deficit/hyperactivity disorder: deficient inhibitory motor control? J Abnorm Psychol 114:216–222

Liu S, Heitz RP, Bradberry CW (2009) A touch screen based Stop Signal Response Task in rhesus monkeys for studying impulsivity associated with chronic cocaine self-administration. J Neurosci Methods 177:67–72

Logan GD (1994) On the ability to inhibit thought and action: a users' guide to the stop signal paradigm. In: Dagenbach D, Carr TH (eds) Inhibitory processes in attention, memory, and language. Academic Press, San Diego, CA US, pp 189–239

Luman M, Tripp G, Scheres A (2010) Identifying the neurobiology of altered reinforcement sensitivity in ADHD: a review and research agenda. Neurosci Biobehav Rev 34:744–754

Marco-Pallarés J, Mohammadi B, Samii A, Münte TF (2010) Brain activations reflect individual discount rates in intertemporal choice. Brain Res 1320:123–129

McClure SM, Laibson DI, Loewenstein G, Cohen JD (2004) Separate neural systems value immediate and delayed monetary rewards. Science 306:503–507

McClure SM, Ericson KM, Laibson DI, Loewenstein G, Cohen JD (2007) Time discounting for primary rewards. J Neurosci 27:5796–5804

Miller EK (2000) The prefrontal cortex and cognitive control. Nat Rev Neurosci 1:59–65

Moeller FG, Barratt ES, Dougherty DM, Schmitz JM, Swann AC (2001) Psychiatric aspects of impulsivity. Am J Psychiatry 158:1783–1793

Nestler EJ (2001) Molecular basis of long-term plasticity underlying addiction. Nat Rev Neurosci 2:119–128

Nigg JT, Wong MM, Martel MM, Jester JM, Puttler LI, Glass JM, Adams KM, Fitzgerald HE, Zucker RA (2006) Poor response inhibition as a predictor of problem drinking and illicit drug use in adolescents at risk for alcoholism and other substance use disorders. J Am Acad Child Adolesc Psychiatry 45:468–475

Ornstein TJ, Iddon JL, Baldacchino AM, Sahakian BJ, London M, Everitt BJ, Robbins TW (2000) Profiles of cognitive dysfunction in chronic amphetamine and heroin abusers. Neuropsychopharmacology 23:113–126

Pearce JM, Hall G (1979) The influence of context-reinforcer associations on instrumental performance. Anim Learn Behav 7:504–508

Perry JL, Carroll ME (2008) The role of impulsive behavior in drug abuse. Psychopharmacology (Berl) 200:1–26

Perry JL, Larson EB, German JP, Madden GJ, Carroll ME (2005) Impulsivity (delay discounting) as a predictor of acquisition of IV cocaine self-administration in female rats. Psychopharmacology (Berl) 178:193–201

Piazza PV, Deminière JM, Le Moal M, Simon H (1989) Factors that predict individual vulnerability to amphetamine self-administration. Science 245:1511–1513

Plessen KJ, Bansal R, Zhu H, Whiteman R, Amat J, Quackenbush GA, Martin L, Durkin K, Blair C, Royal J, Hugdahl K, Peterson BS (2006) Hippocampus and amygdala morphology in attention-deficit/hyperactivity disorder. Arch Gen Psychiatry 63:795–807

Plichta MM, Vasic N, Wolf RC, Lesch K, Brummer D, Jacob C, Fallgatter AJ, Grön G (2009) Neural hyporesponsiveness and hyperresponsiveness during immediate and delayed reward processing in adult attention-deficit/hyperactivity disorder. Biol Psychiatry 65:7–14

Porrino LJ, Smith HR, Nader MA, Beveridge TJ (2007) The effects of cocaine: a shifting target over the course of addiction. Prog Neuropsychopharmacol Biol Psychiatry 31:1593–1600

Reed P, Reilly S (1990) Context extinction following conditioning with delayed reward enhances subsequent instrumental responding. J Exp Psychol Anim Behav Process 16:48–55

Robbins TW (1996) Dissociating executive functions of the prefrontal cortex. Philos Trans R Soc Lond B Biol Sci 351:1463–1470, discussion 1470–1471

Robbins TW (2002) The 5-choice serial reaction time task: behavioural pharmacology and functional neurochemistry. Psychopharmacology (Berl) 163:362–380

Roberts AC, Robbins TW, Weiskrantz L (eds) (1998) The prefrontal cortex: executive and cognitive functions. Oxford University Press, New York, NY US

Robinson TE, Flagel SB (2009) Dissociating the predictive and incentive motivational properties of reward-related cues through the study of individual differences. Biol Psychiatry 65:869–873

Robinson ESJ, Eagle DM, Economidou D, Theobald DEH, Mar AC, Murphy ER, Robbins TW, Dalley JW (2009) Behavioural characterisation of high impulsivity on the 5-choice serial reaction time task: specific deficits in 'waiting' versus 'stopping'. Behav Brain Res 196:310–316

Rogers RD, Robbins TW (2001) Investigating the neurocognitive deficits associated with chronic drug misuse. Curr Opin Neurobiol 11:250–257

Rogers RD, Everitt BJ, Baldacchino A, Blackshaw AJ, Swainson R, Wynne K, Baker NB, Hunter J, Carthy T, Booker E, London M, Deakin JF, Sahakian BJ, Robbins TW (1999) Dissociable deficits in the decision-making cognition of chronic amphetamine abusers, opiate abusers, patients with focal damage to prefrontal cortex, and tryptophan-depleted normal volunteers: evidence for monoaminergic mechanisms. Neuropsychopharmacology 20:322–339

Roiser JP, Cannon DM, Gandhi SK, Taylor Tavares J, Erickson K, Wood S, Klaver JM, Clark L, Zarate CA, Sahakian BJ, Drevets WC (2009) Hot and cold cognition in unmedicated depressed subjects with bipolar disorder. Bipolar Disord 11:178–189

Saunders BT, Robinson TE (2010) A cocaine cue acts as an incentive stimulus in some but not others: implications for addiction. Biol Psychiatry 67:730–736

Schoenbaum G, Roesch MR, Stalnaker TA, Takahashi YK (2009) A new perspective on the role of the orbitofrontal cortex in adaptive behaviour. Nat Rev Neurosci 10:885–892

Shackman AJ, McMenamin BW, Maxwell JS, Greischar LL, Davidson RJ (2009) Right dorsolateral prefrontal cortical activity and behavioral inhibition. Psychol Sci 20:1500–1506

Sher KJ, Bartholow BD, Wood MD (2000) Personality and substance use disorders: a prospective study. J Consult Clin Psychol 68:818–829

Shors TJ (2004) Memory traces of trace memories: neurogenesis, synaptogenesis and awareness. Trends Neurosci 27:250–256

Sikström S, Söderlund G (2007) Stimulus-dependent dopamine release in attention-deficit/hyperactivity disorder. Psychological Rev 114:1047–1075

Solanto MV (2002) Dopamine dysfunction in AD/HD: integrating clinical and basic neuroscience research. Behav Brain Res 130:65–71

Solanto MV, Abikoff H, Sonuga-Barke E, Schachar R, Logan GD, Wigal T, Hechtman L, Hinshaw S, Turkel E (2001) The ecological validity of delay aversion and response inhibition as measures of impulsivity in AD/HD: a supplement to the NIMH multimodal treatment study of AD/HD. J Abnorm Child Psychol 29:215–228

Sonuga-Barke EJ (1994) On dysfunction and function in psychological theories of childhood disorder. J Child Psychol Psychiatry 35:801–815

Sonuga-Barke EJS (2003) The dual pathway model of AD/HD: an elaboration of neuro-developmental characteristics. Neurosci Biobehav Rev 27:593–604

Sonuga-Barke EJS (2005) Causal models of attention-deficit/hyperactivity disorder: from common simple deficits to multiple developmental pathways. Biol Psychiatry 57:1231–1238

Sonuga-Barke EJ, Taylor E, Sembi S, Smith J (1992) Hyperactivity and delay aversion–I. the effect of delay on choice. J Child Psychol Psychiatry 33:387–398

Sonuga-Barke E, Bitsakou P, Thompson M (2010) Beyond the dual pathway model: evidence for the dissociation of timing, inhibitory, and delay-related impairments in attention-deficit/hyperactivity disorder. J Am Acad Child Adolesc Psychiatry 49:345–355

Stephens DN, Duka T, Crombag HS, Cunningham CL, Heilig M, Crabbe JC (2010) Reward sensitivity: issues of measurement, and achieving consilience between human and animal phenotypes. Addict Biol 15:145–168

Taylor JR, Robbins TW (1986) 6-Hydroxydopamine lesions of the nucleus accumbens, but not of the caudate nucleus, attenuate enhanced responding with reward-related stimuli produced by intra-accumbens d-amphetamine. Psychopharmacology (Berl) 90:390–397

van Gaalen MM, Brueggeman RJ, Bronius PFC, Schoffelmeer ANM, Vanderschuren LJMJ (2006) Behavioral disinhibition requires dopamine receptor activation. Psychopharmacology (Berl) 187:73–85

Verdejo-García A, Lawrence AJ, Clark L (2008) Impulsivity as a vulnerability marker for substance-use disorders: review of findings from high-risk research, problem gamblers and genetic association studies. Neurosci Biobehav Rev 32:777–810

Vertes RP (2006) Interactions among the medial prefrontal cortex, hippocampus and midline thalamus in emotional and cognitive processing in the rat. Neuroscience 142:1–20

Volkow ND, Wang G, Newcorn J, Telang F, Solanto MV, Fowler JS, Logan J, Ma Y, Schulz K, Pradhan K, Wong C, Swanson JM (2007) Depressed dopamine activity in caudate and preliminary evidence of limbic involvement in adults with attention-deficit/hyperactivity disorder. Arch Gen Psychiatry 64:932–940

Volkow ND, Wang G, Kollins SH, Wigal TL, Newcorn JH, Telang F, Fowler JS, Zhu W, Logan J, Ma Y, Pradhan K, Wong C, Swanson JM (2009) Evaluating dopamine reward pathway in ADHD: clinical implications. JAMA 302:1084–1091

Voorn P, Vanderschuren LJMJ, Groenewegen HJ, Robbins TW, Pennartz CMA (2004) Putting a spin on the dorsal-ventral divide of the striatum. Trends Neurosci 27:468–474

Williams J (2010) Attention-deficit/hyperactivity disorder and discounting: multiple minor traits and states. In: Madden GJ, Bickel WK (eds) Impulsivity: the behavioral and neurological science of discounting. Washington DC, US. pp 323–357

Wills TA, Vaccaro D, McNamara G (1994) Novelty seeking, risk taking, and related constructs as predictors of adolescent substance use: an application of Cloninger's theory. J Subst Abuse 6:1–20

Winstanley CA, Theobald DEH, Cardinal RN, Robbins TW (2004) Contrasting roles of basolateral amygdala and orbitofrontal cortex in impulsive choice. J Neurosci 24:4718–4722

Winstanley CA, Eagle DM, Robbins TW (2006) Behavioral models of impulsivity in relation to ADHD: translation between clinical and preclinical studies. Clin Psychol Rev 26:379–395

Wong MM, Nigg JT, Zucker RA, Puttler LI, Fitzgerald HE, Jester JM, Glass JM, Adams K (2006) Behavioral control and resiliency in the onset of alcohol and illicit drug use: a prospective study from preschool to adolescence. Child Dev 77:1016–1033

Zeeb FD, Floresco SB, Winstanley CA (2010) Contributions of the orbitofrontal cortex to impulsive choice: interactions with basal levels of impulsivity, dopamine signalling, and reward-related cues. Psychopharmacology (Berl) 211:87–98

Zuckerman M (1990) The psychophysiology of sensation seeking. J Pers 58:313–345

Obesity and ADHD: Clinical and Neurobiological Implications

Samuele Cortese and Brenda Vincenzi

Contents

1 Introduction 201
2 Studies Assessing the Relationship Between ADHD and Obesity 202
    2.1 Studies Examining the Prevalence of ADHD in Obese Subjects (Table 1) 202
    2.2 Studies Assessing the Weight Status of Subjects with ADHD 204
3 Behavioral and Neurobiological Mechanisms Underlying the Association
   Between ADHD and Obesity 205
    3.1 Obesity or Related Factors Manifest as ADHD 208
    3.2 ADHD and Obesity Share Common Biological Mechanisms 209
    3.3 ADHD Contributes to Obesity 211
4 Clinical Management of Patients with ADHD and Obesity 212
    4.1 Screening for Both ADHD and Problematic Eating Behavior 212
    4.2 Therapeutic Strategies 213
5 Conclusions and Future Perspectives 214
References 215

Abstract Although quite overlooked, increasing evidence points to a significant association between attention-deficit/hyperactivity disorder (ADHD) and obesity. Here, we present an updated systematic review and a critical discussion of studies on the relationship between ADHD and obesity, with a particular emphasis on the possible behavioral, neurobiological, and genetics underlying mechanisms.

S. Cortese (✉)
Institute for Pediatric Neuroscience, New York University Child Study Center, New York, NY, USA

215, Lexington Ave, 14th Floor, New York 10016 NY
e-mail: samuele.cortese@gmail.com

B. Vincenzi
Ellern Mede Centre, London, UK

31 Totteridge Common, London N20 8LR,
e-mail: vincenzibrenda@gmail.com

Available empirically based studies indicate that the prevalence of ADHD in clinical samples of patients seeking treatment for their obesity is higher than that in the general population. Moreover, although still limited, current evidence shows that individuals with ADHD have higher-than-average body mass index z-scores and/or significantly higher obesity rates compared with subjects without ADHD. Three mechanisms underlying the association between ADHD and obesity have been proposed: (1) obesity and/or factors associated with it (such as sleep-disordered breathing and deficits in arousal/alertness) manifest as ADHD-like symptoms; (2) ADHD and obesity share common genetics and neurobiological dysfunctions, involving the dopaminergic and, possibly, other systems (e.g., brain-derived neurotropic factor, melanocortin-4-receptor); and (3) impulsivity and inattention of ADHD contribute to weight gain via dysregulated eating patterns. With regards to the possible clinical implications, we suggest that it is noteworthy to screen for ADHD in patients with obesity and to look for abnormal eating behaviors as possible contributing factors of obesity in patients with ADHD. If further studies confirm a causal relationship between ADHD and obesity, appropriate treatment of ADHD may improve eating patterns and, as a consequence, weight status of individuals with both obesity and ADHD.

Keywords ADHD · Binge eating · BMI · Eating disorders · Obesity

Abbreviations

ADD	Attention-deficit disorder
BDNF	Brain-derived neurotropic factor
BMI-SDS	Body mass index standard deviations scores
CAPA	Child and Adolescent Psychiatric Assessment
CI	Confidence interval
CPRS-R:S	Conners Parent Rating Scale-Revised:Short version
DRD2	Dopamine D2 receptor
DRD4	Dopamine D4 receptor
EDS	Excessive daytime sleepiness
MC4-R	Melanocortin-4-receptor
NSCH	National Survey of Children's Health
OR	Odds ratio
SAD	Seasonal affective disorder
SDSC	Sleep Disturbance Scale for Children
WURS-25	Wender Utah Rating Scale-25

1 Introduction

Recent evidence suggests strong links between attention-deficit/hyperactivity disorder (ADHD) and obesity, although findings are not always consistent. A better insight into this association is of relevance for two reasons. First, it may contribute to the understanding of possible psychopathological and pathophysiological mechanisms underlying both ADHD and obesity, at least in a subset of patients. Second, it might have important implications for the management of patients with both obesity and ADHD, suggesting common therapeutic strategies when these two conditions coexist. This seems particularly noteworthy because of the high prevalence and the enormous personal, family, and social burden associated with both obesity and ADHD.

Therefore, the aims of this chapter are to: (1) review the evidence for an association between ADHD and obesity; (2) examine the behavioral and pathophysiological mechanisms that have been proposed to underlie this potential association; and (3) discuss the implications of this newly described potential comorbidity on the clinical management of patients who present with both ADHD and obesity.

Before examining the relationship between ADHD and obesity, we introduce in the following paragraph the definition of obesity and overweight that will be used throughout this chapter. We refer to the definitions available in the World Health Organization (WHO) website: http://www.who.int/en/. Overweight and obesity are characterized by abnormal or excessive fat accumulation that may impair health. Body mass index (BMI) is a simple index of weight-for-height that is commonly used in classifying overweight and obesity in adult populations and individuals. It is defined as the weight in kilograms divided by the square of the height in meters (kg/m^2). BMI provides the most useful population-level measure of overweight and obesity, as it is the same for both sexes and for adults of all ages. However, it should be considered as a rough guide because it may not correspond to the same degree of fatness in different individuals. The WHO defines "overweight" as a BMI greater than or equal to 25 kg/m^2, and "obesity" as a BMI greater than or equal to 30. The new WHO Child Growth Standards, launched in April 2006, include BMI charts for infants and young children up to the age of 5. However, measuring overweight and obesity in children aged 5–14 years is challenging because there is not a standard definition of childhood obesity applied worldwide. WHO is currently developing an international growth reference for school-age children and adolescents. Since BMI does not adequately describe central adiposity, other indices of body fatness are being explored. A recent study showed a significant correlation between BMI and neck circumference in children; the best cutoff that identified high BMI in boys ranged from 28.5 to 39.0 cm; corresponding values in girls ranged to 27.0–34.6 cm (Nafiu et al. 2010). However, since this a recent proposal, no study assessing the relationship between ADHD and obesity has used this index, which, therefore, will not be taken into consideration in this chapter.

2 Studies Assessing the Relationship Between ADHD and Obesity

Our group published in 2008 a systematic review (Cortese et al. 2008a) of the studies exploring the relationship between ADHD and obesity. We searched both for studies assessing the prevalence of ADHD in obese subjects and for those evaluating the weight status of subjects with ADHD. We included methodologically sound studies, published up to January 2007, conducted in children and in adults with ADHD diagnosed according to formal criteria. Since drugs used for the treatment of ADHD may have anorexigenic effects, we did not consider studies examining weight status in treated patients with ADHD or studies which did not control for the effect of ADHD medications. Since that review, an additional three methodologically sound studies (where the diagnosis of ADHD was based on formal criteria) have been published (Braet et al. 2007; Pagoto et al. 2009; Ptacek et al. 2009). In the following section, we present and critically discuss the key results from the studies included in our initial systematic review plus those from the three new additional studies published after January 2007. We also present additional studies that, although do not meet the criteria for inclusion in the systematic review, provide interesting insights on the relationship between ADHD symptoms and obesity.

2.1 Studies Examining the Prevalence of ADHD in Obese Subjects (Table 1)

We located five methodologically sound studies (Agranat-Meged et al. 2005; Altfas 2002; Braet et al. 2007; Erermis et al. 2004; Fleming et al. 2005) conducted in clinical settings, and one survey (Mustillo et al. 2003) in the general population. The five clinical studies were conducted, respectively, in a clinical sample of: 215 obese adults receiving obesity treatment in a clinic specialized in the prevention and treatment of obesity (Altfas 2002); 30 obese adolescents (aged 12–16 years) seeking treatment from a pediatric endocrinology outpatient clinic (Erermis et al. 2004); 26 children (aged 8–17 years) hospitalized in an eating disorder unit for the treatment of their refractory morbid obesity (Agranat-Meged et al. 2005); 75 obese women (mean age: 40.4 years) consecutively referred to a medical obesity clinic (Fleming et al. 2005); and in a clinical sample of 56 overweight children versus 53 normal weight children (10–18 years) (Braet et al. 2007). All these studies, except one (Braet et al. 2007), reported a significantly higher prevalence of ADHD (diagnosed according to formal criteria) in obese patients than in comparable controls or age-adapted reference values. As for the negative study (Braet et al. 2007), we believe that its results cannot lead to a firm conclusion since

Table 1 Studies examining the prevalence of ADHD in obese subjects. Case reports, nonempirical studies, and studies not using standardized ADHD diagnostic criteria were excluded

First author (year)	Subjects	Mean age (SD)/age range (years)	Key results
Altfas (2002)	215 obese adults treated in a bariatric clinic	43.4 (10.9)	27.4% of all patients presented with ADHD. The proportion of patients having ADHD in obesity class III (42.6%) was significantly higher than those in obesity class I–II and overweight class (22.8% and 18.9 %, respectively) ($p = 0.002$). Patients with ADHD had a significantly higher BMI (39.2) than patients without ADHD (34.6) ($p = 0.01$)
Mustillo et al. (2003)	991 youth in the general pediatric population	Age range: 9–16 years	The diagnosis of ADHD was not associated with any of the obesity trajectories (no obesity, childhood obesity, adolescent obesity, and chronic obesity)
Erermis et al. (2004)	30 obese adolescents (clinical), 30 obese adolescents (nonclinical), 30 normal weight adolescents	Clinical group: boys: 13.8 (1.2); girls: 13.8 (1.3)	The prevalence of ADHD was significantly higher in the clinical obese group (13.3%) in comparison to the nonclinical obese group (3.3%) and the control group (3.3%)
Fleming et al. (2005)	75 obese women consecutively referred to a medical obesity clinic	40.4 (10.8)	38.6% of the participants scored at or above the cutoff of 36 on the WURS, significantly higher than the 4% prevalence previously reported in a normal sample ($p < 0.001$); the inattention-memory, the impulsivity-emotionality, the DSM-IV-inattentive, and the ADHD-index subscales scores of the CAARS were significantly higher than the expected frequency; 61% of the patients had scores at the Brown ADD Scale for Adults, suggestive of probable ADHD; 26.6% of the patients were classified as likely cases of ADHD (vs. DSM-IV adult prevalence estimates of 3–5%)
Agranat-Meged et al. (2005)	26 obese children in a pediatric eating disorder unit	13.04 (2.78)	57.7% of children presented with ADHD vs. estimates of 10% in the general population in the same age group ($p < 0.0001$).
Braet et al. (2007)	56 overweight children (clinical) and 53 normal weight children	10–18 years	Overweight boys showed significantly more impulsivity, hyperactivity, and attention-deficit symptoms than control boys on several tests but no higher prevalence of ADHD

neither the details of the interview nor the medication status of the patients was reported.

However, the four positive studies also present some methodological limitations that hamper a firm conclusion from their results, such as: the relatively small sample sizes; the lack of a control group in some of them (Altfas 2002; Fleming et al. 2005); and possible selection biases (e.g., the study by Fleming et al. was conducted in a female sample (Fleming et al. 2005)). However, despite these limitations, they all suggest that obese patients referred to obesity clinics may present with higher than expected prevalence of ADHD.

By contrast, the only epidemiological study reported no association between ADHD and obesity. This study (Mustillo et al. 2003) was conducted in a large sample (991 youths aged 9–16 years) and was based on the state-of-the-art assessment for psychiatric disorder in childhood (Child and Adolescent Psychiatric Assessment, CAPA). However, given the epidemiological nature of the study and the large sample size, it was difficult to obtain information from multiple sources, as required for an appropriate diagnosis of ADHD. The discordance of the results between the four clinical studies and the epidemiological survey may be due to methodological issues. Indeed, clinical settings may favor a higher case-finding rate because the opportunity to observe and assess behavior is greater than the methods of epidemiological surveys.

In summary, current evidence points to higher than expected prevalence of ADHD in clinical samples of obese patients seeking for treatment, while there are no data supporting significantly higher prevalence of ADHD in the general population of obese individuals.

2.2 Studies Assessing the Weight Status of Subjects with ADHD

In our systematic review, we located 12 studies (Anderson et al. 2006; Biederman et al. 2003; Curtin et al. 2005; Faraone et al. 2005; Holtkamp et al. 2004; Hubel et al. 2006; Lam and Yang 2007; Pagoto et al. 2009; Ptacek et al. 2009; Spencer et al. 1996, 2006; Swanson et al. 2006) that examined the weight status of children with ADHD. Eight of these studies (Biederman et al. 2003; Curtin et al. 2005; Faraone et al. 2005; Holtkamp et al. 2004; Ptacek et al. 2009; Spencer et al. 1996, 2006; Swanson et al. 2006) were conducted in clinical settings, while four (Anderson et al. 2006; Hubel et al. 2006; Lam and Yang 2007; Pagoto et al. 2009) examined the general population. These studies showed that children with ADHD presented with higher-than-average body mass index standard deviations scores (BMI-SDS). Interestingly, one of these studies (Ptacek et al. 2009), besides BMI, used also other indices of body mass (percentage of body fat and abdominal circumference): the authors found that these were significantly higher in ADHD compared to controls.

Limitations of these 12 studies included a lack of control group (Curtin et al. 2005; Faraone et al. 2005; Holtkamp et al. 2004; Spencer et al. 2006; Swanson et al. 2006), selection biases (e.g., inclusion of males, but not females (Holtkamp et al. 2004)), lack of control for the effect of comorbid psychiatric disorders, which may have an impact on subjects' BMI (Anderson et al. 2006; Lam and Yang 2007), or retrospective diagnosis of adult ADHD (Pagoto et al. 2009). However, this last study (Pagoto et al. 2009) is the only one that we were able to locate in adults. Interestingly, in this study, adult ADHD remained associated with overweight (OR = 1.57; 95% CI = 0.99, 2.70) and obesity (OR = 1.69; 95% CI = 1.01, 2.82) when controlling for major depressive disorder. However, the association was no longer statistically significant when controlling for binge-eating disorder in the past 12 months (1.41, 95% CI = 0.76–2.53).

Although they did not rely on formal criteria for the diagnosis of ADHD (and, therefore, they are not mentioned in Table 2), we believe it is interesting to mention two additional studies that we located in our search. In the first, a large cohort of 46,707 subjects, aged 10–17 years, collected by the National Survey of Children's Health (NSCH-2003), the authors (Chen et al. 2010) reported that the prevalence of obesity was 18.9% (95% CI: 18.7–19.0) in children with ADHD compared to 12.2% (95% CI: 11.5–13.0) in children without chronic diseases. Differences in the prevalence of obesity in children with and without ADHD remained significant after adjustment for age, sex, race/ethnicity, family income, family structure, parental education, and region. Similarly, in the second study, another large cross-sectional analysis of 62,887 children and adolescents, aged 5–17 years from the 2003–2004 National Survey of Children's Health, the authors (Waring and Lapane 2008) found that children with ADHD, who were not currently using medication, had about 1.5 times the odds of being overweight compared to non-ADHD children. However, as we stated, the major limitation of both these studies was the lack of a formal diagnosis of ADHD, since the assessment of the presence of ADHD was based on the question: "Has a doctor or health professional ever told you [you suffer from ADHD]?" Moreover, the authors did not control for the effect of medication status and psychiatric comorbidity.

In summary, studies both in clinical and epidemiological samples showed that individuals with ADHD have higher than average BMI-SDS and a higher prevalence of obesity compared to non-ADHD subjects. We note that, besides ADHD, other psychological disorders such as depression, low-self esteem, and anxiety have also been found to relate to obesity in children and adolescents (Zametkin et al. 2004).

3 Behavioral and Neurobiological Mechanisms Underlying the Association Between ADHD and Obesity

Since all the studies reviewed above (with the exception of one study, Mustillo et al. (2003)) are cross-sectional, they do not enable an understanding of the causality between ADHD and obesity.

Table 2 Studies on the weight status of subjects with ADHD. Case reports, nonempirical studies, and studies not using standardized ADHD diagnostic criteria were excluded

First author (year)	Subjects	Mean age (SD)/age range (years)	Key results
Spencer et al. (1996)	124 males with ADHD and 109 normal controls	Age range: 6–17	ADHD subjects had greater than average body mass (age- and height-corrected weight index: 109 ± 15), although no significant difference was found between the age- and height-corrected weight index of ADHD and control subjects. Age- and height-corrected weight index of untreated ADHD was 115, indicative of overweight
Biederman et al. (2003)	140 ADHD girls and 122 female controls	Age range: 6–17	The age-and height-corrected weight index was greater than average (1.1), although not indicative of overweight or obesity. No significant differences were found between ADHD girls and controls, as well as between treated and untreated subjects. ADHD girls with comorbid major depression (MD) had a significantly greater average height- and age-corrected weight index relative to ADHD girls without MD ($p = 0.011$)
Holtkamp et al. (2004)	97 inpatient and outpatient boys with ADHD in a Child and Adolescent Psychiatric Department	10 (2)	The mean BMI-SDS of ADHD patients were significantly higher than the age-adapted reference values ($p = 0.038$). The proportion of obese (7.2%) and overweight (19.6%) participants was significantly higher than estimated prevalence ($p = 0.0008$ and $p = 0.0075$, respectively).
Curtin et al. (2005)	98 children with ADHD in a tertiary care clinic for developmental, behavioral, and cognitive disorders	Age range: 3–18	29% of children with ADHD were at-risk for overweight and 17.3% were overweight. No significant difference in comparison to an age-matched reference population. However, the prevalence of at-risk for overweight and overweight in children not treated with ADHD drugs (36% and 23%, respectively) was significantly higher than that found in treated participants (16% and 6.3%, respectively) ($p < 0.05$)
Faraone et al. (2005)	568 children with ADHD enrolled in a study of the safety of mixed amphetamine salts	Age range: 6–12	At baseline, subjects were heavier than average (mean BMI z-score = 0.41)
Hubel et al. (2006)	39 boys with ADHD and 30 healthy controls	Age range: 8–14	BMI-SDS were higher in ADHD than in controls. No significant association between group membership (control vs. ADHD) and obesity or overweight

Anderson et al. (2006)	655 subjects (general population)	Younger than 16.6	Subjects with ADHD had higher mean BMI z-scores at all ages compared with individuals who were not observed with a disruptive disorder
Spencer et al. (2006)	178 children with ADHD receiving OROS methylphenidate	Age range: 6–13	Subjects were slightly overweight compared with that expected for their age (mean BMI z-score = 0.230)
Swanson et al. (2006)	140 children with ADHD	Age range: 3–5.5	The average BMI was 16.9, which corresponds to the 86th percentile at the baseline assessment
Lam and Yang (2007)	1,429 students (general population)	Age range: 13–17	Subjects with high ADHD tendency had an increased risk for obesity of 1.4 times as compared to subjects with low ADHD tendency
Ptacek et al. (2009)	46 nonmedicated ADHD boys	Average age 11.03 years	Boys with ADHD had significantly higher values of percentage of body fat and abdominal circumference
Pagoto et al. (2009)	6,735 US residents	18–44 years	Obesity was more prevalent among persons with adult ADHD (29.4%) than among those with a history of childhood ADHD but no adult symptoms (23.7%) and those with no history of ADHD (21.6%)

From a theoretical point of view, it is possible that: (1) obesity and/or factors associated with obesity (such as sleep-disordered breathing (SDB)) lead to, or manifest as, ADHD/ADHD symptoms; (2) ADHD and obesity are the expression of a common biological dysfunction that manifests itself as both obesity and ADHD in a subset of patients; (3) ADHD contributes to obesity. We discuss these hypotheses in the following sections.

3.1 Obesity or Related Factors Manifest as ADHD

Two factors, potentially associated with obesity, have been proposed that might manifest as ADHD or its symptoms: these are binge eating and excessive daytime sleepiness.

3.1.1 Binge Eating

It is known that a subgroup of patients with obesity, especially those with severe obesity (Hudson et al. 2006), presents with binge eating. One can hypothesize that impulsivity associated with abnormal eating behavior in these patients contributes to, or manifests itself as, impulsivity of ADHD. It is also possible that impulsivity associated with abnormal eating behaviors fosters symptoms of inattention and hyperactivity. It has been reported that patients with bulimic or abnormal eating behaviors may present with repeated and impulsive interruptions of their activities to get food; this behavior results in ADHD symptoms, such as disorganization, inattention, and restlessness (Cortese et al. 2007a). Of note, a research group (Rosval et al. 2006) reported higher rates of motor impulsiveness in patients with bulimia nervosa and anorexia nervosa binge/purge subtype, in comparison to anorexia nervosa, restricting subtype and normal eater control group. This finding further confirms previous studies that linked binge-eating behaviors and behavioral impulsivity (Engel et al. 2005; Nasser et al. 2004). However, other authors (Davis et al. 2009) found evidence against the role of binge eating in explaining the association between ADHD and obesity. After predicting that ADHD symptoms would be more severe in the binge eaters compared to obese controls, using a case–control design (binge eaters, $n = 60$, versus non-binge eaters with normal weight, $n = 61$, and obese subjects, $n = 60$) these authors, contrary to their prediction, reported that symptoms of ADHD were elevated in obese adults, with and without binge eating, but there was no difference between these two groups.

3.1.2 Excessive Daytime Sleepiness

Another hypothesis is based on the link between obesity, hypoarousal (manifested as excessive daytime sleepiness, EDS), and ADHD symptoms. According to this theory, obesity may contribute to excessive daytime sleepiness which, in turn, could lead to ADHD symptoms.

As for the link between EDS and ADHD, EDS, either due to SDB or independent from it, may contribute to ADHD symptoms. According to the "hypoarousal theory" of ADHD (Weinberg and Brumback 1990), subjects with ADHD behaviors (or at least a subgroup of them) might be sleepier than controls and might use motor hyperactivity and impulsivity as a strategy to stay awake and alert. In a recent meta-analysis, we found evidence supporting this proposal: a higher prevalence of excessive daytime sleepiness in children with ADHD versus controls (Cortese et al. 2009). As for the link between obesity and EDS, several studies have reported an association between obesity and SDB or other sleep disorders (Cortese et al. 2008b). These disorders may cause sleep fragmentation, leading to excessive daytime sleepiness. Moreover, as reported in the review of the literature (Vgontzas et al. 2006), recent evidence, based on subjective as well as objective measures, indicates that obesity may be associated with EDS independently of SDB or any other sleep disturbances. It has been suggested (Vgontzas et al. 2006) that, at least in some obese patients, EDS may be related to a metabolic and/or circadian abnormality associated with obesity rather than being a consequence of SDB or other sleep disturbances. Interestingly, in a study by our group (Cortese et al. 2007b) using a subjective measure of sleepiness (a subscale of the Sleep Disturbance Scale for Children, SDSC) in a sample of 70 obese children (age range: 10–16 years), scores of excessive daytime sleepiness on the SDSC were associated with symptoms of inattention, hyperactivity, and impulsivity on both the Conners Parent Rating Scale-R:S and the ADHD-Rating Scale. Clearly, this finding needs to be replicated in studies using objective (i.e., polysomnographic) measures of EDS (hypoarousal).

3.2 ADHD and Obesity Share Common Biological Mechanisms

Another possibility is that obesity and ADHD are different expressions of common underlying biological mechanisms or share common biological underpinnings, at least in a subset of patients with both these conditions. Several biological processes have been proposed, including: altered reinforcement mechanisms, brain-derived neurotropic disorder, and melanocortin-4-receptor (MC4-R) deficiency.

3.2.1 Altered Reinforcement Mechanisms

The "reward deficiency syndrome" may play a significant role. This syndrome is characterized by an insufficient dopamine-related natural reinforcement that leads to the use of "unnatural" immediate rewards, such as: substance use, gambling, risk-taking, and inappropriate eating. Several lines of evidence suggest that patients with ADHD may present with behaviors consistent with the "reward deficiency syndrome" (e.g., Blum et al. 1995; Heiligenstein and Keeling 1995). This syndrome has been reported also in obese patients with abnormal eating behaviors (Comings and Blum 2000).

In a sample of women with seasonal affective disorder (SAD), the 7R allele of DRD4 was associated with higher scores of childhood inattention on the WURS-25 and with higher maximal lifetime BMI. SAD is characterized by marked craving for high-carbohydrate/high-fat foods, resulting in significant weight gain during depressive episodes in the winter. A potential implication of the reward system in the pathophysiology of the disorder has been suggested. It has been hypothesized (Levitan et al. 2004) that childhood attention deficit and adult obesity could be the expression of a common biological dysfunction of the 7R allele of DRD4 associated with a dopamine dysfunction in prefrontal attentional areas and brain circuits involved in the reward pathways. Alterations in the dopamine D2 receptor (DRD2) (Bazar et al. 2006) and, to a lesser extent, DRD4 (Mitsuyasu et al. 2001; Tsai et al. 2004) have been associated with the above-mentioned "reward deficiency syndrome." Dysfunctions of DRD2 and DRD4 have been found in obese patients (Poston et al. 1998). Several studies suggest a role of altered DRD4 and DRD2 in ADHD as well (although the alteration in DRD2 has not been replicated in all studies) (Noble 2003). Therefore, obese patients with abnormal eating behaviors and ADHD may present with common genetically determined dysfunctions in the dopaminergic system, in particular the dopaminergic circuits underlying reward mechanisms and executive functions necessary for appropriate eating behaviors. However, these hypotheses need to be tested in empirically based, methodologically sound studies.

3.2.2 Brain-Derived Neurotropic Factor

Another potential common biological mechanism involves alterations in the brain-derived neurotropic factor (BDNF), a protein member of the family of growth factors. It supports the survival of existing neurons and encourages the growth and differentiation of new neurons and synapses. Preliminary evidence from animal models points to a potential dysfunction of BDNF underlying both ADHD and obesity (Kernie et al. 2000; Lyons et al. 1999). Although one research group (Friedel et al. 2005) failed to find a large role of genetic variation of BDNF in ADHD and obesity, another group (Gray et al. 2006) reported a functional loss of one copy of the BDNF gene in an 8-year old with hyperphagia, severe obesity, impaired cognitive function, and hyperactivity. Moreover, a recent study has reported a significant association among 11p14.1 microdeletions (encompassing the BDNF gene), ADHD, and obesity (Shinawi et al. 2011).

3.2.3 Melanocortin-4-Receptor Deficiency

MC4-R is a 332-amino acid protein encoded by a single exon gene localized on chromosome 18q22 and is a key element in the hypothalamic control of food intake.

MC4-R deficiency has been proposed recently (Agranat-Meged et al. 2008) and has been reported to disrupt neuronal pathways that regulate hunger/satiety and results in abnormal eating behaviors. The authors analyzed 29 subjects (19 males and 10 females) from 5 "proband nuclear families" with morbid obese children (BMI percentile >97%), and found that the prevalence of ADHD was higher than expected only in the groups carrying the homozygous or heterozygous mutation. Recently, Albayrak et al. (2011) reported the case of a 2-year-old child heterozygous for a non-conservative functionally relevant MC4-R mutation (Glu308Lys) and who also showed severe ADHD.

Given the limited and, at least in part, inconsistent findings, the role of common genetic mutations underlying shared neurobiological dysfunctions in ADHD and obesity deserves further investigation.

3.3 ADHD Contributes to Obesity

Finally, behavioral and cognitive features of ADHD may play a significant role in abnormal eating behaviors, which in turn lead to obesity. A study of 110 adult healthy women (age range: 25–46 years) using structural equation modeling (Davis et al. 2006) found that ADHD symptoms and impulsivity correlated with abnormal eating behaviors, including binge eating and emotionally induced eating, which were, in turn, positively associated with BMI. These results have been recently replicated in males (Strimas et al. 2008). In our study, conducted in a clinical sample of 99 severely obese adolescents (aged 12–17 years), we found that, after controlling for potentially confounding depressive and anxiety symptoms, ADHD symptoms were associated with bulimic behaviors (Cortese et al. 2007a). Bearing in mind that loss of control and excessive food intake are important determiners of weight gain, children and adolescents who report binge-eating symptoms gain more weight over time than children and adolescents without such history (Lourenco et al. 2008).

At present, it is not clear which dimension of ADHD (inattention, hyperactivity, or impulsivity) is associated with abnormal eating behaviors. We speculated that both impulsivity and inattention, but not hyperactivity, might lead or contribute to abnormal eating behaviors (Cortese et al. 2007a). We detail this hypothesis in the following sections.

3.3.1 Role of Impulsivity

It has been suggested (Davis et al. 2006) that both *deficient inhibitory control* and *delay aversion*, which are an expression of the impulsivity component of ADHD, may foster abnormal eating behaviors and, as a consequence, obesity. *Deficient inhibitory control* manifests as poor planning and difficulty in monitoring one's behavior effectively. This could lead to over-consumption when not hungry, associated with a relative absence of concern for daily caloric intake. A strong *delay aversion* could favor the tendency to eat high caloric, but readily available "fast food," in preference to less caloric content home-cooked meals, which take longer to prepare. This may contribute to maintenance of a chronic positive energy balance that culminates in obesity.

3.3.2 Role of Inattention

It is also possible that ADHD-related inattention and associated deficits in executive functions cause difficulties in adhering to a regular eating pattern and so favor abnormal eating behaviors. It has been pointed out that patients with ADHD may be relatively inattentive to internal signs of hunger and satiety (Davis et al. 2006). Therefore, they may forget about eating when they are engaged in interesting activities but are more likely to eat when less stimulated.

Another hypothesis is that compulsive eating may be a compensatory mechanism to cope frustration associated with attentional and organizational difficulties (Schweickert et al. 1997). It is also possible that difficulties in initiating activities linked to attentional and organizational problems contribute to decreased caloric expenditure, leading to progressive weight gain (Levitan et al. 2004). Moreover, it has been hypothesized that, since patients with ADHD are susceptible to committing more cognitive effort to take charge of standard mental tasks, it is likely that this cognitive effort accentuates a proneness to an abnormally increased appetite, and the consequent long-term weight gain (Riverin and Tremblay 2009). Finally, other researchers (Waring and Lapane 2008) evoked the role of television, also. Since children with ADHD claim to spend more time watching television, or playing computer or video games to the detriment of exercising, this would contribute to weight gain. However, one cannot exclude the possibility that all the preceding hypotheses hold true, and that the effects on eating behavior may be additive or interactive, at least in certain subjects.

4 Clinical Management of Patients with ADHD and Obesity

4.1 Screening for Both ADHD and Problematic Eating Behavior

From a clinical standpoint, the putative association between ADHD and obesity suggests that it may be useful to screen for ADHD in patients with obesity and also to look for abnormal eating behaviors, as possible contributing factors of obesity. It has been pointed out that clinicians may overlook the need to screen for ADHD in obese patients since "obese individuals are less mobile" and, as a consequence, obesity may mask hyperactivity (Agranat-Meged et al. 2005). However, screening for ADHD in patients with obesity may be relevant since treatment of ADHD might prevent this disorder from exacerbating the personal and social burden of obesity. Conversely, binge eating and other abnormal eating behaviors, which are generally poorly investigated in patients with ADHD, should be screened as well. These abnormal eating behaviors could further compromise the quality of life of ADHD patients and exacerbate their weight gain.

4.2 Therapeutic Strategies

If ADHD contributes to obesity, as hypothesized, then the treatment of ADHD might be expected to prevent obesity. Several reports (Schweickert et al. 1997; Sokol et al. 1999) have suggested that stimulants improve ADHD and restore normal eating behaviors in patients with both conditions. According to some authors (Surman et al. 2006), the treatment of ADHD-related impulsivity could improve abnormal eating behaviors as well. Improvement in attention, leading to more regular eating patterns, may also play a role.

A group of Canadian researchers (Levy et al. 2009) conducted an interesting study in 78 obese patients with a lengthy history of failure to lose weight. After an average of 466 days of continuous ADHD pharmacotherapy, weight loss in treated patients was 15.05 kg, whereas nontreated patients gained weight (3.26 kg). The authors noted the low attrition rate in pharmacologically treated patients and observed that "with the advent of the salutary effects of medication on ADHD symptoms, subjects spoke enthusiastically and in optimistic terms regarding their future plans in life." Moreover, "improvements in daytime energy, restlessness, distractibility, working memory, impulsivity and mood were instrumental in their successful execution of weight loss plans." Of note, appetite suppression almost vanished within 2 months. Therefore, it is unlikely that appetite suppression explains the positive outcome in terms of weight control. Finally, subjects reported feeling more alert and energized with treatment. As a consequence, it is likely that this greater physical movement contributed to successful weight control. These interesting clinical observations should be investigated systematically and thoroughly in the future studies.

Indeed, all the above-mentioned studies on pharmacological treatments support the hypothesis that ADHD and abnormal eating behaviors share common underlying biological mechanisms. These could be the common target of ADHD medications. Alternatively, ADHD medications could act on both the brain pathways involved in ADHD and those that mediate abnormal eating behaviors. Interestingly, a trial reported the efficacy of atomoxetine in weight reduction in obese women (Gadde et al. 2006). The positive results of the trial suggested that this ADHD treatment may act on the noradrenergic synapses in the medial and paraventricular hypothalamus that are thought to play a major role in modulating satiety and feeding behavior.

These preliminary observations suggest that in patients with ADHD and abnormal eating behaviors (associated with obesity), both conditions might improve using the same class of agents.

5 Conclusions and Future Perspectives

Converging empirical evidence from recent literature points to a bidirectional association between ADHD and obesity. Both clinical and nonclinical samples indicate that subjects with ADHD have a higher BMI than average, as well as a higher prevalence of obesity/overweight, compared to non-ADHD controls. On the other hand, obese patients present with a higher prevalence of ADHD. To date, there are no data supporting a higher prevalence of ADHD in obese persons, seeking treatment in specialized clinics, from the general population. Evidently, further epidemiological studies exploring the prevalence of ADHD in obese subjects are needed.

Even if the association between ADHD and obesity holds true only in obese patients who seek treatment in specialized clinics, it would still be noteworthy because this is the group of patients with the highest mortality and morbidity risk and the greatest need for effective treatment.

There are other areas of research that deserve further and deeper investigation. First, the potential role of psychiatric comorbidities in explaining the higher prevalence of obesity in ADHD patients should be better addressed in the future. Second, all available studies are cross-sectional prospective and, therefore, they cannot contribute to an understanding of causality: prospective, longitudinal studies, at present, are lacking but could lead to a better understanding of the causal relationship between ADHD and obesity and shed light on the psychopathological pathways linking the two conditions. Third, the exact neurobiological and psychopathological mechanisms underlying the association deserve further exploration. In particular, both family studies, examining the occurrence of ADHD and obesity, and specific animal model studies (which, at present time, are lacking) could advance our knowledge of any common genetic underpinnings. Fourth, only one study (Levy et al. 2009) has assessed the effects of pharmacological treatment of ADHD on weight control in obese patients: further research in this area is warranted. Moreover, non-pharmacological treatment studies and multimodal (pharmacological plus non-pharmacological) ADHD treatment strategies for these patients are greatly needed to help us find more appropriate and effective therapeutic strategies for their weight control.

In summary, clinical empirically based studies, epidemiological surveys (both cross-sectional and prospective), genetic studies, animal model studies, non-pharmacological treatment studies, and pharmacological trials should all be encouraged since they would advance our knowledge in this field, allowing for better management and quality of life for patients with both obesity and ADHD.

Acknowledgment Dr. Cortese is supported by a grant from European Union "International outgoing fellowship, Marie Curie actions, FP7 Program, # PIOF-GA-2009-253103."

References

Agranat-Meged AN, Deitcher C, Goldzweig G, Leibenson L, Stein M, Galili-Weisstub E (2005) Childhood obesity and attention deficit/hyperactivity disorder: a newly described comorbidity in obese hospitalized children. Int J Eat Disord 37:357–359

Agranat-Meged A, Ghanadri Y, Eisenberg I, Ben NZ, Kieselstein-Gross E, Mitrani-Rosenbaum S (2008) Attention deficit hyperactivity disorder in obese melanocortin-4-receptor (MC4R) deficient subjects: a newly described expression of MC4R deficiency. Am J Med Genet B Neuropsychiatr Genet 147B:1547–1553

Albayrak O, Albrecht B, Scherag S, Barth N, Hinney A, Hebebrand J (2011) Eur J Pharmacol 660 (1):165–70

Altfas JR (2002) Prevalence of attention deficit/hyperactivity disorder among adults in obesity treatment. BMC Psychiatry 2:9

Anderson SE, Cohen P, Naumova EN, Must A (2006) Relationship of childhood behavior disorders to weight gain from childhood into adulthood. Ambul Pediatr 6:297–301

Bazar KA, Yun AJ, Lee PY, Daniel SM, Doux JD (2006) Obesity and ADHD may represent different manifestations of a common environmental oversampling syndrome: a model for revealing mechanistic overlap among cognitive, metabolic, and inflammatory disorders. Med Hypotheses 66:263–269

Biederman J, Faraone SV, Monuteaux MC, Plunkett EA, Gifford J, Spencer T (2003) Growth deficits and attention-deficit/hyperactivity disorder revisited: impact of gender, development, and treatment. Pediatrics 111:1010–1016

Blum K, Sheridan PJ, Wood RC, Braverman ER, Chen TJ, Comings DE (1995) Dopamine D2 receptor gene variants: association and linkage studies in impulsive-addictive-compulsive behaviour. Pharmacogenetics 5:121–141

Braet C, Claus L, Verbeken S, Van VL (2007) Impulsivity in overweight children. Eur Child Adolesc Psychiatry 16:473–483

Chen AY, Kim SE, Houtrow AJ, Newacheck PW (2010) Prevalence of obesity among children with chronic conditions. Obesity (Silver Spring) 18:210–213

Comings DE, Blum K (2000) Reward deficiency syndrome: genetic aspects of behavioral disorders. Prog Brain Res 126:325–341

Cortese S, Isnard P, Frelut ML, Michel G, Quantin L, Guedeney A, Falissard B, Acquaviva E, Dalla BB, Mouren MC (2007a) Association between symptoms of attention-deficit/hyperactivity disorder and bulimic behaviors in a clinical sample of severely obese adolescents. Int J Obes (Lond) 31:340–346

Cortese S, Maffeis C, Konofal E, Lecendreux M, Comencini E, Angriman M, Vincenzi B, Pajno-Ferrara F, Mouren MC, Dalla Bernardina B (2007b) Parent reports of sleep/alertness problems and ADHD symptoms in a sample of obese adolescents. J Psychosom Res 63:587–590

Cortese S, Angriman M, Maffeis C, Isnard P, Konofal E, Lecendreux M, Purper-Ouakil D, Vincenzi B, Bernardina BD, Mouren MC (2008a) Attention-deficit/hyperactivity disorder (ADHD) and obesity: a systematic review of the literature. Crit Rev Food Sci Nutr 48:524–537

Cortese S, Konofal E, Bernardina BD, Mouren MC, Lecendreux M (2008b) Does excessive daytime sleepiness contribute to explaining the association between obesity and ADHD symptoms? Med Hypotheses 70(1):12–16

Cortese S, Faraone SV, Konofal E, Lecendreux M (2009) Sleep in children with attention-deficit/hyperactivity disorder: meta-analysis of subjective and objective studies. J Am Acad Child Adolesc Psychiatry 48:894–908

Curtin C, Bandini LG, Perrin EC, Tybor DJ, Must A (2005) Prevalence of overweight in children and adolescents with attention deficit hyperactivity disorder and autism spectrum disorders: a chart review. BMC Pediatr 5:48

Davis C, Levitan RD, Smith M, Tweed S, Curtis C (2006) Associations among overeating, overweight, and attention deficit/hyperactivity disorder: a structural equation modelling approach. Eat Behav 7:266–274

Davis C, Patte K, Levitan RD, Carter J, Kaplan AS, Zai C, Reid C, Curtis C, Kennedy JL (2009) A psycho-genetic study of associations between the symptoms of binge eating disorder and those of attention deficit (hyperactivity) disorder. J Psychiatr Res 43:687–696

Engel SG, Corneliussen SJ, Wonderlich SA, Crosby RD, le Grange D, Crow S, Klein M, Bardone-Cone A, Peterson C, Joiner T, Mitchell JE, Steiger H (2005) Impulsivity and compulsivity in bulimia nervosa. Int J Eat Disord 38:244–251

Erermis S, Cetin N, Tamar M, Bukusoglu N, Akdeniz F, Goksen D (2004) Is obesity a risk factor for psychopathology among adolescents? Pediatr Int 46:296–301

Faraone SV, Biederman J, Monuteaux M, Spencer T (2005) Long-term effects of extended-release mixed amphetamine salts treatment of attention- deficit/hyperactivity disorder on growth. J Child Adolesc Psychopharmacol 15:191–202

Fleming JP, Levy LD, Levitan RD (2005) Symptoms of attention deficit hyperactivity disorder in severely obese women. Eat Weight Disord 10:e10–e13

Friedel S, Horro FF, Wermter AK, Geller F, Dempfle A, Reichwald K, Smidt J, Bronner G, Konrad K, Herpertz-Dahlmann B, Warnke A, Hemminger U, Linder M, Kiefl H, Goldschmidt HP, Siegfried W, Remschmidt H, Hinney A, Hebebrand J (2005) Mutation screen of the brain derived neurotrophic factor gene (BDNF): identification of several genetic variants and association studies in patients with obesity, eating disorders, and attention-deficit/hyperactivity disorder. Am J Med Genet B Neuropsychiatr Genet 132:96–99

Gadde KM, Yonish GM, Wagner HR, Foust MS, Allison DB (2006) Atomoxetine for weight reduction in obese women: a preliminary randomised controlled trial. Int J Obes (Lond) 30:1138–1142

Gray J, Yeo GS, Cox JJ, Morton J, Adlam AL, Keogh JM, Yanovski JA, El GA, Han JC, Tung YC, Hodges JR, Raymond FL, O'rahilly S, Farooqi IS (2006) Hyperphagia, severe obesity, impaired cognitive function, and hyperactivity associated with functional loss of one copy of the brain-derived neurotrophic factor (BDNF) gene. Diabetes 55:3366–3371

Heiligenstein E, Keeling RP (1995) Presentation of unrecognized attention deficit hyperactivity disorder in college students. J Am Coll Health 43:226–228

Holtkamp K, Konrad K, Muller B, Heussen N, Herpertz S, Herpertz-Dahlmann B, Hebebrand J (2004) Overweight and obesity in children with Attention-Deficit/Hyperactivity Disorder. Int J Obes Relat Metab Disord 28:685–689

Hubel R, Jass J, Marcus A, Laessle RG (2006) Overweight and basal metabolic rate in boys with attention-deficit/hyperactivity disorder. Eat Weight Disord 11:139–146

Hudson JI, Lalonde JK, Berry JM, Pindyck LJ, Bulik CM, Crow SJ, McElroy SL, Laird NM, Tsuang MT, Walsh BT, Rosenthal NR, Pope HG Jr (2006) Binge-eating disorder as a distinct familial phenotype in obese individuals. Arch Gen Psychiatry 63:313–319

Kernie SG, Liebl DJ, Parada LF (2000) BDNF regulates eating behavior and locomotor activity in mice. EMBO J 19:1290–1300

Lam LT, Yang L (2007) Overweight/obesity and attention deficit and hyperactivity disorder tendency among adolescents in China. Int J Obes (Lond) 31:584–590

Levitan RD, Masellis M, Lam RW, Muglia P, Basile VS, Jain U, Kaplan AS, Tharmalingam S, Kennedy SH, Kennedy JL (2004) Childhood inattention and dysphoria and adult obesity associated with the dopamine D4 receptor gene in overeating women with seasonal affective disorder. Neuropsychopharmacology 29:179–186

Levy LD, Fleming JP, Klar D (2009) Treatment of refractory obesity in severely obese adults following management of newly diagnosed attention deficit hyperactivity disorder. Int J Obes (Lond) 33:326–334

Lourenco BH, Arthur T, Rodrigues MD, Guazzelli I, Frazzatto E, Deram S, Nicolau CY, Halpern A, Villares SM (2008) Binge eating symptoms, diet composition and metabolic characteristics of obese children and adolescents. Appetite 50:223–230

Lyons WE, Mamounas LA, Ricaurte GA, Coppola V, Reid SW, Bora SH, Wihler C, Koliatsos VE, Tessarollo L (1999) Brain-derived neurotrophic factor-deficient mice develop aggressiveness and hyperphagia in conjunction with brain serotonergic abnormalities. Proc Natl Acad Sci U S A 96:15239–15244

Mitsuyasu H, Hirata N, Sakai Y, Shibata H, Takeda Y, Ninomiya H, Kawasaki H, Tashiro N, Fukumaki Y (2001) Association analysis of polymorphisms in the upstream region of the human dopamine D4 receptor gene (DRD4) with schizophrenia and personality traits. J Hum Genet 46:26–31

Mustillo S, Worthman C, Erkanli A, Keeler G, Angold A, Costello EJ (2003) Obesity and psychiatric disorder: developmental trajectories. Pediatrics 111:851–859

Nafiu OO, Burke C, Lee J, Voepel-Lewis T, Malviya S, Tremper KK (2010) Neck circumference as a screening measure for identifying children with high body mass index. Pediatrics 126: e306–e310

Nasser JA, Gluck ME, Geliebter A (2004) Impulsivity and test meal intake in obese binge eating women. Appetite 43:303–307

Noble EP (2003) D2 dopamine receptor gene in psychiatric and neurologic disorders and its phenotypes. Am J Med Genet B Neuropsychiatr Genet 116:103–125

Pagoto SL, Curtin C, Lemon SC, Bandini LG, Schneider KL, Bodenlos JS, Ma Y (2009) Association between adult attention deficit/hyperactivity disorder and obesity in the US population. Obesity (Silver Spring) 17:539–544

Poston WS, Ericsson M, Linder J, Haddock CK, Hanis CL, Nilsson T, Astrom M, Foreyt JP (1998) D4 dopamine receptor gene exon III polymorphism and obesity risk. Eat Weight Disord 3:71–77

Ptacek R, Kuzelova H, Paclt I, Zukov I, Fischer S (2009) Anthropometric changes in non-medicated ADHD boys. Neuro Endocrinol Lett 30:377–381

Riverin M, Tremblay A (2009) Obesity and ADHD. Int J Obes (Lond) 33:945

Rosval L, Steiger H, Bruce K, Israel M, Richardson J, Aubut M (2006) Impulsivity in women with eating disorders: problem of response inhibition, planning, or attention? Int J Eat Disord 39:590–593

Schweickert LA, Strober M, Moskowitz A (1997) Efficacy of methylphenidate in bulimia nervosa comorbid with attention-deficit hyperactivity disorder: a case report. Int J Eat Disord 21:299–301

Shinawi M, Sahoo T, Maranda B et al (2011) 11p14.1 microdeletions associated with ADHD, autism, developmental delay, and obesity. Am J Med Genet A 155:1272–1280

Sokol MS, Gray NS, Goldstein A, Kaye WH (1999) Methylphenidate treatment for bulimia nervosa associated with a cluster B personality disorder. Int J Eat Disord 25:233–237

Spencer TJ, Biederman J, Harding M, O'Donnell D, Faraone SV, Wilens TE (1996) Growth deficits in ADHD children revisited: evidence for disorder-associated growth delays? J Am Acad Child Adolesc Psychiatry 35:1460–1469

Spencer TJ, Faraone SV, Biederman J, Lerner M, Cooper KM, Zimmerman B (2006) Does prolonged therapy with a long-acting stimulant suppress growth in children with ADHD? J Am Acad Child Adolesc Psychiatry 45:527–537

Strimas R, Davis C, Patte K, Curtis C, Reid C, McCool C (2008) Symptoms of attention-deficit/ hyperactivity disorder, overeating, and body mass index in men. Eat Behav 9:516–518

Surman CB, Randall ET, Biederman J (2006) Association between attention-deficit/hyperactivity disorder and bulimia nervosa: analysis of 4 case-control studies. J Clin Psychiatry 67:351–354

Swanson J, Greenhill L, Wigal T, Kollins S, Stehli A, Davies M, Chuang S, Vitiello B, Skrobala A, Posner K, Abikoff H, Oatis M, McCracken J, McGough J, Riddle M, Ghuman J, Cunningham C, Wigal S (2006) Stimulant-related reductions of growth rates in the PATS. J Am Acad Child Adolesc Psychiatry 45:1304–1313

Tsai SJ, Hong CJ, Yu YW, Chen TJ (2004) Association study of catechol-O-methyltransferase gene and dopamine D4 receptor gene polymorphisms and personality traits in healthy young chinese females. Neuropsychobiology 50:153–156

Vgontzas AN, Bixler EO, Chrousos GP (2006) Obesity-related sleepiness and fatigue: the role of the stress system and cytokines. Ann N Y Acad Sci 1083:329–344

Waring ME, Lapane KL (2008) Overweight in children and adolescents in relation to attention-deficit/hyperactivity disorder: results from a national sample. Pediatrics 122:e1–e6

Weinberg WA, Brumback RA (1990) Primary disorder of vigilance: a novel explanation of inattentiveness, daydreaming, boredom, restlessness, and sleepiness. J Pediatr 116:720–725

Zametkin AJ, Zoon CK, Klein HW, Munson S (2004) Psychiatric aspects of child and adolescent obesity: a review of the past 10 years. J Am Acad Child Adolesc Psychiatry 43:134–150

Face Processing in Attention Deficit/Hyperactivity Disorder

Daniel P. Dickstein and F. Xavier Castellanos

Contents

1 Introduction 220
2 Faces as Emotional Stimuli 221
3 Attention's Role in Face Processing 223
   3.1 Speed of Face Processing 223
   3.2 Is Face Processing Unconscious? 224
   3.3 Is Face Processing Mandatory? 225
   3.4 Is Face Processing Capacity Free? 225
4 Attentional Face Processing: A Model 226
5 Developmental Effects in Face Processing 227
6 Face Processing in ADHD 228
7 Face Processing in Related Psychopathology: Bipolar Disorder, Oppositional Defiant Disorder, Conduct Disorder 232
8 Conclusion 233
References 233

Abstract ADHD is one of the most common and impairing psychiatric conditions affecting children today. Thus far, much of the phenomenological and neurobiological research has emphasized the core symptoms of inattention, hyperactivity, and impulsivity which are thought to be mediated by frontostriatal alterations. However, increasing evidence suggests that ADHD involves emotional problems in addition to cognitive impairments. Here, we review the neurobiology of face

---

D.P. Dickstein
PediMIND Program, E.P. Bradley Hospital, Alpert Medical School of Brown University
e-mail: daniel_dickstein@brown.edu

F.X. Castellanos
Neidich Professor of Child and Adolescent Psychiatry, Director of the Phyllis Green and Randolph Cöwen Institute for Pediatric Neuroscience, Director of Research, New York University Child Study Center. Research Psychiatrist, Nathan Kline Institute for Psychiatric Research, Orangeburg, NY

processing and suggest that face-processing alterations offer a window into the emotional dysfunction often accompanying ADHD.

Keywords Adolescent · Attention deficit hyperactivity disorder (ADHD) · Child · Emotion · Face processing

Abbreviations

ACC	Anterior cingulate cortex
ADHD	Attention-deficit hyperactivity disorder
ADS-CPT	ADHD diagnostic system continuous performance task
ALE	Activation likelihood estimation
BD	Bipolar disorder
CD	Conduct disorder
DANVA	Diagnostic Analysis of Non-Verbal Accuracy
dlPFC	Dorsolateral PFC
DSM-IV-TR	Diagnostic and Statistical Manual 4th Edition Text Revision
ERP	Event-related potential
FFA	Fusiform face area
fMRI	Functional magnetic resonance imaging
IAPS	International Affective Picture Set
MEG	Magnetoencephalography
MRI	Magnetic resonance imaging
ODD	Oppositional defiant disorder
PFC	Prefrontal cortex
SMD	Severe mood dysregulation
STS	Superior temporal sulcus
vlPFC	Ventrolateral prefrontal cortex
vmPFC	Ventromedial prefrontal cortex

1 Introduction

Attention-deficit hyperactivity disorder (ADHD) is one of the most common and impairing psychiatric disorders affecting children today. Between 6–9% of children and adolescents struggle with ADHD (Polanczyk and Jensen 2008). Since 1980 (American Psychiatric Association 1980), the clinical entity now known as ADHD has been defined in the Diagnostic and Statistical Manual (DSM) by cardinal symptoms of inattention, hyperactivity, and impulsivity (American Psychiatric Association 2000). However, increasing evidence suggests that, in addition to cognitive impairments, ADHD also involves emotional problems in many cases.

By far most ADHD research has focused on the core symptoms of inattention, hyperactivity, and impulsivity. This work has largely implicated frontostriatal alterations in the pathophysiology of ADHD. Early cross-sectional neuroimaging studies suggested volumetric alterations between ADHD versus typically-developing youths, including smaller striatal structures, alterations in typical cerebral asymmetry, and volumetric decreases in other functionally related structures such as the cerebellar vermis (Castellanos et al. 1996, 2001). Decreases in total cerebral and cerebellar volume, global cortical thickness {Shaw, 2006 3752 /id and reduced volumes of specific frontostriatal structures tend to persist throughout childhood and into young adulthood in ADHD (Castellanos et al. 2002). However, a recent analysis of cortical thickness in the same cross-lag longitudinal structural magnetic resonance imaging (MRI) data set emphasized the dramatic developmental lag in the trajectories of increasing followed by decreasing cortical thickness in ADHD {Shaw, 2007}.

Using tasks that tap the classic features of ADHD, such as go/no-go tasks to evaluate response inhibition, functional MRI (fMRI) studies have shown that ADHD children have neural alterations in these same structures (Booth et al. 2005; Dickstein et al. 2006; Pliszka et al. 2006). Such neural alterations have also been shown in the first-degree relatives of ADHD children, suggesting that this may be an endophenotype of ADHD (Durston et al. 2006).

Nevertheless, research has begun to support what many clinicians have long suspected: that ADHD often involves emotional impairments in addition to cognitive impairments (Castellanos et al. 2006). For example, Barkley et al. evaluated the impact of emotional impulsiveness, defined by symptoms such as low frustration tolerance, impatience, quick to anger, hot-temperedness, irritability, and easy emotional excitability, in ADHD children compared to typically-developing controls followed into adulthood (Barkley and Fischer 2010). Those with ADHD persisting into adulthood had more emotional impulsiveness than those with either nonpersistent ADHD or controls. Moreover, emotional impulsivity contributed additional variance in predicting impairments beyond ADHD diagnostic features, including family, peer, relationship, financial, and driving problems. In another study, Martel et al. demonstrated the relationship between ADHD symptoms of hyperactivity/impulsivity and the temperamental dimension of reactive control, as well as the relationship between inattention and temperament dysregulation (Martel and Nigg 2006). Thus, there is a need for greater understanding of the brain and behavior interactions underlying emotional processing in ADHD.

Accordingly, this chapter is focused on face processing in ADHD, since faces are among the most important triggers of emotional responses, and are also the most important means of communicating an emotional response to others. We first begin with a brief review of facial stimuli as the most basic emotionally-evocative stimuli. Second, we discuss the role of attention in face processing. Thirdly, we discuss what is known about face processing in ADHD, as well as in related forms of psychopathology and in typical development. We conclude by discussing future research directions.

2 Faces as Emotional Stimuli

From an affective neuroscience perspective, the human face is perhaps most the important emotional stimulus in the human environment. From this perspective, "emotion" is defined as "an evoked response to an environmental stimulus with motivational salience." (Lang et al. 1998) In turn, "motivational salience" may be defined along two orthogonal axes: "arousal" and "valence." "Arousal" refers to the amount of resources generated in response to an environmental stimulus. "Valence" refers to whether the stimulus is appetitively rewarding, such that the organism will devote energy to acquire it ("positively-valenced"), or aversive, such that the organism will devote energy to avoid it ("negatively-valenced") (Dickstein and Leibenluft 2006) (Fig. 1).

Faces are a nonverbal means of conveying one's emotional state to others (Adolphs et al. 1996; Gobbini and Haxby 2007; LeDoux 2000). Much research has focused on the appraisal and judgment of faces, with the neural response in appraising or judging faces seeming to be hardwired. For example, infants exhibit both gaze preference and increased gamma activity when tracking intact versus inverted faces (Farroni et al. 2002; Goren et al. 1975; Grossmann et al. 2007; Johnson et al. 1991). Furthermore, the neural response to familiar faces represents a biological substrate of attachment, as studies have shown that mothers have greater neural activation when looking at their own child versus a familiar but unrelated child (Leibenluft et al. 2004). Similarly, adults have greater neural response to familiar family members' faces than to either familiar or famous people's faces (Gobbini et al. 2004).

At the most basic level, face processing is akin to the body's response to other visual stimuli. The steps include: (1) sensory input regarding the stimulus, (2) categorizing its emotional salience (including arousal and valence), (3) initiating

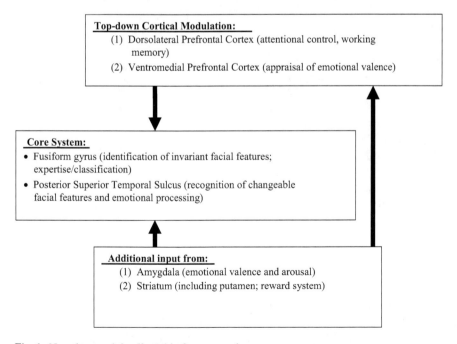

Fig. 1 Neural network implicated in face processing

an emotional response, and (4) mediating higher-order cognitive processing and cognitive controls (Cole et al. 2004; Nelson et al. 2005; Thompson 1994). In particular, regarding faces, face processing begins with receipt of sensory information and categorizing the face's identity, relying primarily on invariant facial features (e.g., eye color, face shape) and then encoding of the changeable aspects of the face, including eye gaze and mouth position that convey its emotional expression (Hoffman and Haxby 2000).

Face processing can be studied using many different techniques, including variations in (a) stimuli, (b) what participants do, and (c) what outcome is being measured. With respect to stimuli, studies can vary basic elements, such as age (e.g., children, adolescents, or adults), gender, race, or emotion (e.g., angry, happy, sad, or neutral). Stimuli can also be morphed, to allow gradations in any of these elements (i.e., to have a series of 10 stimuli going from angry to sad), or masked, to conceal the actual face from conscious processing. Similarly, researchers can vary what participants do during an experiment, such as judging the emotion depicted (either categorically [angry, sad] or dimensionally [how angry?]) or rating nonemotional face elements (e.g., how big is someone's mouth). Finally, numerous outcomes can be measured, including accuracy of ratings or neural activity either when rating faces or after the fact, during postparadigm memory experiments (the latter to evaluate encoding of faces into memory, rather than just the initial response).

3 Attention's Role in Face Processing

With so many naturally occurring visual stimuli in our daily lives, how does the mind regulate which stimuli to process (and when)? Thus the question: to what extent is face processing automatic, requiring minimal attentional resources vs. requiring extensive interplay from areas involved with mediating attention to visual stimuli? Palermo and Rhodes suggest that for face processing in general to be automatic, it must be (a) rapid, (b) nonconscious, (c) mandatory, and (d) capacity free: i.e., requiring no attentional resources (Palermo and Rhodes 2007).

3.1 Speed of Face Processing

With respect to speed, event-related potential (ERP) studies have shown that frontal regions can categorize faces in around 100 ms, whereas 200 ms are required to categorize objects and words (Pegna et al. 2004). Moreover, people have a neural bias toward identifying human faces in context, with studies showing that human faces are detected 10 ms faster than animal faces when these stimuli are embedded in natural scenes (Rousselet et al. 2003).

Additionally, studies suggest that there is an initial frontal response reflecting rapid detection of emotional expression, followed by subsequent higher-order processing, including evaluation of emotional content by more distributed circuits. For example, adults viewing fearful versus neutral faces had a sharp ERP peak at ~110–200 ms, followed by a broader response around 250 ms after the stimulus. Moreover, with inverted faces, the sharp peak was delayed to 150 ms, with the broader response ending at 700 ms (Eimer and Holmes 2002). Studies using a face-matching task have shown that occipitotemporal activity at 100 ms correlated with face categorization and at 170 ms correlated with correct face categorization and identification (Carmel and Bentin 2002; Itier and Taylor 2004).

Lastly, while some studies have suggested that some facial emotions, such as fear, may be processed more rapidly than others (LeDoux 2003), the evidence is mixed (Batty and Taylor 2003). For example, Eimer and Holmes conducted an ERP study using faces depicting anger, disgust, fear, happiness, sadness, surprise, and neutral. Face stimuli were presented in pairs and participants had to decide if the emotional expression displayed by the face pair was emotional or neutral. They found an enhanced positivity across all six emotion types versus neutral starting at 100 ms after stimulus onset and sustained throughout the 1,000 ms recording period, suggesting that this so-called "N170" (ERP activity that is prominent at occipitotemporal recording sites at 170 ms) amplitude and latency were insensitive to emotional face type (Eimer et al. 2003). Thus, in aggregate, the data support the conclusion that the processing of facial stimuli is rapid.

3.2 Is Face Processing Unconscious?

With respect to unconscious processing, even those with acquired prosopagnosia, who cannot consciously recognize previously familiar faces, demonstrate unconscious recognition to familiar versus unfamiliar faces by exhibiting an altered skin conductance response (Bauer 1984; Tranel and Damasio 1985). Adults with right parietal damage, resulting in visual neglect, do not perceive or respond to a stimulus placed in the left visual field (opposite the lesion). However, "unseen" facial stimuli – i.e., those placed in the affected visual field ipsilateral to this damage – depicting fear, result in the same amygdala activation as stimuli placed in the right (unaffected) visual field (Vuilleumier et al. 2002). Another way of studying unconscious perception of facial stimuli is by rapidly masking the face stimuli so that they are not perceived consciously. Using this approach, studies have shown that healthy adults without neurological lesions have similar facial muscle response to happy and angry faces when these faces were presented either with or without masking as measured by facial electromyography (Dimberg et al. 2000). A similar study showed that masked angry, but not happy, faces can evoke a conditioned autonomic response (Ohman 1988).

3.3 Is Face Processing Mandatory?

Palermo and Rhodes also suggested that automatic processes should be mandatory: i.e., it would be impossible not to process faces, regardless of an individual's desires. Here, the data are more mixed. On the one hand, face processing may be mandatory, as indicated by studies showing that participants had delayed recognition of line curvatures when the lines resembled facial silhouettes compared to nonface curvatures (Suzuki and Cavanagh 1998). On the other hand, purposeful attention can alter brain/behavior relationships during face processing. For example, several studies have demonstrated greater fMRI neural activation and ERPs to attended versus unattended faces (Downing et al. 2001; Eimer 2000; O'Craven et al. 1999). Importantly, whether or not face processing is mandatory may vary between brain regions. For example, dual-task studies in which attention is directed away from faces, show that FFA response is reduced, but not eliminated, while the amygdala's response is maintained (Anderson and Phelps 2001; Vuilleumier et al. 2001; Williams et al. 2005).

3.4 Is Face Processing Capacity Free?

Finally, for face processing to be truly automatic, it should be capacity-free, meaning it should occur without any disruption in processing efficiency even when competing or distracting stimuli are present. This attribute can be tested using visual search tasks, which require participants to identify target stimuli amid a background of distractors, akin to the "Where's Waldo" children's books (Handford 1987). Studies have shown that it is more efficient to identify angry faces among a background of either neutral or happy faces, rather than the reverse (Fox et al. 2000; Horstmann and Bauland 2006). However, these differences might reflect low-level visual artifacts, rather than a biological predisposition to detect certain emotional face types. This possibility emerges from a recent study showing that even obliquely oriented lines were detected as efficiently as schematic faces when they had upside down-curved lines (akin to a frown) versus right-side up (akin to a smile) (Coelho et al. 2010).

Another recent study advanced the ecological validity of these visual search tasks by using dynamic, i.e., moving facial stimuli. Angry dynamic faces were efficiently detected versus dynamic friendly faces; yet, when controlling for the degree of movement, there was no strong advantage of dynamic over static faces (Horstmann and Ansorge 2009). Although these data suggest that even very crude representations of facial emotions may be processed with minimal attentional resources, others have shown significant interactions between stimulus valence and attention. For example, Pessoa showed that attended versus unattended faces produced greater amygdala activation across all emotions (fear, happy, neutral), whereas the neural response to unattended faces did not differ (Pessoa et al. 2002).

Thus, while face processing may be rapid and unconscious, it is only partially mandatory or capacity-free. Therefore, face processing is not entirely automatic and does involve a significant contribution of attention.

4 Attentional Face Processing: A Model

Taken as a whole, two attentional mechanisms are likely implicated in face processing: one that involves top-down cortical regulation of competing stimuli via selective attention, and a second subcortical circuit that semi-automatically evaluates emotionally-evocative stimuli, including faces (Fig. 1) (Adolphs and Spezio 2006; Gobbini and Haxby 2007; Palermo and Rhodes 2007).

At its core, face processing involves top-down interaction between cortical structures including the fusiform gyrus (also known as the "fusiform face area" [FFA]) (Kanwisher et al. 1997), inferior occipital gyri, and superior temporal sulcus (STS) (Gobbini and Haxby 2007; Haxby et al. 2002). Whereas the fusiform and inferior occipital gyri mediate identification of invariant facial features, the STS mediates dynamic features required for recognition of particular individuals and emotions (Adolphs et al. 2000; Haxby et al. 2002; Hoffman and Haxby 2000).

Moreover, much has been written about the fusiform's potential functionally specific role in recollection of biographical information about face identity (Kanwisher et al. 1997). However, more recent data have shown that FFA is involved when expertise is required to categorize and to classify items, including nonface stimuli such as an ornithologist's ability to recognize bird species (Bukach et al. 2006; Gauthier et al. 2000). The prefrontal cortex (PFC) plays a role in attentional mediation of this top-down circuit, from the dorsolateral PFC (dlPFC) via its role in attentional set-shifting and working memory, and also from the ventromedial PFC (vmPFC) via its role in appraising the emotional valence of stimuli, such as faces (Rolls 2007; Vuilleumier and Pourtois 2007).

With respect to the subcortical circuit, the amygdala may play a role via its connections to the vmPFC (Adolphs et al. 2002; Adolphs and Tranel 2004; Gorno-Tempini et al. 2001). However, the data are mixed. On the one hand, some studies suggest that the amygdala can respond to fearful faces independently of spatial attention (Vuilleumier et al. 2001). Also, patients with neglect or extinction are more likely to detect emotionally significant versus neutral pictures when presented in the affected visual hemi-field (Vuilleumier and Schwartz 2001a, b). On the other hand, the amygdala's response to fearful or happy faces can be modulated by attention (Pessoa et al. 2002). Moreover, others have shown that the amygdala may not have a speed advantage over other brain regions in processing faces, including fearful versus happy faces (Batty and Taylor 2003; Eimer et al. 2003; Eimer and Holmes 2007). Thus, the amygdala is part of a subcortical circuit mediating face processing with a greater degree of automaticity than the cortical route, although this is not fully independent of attention (Eimer and Holmes 2007).

Putting this all together, Fusar-Poli et al. recently conducted a novel activation likelihood estimation (ALE) meta-analysis of 105 face-processing fMRI studies involving 1,785 neural coordinates and 1,600 healthy control participants. They found a main effect for faces independent of emotional valence consisting of increased neural activation in several areas, including: (a) visual areas (fusiform gyrus, inferior/middle occipital gyri, lingual gyri), (b) limbic areas (amygdala, parahippocampal gyrus, posterior cingulate cortex), (c) parietal lobule, (d) middle temporal gyrus and insula, (e) PFC (medial frontal gyrus), (f) putamen, and (g) cerebellum (declive). They found a main effect of emotional valence, with happy, fearful, and sad faces resulting in increased amygdala activation, whereas angry or disgusted faces had no effect on the amygdala. With respect to attentional effects, they demonstrated that explicit versus implicit face processing was associated with increased neural activation in the amygdala bilaterally, right fusiform gyrus, caudate, and inferior occipital gyrus, and left declive, inferior and medial frontal gyri. In contrast, implicit versus explicit face processing was associated with increased activation of the right fusiform gyrus and insula as well as the left lingual and postcentral gyri (Fusar-Poli et al. 2009).

5 Developmental Effects in Face Processing

Studies suggest that development plays an important role in the neurocircuitry mediating face processing (Monk et al. 2003). For example, adolescents have greater activation in paralimbic regions and less activation in the right vlPFC than adults when attending to nonemotional aspects of faces, suggesting that adults have greater neural modulation based on attentional demands. Additionally, when judging the emotional content of faces, adolescents have decreased activation of the right ACC compared to adults, suggesting the ongoing development of these structures into adulthood (Passarotti et al. 2009). Studies have shown positive correlations between age and attentional bias toward happy faces (Lindstrom et al. 2009). In another study by Nelson et al., adolescents had greater activity in the ACC when viewing subsequently remembered angry faces, and more activity in the right temporal pole when viewing subsequently remembered fear faces than adults. In contrast, adults had greater activity in the subgenual ACC, when viewing subsequently remembered happy faces, and more activity in the right posterior hippocampus when viewing subsequently remembered neutral faces (Nelson et al. 2003).

In sum, developmental differences in the brain/behavior relationships underlying face processing mirror those underlying other cognitive processes, providing further proof that adolescents are not just little adults when it comes to how they understand the most important of emotional stimuli: i.e., emotional faces.

6 Face Processing in ADHD

Although most research on ADHD has focused on neurocognitive deficits in prototypical domains, including attention, working memory, and executive function, only a few studies have evaluated emotional face processing. Given the dearth of such studies, we review each below, in order to stimulate further research in this area (see Table 1).

Table 1 Face processing studies of ADHD participants

Study	Participants	Task	Key findings
Corbett and Glidden (2000)	7–12 yo ADHD vs HC ($N = 37$ per group)	Pictures of facial affect (emotional face identification)	ADHD children had mild/moderate deficits in perception of affect
Da Fonseca et al. (2009)	Adolescents with ADHD vs. HC ($N = 27$ per group)	Emotional face identification task and context task	ADHD less accurate in emotional face identification and in using contextual information
Shin et al. (2008)	6–10 yo boys with ADHD ($N = 42$) vs. HC ($N = 27$)	Emotion recognition task-revised and ADHD diagnostic system continuous performance task	No differences in recognition of positive or negative valenced emotional faces but ADHD had lower contextual understanding
Pelc et al. (2006)	7–12 yo ADHD vs. HC ($N = 30$ per group)	Morphed face emotional recognition	ADHD participants made more errors especially on angry and sad faces
Blaskey et al. (2008)	7–12 yo ADHD combined type, ADHD inattentive type, subthreshold ADHD, or HC	Chimeric face test and stop signal task	No association between ADHD and reactive inhibition measures
Klimkeit et al. (2003)	7–12 yo ADHD ($N = 16$) or HC ($N = 52$)	Face matching, chimeric face task, and grey scale task	ADHD participants did not have expected right-ward bias
Williams et al. (2008)	Adolescents with ADHD or HC ($N = 51$ of each)	Emotional face recognition	ADHD participants had altered ERP measures (including P120 and P300)
Yuill and Lyon (2007)	5–11 yo ADHD boys vs. HC ($N = 19$ of each)	Matching emotional faces to situations and non-face task	ADHD participants had worse performance
Kats-Gold et al. (2007)	4th and 5th graders at risk for ADHD by ratings ($N = 50$) vs. HC ($N = 61$)	Emotional face recognition	At-risk for ADHD performed worse
Rapport et al. (2002)	Adults with ADHD vs. HC ($N = 28$ of each)	DANVA and recognition of human faces vs. animals	ADHD adults performed worse on both tasks, including DANVA child faces

ADHD attention deficit hyperactivity disorder, *HC* typically-developing healthy control, *YO* years old, *DANVA* diagnostic analysis of nonverbal accuracy, *ERP* event-related potential

Corbett et al. administered the Pictures of Facial Affect task that required 7–12 year old participants with either ADHD or typical development ($N = 37$ per group) to identify 21 slides depicting fundamental emotions (such as, anger, disgust, fear, happiness, sadness, surprise, and neutral). Their primary analyses showed that children with ADHD had mild-to-moderate deficits in perception of affect, with data suggesting that attentional issues may be involved in problems encoding stimulus properties (Corbett and Glidden 2000).

DaFonseca et al. evaluated emotional and contextual visual processing in adolescents with ADHD and age-matched controls ($N = 27$ per group). In the first of two experiments, participants had to identify emotional faces. In the second, participants had to utilize contextual information to recognize an emotion (e.g., identifying 26 adult faces and 16 child faces from French magazines as fear, sadness, anger, or happiness) or detect which object was missing from a scene (e.g., viewing a picture of a girl in a kitchen holding her hand near her mouth covered by a white circle, with participants required to select from a tea kettle, pitcher, or cup). Adolescents with ADHD were less accurate than controls at identifying face emotional expressions and in using contextual information (Da Fonseca et al. 2009). Taken as a whole, the authors suggest that this supports an overall deficit in emotional processing in ADHD youths, extending beyond just face-processing problems.

To evaluate the effect of attention on emotional face recognition, Shin et al. evaluated 6–10-year-old boys with either ADHD ($N = 42$) or controls without psychopathology ($N = 27$) using two computerized tasks. The Emotion Recognition Test-Revised required participants to recognize positive (happy and surprise) and negative (anger, sad, disgust, and fear) emotions and to evaluate participant's ability to match facial expressions to cartoons depicting emotional contexts (e.g., happy face and picture of a birthday party). Their second task was the ADHD Diagnostic System Continuous Performance Task (ADS-CPT) to evaluate attention (Shin et al. 2008). While there was no between-group difference in recognition of positive- or negatively-valenced emotional faces, ADHD participants had lower contextual understanding scores than controls. Among ADHD participants, attention (as indexed by the ADS-CPT) was associated with the contextual understanding score, suggesting that attention plays a role in ADHD participants' impaired ability to identify the contextual understanding of faces.

To evaluate the effect of emotional intensity on face processing, Pelc et al. employed a morphed faces paradigm that incrementally attenuated the emotional intensity of angry, sad, happy, and disgust faces. They found that 7–12 year olds with ADHD hyperactive/impulsive type made more errors than controls ($N = 30$ in each group), with specific impairments in identifying angry and sad facial expressions (Pelc et al. 2006). Furthermore, the ADHD group had significant correlations between impaired emotional face recognition and interpersonal problems that was most prominent for angry faces. The authors suggested that this may offer an important new target for treatment to address emotional problems associated with ADHD.

Blaskey et al. explored the relationship between reactive (behavioral) inhibition (thought to be a bottom-up automatic process) versus executive inhibition (thought to be a top-down cognitive process) in ADHD (Blaskey et al. 2008). Reactive inhibition was assessed using Sensation Seeking Scale ratings and performance on the Chimeric Face Test, whereas executive inhibition was evaluated by Stop Signal Task performance. For the Chimeric Face Test, stimuli were created from facial images separated at the midline, with vertical halves represented different emotion types: i.e., left side showing "happy" and right side showing "neutral." Two mirror-image chimeric faces were presented simultaneously, and participants had to answer the question "which face is happier, left or right?" Data suggested that there would be a left visual hemi-space bias in such judgments, with participants selecting the left-sided stimulus in 65–75% of trials. Participants included children ages 7–12 years with either ADHD combined type, ADHD inattentive type, non-ADHD controls, or a subthreshold group who failed to meet full ADHD criteria. All participants were medication free for a minimum of 48 h, for long-acting, and 24 h for short-acting, stimulants. There was no association between ADHD and reactive inhibition measures (including on the Chimeric Face Test). From this, the authors concluded that both types of inhibition were involved in ADHD.

To test left-ward biases in ADHD, Klimkeit et al. administered three face tasks to 7–12-year-olds with ADHD combined type ($N = 16$) and controls without psychopathology ($N = 52$) (Klimkeit et al. 2003). They included a (1) face-matching task, requiring participants to determine if a face composed of left–left or right–right composites was most similar to the original photograph, (2) a chimeric face task, requiring participants to identify whether a composite face with the smile on the left or right looked happier, and (3) a gray scale task, requiring participants to identify which of two rectangles (one with the dark end on the left and the other with the dark end on the right) appeared darker. Control children had a left-ward perceptual bias, indicated by choosing the left–left composite and the happier chimera with the smile on the left, without leftward bias in the grey scale task. ADHD participants did not demonstrate an expected rightward bias, performing instead more like the controls. Post-hoc analyses did not reveal differences between medicated ($N = 8$) and unmedicated ($N = 8$) ADHD participants. The authors suggest that prior work showing perceptual asymmetry in ADHD children may reflect confounding with motor responses or spatial processing that was absent from their purely perceptual tasks.

Williams et al. tested the hypothesis that ADHD would be associated with deficits in recognizing negative emotional expressions in adolescents with ADHD ($N = 51$) and healthy control ($N = 51$) boys using event-related potentials (ERP) while participants categorized faces as expressing fear, anger, sadness, disgust, happiness, or neutral (Williams et al. 2008). ADHD participants were worse at recognizing facial emotions. Also, ADHD participants had a reduced P120 response (waveform occurring between 120–220 ms) at occipital sites during the early perceptual analysis of emotional expression, and subsequent reduction and slowing of the P300 response, an ERP amplitude found in temporal sites at 300–400 ms that is associated with context processing. Stimulant medications normalized both emotion recognition and ERP findings. Thus, this study shows that face processing may represent a biological marker of both ADHD that responds to treatment with stimulant medications.

Yuill et al. found that 5–11 year old boys with ADHD performed poorly on a task of matching emotional faces to situations compared with age-matched controls ($N = 19$ per group) and found it more difficult than a nonface task (Yuill and Lyon 2007). A second experiment used a scaffolding procedure, to discourage impulsive responding, by ensuring that each of six possible answers was attended to prior to the participant answering. ADHD boys ($N = 17$) performed worse than a younger group of controls ($N = 13$), who were selected in an attempt to equalize overall performance based on the first experiment. The authors suggest that this may have implications for why boys with ADHD have peer-relationship problems.

To fully evaluate the possibility that aberrant face processing may represent a behavioral intermediate phenotype for ADHD, face processing has been evaluated in children at elevated risk for ADHD: e.g., first-degree relatives of those with ADHD, or individuals with elevated clinical scores for ADHD symptoms but without actually fully meeting criteria for having ADHD. In one such study, Kats-Gold et al. found deficits in fourth and fifth grade boys at-risk for ADHD based on their ADHD symptom scores. Specifically, this study found that those identified as at-risk for ADHD ($N = 50$) performed worse on a computerized emotional face identification task than controls ($N = 61$). Moreover, only in the at-risk group did impaired face emotion identification correlate with social skills and behavioral problems (Kats-Gold et al. 2007). This suggests that behavioral disinhibition or executive function problems may not solely mediate emotional or social-relationship problems in children with ADHD.

The literature on face processing in adults with ADHD is almost nonexistent. The only exception found that adults with ADHD ($N = 28$; age mean 36.3 years, range 18–64 years) performed worse than controls ($N = 28$; age mean 33.4 years) on the child faces segment of the Diagnostic Analysis of Non-Verbal Accuracy (DANVA) (Nowicki and Duke 1994). Moreover, adults with ADHD performed worse than controls on a task of human faces versus animals using a computerized tachistoscope, but this impairment was unrelated to gross perceptual processing, basic face recognition abilities, or attentional aspects of face perception. This study found that the intensity of experienced emotion moderated emotion recognition, with controls' emotional identification improved by experienced emotion, whereas ADHD adults had the inverse relationship (Rapport et al. 2002).

Thus, current data support aberrant face processing as impaired in ADHD. However, the brain/behavior interactions underlying this deficit remain unknown. Moreover, we need studies that can clarify how development, gender, and treatment play into these alterations. In addition, we need studies that can probe the relationship between face processing and real-world measures of emotional difficulty as well as core ADHD features of inattention, hyperactivity, and impulsivity.

7 Face Processing in Related Psychopathology: Bipolar Disorder, Oppositional Defiant Disorder, Conduct Disorder

Face processing has been studied in disorders that are often comorbid with ADHD. These include oppositional defiant disorder (ODD), conduct disorder (CD), bipolar disorder (BD), and even autistic disorder.

Using International Affective Picture Set (IAPS) images (standardized emotionally-evocative pictures, such as of a plane crash), Krauel et al. found that ADHD participants with comorbid ODD and CD ($N = 16$) made more recollection errors for neutral and positive images versus controls ($N = 25$), whereas ADHD without those comorbid conditions ($N = 14$) performed worse than controls only for neutral images (Krauel et al. 2009).

Cadesky used the DANVA to compare emotional face processing in children with conduct problems alone, ADHD alone, conduct problems plus ADHD, and controls. Using quantitative analyses, they found that children with conduct problems or ADHD were less accurate at emotional face identification than controls. With qualitative analyses, they found that participants with ADHD made random errors, whereas those with conduct problems misinterpreted angry faces (Cadesky et al. 2000). This supports the hypothesis that social problems result from different mechanisms in those with either ADHD or conduct problems.

Given the clinical overlap between ADHD and BD, studies have begun to evaluate the brain/behavior interactions underlying face processing in these disorders. Of note, these studies included not only BD, ADHD, and typically developing controls but also children meeting Leibenluft et al.'s research criteria for "severe mood dysregulation" (SMD), including functionally disabling course of chronic irritability and ADHD-like symptoms of hyperarousal, (Leibenluft et al. 2003). The first by Guyer et al. administered the DANVA to 252 youths ages 7–18 years old, including those with BD, SMD, ADHD, anxiety and/or major depression, and controls. They found that participants in the BD and SMD groups made significantly more errors than the other three groups, but did not differ from one another (Guyer et al. 2007). The second by Brotman et al. conducted an event-related fMRI study to evaluate face processing in children and adolescents with BD, ADHD, SMD, and typically-developing controls. They found that ADHD youth had significantly greater left amygdala activity than the other three groups when rating their fear of neutrally valenced faces, whereas youth with SMD had significantly less left amygdala activity (Brotman et al. 2010).

Sinzig compared children with Autistic disorder ($N = 19$), Autistic disorder plus ADHD ($N = 21$), ADHD alone ($N = 30$), and controls ($N = 29$) in their ability to recognize facial emotions using the Frankfurt Test and Training of Social Affect, which uses 50 face photographs and 40 eye-pair photographs depicting joy, sadness, fear, anger, surprise, disgust, and neutral (Sinzig et al. 2008). Participants with Autism plus ADHD and those with ADHD alone had worse performance than either controls or those with Autism alone. Post-hoc analyses showed that ADHD participants performed worse than controls on joy eye-pairs, whereas those with autism plus ADHD performed worse on surprise faces than controls. These finding suggest that attentional problems should be considered in face processing studies in autism spectrum disorders.

Thus, studies examining the specificity of face processing deficits in ADHD participants versus those with related psychopathology are at an early stage. Further work is necessary to examine important issues including developmental and treatment effects on the brain/behavior interactions underlying face processing in ADHD.

8 Conclusion

Face processing involves attentional circuits: both top-down cortical pathways and subcortical relays involving the amygdala. Although inattention, distractibility, and hyperactivity are considered the cardinal symptoms of ADHD, clinicians and researchers alike are becoming increasingly aware of emotional impairments in ADHD. Face-processing experiments are particularly apt for testing these issues. Thus far, studies suggest that ADHD involves behavior and brain alterations in face processing. However, more work is clearly needed, including work to probe the impact of development, gender, and treatment, as well as correlations between face processing, core ADHD symptoms, and real-world measures of emotional dysfunction. Efforts to improve our understanding of the brain/behavior alterations underlying emotional dysfunction in those with ADHD are likely to yield scientific and clinical benefits.

References

Adolphs R, Spezio M (2006) Role of the amygdala in processing visual social stimuli. Prog Brain Res 156:363–378
Adolphs R, Tranel D (2004) Impaired judgments of sadness but not happiness following bilateral amygdala damage. J Cogn Neurosci 16:453–462
Adolphs R, Damasio H, Tranel D, Damasio AR (1996) Cortical systems for the recognition of emotion in facial expressions. J Neurosci 16:7678–7687
Adolphs R, Damasio H, Tranel D, Cooper G, Damasio AR (2000) A role for somatosensory cortices in the visual recognition of emotion as revealed by three-dimensional lesion mapping. J Neurosci 20:2683–2690
Adolphs R, Baron-Cohen S, Tranel D (2002) Impaired recognition of social emotions following amygdala damage. J Cogn Neurosci 14:1264–1274
American Psychiatric Association (1980) Diagnostic and statistical manual of mental disorders 3 rd edition (DSM-III). American Psychiatric Association Press, Washington, D.C
American Psychiatric Association (2000) Diagnostic and Statistical Manual of Mental Disorders 4th Edition Text Revision (DSM-IV-TR). American Psychiatric Association, Washington, D.C
Anderson AK, Phelps EA (2001) Lesions of the human amygdala impair enhanced perception of emotionally salient events. Nature 411:305–309

Barkley RA, Fischer M (2010) The unique contribution of emotional impulsiveness to impairment in major life activities in hyperactive children as adults. J Am Acad Child Adolesc Psychiatry 49:503–513

Batty M, Taylor MJ (2003) Early processing of the six basic facial emotional expressions. Brain Res Cogn Brain Res 17:613–620

Bauer RM (1984) Autonomic recognition of names and faces in prosopagnosia: a neuropsychological application of the Guilty Knowledge Test. Neuropsychologia 22:457–469

Blaskey LG, Harris LJ, Nigg JT (2008) Are sensation seeking and emotion processing related to or distinct from cognitive control in children with ADHD? Child Neuropsychol 14:353–371

Booth JR, Burman DD, Meyer JR, Lei Z, Trommer BL, Davenport ND, Li W, Parrish TB, Gitelman DR, Mesulam MM (2005) Larger deficits in brain networks for response inhibition than for visual selective attention in attention deficit hyperactivity disorder (ADHD). J Child Psychol Psychiatry 46:94–111

Brotman MA, Rich BA, Guyer AE, Lunsford JR, Horsey SE, Reising MM, Thomas LA, Fromm SJ, Towbin K, Pine DS, Leibenluft E (2010) Amygdala activation during emotion processing of neutral faces in children with severe mood dysregulation versus ADHD or bipolar disorder. Am J Psychiatry 167:61–69

Bukach CM, Gauthier I, Tarr MJ (2006) Beyond faces and modularity: the power of an expertise framework. Trends Cogn Sci 10:159–166

Cadesky EB, Mota VL, Schachar RJ (2000) Beyond words: how do children with ADHD and/or conduct problems process nonverbal information about affect? J Am Acad Child Adolesc Psychiatry 39:1160–1167

Carmel D, Bentin S (2002) Domain specificity versus expertise: factors influencing distinct processing of faces. Cognition 83:1–29

Castellanos FX, Giedd JN, Marsh WL, Hamburger SD, Vaituzis AC, Dickstein DP, Sarfatti SE, Vauss YC, Snell JW, Lange N, Kaysen D, Krain AL, Ritchie GF, Rajapakse JC, Rapoport JL (1996) Quantitative brain magnetic resonance imaging in attention-deficit hyperactivity disorder. Arch Gen Psychiatry 53:607–616

Castellanos FX, Giedd JN, Berquin PC, Walter JM, Sharp W, Tran T, Vaituzis AC, Blumenthal JD, Nelson J, Bastain TM, Zijdenbos A, Evans AC, Rapoport JL (2001) Quantitative brain magnetic resonance imaging in girls with attention-deficit/hyperactivity disorder. Arch Gen Psychiatry 58:289–295

Castellanos FX, Lee PP, Sharp W, Jeffries NO, Greenstein DK, Clasen LS, Blumenthal JD, James RS, Ebens CL, Walter JM, Zijdenbos A, Evans AC, Giedd JN, Rapoport JL (2002) Developmental trajectories of brain volume abnormalities in children and adolescents with attention-deficit/hyperactivity disorder. JAMA 288:1740–1748

Castellanos FX, Sonuga-Barke EJ, Milham MP, Tannock R (2006) Characterizing cognition in ADHD: beyond executive dysfunction. Trends Cogn Sci 10:117–123

Coelho CM, Cloete S, Wallis G (2010) The face-in-the-crowd effect: when angry faces are just cross(es). J Vis 10:7–14

Cole PM, Martin SE, Dennis TA (2004) Emotion regulation as a scientific construct: methodological challenges and directions for child development research. Child Dev 75:317–333

Corbett B, Glidden H (2000) Processing affective stimuli in children with attention-deficit hyperactivity disorder. Child Neuropsychol 6:144–155

Da Fonseca D, Seguier V, Santos A, Poinso F, Deruelle C (2009) Emotion understanding in children with ADHD. Child Psychiatry Hum Dev 40:111–121

Dickstein DP, Leibenluft E (2006) Emotion regulation in children and adolescents: boundaries between normalcy and bipolar disorder. Dev Psychopathol 18:1105–1131

Dickstein SG, Bannon K, Castellanos FX, Milham MP (2006) The neural correlates of attention deficit hyperactivity disorder: an ALE meta-analysis. J Child Psychol Psychiatry 47:1051–1062

Dimberg U, Thunberg M, Elmehed K (2000) Unconscious facial reactions to emotional facial expressions. Psychol Sci 11:86–89

Downing P, Liu J, Kanwisher N (2001) Testing cognitive models of visual attention with fMRI and MEG. Neuropsychologia 39:1329–1342

Durston S, Mulder M, Casey BJ, Ziermans T, Van Engeland H (2006) Activation in ventral prefrontal cortex is sensitive to genetic vulnerability for attention-deficit hyperactivity disorder. Biol Psychiatry 60:1062–1070

Eimer M (2000) Effects of face inversion on the structural encoding and recognition of faces. Evidence from event-related brain potentials. Brain Res Cogn Brain Res 10:145–158

Eimer M, Holmes A (2002) An ERP study on the time course of emotional face processing. Neuroreport 13:427–431

Eimer M, Holmes A (2007) Event-related brain potential correlates of emotional face processing. Neuropsychologia 45:15–31

Eimer M, Holmes A, McGlone FP (2003) The role of spatial attention in the processing of facial expression: an ERP study of rapid brain responses to six basic emotions. Cogn Affect Behav Neurosci 3:97–110

Farroni T, Csibra G, Simion F, Johnson MH (2002) Eye contact detection in humans from birth. Proc Natl Acad Sci U S A 99:9602–9605

Fox E, Lester V, Russo R, Bowles RJ, Pichler A, Dutton K (2000) Facial expressions of emotion: are angry faces detected more efficiently? Cogn Emot 14:61–92

Fusar-Poli P, Placentino A, Carletti F, Landi P, Allen P, Surguladze S, Benedetti F, Abbamonte M, Gasparotti R, Barale F, Perez J, McGuire P, Politi P (2009) Functional atlas of emotional faces processing: a voxel-based meta-analysis of 105 functional magnetic resonance imaging studies. J Psychiatry Neurosci 34:418–432

Gauthier I, Skudlarski P, Gore JC, Anderson AW (2000) Expertise for cars and birds recruits brain areas involved in face recognition. Nat Neurosci 3:191–197

Gobbini MI, Haxby JV (2007) Neural systems for recognition of familiar faces. Neuropsychologia 45:32–41

Gobbini MI, Leibenluft E, Santiago N, Haxby JV (2004) Social and emotional attachment in the neural representation of faces. Neuroimage 22:1628–1635

Goren CC, Sarty M, Wu PY (1975) Visual following and pattern discrimination of face-like stimuli by newborn infants. Pediatrics 56:544–549

Gorno-Tempini ML, Pradelli S, Serafini M, Pagnoni G, Baraldi P, Porro C, Nicoletti R, Umita C, Nichelli P (2001) Explicit and incidental facial expression processing: an fMRI study. Neuroimage 14:465–473

Grossmann T, Johnson MH, Farroni T, Csibra G (2007) Social perception in the infant brain: gamma oscillatory activity in response to eye gaze. Soc Cogn Affect Neurosci 2:284–291

Guyer AE, McClure EB, Adler AD, Brotman MA, Rich BA, Kimes AS, Pine DS, Ernst M, Leibenluft E (2007) Specificity of facial expression labeling deficits in childhood psychopathology. J Child Psychol Psychiatry 48:863–871

Handford M (1987) Where's Waldo. Candlewick Press, Cambridge, MA

Haxby JV, Hoffman EA, Gobbini MI (2002) Human neural systems for face recognition and social communication. Biol Psychiatry 51:59–67

Hoffman EA, Haxby JV (2000) Distinct representations of eye gaze and identity in the distributed human neural system for face perception. Nat Neurosci 3:80–84

Horstmann G, Ansorge U (2009) Visual search for facial expressions of emotions: a comparison of dynamic and static faces. Emotion 9:29–38

Horstmann G, Bauland A (2006) Search asymmetries with real faces: testing the anger-superiority effect. Emotion 6:193–207

Itier RJ, Taylor MJ (2004) Face recognition memory and configural processing: a developmental ERP study using upright, inverted, and contrast-reversed faces. J Cogn Neurosci 16:487–502

Johnson MH, Dziurawiec S, Ellis H, Morton J (1991) Newborns' preferential tracking of face-like stimuli and its subsequent decline. Cognition 40:1–19

Kanwisher N, McDermott J, Chun MM (1997) The fusiform face area: a module in human extrastriate cortex specialized for face perception. J Neurosci 17:4302–4311

Kats-Gold I, Besser A, Priel B (2007) The role of simple emotion recognition skills among school aged boys at risk of ADHD. J Abnorm Child Psychol 35:363–378

Klimkeit EI, Mattingley JB, Sheppard DM, Lee P, Bradshaw JL (2003) Perceptual asymmetries in normal children and children with attention deficit/hyperactivity disorder. Brain Cogn 52:205–215

Krauel K, Duzel E, Hinrichs H, Lenz D, Herrmann CS, Santel S, Rellum T, Baving L (2009) Electrophysiological correlates of semantic processing during encoding of neutral and emotional pictures in patients with ADHD. Neuropsychologia 47:1873–1882

Lang PJ, Bradley MM, Cuthbert BN (1998) Emotion, motivation, and anxiety: brain mechanisms and psychophysiology. Biol Psychiatry 44:1248–1263

LeDoux JE (2000) Emotion circuits in the brain. Annu Rev Neurosci 23:155–184

LeDoux J (2003) The emotional brain, fear, and the amygdala. Cell Mol Neurobiol 23:727–738

Leibenluft E, Charney DS, Towbin KE, Bhangoo RK, Pine DS (2003) Defining clinical phenotypes of juvenile mania. Am J Psychiatry 160:430–437

Leibenluft E, Gobbini MI, Harrison T, Haxby JV (2004) Mothers' neural activation in response to pictures of their children and other children. Biol Psychiatry 56:225–232

Lindstrom KM, Guyer AE, Mogg K, Bradley BP, Fox NA, Ernst M, Nelson EE, Leibenluft E, Britton JC, Monk CS, Pine DS, Bar-Haim Y (2009) Normative data on development of neural and behavioral mechanisms underlying attention orienting toward social-emotional stimuli: an exploratory study. Brain Res 1292:61–70

Martel MM, Nigg JT (2006) Child ADHD and personality/temperament traits of reactive and effortful control, resiliency, and emotionality. J Child Psychol Psychiatry 47:1175–1183

Monk CS, McClure EB, Nelson EE, Zarahn E, Bilder RM, Leibenluft E, Charney DS, Ernst M, Pine DS (2003) Adolescent immaturity in attention-related brain engagement to emotional facial expressions. Neuroimage 20:420–428

Nelson EE, McClure EB, Monk CS, Zarahn E, Leibenluft E, Pine DS, Ernst M (2003) Developmental differences in neuronal engagement during implicit encoding of emotional faces: an event-related fMRI study. J Child Psychol Psychiatry 44:1015–1024

Nelson EE, Leibenluft E, McClure EB, Pine DS (2005) The social re-orientation of adolescence: a neuroscience perspective on the process and its relation to psychopathology. Psychol Med 35:163–174

Nowicki S, Duke M (1994) Individual differences in the nonverbal communication of affect: The Diagnostic Analysis of Nonverbal Accuracy Scale. J Nonverbal Behav 18:9–35

O'Craven KM, Downing PE, Kanwisher N (1999) fMRI evidence for objects as the units of attentional selection. Nature 401:584–587

Ohman A (1988) Nonconscious control of autonomic responses: a role for Pavlovian conditioning? Biol Psychol 27:113–135

Palermo R, Rhodes G (2007) Are you always on my mind? A review of how face perception and attention interact. Neuropsychologia 45:75–92

Passarotti AM, Sweeney JA, Pavuluri MN (2009) Neural correlates of incidental and directed facial emotion processing in adolescents and adults. Soc Cogn Affect Neurosci 4:387–398

Pegna AJ, Khateb A, Michel CM, Landis T (2004) Visual recognition of faces, objects, and words using degraded stimuli: where and when it occurs. Hum Brain Mapp 22:300–311

Pelc K, Kornreich C, Foisy ML, Dan B (2006) Recognition of emotional facial expressions in attention-deficit hyperactivity disorder. Pediatr Neurol 35:93–97

Pessoa L, McKenna M, Gutierrez E, Ungerleider LG (2002) Neural processing of emotional faces requires attention. Proc Natl Acad Sci U S A 99:11458–11463

Pliszka SR, Glahn DC, Semrud-Clikeman M, Franklin C, Perez IR, Xiong J, Liotti M (2006) Neuroimaging of inhibitory control areas in children with attention deficit hyperactivity disorder who were treatment naive or in long-term treatment. Am J Psychiatry 163:1052–1060

Polanczyk G, Jensen P (2008) Epidemiologic considerations in attention deficit hyperactivity disorder: a review and update. Child Adolesc Psychiatr Clin N Am 17:245–260, vii

Rapport LJ, Friedman SR, Tzelepis A, Van Voorhis A (2002) Experienced emotion and affect recognition in adult attention-deficit hyperactivity disorder. Neuropsychology 16:102–110

Rolls ET (2007) The representation of information about faces in the temporal and frontal lobes. Neuropsychologia 45:124–143

Rousselet GA, Mace MJ, Fabre-Thorpe M (2003) Is it an animal? Is it a human face? Fast processing in upright and inverted natural scenes. J Vis 3:440–455

Shin DW, Lee SJ, Kim BJ, Park Y, Lim SW (2008) Visual attention deficits contribute to impaired facial emotion recognition in boys with attention-deficit/hyperactivity disorder. Neuropediatrics 39:323–327

Sinzig J, Morsch D, Lehmkuhl G (2008) Do hyperactivity, impulsivity and inattention have an impact on the ability of facial affect recognition in children with autism and ADHD? Eur Child Adolesc Psychiatry 17:63–72

Suzuki S, Cavanagh P (1998) A shape-contrast effect for briefly presented stimuli. J Exp Psychol Hum Percept Perform 24:1315–1341

Thompson RA (1994) Emotion regulation: a theme in search of definition. Monogr Soc Res Child Dev 59:25–52

Tranel D, Damasio AR (1985) Knowledge without awareness: an autonomic index of facial recognition by prosopagnosics. Science 228:1453–1454

Vuilleumier P, Pourtois G (2007) Distributed and interactive brain mechanisms during emotion face perception: evidence from functional neuroimaging. Neuropsychologia 45:174–194

Vuilleumier P, Schwartz S (2001a) Beware and be aware: capture of spatial attention by fear-related stimuli in neglect. Neuroreport 12:1119–1122

Vuilleumier P, Schwartz S (2001b) Emotional facial expressions capture attention. Neurology 56:153–158

Vuilleumier P, Armony JL, Driver J, Dolan RJ (2001) Effects of attention and emotion on face processing in the human brain: an event-related fMRI study. Neuron 30:829–841

Vuilleumier P, Armony JL, Clarke K, Husain M, Driver J, Dolan RJ (2002) Neural response to emotional faces with and without awareness: event-related fMRI in a parietal patient with visual extinction and spatial neglect. Neuropsychologia 40:2156–2166

Williams MA, McGlone F, Abbott DF, Mattingley JB (2005) Differential amygdala responses to happy and fearful facial expressions depend on selective attention. Neuroimage 24:417–425

Williams LM, Hermens DF, Palmer D, Kohn M, Clarke S, Keage H, Clark CR, Gordon E (2008) Misinterpreting emotional expressions in attention-deficit/hyperactivity disorder: evidence for a neural marker and stimulant effects. Biol Psychiatry 63:917–926

Yuill N, Lyon J (2007) Selective difficulty in recognising facial expressions of emotion in boys with ADHD. General performance impairments or specific problems in social cognition? Eur Child Adolesc Psychiatry 16:398–404

Quantitative and Molecular Genetics of ADHD

Philip Asherson and Hugh Gurling

Contents

1 Introduction 241
2 Quantitative Genetic Studies 241
3 ADHD as a Quantitative Trait 242
4 ADHD in Adults 243
5 ADHD, Inattention and Hyperactivity–Impulsivity 244
6 Cognitive Performance and Brain Function 246
7 Molecular Genetic Studies 248
    7.1 Early Genetic Findings in ADHD 248
    7.2 The Dopamine Transporter Gene (DAT1/SLC6A3) 249
    7.3 Other Candidate Gene Associations 250
    7.4 Dopamine D4 Receptor Gene (DRD4) 252
    7.5 The Tachykinin Receptor 1 252
    7.6 The Serotonin 1B Receptor Gene (HTR1B) 253
    7.7 Serotonin Transporter Gene (SLC6A4/5-HTT) 253
    7.8 Synaptosomal-Associated Protein 25 Isoform Gene 254
    7.9 ADHD and Co-occurring Disorders 254
    7.10 ADHD in Adults 255
8 Whole Genome Approaches 256
    8.1 Linkage Studies 256
    8.2 Genome-Wide Association 257
    8.3 Copy Number Variants 259
    8.4 Endophenotypes and Intermediate Phenotypes 260
9 Research Gaps and Future Prospects 263
References 264

P. Asherson
MRC Social Genetic and Developmental Psychiatry, Institute of Psychiatry, Kings College London, UK
e-mail: philip.asherson@kcl.ac.uk

H. Gurling
Mental Health Sciences Unit, Faculty of Brain Sciences, University College London, UK
e-mail: h.gurling@ucl.ac.uk

Abstract ADHD is a common and highly heritable disorder. Family, twin, and adoption studies confirm a strong genetic influence in risk for ADHD and there has been a great deal of interest in identifying the genetic factors involved. Quantitative genetic studies find that genetic risk for ADHD is continuously distributed throughout the population, that there are both shared and unique genetic influences on inattention and hyperactivity-impulsivity, and that ADHD shares genetic risk factors with commonly co-occurring clinical syndromes and traits. ADHD is found at all ages and the underlying genetic architecture is similar across the lifespan. In terms of specific genetic findings, there is consistent evidence of monoamine neurotransmitter involvement with the best evidence coming from genetic markers in or near the dopamine D4 and D5 receptor genes. Recent genome-wide association studies have identified new association findings, including genes involved in cell division, cell adhesion, neuronal migration, and neuronal plasticity. However, as yet, none of these pass genome-wide levels of significance. Finally, recent data confirm an important role for rare copy number variants, including those that are found in schizophrenia and autism. Future work should use genetic association data to determine the nature of the cognitive, neuronal and cellular processes that mediate genetic risks on behaviour, and identify environmental factors that interact with genetic risks for ADHD.

Keywords ADHD · Association studies · Copy number variants · Genes · Heritability · Twin studies

Abbreviations

ADHD	Attention deficit hyperactivity disorder
CEPH	Centre d'Etude du Polymorphisme Humain
CNV	Copy number variants
DAT1	Dopamine transporter (SLC/6A3)
DZ	Dizygotic
EAA	Equal environments assumption
GWAS	Genomewide association studies
IMAGE	International Multisite ADHD Genetics
Met	Methionine
MV	Multivariate
MZ	Monozygotic
NICE	National Institute for Clinical Health and Excellence (UK)
RD	Reading disability
RT	Response time
SNP	Single nucleotide polymorphism
TDT	Transmission disequilibrium test
Val	Valine
VNTR	Variable number tandem repeat

1 Introduction

ADHD is a common childhood onset disorder that affects around 5% of the childhood population (Polanczyk et al. 2007). The disorder persists into adulthood in around two-thirds of cases, either as the full-blown form or in partial remission, with impairment from continued symptoms of the disorder (Faraone et al. 2006). ADHD is also associated with a wide range of cognitive and functional deficits as well as neurodevelopmental, psychiatric and behavioural comorbidities. The precise causes of ADHD are not well understood although family, twin, and adoption studies define a condition that is largely influenced by genetic factors. ADHD aggregates within families and estimates from numerous twin studies indicate an average heritability (the proportion of phenotypic variance explained by genetic factors) of around 76% (Faraone et al. 2005). As such, there has been a great deal of interest in identifying the genetic factors underlying susceptibility for the disorder and understanding the ways in which genetic factors influence ADHD and associated cognitive and behavioural symptoms, syndromes and traits. While the finding of high heritability for ADHD has naturally focused research on delineating the molecular mechanisms involved, it is unlikely that genetic risk factors act independently of the environment. Genetically sensitive designs are also required for a full understanding of the role of the environment and the interplay between genetic and environmental risks.

Genetic studies of ADHD are relevant in two main ways. First, quantitative genetic studies enable the investigation of the extent of genetic effects on ADHD and the extent to which these are shared with associated cognitive impairments, brain function deficits and co-occurring disorders and traits. This is important to our understanding of the relationship between ADHD and associated behavioural and cognitive functions. Secondly, molecular genetic studies enable the identification of the specific genetic risk factors involved, allowing for a detailed understanding of the underlying molecular and neurobiological processes that underlie the disorder. Identifying specific risk alleles is an important step towards a complete understanding of the aetiological processes involved in the development and persistence of ADHD throughout the lifespan. Taken together these two main approaches provide a better understanding of the aetiology of ADHD with the potential for improved prediction of clinical outcomes and development of novel pharmacological and non-pharmacological treatments

2 Quantitative Genetic Studies

Family, adoption and twin studies delineate a disorder that tends to run in families with a risk to first degree relatives in the order of five- to tenfold the population rate (Chen et al. 2008; Faraone et al. 2000). The proportion of phenotypic variance explained by genetic factors (heritability) averages around 76% (Faraone et al. 2005).

Most information on the heritability of ADHD comes from population twin studies that compare monozygotic twin pairs (MZ: identical twins), who share 100% of genetic variation, with dizygotic (DZ: non-identical twins) twin pairs, who share on average 50% of genetic variation. The main assumption of twin studies is that MZ and DZ twins have an equal exposure to environments that influence the trait being studied: the so-called equal environment assumption (EEA). The EEA has been directly tested by Kendler and colleagues for twin studies of psychiatric disorders. Using a subset of twin pairs for whom perceived zygosity differed from tested zygosity, they modelled perceived zygosity as a form of specific familial environment and found that for five common psychiatric disorders there was no influence of perceived zygosity status on twin similarity (Kendler et al. 1993), supporting the validity of the EAA in twin studies of psychiatric disorders.

Twin studies of ADHD have mainly used continuous rating scale scores of ADHD symptoms as reported by parents and teachers and found substantial differences in the phenotypic correlation between MZ and DZ pairs, with MZ twin correlations being twice, or more than twice, DZ correlations. Most studies show little or no evidence for shared environmental influences, so that the general conclusion has been that familial effects are entirely due to genetic factors, while unique environmental effects, including measurement error and non-inherited epigenetic factors, explain non-familial effects on the clinical phenotype. For example, in a recent meta-analysis and review there was no overall evidence for shared environmental influences on ADHD, while there were for some other behavioural phenotypes, including conduct disorder, oppositional defiant disorder, anxiety and depression (Burt 2009). The finding that there are no shared environmental effects on ADHD is important for the interpretation of family studies because we can infer that familial influences that are shared between ADHD and other disorders or traits are likely to result from shared genetic factors rather than the familial environment.

3 ADHD as a Quantitative Trait

Family and twin studies both support the concept of ADHD as representing the extreme of one or more quantitative traits that show variation throughout the population. There are several lines of evidence for this conclusion (Chen et al. 2008). First, twin studies that report on concordance rates for the diagnosis of ADHD estimate similar high heritabilities to that estimated from the numerous studies that report on correlations of ADHD symptoms in population twin samples (Sherman et al. 1997). Secondly, twin studies that define ADHD dimensionally have shown that the genetic contribution to ADHD operates across the continuum and exerts similar influences to those acting on individuals with extreme ADHD scores (Levy et al. 1997). Thirdly, environmental risk measures also correlate with trait scores for ADHD: for example, the impact of food additives (Stevenson et al. 2010) and maternal smoking during pregnancy (Becker et al. 2008; Kahn et al. 2003). Fourthly, as we shall see, numerous studies report on the association of

ADHD trait scores with other dimensional measures such as, reading ability, conduct problems and cognitive performance, that are also reflected in ADHD case-control differences.

These findings promoted an investigation of the relationship between DSM-IV combined type ADHD and quantitative trait scores of ADHD among siblings. Using a sample of 894 ADHD probands with a research diagnosis of combined type ADHD and 1,135 of their siblings, who were unselected for ADHD, a sibling correlation of 0.30 for teacher ratings and 0.21 for parents' ratings was estimated, which was comparable to the average DZ correlations estimated from review of the twin literature (Chen et al. 2008). Furthermore, there was no evidence for bimodality in the sibling data. This study also estimated the sibling risk for combined type ADHD to be about ninefold that of the general population rate. Overall, this study leads to the conclusion that genetic liability for ADHD is distributed throughout the population and is reasonably well measured using parent or teacher rated DSM-IV rating scales.

From a clinical perspective, there is a clear implication that there is no natural boundary between those with and without ADHD based on symptom count alone. For this reason, the National Institute for Clinical Health and Excellence in their recent review (NICE 2008) emphasized the importance of linking high levels of symptoms to impairment when defining ADHD as a clinical disorder. However, for aetiological investigations, both categorical and quantitative approaches to the ADHD phenotype appear to be valid.

4 ADHD in Adults

Until recently, the evidence for familial effects on ADHD in adults came from family and adoption studies, which reported rates of ADHD among the parents of children with ADHD and rates of ADHD among the offspring of adults with ADHD. Early reports found that adoptive parents of hyperactive children were less likely than biological parents to have hyperactivity or associated disorders (Cantwell 1972; Morrison and Stewart 1971). Subsequent studies confirmed these findings (Alberts-Corush et al. 1986; Epstein et al. 2000; Sprich et al. 2000). For example, Sprich and colleagues reported a rate of ADHD of 18% among the biological parents of children with ADHD compared to 6% in adoptive parents and 3% in controls (Sprich et al. 2000). As expected, family studies also find differences among the parents of children with and without ADHD (Curko Kera et al. 2004). The study from Faraone and colleagues also found that parents of children with ADHD were far more likely to have ADHD than control parents, particularly among the parents of children who retained the diagnosis of ADHD into adolescence at 4-year follow up (Faraone et al. 2000). Other studies investigated the rates of ADHD among the relatives of adults with ADHD and found high familial risks. For example, Biederman et al. reported 57% of the offspring of adults with ADHD

(Biederman et al. 1995), while Manshadi et al. found 41% of the siblings of adults with attention deficit disorder (ADD) had ADHD (Manshadi et al. 1983).

These studies appear to demonstrate that familial effects on ADHD are found throughout the lifespan. Furthermore, the high rates of ADHD reported in some family studies suggest that persistent forms of ADHD may be associated with a greater familial loading than remitting forms of the disorder (Faraone 2004). However, there is a discrepancy within the data that needs further investigation to understand fully the familial influences on ADHD symptoms in adults. Where family studies have used cross-sectional data, they have found that self-report for current ADHD symptoms does not always show the expected familial effects, but rather differences were found with informant report or on tests of cognitive performance.

Adult twin studies that rely entirely on cross-section self-report data of ADHD symptoms are consistent with the family studies by finding low familial risks and heritabilities for self-report measures of ADHD symptoms. In a large Dutch adult population twin study, estimates of heritability were in the region of 0.30 (Boomsma et al. 2010), which is close to the estimate of 0.35 reported in a similar Swedish study (Larsson et al. unpublished data). There are several potential explanations for this finding. First, self-rating for current level of ADHD symptoms appears to be a poor index of genetic liability, since such low heritabilities are also found when using self-rated data in adolescent twin samples (Ehringer et al. 2006; Martin et al. 2002). Secondly, unreliability of self-ratings may contribute to the low estimates; related to this, the use of self-ratings means that two raters are involved in the evaluation of each twin pair, whereas traditional twin studies of ADHD in children use the same parent or teacher to report on both members of a twin pair. This is a potentially important cause of low heritability estimates, since we observed such a rater effect in children when comparing the heritability estimated from same teacher for both members of a twin pair (74%) versus different teachers (46%) in a large twin sample of 12-year-olds (Merwood and Asherson, unpublished data). Thirdly, adult-onset conditions may generate ADHD-like symptoms and these would not be expected to share genetic effects with ADHD. A recent twin study found that heritability for attention problems in 19–20 year olds was around 80% when self and parent data were combined, suggesting measurement error as the main explanation for the lower heritabilities seen in the previous adult studies that used self ratings alone (Larsson et al., unpublished data). Further work is needed to fully understand the observed discrepancies in heritability and their relationship to rater and developmental effects on the measures used.

5 ADHD, Inattention and Hyperactivity–Impulsivity

An important question that can be addressed by quantitative genetic approaches is the aetiological relationship between inattention and hyperactivity–impulsivity. Twin studies that have investigated this have found that these are partially separable measures with only partially overlapping genetic influences. While the correlation between the two domains is largely accounted for by shared genetic effects, there are additional significant unique genetic effects acting on each of the two domains (McLoughlin et al. 2007). Furthermore, a similar pattern of findings was found for ADHD in adults, despite low overall heritability of the self-rating scale scores (Larsson et al. in review), indicating that the underlying genetic architecture of ADHD is similar in children, adolescents and adults.

Twin data have also been used to investigate the DSM subtypes of ADHD and comparison with empirically derived latent class analyses of DSM symptoms that identify groups based on the clustering of symptoms. Such statistical approaches found that most DSM-IV inattentive subtype cases belong to the combined type latent class and make a clear separation from the purely inattentive subtype with only one or two symptoms of hyperactivity–impulsivity; a distinction that is reflected in the revised version of the DSM (www.dsm5.org). These findings that use empirical approaches to define clinical subtypes also lend support for the separation of the two symptom domains. The studies from Todd et al. (2001) showed that latent class subtypes tended to co-segregate within twin pairs, whereas this was not the case for the DSM-IV subtypes, with cross concordance identified between the DSM-IV combined and inattentive subtypes. The DSM-IV predominantly hyperactive–impulsive cases were, however, genetically distinct (Todd et al. 2001).

Molecular genetic investigations of the two symptom domains are therefore expected to identify genetic variants that confer unique, as well as shared, risks and imply the existence of two or more molecular and neurobiological pathways. In support of this general conclusion, we see differential relationships between each of the two domains and comorbid traits. One example is reading disability (RD) and inattention where the correlation between ADHD and RD was found to be largely driven by genetic factors not shared with hyperactivity–impulsivity (Paloyelis et al. 2010). A contrasting example is a study of ADHD and oppositional behaviour that found high phenotypic and genetic correlations with hyperactivity–impulsivity but much lower with inattention (Wood et al. 2009). Since both the genetic and environmental correlations between hyperactivity–impulsivity and oppositional behaviour were around 95% in this study, the authors concluded that they represented the same underlying liability and that this was distinct from the inattentive domain.

Twin studies have found that there are many other examples of overlapping genetic influences between ADHD symptoms and co-occurring disorders and traits, including autism spectrum disorder (Ronald et al. 2008), motor coordination (Francks et al. 2003), conduct disorder (Thapar et al. 2001), and depression (Cole et al. 2009). These studies have yet to investigate whether there are differential effects between the inattentive and hyperactive–impulsive components of ADHD. However, they clearly demonstrate the complexity of shared and unique genetic and other aetiological influences among many mental health disorders and comorbid traits. Further work is needed to understand fully the causal pathways involved. At a basic level, longitudinal twin designs can further delineate the causal relationships between ADHD and co-occurring traits, which include pleiotropic effects (multiple effects of individual sets of genes), mediating effects (brain functions or behaviours that mediate genetic effects on ADHD) and risk models (one disorder leading to another). Finally, some symptoms/traits traditionally regarded as separate from ADHD, such as mood instability in adult ADHD, are being re-evaluated and may be part of the core syndrome (Skirrow et al. 2009).

6 Cognitive Performance and Brain Function

Family and twin data have also addressed the phenotypic and aetiological relationships between ADHD and cognitive performance and other measures of brain function. Such approaches have been used to identify endophenotypes for ADHD, which are measures of brain function that index genetic risk for a disorder. The basic idea behind endophenotype research is that there are measurable brain functions that are closer to molecular mechanisms and, as such, are likely to represent simpler clues to genetic underpinnings than the disease syndrome itself (Gottesman and Gould 2003). The distinction needs to be made from biological markers, which is a more general term that can include state markers for the disorder as well as trait markers for underlying liability.

Even where there is evidence for shared aetiological relationships between measures of brain function and a disorder such as ADHD, it is necessary to further clarify the distinction between pleiotropic effects and intermediate phenotypes (Kendler and Neale 2010). Pleiotropy refers to the effects of genes on multiple phenotypes, such as those that account for the association between ADHD and neurodevelopmental disorders such as dyslexia and autism spectrum disorder. These can be difficult to differentiate from "intermediate phenotypes" which, according to the nomenclature recommended by Kendler and Neale, should be reserved for the subgroup of "endophenotypes" that mediate genetic effects on behaviour. Both endophenotypes and intermediate phenotypes can be used to map some of the genes that increase risk for ADHD. However, only intermediate phenotypes represent direct causal pathways from genes to ADHD, which could therefore be targeted for treatment interventions of ADHD.

Key criteria when considering a measure as a putative endophenotype for ADHD are association with the clinical disorder or correlation with the quantitative trait; and shared genetic influences with the clinical disorder or quantitative trait. The endophenotype must itself be heritable, although the size of heritability does not provide information on the expected genetic effects at individual genetic loci. Genetic correlation with ADHD is often inferred from family studies that show significant deficits among the unaffected siblings of affected ADHD probands as compared to healthy controls: the key test being whether the endophenotype measure is significantly different in siblings of proband compared to controls. Estimates of sibling familiality, which can be compared to concordance rates or correlations between DZ twins, are best derived by comparing the means among probands, with sibling and population controls unselected for phenotype

(Andreou et al. 2007). Since family studies cannot separate out the effects of the familial environment from genetic influences, it should not be assumed that any observed familial effects are genetic in origin. Twin designs are far better, since they decompose the familial genetic and environmental influences. However where previous twin studies have found no familial environmental effects on both phenotypes, as is the case for ADHD and most endophenotypes of ADHD, it is reasonable to assume that only genetic and unique environmental factors play a role in the ADHD-endophenotype familial association.

The family design has now been used extensively to nominate putative endophenotypes for ADHD including cognitive-performance, electrophysiological and brain neuroimaging measures. However, it is not clear what criteria can be applied to pick out the most promising measures from family study data. One approach is to take advantage of the power of multivariate (MV) genetic model fitting to consider the phenotypic and genetic correlations between several putative endophenotype measures, as well as estimating the extent to which the measures being studied share genetic influences with ADHD. Furthermore, MV genetic methods can separate out one or more factors marking distinct aetiological pathways.

Kuntsi and colleagues recently completed such analyses using a large ADHD proband sibling pair sample collected as part of the International Multisite ADHD Genetics (IMAGE) project, and control sibling pair sample, to study the aetiological relationship of ADHD with cognitive performance measures (Kuntsi et al. 2010). They were also able to compare the findings from the clinical sample with data from a population twin sample that measured ADHD and included the same measures of cognitive performance (Kuntsi and Klein 2011). The main measures included in these analyses were reaction time mean and variability derived from a speeded reaction-time task (the Fast Task) and a Go/NoGo task, omission and commission on the Go/NoGo task, impulsive choice and delay aversion, IQ and digit span forwards and backwards. Heritabilities for the main cognitive performance variables were reported and found to be in the range of 50–60%, but this depended on the measure used and on whether theoretically similar measures were combined together, which has the impact of reducing error and increasing estimated heritability (Kuntsi et al. 2006b). These data highlight the importance of using stable measures that are reliable over time (i.e., high test–retest reliability) because "heritability" cannot be greater than retest reliability. The same considerations would also impact on the power of the measures being used to map individual genes so that, in general, it is thought better to combine several performance measures to generate a more stable cognitive index, as for example in the generation of IQ.

MV analysis was applied to the IMAGE and control sibling pair samples to address the question of whether one or familial factors underlie the slow and variable response times (RT), impaired response inhibition, sustained attention and choice impulsivity that are associated with ADHD (Kuntsi et al. 2010). The final model consisted of two familial factors. The larger factor, reflecting 85% of the familial variance of ADHD, captured all familial influences on mean RT and RT variability. The second, smaller factor, reflecting 13% of the familial variance of

ADHD, captured all familial influences on omission errors and 60% on commission errors (Kuntsi et al. 2010). Further work from the same group, using the same cognitive performance measures in a population twin sample, confirmed that the two key familial cognitive factors associated with ADHD (RT factor and commission error factor) are influenced by distinct sets of genes with a very low and non-significant genetic correlation for the two factors (Kuntsi et al. unpublished data). Furthermore, the association between the two familial cognitive factors and ADHD was largely independent of any contribution from the familial influences shared between ADHD and IQ, whereas this was not the case for choice impulsivity where 50% of the overlap with ADHD could be accounted by familial variance underlying IQ (Wood et al. 2010). These data suggest that lower IQ does not account for the key cognitive impairments that show overlapping genetic influences with ADHD. Overall, these data demonstrate the power of MV genetic approaches to dissect the genetic influences and neurobiological measure associated with ADHD. Future work should extend such analyses to additional phenotypes and further aim to understand the core processes that underline the two main familial cognitive factors identified to date.

7 Molecular Genetic Studies

7.1 Early Genetic Findings in ADHD

Molecular genetic studies on ADHD started with candidate gene association studies in the mid-1990s with the first two reported associations between genetic variants in the dopamine D4 receptor (DRD4) (LaHoste et al. 1996; Swanson et al. 1998) and dopamine transporter (DAT1) genes (Cook et al. 1995). Subsequently, association was reported with a microsatellite marker near to the dopamine D5 receptor gene (DRD5) (Lowe et al. 2004a). Since then, there have been numerous replication studies with relatively few independent replications. However, meta-analysis of available data reported strong evidence for the association of dopamine system genes, in particular with the 7-repeat allele of a variable number tandem repeat (VNTR) polymorphism within DRD4 and a microsatellite which lies upstream of DRD5. These two associations are significant because they are as yet the only genetic association findings to reach genome-wide levels of significance in the meta-analytic study of Li and colleagues (Li et al. 2006). This level of significance, in the region of 5×10^{-8} (Dudbridge and Gusnanto 2008), is important because it means that, in a systematic screen of the genome for association, these findings would occur less than 5% of the time by chance alone. Conventional levels of significance, such as 0.05, would be detected by chance every 20th selection of a random independent genetic polymorphism, of which there are hundreds of thousands across the genome. We can therefore be confident that these two findings are truly associated with ADHD.

7.2 The Dopamine Transporter Gene (DAT1/SLC6A3)

One of the most commonly cited associations with ADHD is with the 10-repeat allele of a VNTR within the 3'-intranslated region of DAT1. However, the evidence for this association from meta-analytic studies is not yet convincing, with three negative reports (Li et al. 2006; Maher et al. 2002; Purper-Ouakil et al. 2005) while three others found only weak evidence of association (Faraone et al. 2005; Gizer et al. 2009; Yang et al. 2007), far below the stringent criteria reached by DRD4 and DRD5. Whether DAT1 is associated with ADHD therefore remains a contentious question. Yet, there is evidence for heterogeneity across the datasets included in the meta-analytic studies suggesting that there may be identifiable sources for differences in the strength of the association across the various studies (Gizer et al. 2009; Li et al. 2006). There are as yet insufficient studies to clarify which of the various sources of heterogeneity could convincingly explain the meta-analytic findings for DAT1, although potential sources of heterogeneity have been identified in the literature.

One potential source of error in the analysis of DAT1 comes from undetected genotyping errors that cause apparent over transmission of common alleles when using the transmission disequilibrium test (TDT), a test of unequal transmission of alleles from parents to their affected offspring. Mitchell and colleagues found that, in studies reporting a positive association using the TDT, 87% identified an association with alleles that were present at a frequency of >50% whereas, in case–control studies, the proportion of common allele associations was 40% (Mitchell et al. 2003). This highly significant difference could only be explained by systematic genotype error related to the method, especially with high-frequency alleles such as the DAT1 10-repeat, which occurs on around 70% of chromosomes from people of European ancestry.

However, there are more interesting sources of heterogeneity for DAT1. First, recent studies of ADHD in adults find evidence for association with the 9-repeat rather than the more common 10-repeat allele, suggesting that the effects of DAT1 may differ depending on the developmental age; this indicates a role for DAT1 in the modulation rather than aetiology of ADHD (Franke et al. 2008).

Secondly, the DAT1 VNTR interacts with measures of the pre-natal environment, including maternal use of tobacco during pregnancy on oppositional and hyperactive-impulsive symptoms (Becker et al. 2008; Kahn et al. 2003) and maternal use of alcohol during pregnancy (Brookes et al. 2006). Interestingly, recent research has shown that, when genetic factors are controlled for in studies of the association between mothers smoking during pregnancy and ADHD, the association appears to be mediated by genetic factors, suggesting that the environmental measure may be correlated with maternal genes (Thapar et al. 2009). This is an example of gene–environment correlation and, in this case, indicates that the observed interactions with DAT1 may be explained by gene–gene interactions, also known as epistatic interactions.

Thirdly, there is evidence that there may be an association of two or more haplotypes of DAT1 with ADHD, indicating that more than one genetic variant is likely to be involved. Further, the known VNTR may not be the primary functional variant but may tag other genetic variants; and there appears to be more than one associated region with the gene (Asherson et al. 2007; Brookes et al. 2006, 2008). Strength of the association would then depend on the relationships between the various functional genetic variants involved, which may differ across sample populations.

Finally, there is evidence that the association with genetic variants across DAT1 may be specific to a subgroup of cases with ADHD alone, but is not seen among cases of ADHD with comorbid conduct disorder (Zhou et al. 2008a). This finding was more convincing because the association with the subgroup of cases with ADHD without conduct disorder was found in two entirely separate regions of the gene. Since these two regions are not in linkage disequilibrium with each other (i.e. the findings are uncorrelated), this implicates two independent functional genetic variants at the 3′ and 5′ ends of the gene. Evidence for genetic variants associated with ADHD in the 5′ end, within or close to the promoter region, has also been reported by other groups (Brookes et al. 2008; Friedel et al. 2007; Genro et al. 2007, 2008).

7.3 Other Candidate Gene Associations

There have now been numerous candidate gene association studies in ADHD and these are best summarized in the most recent review and meta-analysis from Gizer and colleagues (Gizer et al. 2009). A summary of key findings from this study is listed in Table 1. They conducted a systematic review of the association literature with the aim of identifying the most consistent findings in ADHD research to date. They further tested for heterogeneity across studies because other genes, like DAT1, might show significant sources of variability due to identifiable moderating factors. Overall, they concluded that the following genes were significantly associated with ADHD: DAT1 (dopamine transporter gene), DRD4 dopamine D4 receptor gene), DRD5 (dopamine D5 receptor gene), 5HTT (serotonin transporter gene), HTR1B (serotonin 1B receptor gene) and SNAP-25 (synaptosomal associated protein-25). In their analysis, none of the reported meta-analytic findings passed more stringent genome-wide tests of association, so there remains a level of uncertainty for each of the specific findings reported in this study. In addition, they identified the following genes that showed significant evidence of heterogeneity: DAT1, DRD4, DRD5, DBH (dopamine beta-hydroxylase gene), ADRA2A (adrenergic 2A receptor gene), 5HTT, TPH2 (tryptophan hydroxylase 2 gene), MAOA (monoamine oxidase A gene) and SNAP-25. For these later genes showing heterogeneity, future studies could usefully investigate the potential moderating factors that lead to variability in effect size across studies.

Despite some significant progress from the candidate gene studies, an estimate of the overall impact of the most replicated gene findings suggested that 3.3% of the phenotypic variance for ADHD was explained by the additive effects of these genes, accounting for 4.3% of the estimated average heritability of ADHD of 76% (Kuntsi et al. 2006a). In this analysis, the average odds ratios for the individual genetic variants were taken from a previous meta-analysis (Faraone et al. 2005) and found to be between 1.1 and 1.4, equivalent to 0.1% to 1.0% of the variance in ADHD symptoms (Table 2). Further work is therefore needed to explain the rest of the genetic influences on ADHD.

Table 1 Data from Gizer et al. (2009) summarizing meta-analytic findings of candidate gene studies that reached nominal significance ($P = 0.05$ or better). Risk alleles were included in this analysis if they had been reported in four or more independent studies of clinically diagnosed cases of ADHD following DSM-IV criteria, including both case-control and within family tests of association. Note that risk allele associations should be treated with caution unless they pass the genomewide threshold of significance which is around 5×10^{-8}

Gene	Location	Polymorphism	Risk allele	Number of studies	Odds ratio	95% confidence interval	p-value	Evidence of heterogeneity (P-value)
DAT1	3′ UTR	VNTR	10-repeat	34	1.12	1.00–1.27	0.03	<0.00001
DAT1	Intron 8	VNTR	6-repeat	5	1.25	0.98–1.58	0.03	0.01
DAT1	3′ UTR	Rs27072	G	7	1.20	1.04–1.38	0.006	NS
DRD4	Exon 3	VNTR	7-repeat	26	1.33	1.15–1.54	0.000007	0.00006
DRD4	Promoter	Rs180095	T	5	1.21	1.04–1.41	0.007	0.03
DRD5	5′ flank	Dinucleotide	148-bp	9	1.23	1.06–1.43	0.003	NS
5HTT	Promoter	VNTR	Long	19	1.17	1.02–1.33	0.01	NS
SNAP25	3′ UTR	Rs3746544	Unknown	7	1.15	1.01–1.31	0.03	NS

Table 2 Summary of effect sizes for replicated candidate gene associations reported by Kuntsi et al. (2006a), using data from Faraone et al. (2005). The percentage of the variance explained by each risk allele has been estimated assuming a liability threshold risk model and additive genetic effects. The total phenotypic variance explained sums to 3.2%, which is 4.2% of the additive genetic effects if we using the average heritability estimate of 76% reported in Faraone et al. (2005)

Gene	Frequency of risk allele	Odds ratio	95% confidence interval	QTL effect size (% variance)	Number of families to replicate with 80% power and nominal significance (0.05)
DRD4	0.12	1.16	1.03–1.31	0.1	3,196
DRD5	0.35	1.24	1.12–1.65	0.4	728
DAT1	0.73	1.13	1.03–1.24	0.1	2,748
DBH	0.50	1.33	1.11–1.59	0.7	391
SNAP25	0.50	1.19	1.03–1.38	0.3	1,043
SERT	0.06	1.31	1.09–1.59	0.6	466
HTR1B	0.71	1.44	1.14–1.83	1.0	315

7.4 Dopamine D4 Receptor Gene (DRD4)

In addition to the association with the 7-repeat allele, described above, other genetic variants have been investigated within this gene. A family-based association analysis of the DRD4 120-bp insertion/deletion promoter, 1.2 kb upstream of the transcriptional start site, showed a significant association with ADHD in 372 ADHD cases and their parents (McCracken et al. 2000). However, several studies have failed to replicate these findings, including a meta-analysis of the data from this polymorphism, which found no evidence for association (Gizer et al. 2009). The SNP (single nucleotide polymorphism), rs1800955, located 521 base pairs upstream of the DRD4 transcriptional start site, has recently been found to show association with ADHD in a Korean sample ($P = 0.01$: Yang et al. 2008). Previous studies had not found evidence for association with this polymorphism. However, meta-analysis of the data (Barr et al. 2001; Kereszturi et al. 2007; Lowe et al. 2004b; Payton et al. 2001) suggests that this polymorphism may have a role in ADHD ($P = 0.007$: Gizer et al. 2009). Alleles of both of these upstream polymorphisms have been reported to alter promoter activity (D'Souza et al. 2004; Okuyama et al. 1999).

7.5 The Tachykinin Receptor 1

Another gene strongly related to dopaminergic function in the prefrontal cortex is the neurokinin (substance P-preferring) receptor (NK1R) also called the tachykinin receptor 1 (TACR1). In a knockout mouse lacking the TACR1 gene (NK1R−/−), mice were found to be hyperactive (Yan et al. 2010, 2011) and this was ameliorated by psychostimulants (*d*-amphetamine and methylphenidate). The mice had reduced

(>50%) spontaneous dopamine efflux in the prefrontal cortex and displayed lack of the striatal dopamine response to d-amphetamine. These behavioural and neurochemical abnormalities in NK1R$-$/$-$ mice, together with their atypical response to psychostimulants, are similar to clinical characteristics of ADHD in humans (Yan et al. 2010). An allelic association study of 450 ADHD cases and 600 controls found that four TACR1 SNPs previously associated with bipolar disorder and alcoholism were also associated with ADHD (Yan et al. 2010). If confirmed, TACR1 may turn out to be another dopamine-related gene that is more strongly associated with an affective disorder subgroup of ADHD.

7.6 The Serotonin 1B Receptor Gene (HTR1B)

Evidence of significant genetic association of the 861G allele of HTR1B was shown in 273 European families ($P = 0.007$; Hawi et al. 2002), as well as in 115 Canadian families ($P = 0.03$; Quist et al. 2003). Similarly, paternal overtransmission of the 5-HTR1B G861 allele to offspring with the inattentive ADHD subtype was observed in 12 multi-generational Centre d'Etude du Polymorphisme Humain (CEPH) pedigrees, comprising 229 families of ADHD probands (Smoller et al. 2006). A haplotype block, encompassing the gene, was associated with the inattentive ADHD subtype. In addition, three polymorphisms in this block were nominally associated with this subtype but did not remain significant after correction for multiple testing (Smoller et al. 2006). Meta-analysis from nine studies yielded significant evidence of association between ADHD and the HTR1B 861G allele (fixed effects $P = 0.01$) with no heterogeneity in study effect sizes (Gizer et al. 2009).

7.7 Serotonin Transporter Gene (SLC6A4/5-HTT)

A functional 44 bp deletion/insertion polymorphism in the promoter region of SLC6A4 has been associated with depression and anxiety (Acosta et al. 2004; Lotrich and Pollock 2004; Munafo et al. 2006; Schinka et al. 2004) and implicated in stress sensitivity leading to phenotypes such as depression in adults and emotional problems in children (Caspi et al. 2010; Sugden et al. 2010). The long promoter variant is associated with more rapid reuptake of serotonin than the short allele (Lesch et al. 1996). Meta-analysis of ten Transmission Disequilibrium Test (TDT) studies and nine case-control and haplotype relative-risk studies shows a significant association with the long allele and ADHD (fixed effects $P = 0.004$, random effects $P = 0.010$) with significant heterogeneity across studies ($P = 0.00003$; Gizer et al. 2009). Other polymorphisms tested within this gene such as the 12-repeat of an intron 2 VNTR and SNP markers were not associated with ADHD in the meta-analytic study.

7.8 Synaptosomal-Associated Protein 25 Isoform Gene

The Synaptosomal-Associated Protein 25 (SNAP25) gene was suggested as a candidate gene for ADHD based on the hyperactivity phenotype of the mouse strain, *Coloboma* (Hess et al. 1995). SNAP-25 is interesting because it is involved in a number of processes, including axonal growth, synaptic plasticity and the vesicular release of neurotransmitters. Four polymorphisms in SNAP-25 were significantly associated with ADHD in 186 Canadian families with 234 ADHD children ($P = 0.04$–0.005; Feng et al. 2005). However, these results were not replicated in an independent sample of 99 families with 102 ADHD children from southern California, possibly due to differences in selection criteria, ethnicity, medication response and other clinical characteristics of the samples (Feng et al. 2005). Analysis of the DSM-IV subtypes in the Toronto sample indicated that the differential results were not attributable to ADHD subtype. Quantitative trait analyses of the dimensions of hyperactivity/impulsivity and inattention in the Toronto sample found that both behavioural traits were associated with SNAP-25. Subsequent investigations used several SNPs spanning the gene with 7 SNPs included in four or more studies. Meta-analysis of four of these SNPs found that only one (rs3746544) was significantly associated with ADHD across seven studies with heterogeneity and sensitivity analyses indicting that significance did not depend on any particular study (Gizer et al. 2009).

7.9 ADHD and Co-occurring Disorders

Another line of enquiry that has been applied to candidate gene studies is the search for genes that might modify the developmental course of ADHD, or be associated with the development of co-occurring behavioural phenotypes, such as conduct problems or emotional reactivity. One behavioural phenotype that is being investigated is the co-occurrence of ADHD with conduct problems and the later development of conduct disorder and antisocial behaviour. Twin studies suggest that conduct problems and ADHD share a common genetic aetiology and that ADHD plus conduct problems appears to be a more severe subtype in terms of genetic loading for ADHD as well as clinical severity (Thapar et al. 2001). Evidence has now accumulated that a functional variant of the catechol-*O*-methyl transferase gene (COMT) is associated with conduct disorder and antisocial behaviour in children with ADHD, and therefore plays a role in developmental outcomes associated with ADHD.

The COMT gene contains a functional SNP that results in a valine (val) to methionine (met) mutation with altered enzymatic activity (Lotta et al. 1995). Since the val variant catabolises dopamine at up to four times the met variant, resulting in significantly lower synaptic dopamine in the prefrontal cortex, this polymorphism was a good candidate for association with ADHD. However, meta-analysis of 16 studies found no significant association between the COMT val/met polymorphism and ADHD (Gizer et al. 2009). On the other hand, in a sample of 240 British children with ADHD or hyperkinetic disorder, the COMT val/met variation was associated with increased symptoms of conduct disorder, in addition to a significant gene–environment interaction between the COMT polymorphism and birth weight (Thapar et al. 2005). Subsequent studies using population samples replicated the association of the COMT polymorphism with antisocial behaviour in groups of children, with high levels of ADHD symptoms, but not in the rest of the population, indicating that there is a gene-by-ADHD interaction on the risk for developing conduct problems during development (Caspi et al. 2008). These findings demonstrate the complexity of aetiological influences on human behaviour and confirm the role of ADHD as a risk factor for development of conduct problems when combined with exposure to other risk factors: in this case changes in dopamine metabolism due to genetic variation of the COMT gene.

7.10 ADHD in Adults

One approach to understanding developmental changes in ADHD is the comparison of association findings from ADHD in childhood with those from ADHD in adults. The idea underlying this approach is that there will be some genes that are specific to ADHD in adults and are closely related to processes that lead to persistence or remission during the transition from childhood and adolescence into adulthood. For example, Halperin and Schultz (Halperin and Schulz 2006) hypothesize a developmental model that takes into account the two main familial cognitive "variability" and "error" factors, identified by Kuntsi et al. (2010), and predicted by previous cognitive models for ADHD, including the arousal-attention model (Johnson et al. 2009; O'Connell et al. 2008, 2009). In the Halperin developmental model (Halperin and Schulz 2006; Halperin et al. 2008), RT variability is proposed to reflect poor state regulation, perceptual sensitivity and/or weak arousal mechanisms. The model proposes a distinction between two neurocognitive processes: (1) the proposed subcortical dysfunction linked to the aetiology of ADHD and (2) prefrontal mediated executive control, linked to persistence or desistence of ADHD during adolescence. As such, one possible interpretation of the two familial cognitive factors is that the first factor (RT) represents the core, enduring deficit and the second factor (errors) represents prefrontally mediated executive control. The model predicts that the extent to which executive control functions, which develop throughout childhood and adolescence, can compensate for the more primary and enduring subcortical deficits, can determine the degree of recovery from ADHD symptoms.

The involvement of two or more processes in the developmental outcomes of ADHD suggests that an informative strategy will be the direct comparison of childhood and adult ADHD samples, or comparison of persistent *versus* remitting forms of ADHD. This will help to identify genes that have a stable influence on ADHD throughout the lifespan as well as those that influence the recovery from, or persistence of, ADHD during adolescent and early adult life. Two studies that adopt this approach investigated genetic associations with 19 genes involved in the regulation of serotonin pathways (Ribases et al. 2009) and 10 genes that encode for neurotrophins (Ribases et al. 2009). Using two clinical samples of adult and child cases with ADHD, they found significant associations for SNPs with the dopamine decarboxylase and serotonin 2A receptor genes in both the adult and childhood ADHD samples, whereas there was evidence of association with the monoamine oxidase B gene only in the adult sample. In the second study, they found evidence of association for CNTFR (ciliary neurotrophic factor receptor gene) in both child and adult samples and for NTF3 (Neurotrophin 3 gene) and NTRK2 (neurotrophic tyrosine kinase gene) in the childhood samples, only. While these findings remain provisional until they have been replicated in independent samples and the significance levels approach genome-wide levels of significance, they represent an important approach to identifying both stable and novel genetic risk factors at different stage of development.

8 Whole Genome Approaches

8.1 Linkage Studies

Linkage analysis is a method for identifying the presence of susceptibility genes for genetic disorders within relatively large chromosomal regions (of up to forty million bases of DNA) possibly containing thousands of genes. Although highly successful in the identification of genes for numerous single gene (Mendelian) disorders, linkage has been far less successful for complex disorders, where numerous genes are involved. This is because the method lacks power under conditions of high-genetic heterogeneity or where genetic effects are relatively small (Dawn Teare and Barrett 2005). The success of linkage studies in psychiatric genetics remains controversial because it has been difficult to replicate key findings and, even when linkage has been identified, there are few examples of genes that have been identified that account for linkage signals (Cichon et al. 2009).

Multiple putative linkage regions were identified in genome-wide linkage scans of ADHD, but few of these passed genome-wide levels of significance or were ever replicated in subsequent studies; they could therefore only be described as potential, rather than confirmed, linkage regions (Acosta et al. 2004; Arcos-Burgos et al. 2004; Asherson et al. 2008; Bakker et al. 2003; Faraone et al. 2008; Fisher et al. 2002; Hebebrand et al. 2006a; Loo et al. 2004; Ogdie et al. 2003; Romanos et al. 2008; Smalley et al. 2002).

The data from seven available linkage studies were summarized by Zhou and colleagues (Zhou et al. 2008b). They concluded that no chromosomal region had been consistently identified across the studies and that the majority of findings were unique to each study. They then went on to perform a meta-analysis to identify the regions showing some consistency between the various findings and concluded that one region on chromosome 16, between 16q21 and 16q24, showed genome wide significance and could therefore be considered as confirmed linkage region, whereas ten other chromosomal regions on 5, 6, 7, 8, 9, 15, 16 and 17 showed evidence suggesting linkage with ADHD. As we shall see, the 16q21–16q24 region obtained further support from genome-wide association studies that had identified a gene called CDH13 (Cadherin 13) that lies within this region, although it has yet to be shown that genetic variation of this gene leads directly to the linkage signal. Other regions, such as 5p13, which had been indicated as potential linkage regions in two previous studies (Hebebrand et al. 2006b; Ogdie et al. 2004), and spans the region containing the dopamine transporter gene (Friedel et al. 2007), were not supported by the meta-analysis, although such regions might still be relevant to the specific populations involved.

One gene that has been specifically identified following an initial linkage strategy is the latrophilin 3 gene (LPHN3) (Arcos-Burgos et al. 2010; Ribases et al. 2011). The original studies used large multigenerational families from the genetically isolated Paisa population in Columbia. A genome linkage study of 16 families found significant linkage on chromosome 4q13 (Arcos-Burgos et al. 2004). Fine mapping applied to nine of the families narrowed a critical region of around 20,000,000 base pairs. Subsequent studies identified a significant region of association within exons 4 through to 19 of LPHN3 that was replicated in US, German, Spanish and Norwegian samples, with an average odds ratio of around 1.2 (Arcos-Burgos et al. 2010). Finally, a further study of 334 adults with ADHD and 334 controls from Spain found additional evidence for the association, indicating an association between genetic variants of LPHN3 and ADHD throughout the lifespan (Ribases et al. 2011). The role of LPHN3 is not well understood, but it is a G-protein coupled receptor that is thought to be involved in neurotransmission and maintenance of neuronal viability.

8.2 Genome-Wide Association

The main approach currently being taken to the identification of novel genes for ADHD include genome-wide association studies, which have the potential to detect entirely novel associations where there is no previous *a priori* hypothesis. Given that we know so little about the function of the brain and how molecular processes lead to ADHD, it makes sense for molecular genetic studies to focus efforts on empirical approaches that systematically screen the entire genome for associations.

Genome-wide association studies (GWAS) take advantage of SNP arrays that enable genotyping of genetically informative markers across the entire human genome. Depending on the density of the arrays these may account for 80% or more of common genetic variation. In ADHD genetic research, however, GWAS have yet to establish confirmed novel associations, since no individual SNP has yet to reach genome-wide levels of significance. The problem is that conventional levels of significance in the region of 0.05 to 0.001 would be found by chance with SNPs throughout the genome, due the large number independent haplotypes (correlated sequences of correlated genetic variation) across the genome. As discussed above, this means that higher levels of significance, in the region of 5×10^{-8}, are recommended to adjust for the low prior odds of association (Dudbridge and Gusnanto 2008). This has meant that, for most common complex disorders, 12,000 or more samples are needed to identify reliably a few associated SNPs (much larger samples are needed to detect appreciable numbers of common risk alleles) (Park et al. 2010) since, in nearly all cases, only small genetic risks have been identified for specific risk alleles with odds ratios in the region of 1.1–1.4 or less.

The first GWAS study of ADHD investigated 438,784 SNPs in 958 combined type ADHD proband–parent trios. No genes of moderate to large effect were identified (Neale et al. 2008) and no findings passed genome-wide levels of significance. Although there were potentially interesting genes among the top 25 SNPs, including the cannabinoid 1 receptor gene (CNR1), none of these has been replicated in subsequent studies. However, when a set of 51 candidate genes was investigated, there was significant evidence at the group level that positive association signals were emerging from the selected SNPs, implicating mainly dopamine, noradrenaline and serotonin neurotransmitter genes and providing further support from traditional sets of candidate genes. Analysis using a quantitative approach in the same sample found some degree of association in monoamine-related genes (SLC9A9, DDC, SNAP25, SLC6A1, ADRB2, HTR1E, ADRA1A, DBH, BDNF, DRD2, TPH2, HTR2A, SLC6A2, PER1, CHRNA4, COMPT and SYT1) (Lasky-Su et al. 2008). Similar findings were subsequently reported in a study that combined genome-wide association data from several studies, with a total sample size of 2,064 trios, 896 cases and 2,455 controls (Neale et al. 2010). As we can see from comparison with other complex genetic disorders, including psychiatric syndromes, such as schizophrenia and bipolar disorder (O'Donovan et al. 2009), the total number of samples analysed in GWAS studies of ADHD to date remains relatively small and the disappointing findings so far are, in fact, not unexpected. Further work is required to collate DNA and obtain GWAS information on much larger sets of samples for significant progress in the identification of common genetic variants associated with ADHD.

However, there are potentially novel findings that have emerged from the GWAS studies of ADHD (reviewed in: Franke et al. 2009). Of particular interest is the Cadherin 13 gene (CDH13) which was associated with ADHD in two out of five GWAS studies completed to date (Lasky-Su et al. 2008; Lesch et al. 2008) and lies within the only region that reached genome-wide significance in a meta-analysis of linkage studies of ADHD (Zhou et al. 2008b). Cadherin 13 is a cell adhesion protein (Patel et al. 2003) and also acts as a negative regulator of neural cell growth (Takeuchi et al. 2000). In the IMAGE project sample, CDH13 reached genome-wide levels of significance using a method that was designed to maximize the power of the sample, by using quantitative phenotypes and a two-stage procedure (Lasky-Su et al. 2008); but was also detected within the top 25 SNPs from the more conventional TDT analysis of the same sample (Neale et al. 2008), as well as an independent study from Germany that used a DNA-pooling approach (Lesch et al. 2008). Furthermore, CDH13 is among the most consistent findings on a wide range of phenotypes related to drug abuse and dependence (Uhl et al. 2008).

Other genes that emerged as potentially interesting include the glucose–fructose oxidoreductase-domain containing 1 gene (GFOD1), which may be involved in electron transport. A SNP in this gene was associated with inattentive symptoms using the same quantitative approach that identified the association with CDH13 (Lasky-Su et al. 2008). In addition, there has also been replication of genetic association findings between the brain expressed metalloprotease genes TLL1 and TLL2 genes with ADHD (Lasky-Su et al. 2008; Lesch et al. 2008). The genes HAS3, SPOCK3, MAN2A2, GPC6, MMP24, KCNIP1, KCNIP4, and DPP10 all show some level of association with ADHD in three studies (Lasky-Su et al. 2008; Lesch et al. 2008; Neale et al. 2008). Finally, in the GWAS study using pooled DNA from Lesch and colleagues, none of the findings achieved genome-wide significance but several genes expressed in the brain, which are of potential interest, were identified among the top hits that are involved in neuronal plasticity, cell–cell communication and/or adhesion (Lesch et al. 2008). These findings and other hints from GWAS indicate that genes involved in cell division, cell adhesion, neuronal migration and neuronal plasticity may confer risk for ADHD (Franke et al. 2009; Lesch et al. 2008).

Overall, there is still a long way to go to delineate the specific genetic factors that explain the high heritability of the disorder. However, this is a common phenomenon in research of common disorders and several possible explanations for the so-called "dark-matter" of (missing) heritability have been put forward: dark matter in the sense that we know it exists, we can detect its influences, but simply cannot see it (Manolio et al. 2009). Potential reasons include: numerous genes of small effect; genetic heterogeneity with risk conferred by many different genes and variants within genes; higher order interactions between genes and with environment; and aetiological heterogeneity (Manolio et al. 2009). In addition, we do not yet understand the contribution made to ADHD from rare copy number variants (CNVs) or other types of rare genetic variation; although as we shall see, recent data suggest that CNVs can exert moderate to large influences on risk for ADHD.

8.3 Copy Number Variants

Genome-wide SNP arrays can also be used to investigate copy number variants or CNVs. These arise through non-allelic homologous recombination and are now known to play an important role in the aetiology of psychiatric disorders including autism and schizophrenia (Cichon et al. 2009). In ADHD, the evidence for aetiologically significant CNVs has been accumulating. It is now clear that not only do CNVs play a significant role in the overall risk for ADHD but also that the regions involved appear to be shared with those that give rise to two neurodevelopmental disorders, schizophrenia and autism (Williams et al. 2010).

There are a few studies that implicate CNVs in ADHD. Some represent reports of individual cases, such as the example of a girl with ADHD, who has a *de novo* 600 kb deletion on a maternal copy of chromosome 16p11.2, encompassing

CORO1A (coronin-1A, which is essential for T cell release from the thymus), flanked by 146 kb segmental duplications (Shiow et al. 2009). Deletion, as well as duplication, at the 16q11.2 locus has also been associated with autism spectrum disorder and neurodevelopmental disorders (Ghebranious et al. 2007). One recent study identified 222 structural variants in 335 children with ADHD and their parents that were not detected in 2,026 unrelated healthy controls (Elia et al. 2009). Although no overall excess of deletions or insertions were found in the ADHD cases, inherited rare CNVs were significantly enriched within genes that had previously been implicated in autism, schizophrenia and Tourettes syndrome. Some of these, such as GRM7, DPP6 and TACR3, have been nominally associated with bipolar disorder in recent GWAS studies (Ferreira et al. 2008). The deletion in the glutamate receptor gene (GRM5) was identified in an affected parent and three affected offspring, who displayed problems with spatial orientation (Elia et al. 2009), a behavioural trait in the GRM5 knock-out mouse (Lu et al. 1997). The CNV in GRM7 was present in an ADHD proband who presented with an anxiety disorder. Additional genes disrupted by CNVs in ADHD that are also thought to be involved in autism include AUTS2 and IMMP2L. Furthermore, other ADHD, CNV-associated genes are implicated in learning, behaviour, synaptic transmission and neuronal development. Four separate deletions were found in the protein tyrosine phosphatase gene (PTPRD), a gene thought to play a role in restless leg syndrome, which is a common symptom in ADHD (Elia et al. 2009). It is of note that a CNV in the tachykinin receptor 3 (TACR3) gene was found to be associated with ADHD.

The Williams et al. study (Williams et al. 2010), published in the Lancet, received a considerable amount of publicity as the first major study that convincingly showed the impact of structural genetic variation on the overall risk for the disorder. They analysed a group of 410 children with ADHD and 1,156 controls and focused on a subset of large CNVs, greater than 500,000 base pairs that could be accurately and confidently called. Overall, they found an increased burden of CNVs in cases compared to controls (15.6% versus 7.5%: odds ratio = 2.1); which was particularly high in those with an IQ below 70 (42.4%: odds ratio = 5.7) compared with the group with higher IQs (12.5%; odds ratio = 1.7). They further identified chromosome 16p13.11 duplications as a particular risk factor, a finding that they were able to replicate in a set of 825 ADHD cases and 35,243 controls from Iceland. Finally, they confirmed the findings from the Elia study for a significant enrichment of loci previously reported in schizophrenia and autism. Further work is needed to clarify the extent to which CNVs associated with ADHD arise *de novo* or are inherited and contribute to the overall heritability of the disorder.

8.4 Endophenotypes and Intermediate Phenotypes

Finally, the focus of much of the genetic research on ADHD has moved to the identification of endophenotypes and intermediate phenotypes. The distinction between the two was outlined by Kendler and Neale (2010) who clarify that intermediate phenotypes reflect neurobiological processes that mediate between genes and behaviour, whereas endophenotypes also reflect the pleiotropic (multiple outcome) effects of genes, which are genetically correlated with, but may not mediate genetic effects on, behaviour. For example, it is hypothesized that the overlap between ADHD and autism spectrum disorder reflects pleiotropic effects (Ronald et al. 2010), whereas the overlap with cognitive performance measures, such as reaction time variability and omission and commission errors, reflects neurobiological processes that mediate between genes and behaviour (Kuntsi et al. 2010). Here, we refer to endophenotypes as encompassing any measure of a neurobiological process that shares genetic effects with ADHD, whereas we reserve the term intermediate phenotype to those that specifically mediate genetic associations with ADHD. The study of intermediate phenotypes is developing rapidly and provides the exciting prospect of linking molecular and cellular processes to neurobiological and psychological processes that underlie complex behavioural disorders such as ADHD. However, as we shall discuss, tests of mediation are required to differentiate intermediate form more general endophenotype effects.

Whether the endophenotype approach will be successful in speeding up the identification of susceptibility genes for ADHD depends critically on the size of the genetic effects from individual genes and size of the samples for molecular genetic analysis. There are a few examples from fMRI studies that indicate that at least for some imaging phenotypes greater effect sizes exist. For example, the association between the serotonin transporter promoter polymorphism (5-HTTLPR) and amygdala activation is reported, from meta-analysis of available data, to account for as much as 10% of the phenotypic variance (Munafo et al. 2008). Were such large genetic effects to be found for individual genetic variants with other neurobiological phenotypes, then considerable progress would be expected. However, it has been argued that, for many measures of biological function, the genetic effects may be similar to behavioural phenotypes (Flint and Munafo 2007) and it is not clear how often brain phenotypes will be more penetrant than behavioural phenotypes. In general, current data point towards a few cases of neuroimaging phenotypes showing large effects from individual genes but it cannot be assumed that this will be the case for all neuroimaging phenotypes.

As mentioned above, genetic influences on ADHD appear to be indexed by two main familial pathways (Kuntsi et al. 2010), so that intermediate phenotype approaches could usefully focus on the processes that underlie these cognitive performance impairments in ADHD. However, association studies, using cognitive and other intermediate neurobiological phenotypes, have often used small samples and assumed the existence of large genetic effects. As a consequence, these studies risk repeating the mistakes learnt from years of human genetic studies that produced multiple false positive findings. To some extent, the potential for misleading findings is even greater with neurobiological phenotypes because of the numerous potential phenotypes that can be derived from test procedures: different brain regions; large number of voxels; methods of analysis and so on. Under these circumstances, controlling the multiple test problem represents a considerable challenge. Although many reasonable hypotheses can link genes to patterns of cognitive and brain function, it is not at all clear how these should be prioritized. As such, the same stringent criteria of replication and genome-wide significance levels remain the gold standard for confirming a positive association with neurobiological and behavioural phenotypes (Castellanos et al. 2008). To take the next steps in this exciting field of research, large samples (or multiple smaller samples that can be combined together) with comparable neurobiological data will be required to have the power for molecular genetic analysis.

Interestingly, the most replicated genetic association with cognitive performance measures in ADHD is with the DRD4 7-repeat allele risk allele for ADHD (see: (Kebir et al. 2009) for review). Among children with ADHD, the high-risk 7-repeat allele is associated with normal cognitive impairment compared to controls, whereas carrying non-risk alleles for ADHD show significant cognitive impairments. This unexpected finding has also been found with the gene ZNF804A and schizophrenia (Walters et al. 2010) where it was found that those with schizophrenia who do not carry the ZNF804A risk allele show greater cognitive impairments, suggesting that this pattern of findings may be common in neuropsychiatric disorders. These findings suggest that some genes that increase risk for disorders, such as ADHD, may also be important sources of heterogeneity among affected individuals; identifying groups with high and low levels of cognitive impairments with distinct genetic or other causal risk factors. This is an entirely different from the intermediate phenotype model in which cognitive impairments are proposed to mediate genetic effects on clinical phenotypes such as ADHD.

Other findings highlighted in the Kebir review (Kebir et al. 2009) include speed of processing, set shifting and cognitive impulsiveness, which were impaired in DRD4 7-repeat carriers; and four studies with conflicting findings for the 10-repeat allele of DAT1 on commission and omission errors from CPT and similar tasks. There was also some evidence for the association of reaction time variability with the DAT1 10-repeat allele. In relation to other phenotypes, both the DAT1 10-repeat allele and the DRD4 7-repeat allele have been associated with structural, functional and electrophysiological changes in the brain (Durston 2010; Loo et al. 2003). However, no clear pattern of findings has yet emerged. This is not surprising given the limited number of studies reported to date, the small sample sizes included in the reported studies, differences in the clinical samples used and methods for obtaining brain-derived phenotypes. Further progress in this area is expected to arise from the large datasets, such as a consortium project that is currently completing genome-wide association studies with large datasets of structural brain imaging.

Finally, we should address the methods needed to establish the measures of cognitive performance or neurobiological function that mediate genetic effects on behaviour. One approach using genetic data is to test for mediation, using regression models, when a particular genetic variant has been found to be associated with both the endophenotype and the clinical disorder. This approach was recently adopted in a study of the association between the high activity, COMT genotype and the risk for developing antisocial behaviour in children with ADHD (Langley et al. 2010). Using a longitudinal population study sample, they were able to replicate the original observation, for the third time, making this one of the most consistent findings in ADHD genetic research to date. The high-activity COMT polymorphism was further associated with measures of executive function and impaired social understanding. This enabled tests of mediation to be performed, which found that controlling for executive function deficits in the analysis made no difference to the size of the association between COMT and antisocial behaviour. However, the strength of the association dropped when social understanding was included in the model, suggesting a mediating role for social understanding but not for executive function. Overall, these studies show the importance of testing for mediating effects and the continued need to stick with the behavioural and clinical phenotypes of primary interest. Both the study of clinical and neurobiological measures are needed to determine the neurobiological processes involved in the aetiology of ADHD and not just those that reflect the multiple outcomes of genetic effects on brain structure and function.

9 Research Gaps and Future Prospects

Further work is needed to identify both common and rare genetic variants that account for the heritability of ADHD, using large samples and, in the near future, whole genome-sequencing technologies. Neurobiological research needs to focus on measures that are genetically correlated with ADHD and use genetic association data to determine the nature of the cognitive, neuronal and cellular processes that mediate genetic risks on behaviour. Genetic studies of ADHD in adults are only just beginning, but it is expected that some genetic factors will influence risk for persistence and remission of the disorder during the transitional years from childhood to adulthood. Finally, further work is needed to identify environmental risks that act in an additive or interactive way with genetic risks for ADHD.

Family, twin and adoption studies have had a major influence on the way that we perceive ADHD and this, in turn, has influenced clinical decision making. We know that the disorder is largely inherited and that the genetic influences account for stability of ADHD over time. Furthermore, genetic studies have helped our understanding of the development of comorbid disorders. Future work will use genetic data to identify aetiologically distinct sub-groups with the aim of improving the prediction of clinical outcome and developing novel targeted-intervention strategies to treat the disorder and prevent its progression into adulthood. These are critical strategies because of the high personal and societal costs of ADHD including education and employment problems, high accident rates and risk for the development of anxiety, depression, drug and alcohol addiction and antisocial behaviour associated with ADHD.

References

Acosta MT, Arcos-Burgos M, Muenke M (2004) Attention deficit/hyperactivity disorder (ADHD): complex phenotype, simple genotype? Genet Med 6:1–15

Alberts-Corush J, Firestone P, Goodman JT (1986) Attention and impulsivity characteristics of the biological and adoptive parents of hyperactive and normal control children. Am J Orthopsychiatry 56:413–423

Andreou P, Neale BM, Chen W, Christiansen H, Gabriels I, Heise A, Meidad S, Muller UC, Uebel H, Banaschewski T, Manor I, Oades R, Roeyers H, Rothenberger A, Sham P, Steinhausen HC, Asherson P, Kuntsi J (2007) Reaction time performance in ADHD: improvement under fast-incentive condition and familial effects. Psychol Med 37:1703–1715

Arcos-Burgos M, Castellanos FX, Pineda D, Lopera F, Palacio JD, Palacio LG, Rapoport JL, Berg K, Bailey-Wilson JE, Muenke M (2004) Attention-deficit/hyperactivity disorder in a population isolate: linkage to loci at 4q13.2, 5q33.3, 11q22, and 17p11. Am J Hum Genet 75:998–1014

Arcos-Burgos M, Jain M, Acosta MT, Shively S, Stanescu H, Wallis D, Domene S, Velez JI, Karkera JD, Balog J, Berg K, Kleta R, Gahl WA, Roessler E, Long R, Lie J, Pineda D, Londono AC, Palacio JD, Arbelaez A, Lopera F, Elia J, Hakonarson H, Johansson S, Knappskog PM, Haavik J, Ribases M, Cormand B, Bayes M, Casas M, Ramos-Quiroga JA, Hervas A, Maher BS, Faraone SV, Seitz C, Freitag CM, Palmason H, Meyer J, Romanos M, Walitza S, Hemminger U, Warnke A, Romanos J, Renner T, Jacob C, Lesch KP, Swanson J, Vortmeyer A, Bailey-Wilson JE, Castellanos FX, Muenke M (2010) A common variant of the latrophilin 3 gene, LPHN3, confers susceptibility to ADHD and predicts effectiveness of stimulant medication. Mol Psychiatry 15:1053–1066

Asherson P, Brookes K, Franke B, Chen W, Gill M, Ebstein RP, Buitelaar J, Banaschewski T, Sonuga-Barke E, Eisenberg J, Manor I, Miranda A, Oades RD, Roeyers H, Rothenberger A, Sergeant J, Steinhausen HC, Faraone SV (2007) Confirmation that a specific haplotype of the dopamine transporter gene is associated with combined-type ADHD. Am J Psychiatry 164:674–677

Asherson P, Zhou K, Anney RJ, Franke B, Buitelaar J, Ebstein R, Gill M, Altink M, Arnold R, Boer F, Brookes K, Buschgens C, Butler L, Cambell D, Chen W, Christiansen H, Feldman L, Fleischman K, Fliers E, Howe-Forbes R, Goldfarb A, Heise A, Gabriels I, Johansson L, Lubetzki I, Marco R, Medad S, Minderaa R, Mulas F, Muller U, Mulligan A, Neale B, Rijsdijk F, Rabin K, Rommelse N, Sethna V, Sorohan J, Uebel H, Psychogiou L, Weeks A, Barrett R, Xu X, Banaschewski T, Sonuga-Barke E, Eisenberg J, Manor I, Miranda A, Oades RD, Roeyers H, Rothenberger A, Sergeant J, Steinhausen HC, Taylor E, Thompson M, Faraone SV (2008) A high-density SNP linkage scan with 142 combined subtype ADHD sib pairs identifies linkage regions on chromosomes 9 and 16. Mol Psychiatry 13:514–521

Bakker SC, van der Meulen EM, Buitelaar JK, Sandkuijl LA, Pauls DL, Monsuur AJ, van 't Slot R, Minderaa RB, Gunning WB, Pearson PL, Sinke RJ (2003) A whole-genome scan in 164 Dutch sib pairs with attention-deficit/hyperactivity disorder: suggestive evidence for linkage on chromosomes 7p and 15q. Am J Hum Genet 72:1251–1260

Barr CL, Feng Y, Wigg KG, Schachar R, Tannock R, Roberts W, Malone M, Kennedy JL (2001) 5'-untranslated region of the dopamine D4 receptor gene and attention-deficit hyperactivity disorder. Am J Med Genet 105:84–90

Becker K, El-Faddagh M, Schmidt MH, Esser G, Laucht M (2008) Interaction of dopamine transporter genotype with prenatal smoke exposure on ADHD symptoms. J Pediatr 152:263–269

Biederman J, Faraone SV, Mick E, Spencer T, Wilens T, Kiely K, Guite J, Ablon JS, Reed E, Warburton R (1995) High risk for attention deficit hyperactivity disorder among children of parents with childhood onset of the disorder: a pilot study. Am J Psychiatry 152:431–435

Boomsma DI, Saviouk V, Hottenga JJ, Distel MA, de Moor MH, Vink JM, Geels LM, van Beek JH, Bartels M, de Geus EJ, Willemsen G (2010) Genetic epidemiology of attention deficit hyperactivity disorder (ADHD index) in adults. PLoS One 5:e10621

Brookes KJ, Mill J, Guindalini C, Curran S, Xu X, Knight J, Chen CK, Huang YS, Sethna V, Taylor E, Chen W, Breen G, Asherson P (2006) A common haplotype of the dopamine transporter gene associated with attention-deficit/hyperactivity disorder and interacting with maternal use of alcohol during pregnancy. Arch Gen Psychiatry 63:74–81

Brookes KJ, Xu X, Anney R, Franke B, Zhou K, Chen W, Banaschewski T, Buitelaar J, Ebstein R, Eisenberg J, Gill M, Miranda A, Oades RD, Roeyers H, Rothenberger A, Sergeant J, Sonuga-Barke E, Steinhausen HC, Taylor E, Faraone SV, Asherson P (2008) Association of ADHD with genetic variants in the 5'-region of the dopamine transporter gene: evidence for allelic heterogeneity. Am J Med Genet B Neuropsychiatr Genet 147B:1519–1523

Burt SA (2009) Rethinking environmental contributions to child and adolescent psychopathology: a meta-analysis of shared environmental influences. Psychol Bull 135:608–637

Cantwell DP (1972) Psychiatric illness in the families of hyperactive children. Arch Gen Psychiatry 27:414–417

Caspi A, Langley K, Milne B, Moffitt TE, O'Donovan M, Owen MJ, Polo Tomas M, Poulton R, Rutter M, Taylor A, Williams B, Thapar A (2008) A replicated molecular genetic basis for subtyping antisocial behavior in children with attention-deficit/hyperactivity disorder. Arch Gen Psychiatry 65:203–210

Caspi A, Hariri AR, Holmes A, Uher R, Moffitt TE (2010) Genetic sensitivity to the environment: the case of the serotonin transporter gene and its implications for studying complex diseases and traits. Am J Psychiatry 167:509–527

Castellanos FX, Margulies DS, Kelly C, Uddin LQ, Ghaffari M, Kirsch A, Shaw D, Shehzad Z, Di Martino A, Biswal B, Sonuga-Barke EJ, Rotrosen J, Adler LA, Milham MP (2008) Cingulate-precuneus interactions: a new locus of dysfunction in adult attention-deficit/hyperactivity disorder. Biol Psychiatry 63:332–337

Chen W, Zhou K, Sham P, Franke B, Kuntsi J, Campbell D, Fleischman K, Knight J, Andreou P, Arnold R, Altink M, Boer F, Boholst MJ, Buschgens C, Butler L, Christiansen H, Fliers E, Howe-Forbes R, Gabriels I, Heise A, Korn-Lubetzki I, Marco R, Medad S, Minderaa R, Muller UC, Mulligan A, Psychogiou L, Rommelse N, Sethna V, Uebel H, McGuffin P, Plomin R, Banaschewski T, Buitelaar J, Ebstein R, Eisenberg J, Gill M, Manor I, Miranda A, Mulas F, Oades RD, Roeyers H, Rothenberger A, Sergeant J, Sonuga-Barke E, Steinhausen HC, Taylor E, Thompson M, Faraone SV, Asherson P (2008) DSM-IV combined type ADHD shows familial association with sibling trait scores: a sampling strategy for QTL linkage. Am J Med Genet B Neuropsychiatr Genet 147B:1450–1460

Cichon S, Craddock N, Daly M, Faraone SV, Gejman PV, Kelsoe J, Lehner T, Levinson DF, Moran A, Sklar P, Sullivan PF (2009) Genomewide association studies: history, rationale, and prospects for psychiatric disorders. Am J Psychiatry 166:540–556

Cole J, Ball HA, Martin NC, Scourfield J, McGuffin P (2009) Genetic overlap between measures of hyperactivity/inattention and mood in children and adolescents. J Am Acad Child Adolesc Psychiatry 48:1094–1101

Cook EH Jr, Stein MA, Krasowski MD, Cox NJ, Olkon DM, Kieffer JE, Leventhal BL (1995) Association of attention-deficit disorder and the dopamine transporter gene. Am J Hum Genet 56:993–998

Curko Kera EA, Marks DJ, Berwid OG, Santra A, Halperin JM (2004) Self-report and objective measures of ADHD-related behaviors in parents of preschool children at risk for ADHD. CNS Spectr 9:639–647

Dawn Teare M, Barrett JH (2005) Genetic linkage studies. Lancet 366:1036–1044

D'Souza UM, Russ C, Tahir E, Mill J, McGuffin P, Asherson PJ, Craig IW (2004) Functional effects of a tandem duplication polymorphism in the 5'flanking region of the DRD4 gene. Biol Psychiatry 56:691–697

Dudbridge F, Gusnanto A (2008) Estimation of significance thresholds for genomewide association scans. Genet Epidemiol 32:227–234

Durston S (2010) Imaging genetics in ADHD. Neuroimage 53:832–838

Ehringer MA, Rhee SH, Young S, Corley R, Hewitt JK (2006) Genetic and environmental contributions to common psychopathologies of childhood and adolescence: a study of twins and their siblings. J Abnorm Child Psychol 34:1–17

Elia J, Gai X, Xie HM, Perin JC, Geiger E, Glessner JT, D'Arcy M, Deberardinis R, Frackelton E, Kim C, Lantieri F, Muganga BM, Wang L, Takeda T, Rappaport EF, Grant SF, Berrettini W, Devoto M, Shaikh TH, Hakonarson H, White PS (2009) Rare structural variants found in attention-deficit hyperactivity disorder are preferentially associated with neurodevelopmental genes. Mol Psychiatry 15:637–646

Epstein JN, Conners CK, Erhardt D, Arnold LE, Hechtman L, Hinshaw SP, Hoza B, Newcorn JH, Swanson JM, Vitiello B (2000) Familial aggregation of ADHD characteristics. J Abnorm Child Psychol 28:585–594

Faraone SV (2004) Genetics of adult attention-deficit/hyperactivity disorder. Psychiatr Clin North Am 27:303–321

Faraone SV, Biederman J, Monuteaux MC (2000) Toward guidelines for pedigree selection in genetic studies of attention deficit hyperactivity disorder. Genet Epidemiol 18:1–16

Faraone SV, Perlis RH, Doyle AE, Smoller JW, Goralnick JJ, Holmgren MA, Sklar P (2005) Molecular genetics of attention-deficit/hyperactivity disorder. Biol Psychiatry 57:1313–1323

Faraone SV, Biederman J, Mick E (2006) The age-dependent decline of attention deficit hyperactivity disorder: a meta-analysis of follow-up studies. Psychol Med 36:159–165

Faraone SV, Doyle AE, Lasky-Su J, Sklar PB, D'Angelo E, Gonzalez-Heydrich J, Kratochvil C, Mick E, Klein K, Rezac AJ, Biederman J (2008) Linkage analysis of attention deficit hyperactivity disorder. Am J Med Genet B Neuropsychiatr Genet 147B:1387–1391

Feng Y, Crosbie J, Wigg K, Pathare T, Ickowicz A, Schachar R, Tannock R, Roberts W, Malone M, Swanson J, Kennedy JL, Barr CL (2005) The SNAP25 gene as a susceptibility gene contributing to attention-deficit hyperactivity disorder. Mol Psychiatry 10(998–1005):973

Ferreira MA, O'Donovan MC, Meng YA, Jones IR, Ruderfer DM, Jones L, Fan J, Kirov G, Perlis RH, Green EK, Smoller JW, Grozeva D, Stone J, Nikolov I, Chambert K, Hamshere ML, Nimgaonkar VL, Moskvina V, Thase ME, Caesar S, Sachs GS, Franklin J, Gordon-Smith K, Ardlie KG, Gabriel SB, Fraser C, Blumenstiel B, Defelice M, Breen G, Gill M, Morris DW, Elkin A, Muir WJ, McGhee KA, Williamson R, MacIntyre DJ, MacLean AW, St CD, Robinson M, Van Beck M, Pereira AC, Kandaswamy R, McQuillin A, Collier DA, Bass NJ, Young AH, Lawrence J, Ferrier IN, Anjorin A, Farmer A, Curtis D, Scolnick EM, McGuffin P, Daly MJ, Corvin AP, Holmans PA, Blackwood DH, Gurling HM, Owen MJ, Purcell SM, Sklar P, Craddock N (2008) Collaborative genome-wide association analysis supports a role for ANK3 and CACNA1C in bipolar disorder. Nat Genet 40:1056–1058

Fisher SE, Francks C, McCracken JT, McGough JJ, Marlow AJ, MacPhie IL, Newbury DF, Crawford LR, Palmer CG, Woodward JA, Del'Homme M, Cantwell DP, Nelson SF, Monaco AP, Smalley SL (2002) A genomewide scan for loci involved in attention-deficit/hyperactivity disorder. Am J Hum Genet 70:1183–1196

Flint J, Munafo MR (2007) The endophenotype concept in psychiatric genetics. Psychol Med 37:163–180

Francks C, Fisher SE, Marlow AJ, MacPhie IL, Taylor KE, Richardson AJ, Stein JF, Monaco AP (2003) Familial and genetic effects on motor coordination, laterality, and reading-related cognition. Am J Psychiatry 160:1970–1977

Franke B, Hoogman M, Arias Vasquez A, Heister JG, Savelkoul PJ, Naber M, Scheffer H, Kiemeney LA, Kan CC, Kooij JJ, Buitelaar JK (2008) Association of the dopamine transporter (SLC6A3/DAT1) gene 9–6 haplotype with adult ADHD. Am J Med Genet B Neuropsychiatr Genet 147B:1576–1579

Franke B, Neale BM, Faraone SV (2009) Genome-wide association studies in ADHD. Hum Genet 126:13–50

Friedel S, Saar K, Sauer S, Dempfle A, Walitza S, Renner T, Romanos M, Freitag C, Seitz C, Palmason H, Scherag A, Windemuth-Kieselbach C, Schimmelmann BG, Wewetzer C, Meyer J, Warnke A, Lesch KP, Reinhardt R, Herpertz-Dahlmann B, Linder M, Hinney A, Remschmidt H,

Schafer H, Konrad K, Hubner N, Hebebrand J (2007) Association and linkage of allelic variants of the dopamine transporter gene in ADHD. Mol Psychiatry 12:923–933

Genro JP, Zeni C, Polanczyk GV, Roman T, Rohde LA, Hutz MH (2007) A promoter polymorphism (−839 C > T) at the dopamine transporter gene is associated with attention deficit/hyperactivity disorder in Brazilian children. Am J Med Genet B Neuropsychiatr Genet 144B:215–219

Genro JP, Polanczyk GV, Zeni C, Oliveira AS, Roman T, Rohde LA, Hutz MH (2008) A common haplotype at the dopamine transporter gene 5' region is associated with attention-deficit/hyperactivity disorder. Am J Med Genet B Neuropsychiatr Genet 147B:1568–1575

Ghebranious N, Giampietro PF, Wesbrook FP, Rezkalla SH (2007) A novel microdeletion at 16p11.2 harbors candidate genes for aortic valve development, seizure disorder, and mild mental retardation. Am J Med Genet A 143A:1462–1471

Gizer IR, Ficks C, Waldman ID (2009) Candidate gene studies of ADHD: a meta-analytic review. Hum Genet 126:51–90

Gottesman II, Gould TD (2003) The endophenotype concept in psychiatry: etymology and strategic intentions. Am J Psychiatry 160:636–645

Halperin JM, Schulz KP (2006) Revisiting the role of the prefrontal cortex in the pathophysiology of attention-deficit/hyperactivity disorder. Psychol Bull 132:560–581

Halperin JM, Trampush JW, Miller CJ, Marks DJ, Newcorn JH (2008) Neuropsychological outcome in adolescents/young adults with childhood ADHD: profiles of persisters, remitters and controls. J Child Psychol Psychiatry 49:958–966

Hawi Z, Dring M, Kirley A, Foley D, Kent L, Craddock N, Asherson P, Curran S, Gould A, Richards S, Lawson D, Pay H, Turic D, Langley K, Owen M, O'Donovan M, Thapar A, Fitzgerald M, Gill M (2002) Serotonergic system and attention deficit hyperactivity disorder (ADHD): a potential susceptibility locus at the 5-HT(1B) receptor gene in 273 nuclear families from a multi-centre sample. Mol Psychiatry 7:718–725

Hebebrand J, Dempfle A, Saar K, Thiele H, Herpertz-Dahlmann B, Linder M, Kiefl H, Remschmidt H, Hemminger U, Warnke A, Knolker U, Heiser P, Friedel S, Hinney A, Schafer H, Nurnberg P, Konrad K (2006) A genome-wide scan for attention-deficit/hyperactivity disorder in 155 German sib-pairs. Mol Psychiatry 11:196–205

Hess EJ, Rogan PK, Domoto M, Tinker DE, Ladda RL, Ramer JC (1995) Absence of linkage of apparently single gene mediated ADHD with the human syntenic region of the mouse mutant Coloboma. Am J Med Genet 60:573–579

Johnson KA, Wiersema JR, Kuntsi J (2009) What would Karl Popper say? Are current psychological theories of ADHD falsifiable? Behav Brain Funct 5:15

Kahn RS, Khoury J, Nichols WC, Lanphear BP (2003) Role of dopamine transporter genotype and maternal prenatal smoking in childhood hyperactive-impulsive, inattentive, and oppositional behaviors. J Pediatr 143:104–110

Kebir O, Tabbane K, Sengupta S, Joober R (2009) Candidate genes and neuropsychological phenotypes in children with ADHD: review of association studies. J Psychiatry Neurosci 34:88–101

Kendler KS, Neale MC (2010) Endophenotype: a conceptual analysis. Mol Psychiatry 15:789–797

Kendler KS, Neale MC, Kessler RC, Heath AC, Eaves LJ (1993) A test of the equal-environment assumption in twin studies of psychiatric illness. Behav Genet 23:21–27

Kereszturi E, Kiraly O, Csapo Z, Tarnok Z, Gadoros J, Sasvari-Szekely M, Nemoda Z (2007) Association between the 120-bp duplication of the dopamine D4 receptor gene and attention deficit hyperactivity disorder: genetic and molecular analyses. Am J Med Genet B Neuropsychiatr Genet 144B:231–236

Kuntsi J, Neale BM, Chen W, Faraone SV, Asherson P (2006a) The IMAGE project: methodological issues for the molecular genetic analysis of ADHD. Behav Brain Funct 2:27

Kuntsi J, Rogers H, Swinard G, Borger N, van der Meere J, Rijsdijk F, Asherson P (2006b) Reaction time, inhibition, working memory and 'delay aversion' performance: genetic influences and their interpretation. Psychol Med 36:1613–1624

Kuntsi J, Wood AC, Rijsdijk F, Johnson KA, Andreou P, Albrecht B, Arias-Vasquez A, Buitelaar JK, McLoughlin G, Rommelse NN, Sergeant JA, Sonuga-Barke EJ, Uebel H, van der Meere JJ, Banaschewski T, Gill M, Manor I, Miranda A, Mulas F, Oades RD, Roeyers H, Rothenberger A, Steinhausen HC, Faraone SV, Asherson P (2010) Separation of cognitive impairments in attention-deficit/hyperactivity disorder into 2 familial factors. Arch Gen Psychiatry 67:1159–1167

Kuntsi J, Klein C (2011) Intraindividual variability in ADHD and its implications for research of causal links. Curr Topics Behav Neurosci. doi:10.1007/7854_2011_145

LaHoste GJ, Swanson JM, Wigal SB, Glabe C, Wigal T, King N, Kennedy JL (1996) Dopamine D4 receptor gene polymorphism is associated with attention deficit hyperactivity disorder. Mol Psychiatry 1:121–124

Langley K, Heron J, O'Donovan MC, Owen MJ, Thapar A (2010) Genotype link with extreme antisocial behavior: the contribution of cognitive pathways. Arch Gen Psychiatry 67:1317–1323

Lasky-Su J, Neale BM, Franke B, Anney RJ, Zhou K, Maller JB, Vasquez AA, Chen W, Asherson P, Buitelaar J, Banaschewski T, Ebstein R, Gill M, Miranda A, Mulas F, Oades RD, Roeyers H, Rothenberger A, Sergeant J, Sonuga-Barke E, Steinhausen HC, Taylor E, Daly M, Laird N, Lange C, Faraone SV (2008) Genome-wide association scan of quantitative traits for attention deficit hyperactivity disorder identifies novel associations and confirms candidate gene associations. Am J Med Genet B Neuropsychiatr Genet 147B:1345–1354

Lesch KP, Bengel D, Heils A, Sabol SZ, Greenberg BD, Petri S, Benjamin J, Muller CR, Hamer DH, Murphy DL (1996) Association of anxiety-related traits with a polymorphism in the serotonin transporter gene regulatory region. Science 274:1527–1531

Lesch KP, Timmesfeld N, Renner TJ, Halperin R, Roser C, Nguyen TT, Craig DW, Romanos J, Heine M, Meyer J, Freitag C, Warnke A, Romanos M, Schafer H, Walitza S, Reif A, Stephan DA, Jacob C (2008) Molecular genetics of adult ADHD: converging evidence from genome-wide association and extended pedigree linkage studies. J Neural Transm 115:1573–1585

Levy F, Hay DA, McStephen M, Wood C, Waldman I (1997) Attention-deficit hyperactivity disorder: a category or a continuum? Genetic analysis of a large-scale twin study. J Am Acad Child Adolesc Psychiatry 36:737–744

Li D, Sham PC, Owen MJ, He L (2006) Meta-analysis shows significant association between dopamine system genes and attention deficit hyperactivity disorder (ADHD). Hum Mol Genet 15:2276–2284

Loo SK, Specter E, Smolen A, Hopfer C, Teale PD, Reite ML (2003) Functional effects of the DAT1 polymorphism on EEG measures in ADHD. J Am Acad Child Adolesc Psychiatry 42:986–993

Loo SK, Fisher SE, Francks C, Ogdie MN, MacPhie IL, Yang M, McCracken JT, McGough JJ, Nelson SF, Monaco AP, Smalley SL (2004) Genome-wide scan of reading ability in affected sibling pairs with attention-deficit/hyperactivity disorder: unique and shared genetic effects. Mol Psychiatry 9:485–493

Lotrich FE, Pollock BG (2004) Meta-analysis of serotonin transporter polymorphisms and affective disorders. Psychiatr Genet 14:121–129

Lotta T, Vidgren J, Tilgmann C, Ulmanen I, Melen K, Julkunen I, Taskinen J (1995) Kinetics of human soluble and membrane-bound catechol O-methyltransferase: a revised mechanism and description of the thermolabile variant of the enzyme. Biochemistry 34:4202–4210

Lowe N, Kirley A, Hawi Z, Sham P, Wickham H, Kratochvil CJ, Smith SD, Lee SY, Levy F, Kent L, Middle F, Rohde LA, Roman T, Tahir E, Yazgan Y, Asherson P, Mill J, Thapar A, Payton A, Todd RD, Stephens T, Ebstein RP, Manor I, Barr CL, Wigg KG, Sinke RJ, Buitelaar JK, Smalley SL, Nelson SF, Biederman J, Faraone SV, Gill M (2004a) Joint analysis of the DRD5 marker concludes association with attention-deficit/hyperactivity disorder confined to the predominantly inattentive and combined subtypes. Am J Hum Genet 74:348–356

Lowe N, Kirley A, Mullins C, Fitzgerald M, Gill M, Hawi Z (2004b) Multiple marker analysis at the promoter region of the DRD4 gene and ADHD: evidence of linkage and association with the SNP −616. Am J Med Genet B Neuropsychiatr Genet 131B:33–37

Lu YM, Jia Z, Janus C, Henderson JT, Gerlai R, Wojtowicz JM, Roder JC (1997) Mice lacking metabotropic glutamate receptor 5 show impaired learning and reduced CA1 long-term potentiation (LTP) but normal CA3 LTP. J Neurosci 17:5196–5205

Maher BS, Marazita ML, Ferrell RE, Vanyukov MM (2002) Dopamine system genes and attention deficit hyperactivity disorder: a meta-analysis. Psychiatr Genet 12:207–215

Manolio TA, Collins FS, Cox NJ, Goldstein DB, Hindorff LA, Hunter DJ, McCarthy MI, Ramos EM, Cardon LR, Chakravarti A, Cho JH, Guttmacher AE, Kong A, Kruglyak L, Mardis E, Rotimi CN, Slatkin M, Valle D, Whittemore AS, Boehnke M, Clark AG, Eichler EE, Gibson G, Haines JL, Mackay TF, McCarroll SA, Visscher PM (2009) Finding the missing heritability of complex diseases. Nature 461:747–753

Manshadi M, Lippmann S, O'Daniel RG, Blackman A (1983) Alcohol abuse and attention deficit disorder. J Clin Psychiatry 44:379–380

Martin N, Scourfield J, McGuffin P (2002) Observer effects and heritability of childhood attention-deficit hyperactivity disorder symptoms. Br J Psychiatry 180:260–265

McCracken JT, Smalley SL, McGough JJ, Crawford L, Del'Homme M, Cantor RM, Liu A, Nelson SF (2000) Evidence for linkage of a tandem duplication polymorphism upstream of the dopamine D4 receptor gene (DRD4) with attention deficit hyperactivity disorder (ADHD). Mol Psychiatry 5:531–536

McLoughlin G, Ronald A, Kuntsi J, Asherson P, Plomin R (2007) Genetic support for the dual nature of attention deficit hyperactivity disorder: substantial genetic overlap between the inattentive and hyperactive-impulsive components. J Abnorm Child Psychol 35:999–1008

Mitchell AA, Cutler DJ, Chakravarti A (2003) Undetected genotyping errors cause apparent overtransmission of common alleles in the transmission/disequilibrium test. Am J Hum Genet 72:598–610

Morrison JR, Stewart MA (1971) A family study of the hyperactive child syndrome. Biol Psychiatry 3:189–195

Munafo MR, Clark TG, Roberts KH, Johnstone EC (2006) Neuroticism mediates the association of the serotonin transporter gene with lifetime major depression. Neuropsychobiology 53:1–8

Munafo MR, Brown SM, Hariri AR (2008) Serotonin transporter (5-HTTLPR) genotype and amygdala activation: a meta-analysis. Biol Psychiatry 63:852–857

Neale BM, Lasky-Su J, Anney R, Franke B, Zhou K, Maller JB, Vasquez AA, Asherson P, Chen W, Banaschewski T, Buitelaar J, Ebstein R, Gill M, Miranda A, Oades RD, Roeyers H, Rothenberger A, Sergeant J, Steinhausen HC, Sonuga-Barke E, Mulas F, Taylor E, Laird N, Lange C, Daly M, Faraone SV (2008) Genome-wide association scan of attention deficit hyperactivity disorder. Am J Med Genet B Neuropsychiatr Genet 147B:1337–1344

Neale BM, Medland SE, Ripke S, Asherson P, Franke B, Lesch KP, Faraone SV, Nguyen TT, Schafer H, Holmans P, Daly M, Steinhausen HC, Freitag C, Reif A, Renner TJ, Romanos M, Romanos J, Walitza S, Warnke A, Meyer J, Palmason H, Buitelaar J, Vasquez AA, Lambregts-Rommelse N, Gill M, Anney RJ, Langely K, O'Donovan M, Williams N, Owen M, Thapar A, Kent L, Sergeant J, Roeyers H, Mick E, Biederman J, Doyle A, Smalley S, Loo S, Hakonarson H, Elia J, Todorov A, Miranda A, Mulas F, Ebstein RP, Rothenberger A, Banaschewski T, Oades RD, Sonuga-Barke E, McGough J, Nisenbaum L, Middleton F, Hu X, Nelson S (2010) Meta-analysis of genome-wide association studies of attention-deficit/hyperactivity disorder. J Am Acad Child Adolesc Psychiatry 49:884–897

NICE (2008) Attention deficit hyperactivity disorder: The NICE guideline on diagnosis and managment of ADHD in children, young people and adults. The British Psychological Society and The Royal College of Psychiatrists, London

O'Connell RG, Bellgrove MA, Dockree PM, Lau A, Fitzgerald M, Robertson IH (2008) Self-Alert Training: volitional modulation of autonomic arousal improves sustained attention. Neuropsychologia 46:1379–1390

O'Connell RG, Dockree PM, Bellgrove MA, Turin A, Ward S, Foxe JJ, Robertson IH (2009) Two types of action error: electrophysiological evidence for separable inhibitory and sustained attention neural mechanisms producing error on go/no-go tasks. J Cogn Neurosci 21:93–104

O'Donovan MC, Craddock NJ, Owen MJ (2009) Genetics of psychosis; insights from views across the genome. Hum Genet 126:3–12

Ogdie MN, Macphie IL, Minassian SL, Yang M, Fisher SE, Francks C, Cantor RM, McCracken JT, McGough JJ, Nelson SF, Monaco AP, Smalley SL (2003) A genomewide scan for attention-deficit/hyperactivity disorder in an extended sample: suggestive linkage on 17p11. Am J Hum Genet 72:1268–1279

Ogdie MN, Fisher SE, Yang M, Ishii J, Francks C, Loo SK, Cantor RM, McCracken JT, McGough JJ, Smalley SL, Nelson SF (2004) Attention deficit hyperactivity disorder: fine mapping supports linkage to 5p13, 6q12, 16p13, and 17p11. Am J Hum Genet 75:661–668

Okuyama Y, Ishiguro H, Toru M, Arinami T (1999) A genetic polymorphism in the promoter region of DRD4 associated with expression and schizophrenia. Biochem Biophys Res Commun 258:292–295

Paloyelis Y, Rijsdijk F, Wood AC, Asherson P, Kuntsi J (2010) The genetic association between ADHD symptoms and reading difficulties: the role of inattentiveness and IQ. J Abnorm Child Psychol 38:1083–1095

Park JH, Wacholder S, Gail MH, Peters U, Jacobs KB, Chanock SJ, Chatterjee N (2010) Estimation of effect size distribution from genome-wide association studies and implications for future discoveries. Nat Genet 42:570–575

Patel SD, Chen CP, Bahna F, Honig B, Shapiro L (2003) Cadherin-mediated cell-cell adhesion: sticking together as a family. Curr Opin Struct Biol 13:690–698

Payton A, Holmes J, Barrett JH, Hever T, Fitzpatrick H, Trumper AL, Harrington R, McGuffin P, O'Donovan M, Owen M, Ollier W, Worthington J, Thapar A (2001) Examining for association between candidate gene polymorphisms in the dopamine pathway and attention-deficit hyperactivity disorder: a family-based study. Am J Med Genet 105:464–470

Polanczyk G, de Lima MS, Horta BL, Biederman J, Rohde LA (2007) The worldwide prevalence of ADHD: a systematic review and metaregression analysis. Am J Psychiatry 164:942–948

Purper-Ouakil D, Wohl M, Mouren MC, Verpillat P, Ades J, Gorwood P (2005) Meta-analysis of family-based association studies between the dopamine transporter gene and attention deficit hyperactivity disorder. Psychiatr Genet 15:53–59

Quist JF, Barr CL, Schachar R, Roberts W, Malone M, Tannock R, Basile VS, Beitchman J, Kennedy JL (2003) The serotonin 5-HT1B receptor gene and attention deficit hyperactivity disorder. Mol Psychiatry 8:98–102

Ribases M, Ramos-Quiroga JA, Hervas A, Bosch R, Bielsa A, Gastaminza X, Artigas J, Rodriguez-Ben S, Estivill X, Casas M, Cormand B, Bayes M (2009) Exploration of 19 serotoninergic candidate genes in adults and children with attention-deficit/hyperactivity disorder identifies association for 5HT2A, DDC and MAOB. Mol Psychiatry 14:71–85

Ribases M, Antoni Ramos-Quiroga J, Sanchez-Mora C, Bosch R, Richarte V, Alvarez I, Gastaminza X, Bielsa A, Arcos-Burgos M, Muenke M, Castellanos FX, Cormand B, Bayes M, Casas M (2011) Contribution of latrophilin 3 (LPHN3) to the genetic susceptibility to ADHD in adulthood: a replication study. Genes Brain Behav 10:149–157

Romanos M, Freitag C, Jacob C, Craig DW, Dempfle A, Nguyen TT, Halperin R, Walitza S, Renner TJ, Seitz C, Romanos J, Palmason H, Reif A, Heine M, Windemuth-Kieselbach C, Vogler C, Sigmund J, Warnke A, Schafer H, Meyer J, Stephan DA, Lesch KP (2008) Genome-wide linkage analysis of ADHD using high-density SNP arrays: novel loci at 5q13.1 and 14q12. Mol Psychiatry 13:522–530

Ronald A, Simonoff E, Kuntsi J, Asherson P, Plomin R (2008) Evidence for overlapping genetic influences on autistic and ADHD behaviours in a community twin sample. J Child Psychol Psychiatry 49:535–542

Ronald A, Edelson LR, Asherson P, Saudino KJ (2010) Exploring the relationship between autistic-like traits and ADHD behaviors in early childhood: findings from a community twin study of 2-year-olds. J Abnorm Child Psychol 38:185–196

Schinka JA, Busch RM, Robichaux-Keene N (2004) A meta-analysis of the association between the serotonin transporter gene polymorphism (5-HTTLPR) and trait anxiety. Mol Psychiatry 9:197–202

Sherman DK, McGue MK, Iacono WG (1997) Twin concordance for attention deficit hyperactivity disorder: a comparison of teachers' and mothers' reports. Am J Psychiatry 154:532–535

Shiow LR, Paris K, Akana MC, Cyster JG, Sorensen RU, Puck JM (2009) Severe combined immunodeficiency (SCID) and attention deficit hyperactivity disorder (ADHD) associated with a Coronin-1A mutation and a chromosome 16p11.2 deletion. Clin Immunol 131:24–30

Skirrow C, McLoughlin G, Kuntsi J, Asherson P (2009) Behavioral, neurocognitive and treatment overlap between attention-deficit/hyperactivity disorder and mood instability. Expert Rev Neurother 9:489–503

Smalley SL, Kustanovich V, Minassian SL, Stone JL, Ogdie MN, McGough JJ, McCracken JT, MacPhie IL, Francks C, Fisher SE, Cantor RM, Monaco AP, Nelson SF (2002) Genetic linkage of attention-deficit/hyperactivity disorder on chromosome 16p13, in a region implicated in autism. Am J Hum Genet 71:959–963

Smoller JW, Biederman J, Arbeitman L, Doyle AE, Fagerness J, Perlis RH, Sklar P, Faraone SV (2006) Association between the 5HT1B receptor gene (HTR1B) and the inattentive subtype of ADHD. Biol Psychiatry 59:460–467

Sprich S, Biederman J, Crawford MH, Mundy E, Faraone SV (2000) Adoptive and biological families of children and adolescents with ADHD. J Am Acad Child Adolesc Psychiatry 39:1432–1437

Stevenson J, Sonuga-Barke E, McCann D, Grimshaw K, Parker KM, Rose-Zerilli MJ, Holloway JW, Warner JO (2010) The role of histamine degradation gene polymorphisms in moderating the effects of food additives on children's ADHD symptoms. Am J Psychiatry 167:1108–1115

Sugden K, Arseneault L, Harrington H, Moffitt TE, Williams B, Caspi A (2010) Serotonin transporter gene moderates the development of emotional problems among children following bullying victimization. J Am Acad Child Adolesc Psychiatry 49:830–840

Swanson JM, Sunohara GA, Kennedy JL, Regino R, Fineberg E, Wigal T, Lerner M, Williams L, LaHoste GJ, Wigal S (1998) Association of the dopamine receptor D4 (DRD4) gene with a refined phenotype of attention deficit hyperactivity disorder (ADHD): a family-based approach. Mol Psychiatry 3:38–41

Takeuchi T, Misaki A, Liang SB, Tachibana A, Hayashi N, Sonobe H, Ohtsuki Y (2000) Expression of T-cadherin (CDH13, H-Cadherin) in human brain and its characteristics as a negative growth regulator of epidermal growth factor in neuroblastoma cells. J Neurochem 74:1489–1497

Thapar A, Harrington R, McGuffin P (2001) Examining the comorbidity of ADHD-related behaviours and conduct problems using a twin study design. Br J Psychiatry 179:224–229

Thapar A, Langley K, Fowler T, Rice F, Turic D, Whittinger N, Aggleton J, Van den Bree M, Owen M, O'Donovan M (2005) Catechol O-methyltransferase gene variant and birth weight predict early-onset antisocial behavior in children with attention-deficit/hyperactivity disorder. Arch Gen Psychiatry 62:1275–1278

Thapar A, Rice F, Hay D, Boivin J, Langley K, van den Bree M, Rutter M, Harold G (2009) Prenatal smoking might not cause attention-deficit/hyperactivity disorder: evidence from a novel design. Biol Psychiatry 66:722–727

Todd RD, Rasmussen ER, Neuman RJ, Reich W, Hudziak JJ, Bucholz KK, Madden PA, Heath A (2001) Familiality and heritability of subtypes of attention deficit hyperactivity disorder in a population sample of adolescent female twins. Am J Psychiatry 158:1891–1898

Uhl GR, Drgon T, Liu QR, Johnson C, Walther D, Komiyama T, Harano M, Sekine Y, Inada T, Ozaki N, Iyo M, Iwata N, Yamada M, Sora I, Chen CK, Liu HC, Ujike H, Lin SK (2008) Genome-wide association for methamphetamine dependence: convergent results from 2 samples. Arch Gen Psychiatry 65:345–355

Walters JT, Corvin A, Owen MJ, Williams H, Dragovic M, Quinn EM, Judge R, Smith DJ, Norton N, Giegling I, Hartmann AM, Moller HJ, Muglia P, Moskvina V, Dwyer S, O'Donoghue T, Morar B, Cooper M, Chandler D, Jablensky A, Gill M, Kaladjieva L, Morris DW, O'Donovan MC, Rujescu D, Donohoe G (2010) Psychosis susceptibility gene ZNF804A and cognitive performance in schizophrenia. Arch Gen Psychiatry 67:692–700

Williams NM, Zaharieva I, Martin A, Langley K, Mantripragada K, Fossdal R, Stefansson H, Stefansson K, Magnusson P, Gudmundsson OO, Gustafsson O, Holmans P, Owen MJ, O'Donovan M, Thapar A (2010) Rare chromosomal deletions and duplications in attention-deficit hyperactivity disorder: a genome-wide analysis. Lancet 376:1401–1408

Wood AC, Rijsdijk F, Asherson P, Kuntsi J (2009) Hyperactive-impulsive symptom scores and oppositional behaviours reflect alternate manifestations of a single liability. Behav Genet 39:447–460

Wood AC, Rijsdijk F, Johnson KA, Andreou P, Albrecht B, Arias-Vasquez A, Buitelaar JK, McLoughlin G, Rommelse NN, Sergeant JA, Sonuga-Barke EJ, Uebel H, van der Meere JJ, Banaschewski T, Gill M, Manor I, Miranda A, Mulas F, Oades RD, Roeyers H, Rothenberger A, Steinhausen HC, Faraone SV, Asherson P, Kuntsi J (2011) The relationship between ADHD and key cognitive phenotypes is not mediated by shared familial effects with IQ. Psychol Med 41(4):861–871

Yan TC, McQuillin A, Thapar A, Asherson P, Hunt SP, Stanford SC, Gurling H (2010) NK1 (TACR1) receptor gene 'knockout' mouse phenotype predicts genetic association with ADHD. J Psychopharmacol 24:27–38

Yan TC, Dudley JA, Weir RK, Grabowska EM, Pena-Oliver Y, Ripley TL, Hunt SP, Stephens DN, Stanford SC (2011) Performance deficits of NK1 receptor knockout mice in the 5-choice serial reaction-time task: effects of d-amphetamine, stress and time of day. PLoS One 6:e17586

Yang B, Chan RC, Jing J, Li T, Sham P, Chen RY (2007) A meta-analysis of association studies between the 10-repeat allele of a VNTR polymorphism in the 3'-UTR of dopamine transporter gene and attention deficit hyperactivity disorder. Am J Med Genet B Neuropsychiatr Genet 144B:541–550

Yang JW, Jang WS, Hong SD, Ji YI, Kim DH, Park J, Kim SW, Joung YS (2008) A case-control association study of the polymorphism at the promoter region of the DRD4 gene in Korean boys with attention deficit-hyperactivity disorder: evidence of association with the −521 C/T SNP. Prog Neuropsychopharmacol Biol Psychiatry 32:243–248

Zhou K, Chen W, Buitelaar J, Banaschewski T, Oades RD, Franke B, Sonuga-Barke E, Ebstein R, Eisenberg J, Gill M, Manor I, Miranda A, Mulas F, Roeyers H, Rothenberger A, Sergeant J, Steinhausen HC, Lasky-Su J, Taylor E, Brookes KJ, Xu X, Neale BM, Rijsdijk F, Thompson M, Asherson P, Faraone SV (2008a) Genetic heterogeneity in ADHD: DAT1 gene only affects probands without CD. Am J Med Genet B Neuropsychiatr Genet 147B:1481–1487

Zhou K, Dempfle A, Arcos-Burgos M, Bakker SC, Banaschewski T, Biederman J, Buitelaar J, Castellanos FX, Doyle A, Ebstein RP, Ekholm J, Forabosco P, Franke B, Freitag C, Friedel S, Gill M, Hebebrand J, Hinney A, Jacob C, Lesch KP, Loo SK, Lopera F, McCracken JT, McGough JJ, Meyer J, Mick E, Miranda A, Muenke M, Mulas F, Nelson SF, Nguyen TT, Oades RD, Ogdie MN, Palacio JD, Pineda D, Reif A, Renner TJ, Roeyers H, Romanos M, Rothenberger A, Schafer H, Sergeant J, Sinke RJ, Smalley SL, Sonuga-Barke E, Steinhausen HC, van der Meulen E, Walitza S, Warnke A, Lewis CM, Faraone SV, Asherson P (2008b) Meta-analysis of genome-wide linkage scans of attention deficit hyperactivity disorder. Am J Med Genet B Neuropsychiatr Genet 147B:1392–1398

Rodent Models of ADHD

Xueliang Fan, Kristy J. Bruno, and Ellen J. Hess

Contents

1 Introduction 274
2 Neonatal 6-Hydroxydopamine-Lesioned Rat Model 276
    2.1 Validity 276
    2.2 Monoaminergic Regulation 276
    2.3 Therapeutic Mechanisms 278
3 *Coloboma* Mutant Mouse Model 279
    3.1 Validity 279
    3.2 Monoaminergic Regulation 280
    3.3 Therapeutic Mechanisms 281
4 Dopamine Transporter Knockout Mouse Model 282
    4.1 Validity 282
    4.2 Monoaminergic Regulation 282
    4.3 Therapeutic Mechanisms 283
5 Spontaneously Hypertensive Rat Model 284
    5.1 Validity 284
    5.2 Monoaminergic Regulation 284
    5.3 Therapeutic Mechanisms 287
6 Neurokinin 1 Receptor Knockout Mice ($NK1^{-/-}$) 287
    6.1 Validity 287
    6.2 Monoaminergic Regulation 287
    6.3 Therapeutic Mechanisms 288
7 What Have We Learned from Animal Models of ADHD? 288
    7.1 Animal Models of ADHD Exhibit Hyperdopaminergic Neurotransmission 288
    7.2 Norepinephrine Transmission Plays a Dual Role in ADHD 290
    7.3 Serotonin Transmission Plays an Inhibitory Role in ADHD 291
    7.4 Dopamine and Serotonin Receptors Contribute to the Therapeutic Effects
        of Psychostimulants 291
8 Summary 292
References 292

X. Fan, K.J. Bruno, and E.J. Hess (✉)
Departments of Pharmacology and Neurology, Emory University School of Medicine, Atlanta, GA 30322, USA
e-mail: ejhess@emory.edu

C. Stanford and R. Tannock (eds.), *Behavioral Neuroscience of Attention Deficit Hyperactivity Disorder and Its Treatment,* Current Topics in Behavioral Neurosciences 9, DOI 10.1007/7854_2011_121, © Springer-Verlag Berlin Heidelberg 2011, published online 15 March 2011

Abstract The neonatal 6-OHDA-lesioned rat, *coloboma* mouse, DAT-KO mouse, and spontaneously hypertensive rat (SHR) models all bear a phenotypic resemblance to ADHD in that they express hyperactivity, inattention, and/or impulsivity. The models also illustrate the heterogeneity of ADHD: the initial cause (chemical depletion or genetic abnormality) of the ADHD-like behaviors is different for each model. Neurochemical and behavioral studies of the models indicate aberrations in monoaminergic neurotransmission. Hyperdopaminergic neurotransmission is implicated in the abnormal behavior of all models. Norepinephrine has a dual enhancing/inhibitory role in ADHD symptoms, and serotonin acts to inhibit abnormal dopamine and norepinephrine signaling. It is unlikely that symptoms arise from a single neurotransmitter dysfunction. Rather, studies of animal models of ADHD suggest that symptoms develop through the complex interactions of monoaminergic neurotransmitter systems.

Keywords 6-Hydroxydopamine norepinephrine · Animal models · *Coloboma* mouse · Dopamine · Dopamine transporter knockout mouse · Serotonin · Spontaneously hypertensive rat

Abbreviations

5,7-DHT	5,7-Dihydroxytryptamine
6-OHDA	6-Hydroxydopamine
ADHD	Attention-deficit hyperactivity disorder
DAT-KD	Dopamine transporter knockdown
DAT-KO	Dopamine transporter knockout
DSM-IV	Diagnostic and Statistical Manual (Edition IV)
m-CPP	*m*-Chlorophenylpiperazine
NET	Norepinephrine transporter
SERT	Serotonin transporter
SHR	Spontaneously hypertensive rat
SNAP-25	Synaptosomal associated peptide (-25)
TH	Tyrosine hydroxylase
WKY	Wistar Kyoto (rat)

1 Introduction

By manipulating animal models of human diseases in the laboratory, researchers can elucidate their pathogenesis, leading to disease prevention in humans, as well as improvements in diagnosis and treatment. Because animal models play a vital role in the understanding of human diseases, it is important to characterize each model, with a focus on recognizing its advantages and limitations.

What criteria make an animal model suitable for the study of a human disease? A fundamental consideration for an animal model of ADHD is its phenotypic resemblance to the human disease, termed "face validity". Identifying face validity has been the primary goal for establishing animal models of ADHD since the first proposed animal model of ADHD: the 6-hydroxydopamine-lesioned rat (Shaywitz et al. 1976a).

While ADHD is characterized by inattention, impulsivity, and hyperactivity, the three symptoms do not affect all ADHD patients equally. In fact, DSM-IV divides classification of ADHD patients into a Predominantly Inattentive type, Predominantly Hyperactive-Impulsive type, and Combined type. The dichotomy in the presentation of symptoms associated with ADHD in humans, combined with the results of animal studies suggesting that inattention, impulsivity, and hyperactivity may not share a common substrate (Sagvolden et al. 1998), strongly suggests different neurobiological mechanisms underlying the three symptoms. Therefore, in establishing the face validity of an animal model of ADHD, the model need not exhibit all the behavioral symptoms of ADHD.

An animal model that has face validity for ADHD may meet other criteria that make that model amenable to the study of ADHD. A model that exhibits a behavioral response to pharmacological intervention that is similar to the human disease response has predictive validity. The psychostimulants, methylphenidate and amphetamine, are used to treat ADHD. These drugs normally produce an increase in motor activity but ameliorate symptoms in ADHD patients. Both drugs increase synaptic dopamine and norepinephrine, but they do so through different mechanisms. Amphetamine disrupts vesicular stores of dopamine and norepinephrine, causing reversal of the reuptake transporters and, consequently, efflux of catecholamines (Sulzer et al. 1995). In contrast, the methylphenidate class of psychostimulants (e.g., methylphenidate or cocaine) blocks reuptake (Butcher et al. 1991). The increase in extracellular dopamine induced by these drugs drives the increase in locomotor activity in normosensitive animals, although both drugs also increase extracellular norepinephrine. Animal models are essential for understanding the therapeutic effects of these compounds in ADHD. In addition to psychostimulants, norepinephrine reuptake inhibitors, such as atomoxetine, have been introduced more recently to treat ADHD. An animal model of ADHD that shows a similar reduction in expression of symptoms after administration of these drugs has "predictive validity". An animal model of ADHD achieves "etiological validity" when the cause(s)—such as gene mutations or environmental toxins of the ADHD-like behaviors—are the same as those in humans.

Based on their ability to fulfill one or more of these three criteria – face validity, predictive validity, etiological validity – nearly ten animal models of ADHD have emerged since the inception of the 6-hydroxydopamine (6-OHDA) rat model. Because ADHD is a heterogeneous disorder associated with multiple genetic abnormalities and multiple environmental risk factors, each of these animal models is useful in deciphering the pathogenesis of ADHD.

The four most extensively studied animal models of ADHD are the neonatal 6-hydroxydopamine-lesioned rat, the *coloboma* mutant mouse, the dopamine transporter (DAT) knockout/down mouse, and the spontaneously hypertensive rat (SHR). In this chapter, we discuss the behavioral and neurochemical features of each of these models plus a promising new model, and we explore the features shared by the models, which may reveal final common pathways in ADHD.

2 Neonatal 6-Hydroxydopamine-Lesioned Rat Model

2.1 Validity

Shaywitz and colleagues produced the neonatal 6-OHDA-lesioned rat model of ADHD by selective chemical lesion of dopaminergic neurons in 5-day-old rats. Ten to seventeen days after administration of 6-OHDA (to lesion the midbrain dopamine neurons) in combination with desmethylimipramine (to preserve noradrenergic neurons), investigators observed hyperactivity. Subsequent studies revealed that the timing of the lesion, in the context of rat development, and the 6-OHDA dose used to induce the chemical lesion influence the degree and duration of hyperactivity, respectively (Erinoff et al. 1979; Miller et al. 1981). Later, others reported not only hyperactivity in 6-OHDA-lesioned rats but also inattention (Oke and Adams 1978; Archer et al. 1988). Therefore, the neonatal 6-OHDA-lesioned rat has face validity as a model of ADHD.

Amphetamine and methylphenidate reduce hyperactivity of the neonatal 6-OHDA-lesioned rat (Shaywitz et al. 1976b, 1978; Heffner and Seiden 1982; Luthman et al. 1989a; Archer et al. 2002). Because both drugs ameliorate the behavioral deficits, the neonatal 6-OHDA-lesioned rat also has predictive validity as a model of ADHD.

2.2 Monoaminergic Regulation

2.2.1 Dopamine

Intracranial injection of 6-OHDA in neonatal rats causes extensive degeneration of dopaminergic neurons. Consequently, in many dopamine-rich brain regions, such as striatum, nucleus accumbens, and substantia nigra, there is a decrease in the following: the number of dopaminergic neurons, the tissue concentration of dopamine and its metabolites, dopamine transporter expression, activity and expression of tyrosine hydroxylase, and K^+-stimulated dopamine release and clearance (Shaywitz et al. 1976a, b; Oke and Adams 1978; Miller et al. 1981; Castaneda et al. 1990; Luthman et al. 1990a, 1993a, b, 1995; Herrera-Marschitz et al. 1994; Masuo et al. 2004; Dal Bo et al. 2008).

The 6-OHDA-induced depletion of dopamine results in a compensatory increase in basal dopamine efflux. Several investigators demonstrated that, despite the 95% depletion of dopamine content in the striatum of 6-OHDA-treated rats, basal extracellular dopamine as assessed by microdialysis is still 33–85% of the control level (Castaneda et al. 1990; Loupe et al. 2002; Nowak et al. 2007). Likewise, the basal extracellular dopamine level in the nucleus accumbens is not reduced, although a ~75% loss in dopamine tissue content is observed in treated rats (Ikegami et al. 2006). These compensatory effects may result from both increased transmitter release from surviving neurons and a loss of transporter to clear the extracellular dopamine. Overall, although total dopamine is reduced, these results suggest that extracellular dopamine is increased at individual residual terminals.

Studies of D_1- and D_2-like dopamine receptor expression in the 6-OHDA-lesioned rat model of ADHD have generated conflicting results, likely due to the use of rats of different ages in different studies (Broaddus and Bennett 1990; Dewar et al. 1990; Luthman et al. 1990b). However, behavioral studies have consistently demonstrated both D_1 and D_2 dopamine receptor supersensitivity in response to dopamine agonist challenge (Breese et al. 1985; Gong et al. 1993; Brus et al. 1994; Bishop et al. 2005; Archer and Fredriksson 2007). Neonatal 6-OHDA-lesioned rats are generally unresponsive to D_1- and D_2-like dopamine receptor antagonists (Duncan et al. 1987; Johnson and Bruno 1990), whereas dopamine receptor antagonists induce immobility (catalepsy) in normosensitive rats. This lack of response to antagonists is consistent with the supersensitive response to agonists, and it is likely a compensatory response caused by gross dopamine depletion. Interestingly, specific antagonism, or the absence, of D_4 dopamine receptors, which are in the D_2-like dopamine receptor family, prevents hyperactivity in this model, suggesting that this receptor is integral to the behavioral deficit (Zhang et al. 2002b; Avale et al. 2004a).

2.2.2 Norepinephrine

Because the norepinephrine reuptake inhibitor, desmethylimipramine, is used before 6-OHDA injection, brain norepinephrine content is not dramatically altered (Erinoff et al. 1979; Luthman et al. 1989b). Indeed, Luthman et al. (1989a, b) demonstrated that lesioning the noradrenergic system does not contribute to the locomotor hyperactivity. Future studies examining adrenergic receptor regulation will be helpful for understanding the interaction among the monoamines in this model.

2.2.3 Serotonin

Although dopamine neurotransmission is the primary target in the 6-OHDA-lesioned rat model, this model also shows marked changes in serotonin neurotransmission. In the striatum, there are increases in the following: tissue concentration of serotonin and its metabolites, serotonin transporter density, the number of serotonin-containing neurons and fibers, and K^+-stimulated serotonin release and clearance (Stachowiak et al. 1984; Snyder et al. 1986; Bruno et al. 1987; Towle et al. 1989; Luthman et al. 1990a, 1997; Molina-Holgado et al. 1994; Raison et al. 1995; Zhang et al. 2002a; Avale et al. 2004b). Despite serotonergic hyperinnervation, basal extracellular serotonin in the striatum is not altered in this model, likely because serotonin clearance is also increased (Jackson and Abercrombie 1992; Nowak et al. 2007). Indeed, blocking serotonin reuptake elicits greater extracellular serotonin in the striatum of 6-OHDA-treated rats than in control rats (Jackson and Abercrombie 1992).

An increase in the expression of 5-HT$_1$ and 5-HT$_2$ serotonin receptors occurs in the striatum and in other areas of severe dopamine denervation (Radja et al. 1993; Basura and Walker 1999). In fact, the changes in serotonin receptor expression and neurochemistry impact the hyperactivity in this model: Kostrzewa et al. (1994) found that elimination of serotonergic hyperinnervation, by administration of the selective serotonergic toxin 5,7-DHT 10 weeks after 6-OHDA injection, augments the hyperactivity of lesioned rats compared to 6-OHDA injection alone. Furthermore, in these 6-OHDA/5,7-DHT-lesioned rats, administration of the 5-HT$_2$ serotonin receptor agonist, m-CPP (m-chlorophenylpiperazine), reduces their hyperactivity (Kostrzewa et al. 1994; Brus et al. 2004), suggesting that serotonergic hyperinnervation inhibits hyperactivity, possibly through activation of 5-HT$_2$ serotonin receptors.

In addition to the direct effect of serotonin transmission on locomotor activity in the 6-OHDA model, serotonin also indirectly regulates dopamine-mediated behaviors. 5-HT$_{2A}$/$_{2C}$ serotonin receptor antagonists block D$_1$ dopamine receptor agonist-induced oral activity and locomotor hyperactivity (Gong et al. 1992; Kostrzewa et al. 1993; Bishop et al. 2005). Furthermore, disruption of neonatal 6-OHDA-induced serotonergic hyperinnervation by 5,7-DHT eliminates supersensitivity to both D$_1$ and D$_2$ dopamine receptor agonists (Brus et al. 1994, 1995), suggesting that serotonergic hyperinnervation facilitates dopaminergic function.

2.3 Therapeutic Mechanisms

Monoaminergic drug challenges are useful in determining the direct impact of receptor subtypes on behaviors and the indirect impact of receptor subtypes on other neurotransmitter systems. They also help to determine the neurotransmitters involved in the psychostimulant-mediated reduction in locomotor activity in the 6-OHDA model. Surprisingly, amphetamine does not augment dopamine efflux in this model (Castaneda et al. 1990; Herrera-Marschitz et al. 1994; Nowak et al. 2007), despite the disproportionate sparing of extracellular dopamine concentrations. That amphetamine does not increase dopamine efflux might be explained by the loss of transporters. Consistent with this finding, Heffner and Seiden (1982) found that the D$_2$ dopamine receptor antagonist, spiroperidol, dose-dependently reduced the amphetamine-induced increase in locomotor activity in control rats but failed to block the amphetamine-induced reduction in locomotor activity in

6-OHDA-treated rats. In contrast, the serotonin receptor antagonist, methysergide, dose-dependently blocked the amphetamine-induced decrease in locomotor activity in 6-OHDA-treated rats without affecting amphetamine-induced hyperactivity in normal rats. Similar to amphetamine and methylphenidate, the selective norepinephrine uptake inhibitors, desipramine and nisoxetine, and the serotonin uptake inhibitors, fluoxetine and citalopram, reduced hyperactivity in this model. Conversely, the selective dopamine uptake inhibitors, GBR12909 and amfonelic acid, did not influence hyperactivity (Davids et al. 2002). These experiments suggest that the therapeutic effects of amphetamine and methylphenidate in this model may be not through the dopamine system, but through norepinephrine and serotonin transmission, instead.

3 *Coloboma* Mutant Mouse Model

3.1 Validity

The *coloboma* mouse has a ~2 cM deletion mutation that encompasses *Snap25*, a gene that encodes SNAP-25 (Hess et al. 1994). SNAP-25 is a neuron-specific protein that plays a key role in neurotransmitter release. The deletion mutation causes spontaneous locomotor hyperactivity. Transgenic replacement of *Snap25* ameliorates the hyperactivity of *coloboma* mice, suggesting that this gene is central to the phenotype (Hess et al. 1996). Further investigations of *coloboma* mice revealed other behaviors that resemble ADHD.

Latent inhibition – a measure of a subject's ability to attend to and adjust behaviors for relevant and irrelevant stimuli – is a measure of attention in species ranging from the mouse to the human. Both ADHD patients and *coloboma* mice exhibit a disruption of latent inhibition, indicating inattention (Lubow and Josman 1993; Bruno et al. 2007).

Delayed reward paradigms can be used to assess impulsivity in humans and mice. Such paradigms require subjects to choose between two rewards: an immediately available but less desirable small reward or a delayed greater reward. Unlike control littermates, *coloboma* mice are unable to wait for the delayed greater reward, suggesting impulsivity (Bruno et al. 2007). Behavioral characterization of *coloboma* mice has revealed all three signs of ADHD – hyperactivity, inattention, and impulsivity – giving this model face validity as a model of ADHD (Bruno et al. 2007).

Coloboma mice exhibit a mixed response to psychostimulants. Amphetamine reduces hyperactivity in *coloboma* mice, but the same doses increase locomotor activity in normal mice. In contrast, methylphenidate increases locomotor activity in both *coloboma* and normal mice (Hess et al. 1996). A differential response to amphetamine *versus* methylphenidate is also observed in a subset of ADHD patients (Elia et al. 1991; Efron et al. 1997). Genetic studies revealed similarities between the *coloboma* mouse model and ADHD. Identification of the *Snap25* gene defect in these mice provided a candidate gene for human linkage studies that consistently demonstrate an association between *SNAP25* and ADHD (Barr et al. 2000; Mill et al. 2002; Kustanovich et al. 2003). Therefore, in addition to face validity, *coloboma* mice have predictive and etiological validity as a model of ADHD.

3.2 Monoaminergic Regulation

3.2.1 Dopamine

Studies of the neurochemistry of *coloboma* mice revealed no changes in dopamine concentrations of *coloboma* mice compared to wild-type mice (Jones et al. 2001a). However, a study of transmitter release in vitro revealed that K^+-evoked dopamine release from synaptosomes is decreased in the dorsal striatum of *coloboma* mice, relative to control mice (Raber et al. 1997). In contrast, basal extracellular dopamine as assessed by microdialysis in the striatum of alert, freely moving *coloboma* mice is nearly twice that of normal mice (Fan and Hess 2007). The discrepancy between in vitro and in vivo measures of striatal dopamine release in *coloboma* mice might be explained by the loss of feedback loops in vitro, where the direct biochemical consequence of the SNAP-25 deficit would be reflected in synaptosomes by a reduction in dopamine release. These release studies suggest that increased dopamine efflux in *coloboma* mice, compared to control mice, may play an important role in their hyperactivity, just as increased dopaminergic tone causes hyperactivity in normal rodents and humans.

Receptor agonists and antagonists have also been used to determine the role of dopaminergic signaling in the *coloboma* mouse phenotype. *Coloboma* mice show behavioral supersensitivity to a D_2 dopamine receptor agonist challenge, whereby *coloboma* mice respond to the D_2-like dopamine receptor agonist, quinpirole, with a greater reduction in locomotor activity than control mice. In contrast, the *coloboma* response to D_2 dopamine receptor antagonist is attenuated. By creating *coloboma* mice that lacked the D_2, D_3, or D_4 dopamine receptor, Fan et al. (2010) discovered that the D_2 dopamine receptor subtype mediates the hyperactivity and elevates extracellular dopamine concentration in *coloboma* mice. The contribution of D_3 and D_4 dopamine receptors is minor, demonstrating that a single dopamine receptor – the D_2 dopamine receptor – plays a central role in the hyperactivity of this model (Fan et al. 2010).

3.2.2 Norepinephrine

Coloboma mice show an increase in the tissue concentrations of norepinephrine in the dorsal striatum and nucleus accumbens (Jones et al. 2001a; Jones and Hess 2003). Consistent with these findings, tyrosine hydroxylase (TH) mRNA levels are

83% higher in the locus coeruleus of *coloboma* mice compared to control mice. Furthermore, depletion of the excess tissue concentrations of norepinephrine, using the selective noradrenergic neurotoxin DSP-4, reduces locomotor activity and ameliorates inattention in *coloboma* mice, suggesting that norepinephrine plays an important role in *coloboma* mouse behavior (Jones and Hess 2003; Bruno and Hess 2006).

Because *coloboma* mice show abnormalities in norepinephrine neurotransmission, they were challenged with various adrenergic agents to determine the role of adrenergic receptors in their hyperactivity. The nonselective α_2-adrenergic receptor antagonist, yohimbine, reduces locomotor activity in *coloboma* mice but not in wild-type mice. This is consistent with the reduction in locomotor activity observed after noradrenergic lesion and suggests that norepinephrine promotes hyperactivity of *coloboma* mice. Further investigation revealed that the selective α_{2C}-adrenergic receptor antagonist, MK-912, but not the α_{2A}-adrenergic receptor antagonist, BRL44208, or the α_{2B}-adrenergic receptor antagonist, ARC 239, dose-dependently inhibits hyperactivity in *coloboma* mice without affecting wild-type mice (Bruno and Hess 2006). These findings indicate that the α_{2C}-adrenergic receptor mediates hyperactivity in this model.

3.2.3 Serotonin

The contribution of serotonin to the *coloboma* mouse phenotype has not yet been studied in detail. Tissue concentrations of serotonin are normal in *coloboma* mice (Jones et al. 2001a). However, in vitro serotonin release is decreased in the dorsal striatum of *coloboma* mice, relative to control mice (Raber et al. 1997), suggesting that, like dopamine and norepinephrine, serotonin regulation is abnormal in these mice.

3.3 Therapeutic Mechanisms

Amphetamine reduces locomotor activity, but enhances striatal dopamine overflow, in *coloboma* mice. In fact, the magnitude of the increase in amphetamine-induced dopamine efflux is higher in *coloboma* mice than in wild-type mice (Fan and Hess 2007). Blocking D_2 dopamine receptors using the nonselective antagonists, haloperidol and raclopride, or the D_2 dopamine receptor-selective antagonist, L-741,626, prevents the amphetamine-mediated reduction in hyperactivity. Importantly, amphetamine-induced dopamine efflux is unaffected by blocking D_2 dopamine receptors with haloperidol, suggesting that this is not a D_2 dopamine autoreceptor-mediated effect. Instead, it is likely that postsynaptic D_2 dopamine receptors are involved. D_3 and D_4 dopamine receptor-selective antagonists have no effect on the amphetamine-induced reduction in locomotor activity in *coloboma* mice (Fan et al. 2010). Likewise, blocking D_1-like dopamine receptors with the antagonist, SCH23390, has no effect. It appears that the D_2 dopamine receptor subtype mediates both the hyperactivity and response to amphetamine, suggesting a target for therapeutics in ADHD.

4 Dopamine Transporter Knockout Mouse Model

4.1 Validity

In 1996, Giros and colleagues reported hyperactivity in dopamine transporter knockout (DAT-KO) mice. These mice express no DAT and are three to five times more active than control mice. Further studies demonstrated impaired learning ability and memory as well as impulsivity in DAT-KO mice (Gainetdinov et al. 1999; Li et al. 2010). As in ADHD patients, both methylphenidate and amphetamine reduce hyperactivity of DAT-KO mice (Gainetdinov et al. 1999). Therefore, these mice have face validity and predictive validity as a model of ADHD.

Mice that express low levels of DAT also exhibit hyperactivity. DAT knockdown mice (DAT-KD), which express only 10% of the DAT typically produced by this strain of mice, exhibit hyperactivity in a novel environment (Zhuang et al. 2001). Amphetamine reduces hyperactivity of DAT-KD mice; so the DAT-KD model, like the DAT-KO model, has face validity and predictive validity as a model of ADHD.

4.2 Monoaminergic Regulation

4.2.1 Dopamine

DAT-KO mice have reduced tyrosine hydroxylase (TH) mRNA expression, TH protein expression, and TH-positive neurons (Giros et al. 1996; Sora et al. 1998; Jaber et al. 1999). Consistent with decreased TH expression, dopamine content in the caudate and striatum is only about 5% of that in wild types (Jones et al. 1998). These results suggest a compensatory decrease in presynaptic dopamine transmission.

Because DAT-KO mice lack the dopamine reuptake transporter, synaptic dopamine clearance is about 300-fold slower than in control mice. Extracellular dopamine efflux in the striatum is increased by fivefold in homozygous DAT-KO mice and by twofold in heterozygous DAT-KO mice compared to wild-type mice (Gainetdinov et al. 1998; Jones et al. 1998). Despite the elevated dopamine in heterozygous mice, locomotor activity is essentially normal, suggesting that gross changes in extracellular dopamine are necessary to produce the hyperactivity (Giros et al. 1996; Spielewoy et al. 2000). Elevated basal dopamine efflux also occurs in the nucleus accumbens (Carboni et al. 2001), but not in the frontal cortex (Shen et al. 2004).

D_1 and D_2 dopamine receptor mRNA levels are about 50% lower in DAT-KO mice than in control mice (Giros et al. 1996; Jones et al. 1999). The reduction in D_2

dopamine receptor mRNA in the substantia nigra and VTA suggests that D_2 dopamine autoreceptor function is attenuated. Indeed, Jones et al. (1999) found that the dopamine autoreceptor, which provides inhibitory feedback to regulate dopamine release and synthesis, is not functional in DAT-KO mice, which may also contribute to elevated dopamine efflux in this model. While D_1 and D_2 dopamine receptor mRNA levels are decreased, D_3 dopamine receptor mRNA levels are increased by 40–100% in multiple brain regions (Fauchey et al. 2000). It is difficult to perform dopamine receptor agonist and antagonist behavioral challenges in DAT-KO mice because the extraordinarily high levels of extracellular dopamine compete with the drugs for binding to the receptor and so mask their effects. However, in DAT-KD mice, where the extracellular dopamine concentration is lower than in DAT-KO mice, D_2 dopamine receptors are supersensitive: the D_2 dopamine receptor agonist, quinpirole, and the mixed D_1/D_2 dopamine receptor agonist, apomorphine, cause a greater decrease in locomotor activity of DAT-KD mice than control mice (Zhuang et al. 2001).

4.2.2 Norepinephrine

Amphetamine increases dopamine efflux in the nucleus accumbens of DAT-KO mice, despite the lack of dopamine reuptake transporters (Carboni et al. 2001; Budygin et al. 2004). The selective norepinephrine transporter (NET) blocker, reboxetine, but not the selective DAT blocker, GBR12909, also increases extracellular dopamine in the nucleus accumbens (Carboni et al. 2001). These studies suggest that the NET participates in the regulation of dopamine efflux in DAT-KO mice in the nucleus accumbens, a brain region rich in noradrenergic terminals. Therefore, it is possible that amphetamine acts through the NET in DAT-KO mice to mediate extracellular dopamine.

4.2.3 Serotonin

Serotonin efflux is normal in the striatum, nucleus accumbens, and prefrontal cortex of DAT-KO mice (Shen et al. 2004). However, 5-HT_{2A} serotonin receptors may be involved in the hyperactivity of DAT-KO mice: The 5-HT_{2A} serotonin receptor antagonist, M100907, blocks hyperactivity of DAT-KO mice but does not affect activity of control mice (Barr et al. 2004).

4.3 Therapeutic Mechanisms

Although both methylphenidate and amphetamine reduce locomotor activity of DAT-KO mice, extracellular striatal dopamine concentrations are not changed by methylphenidate or amphetamine treatment; this is not surprising in light of the high basal extracellular dopamine concentration. The results suggest that psychostimulants modulate locomotor activity through a non-dopaminergic mechanism in this model. The selective serotonin reuptake transporter (SERT) blocker, fluoxetine, but not the NET blocker, nisoxetine, significantly reduces hyperactivity of DAT-KO mice without affecting locomotor activity of wild-type mice. Furthermore, both quipazine, a nonselective serotonin receptor agonist, and an increase in the endogenous agonist serotonin through dietary manipulation also reduced hyperactivity in DAT-KO mice (Gainetdinov et al. 1999). These data suggest that potentiating serotonin neurotransmission reduces locomotor activity of DAT-KO mice, and that the effects of psychostimulants may be mediated through serotonin neurotransmission, although selective serotonin reuptake inhibitors are not commonly used for the treatment of ADHD.

5 Spontaneously Hypertensive Rat Model

5.1 Validity

The SHR model is the subject of a comprehensive analysis in Chap. 10 of this volume. Information that will help the reader to compare and contrast the models is included here. In the 1990s, Sagvolden and colleagues proposed the use of SHRs as a model of ADHD because they are hyperactive (Wultz et al. 1990; Sagvolden et al. 1992) compared to the Wistar Kiyoto (WKY) control rat strain. Subsequent studies demonstrated inattention and impulsivity in SHRs using a variety of behavioral tests (Evenden and Meyerson 1999; De Bruin et al. 2003; Jentsch 2005; Bizot et al. 2007; Fox et al. 2008). Because SHRs exhibit hyperactivity, inattention, and impulsivity, they have face validity as a model of ADHD. Drugs used clinically to treat ADHD, including psychostimulants and guanfacine, an α_2-adrenoceptor agonist, also ameliorate the behavioral deficits in the SHR model, giving this model predictive validity as a model of ADHD (Myers et al. 1982; Sagvolden 2006).

5.2 Monoaminergic Regulation

5.2.1 Dopamine

In SHRs, TH expression is reduced in the neostriatum and nucleus accumbens (Akiyama et al. 1992), although striatal dopamine content and dopamine metabolites do not differ between SHRs and control WKY rats (Fuller et al. 1983). Electrically and/or K^+-stimulated dopamine release is decreased in the caudate putamen, nucleus accumbens, striatum, and frontal cortex of SHRs compared to WKY rats (Linthorst et al. 1990; van den Buuse et al. 1991; de Villiers et al. 1995; Russell et al. 1998;

Russell 2000; Yousfi-Alaoui et al. 2001). In parallel with the reduction in stimulated release, dopamine uptake is decreased in the frontal cortex and striatum of SHRs. While some studies find no differences in basal extracellular dopamine concentration in SHRs compared to WKY rats across different ages (Kirouac and Ganguly 1995; Ferguson et al. 2003), others report decreased basal extracellular dopamine in the striatum, caudate nucleus, and nucleus accumbens in SHRs compared to WKY rats at 8–9 weeks of age (Linthorst et al. 1991; Fujita et al. 2003). Along these same lines, both in vitro studies using slice preparations and in vivo microdialysis studies suggest an increased influence of D_2 dopamine autoreceptor-mediated inhibition of dopamine release in SHRs compared to control WKY rats (Linthorst et al. 1991; van den Buuse et al. 1991; Russell 2000). However, one study examining perfused synaptosomes did not find any difference in D_2 dopamine receptor-mediated inhibition of dopamine release in SHRs *versus* WKY rats (Yousfi-Alaoui et al. 2001). Overall, these studies suggest hypodopaminergic function in the SHR model.

The results of more recent studies, on the other hand, suggest an increase in dopaminergic transmission. Because SHRs develop hypertension with age, a study that was performed using 6-week-old hyperactive SHRs, which do not yet have complications associated with hypertension, demonstrated increased dopamine overflow in the shell of the nucleus accumbens in SHRs compared to control rats (Carboni et al. 2003). Similarly, using more modern methods of microdialysis, Heal et al. (2008) also found increased striatal dopamine overflow in SHRs compared to control rats, although in these experiments the controls were Sprague Dawleys, not WKY rats; so the results of these studies are not directly comparable to previous experiments.

While some studies indicate no differences in D_1 and D_2 dopamine receptor expression in SHRs compared to control WKY rats (Fuller et al. 1983; Van den Buuse et al. 1992; Linthorst et al. 1993), many studies indicate upregulation of D_1 and D_2 dopamine receptors in several brain areas of SHRs, including the nucleus accumbens, striatum, and frontal cortex (Chiu et al. 1982, 1984; Le Fur et al. 1983; Lim et al. 1989; Kirouac and Ganguly 1993; Sadile 2000; Papa et al. 2002), again suggesting an increase in dopaminergic tone.

Studies of the behavioral effects of D_1 and D_2 dopamine receptor agonists and antagonists also indicate functional changes in dopamine receptors in SHRs. The D_2 dopamine receptor antagonists, haloperidol and sulpiride, had little effect on SHRs but suppressed motor activity in WKY control rats. The reduced response to D_2 dopamine receptor antagonists suggests that D_2 dopamine receptors are supersensitive in SHRs (van den Buuse and de Jong 1989; Van den Buuse et al. 1992). In contrast, low doses of D_2 dopamine receptor agonists decreased locomotor activity in both SHRs and WKY rats, but high doses of agonist increased locomotor activity only in WKY rats (Fuller et al. 1983; Hynes et al. 1985). In these studies, the stimulatory effect of D_2 dopamine receptor activation on locomotor activity appears to be attenuated in SHRs, but these results must be interpreted in the context of the high baseline activity of SHRs: an increase in locomotor activity may be obscured by a ceiling effect. In a study of grooming behavior in a novel environment,

Linthorst et al. (1992) demonstrated that SHRs are less sensitive to D_1 dopamine receptor antagonist SCH23390 and more sensitive to D_2 dopamine receptor agonist quinpirole than WKY rats, suggesting both D_1 and D_2 dopamine receptor supersensitivity.

5.2.2 Norepinephrine

Because hypertension in SHRs may also influence noradrenergic signaling, changes in noradrenergic neurotransmission in SHRs must be interpreted cautiously in the context of informing the underlying mechanisms of ADHD. In vitro, basal norepinephrine release in the locus coeruleus is increased in SHRs compared to WKY rats (de Villiers et al. 1995), but stimulus-evoked release does not differ between the groups (Russell et al. 2000). Furthermore, inhibition of K^+-evoked norepinephrine release by α_2-adrenergic receptor agonists is decreased, suggesting reduced autoreceptor inhibitory feedback and, therefore, hypernoradrenergic function (Tsuda et al. 1990; Russell 2002). Norepinephrine content is increased in lower brainstem, midbrain, cerebellum, and striatum of SHRs at 6–8 weeks old (Howes et al. 1984; Dawson et al. 1987). However, others found that norepinephrine reuptake is increased (Myers et al. 1981), and in vivo norepinephrine efflux is decreased in the frontal cortex (Heal et al. 2008) of SHRs compared to control rats, suggesting a reduction in norepinephrine transmission.

Adrenergic receptor regulation is also altered in SHRs. α_1-Adrenergic receptor density is increased in the hypothalamus (Yamada et al. 1989). α_2-Adrenergic receptor binding is increased in the locus coeruleus (Luque et al. 1991) but decreased in the medulla oblongata (Yamada et al. 1989). β-Adrenergic receptors are decreased in the cortex (Myers et al. 1981) but increased in the cerebellum (Jones et al. 1990). These changes in adrenoceptor expression may contribute to the spontaneous hypertension or may be involved in ADHD-like symptoms exhibited by SHRs. Overall, it is not clear whether norepinephrine transmission is hyper- or hypofunctional in SHRs.

5.2.3 Serotonin

Serotonergic neurotransmission has not yet been the subject of focused experimentation in SHRs. Although striatal serotonin content does not differ between SHRs and WKY rats (Fuller et al. 1983), serotonin content is increased in frontal cortex, cerebellum, midbrain, locus coeruleus, and hypothalamus (Dawson et al. 1987; de Villiers et al. 1995), suggesting hyperfunctional serotonergic systems in some brain areas of SHRs. Clearly, both noradrenergic and serotonergic regulations are altered in the SHRs model, but the role of these changes in the hyperactivity, inattention, and impulsivity of SHRs is as yet unclear.

5.3 Therapeutic Mechanisms

Compared to effects of psychostimulants on extracellular dopamine in control rats, both methylphenidate and amphetamine greatly augment dopamine release in SHRs (Carboni et al. 2003; Heal et al. 2008). However, little is known regarding mechanism of action in this model.

6 Neurokinin 1 Receptor Knockout Mice (NK1$^{-/-}$)

6.1 Validity

The NK1R$^{-/-}$ mouse is a promising new model that also implicates abnormal monoaminergic regulation in ADHD-like behaviors. De Felipe et al. (1998) produced null mutants for *Tacr1*, a gene that encodes the substance P-preferring tachykinin receptor NK1, to explore substance P-mediated nociception. It was only recently noted that these mice are hyperactive (Herpfer et al. 2005). Because this model is relatively young, there are not yet assessments of impulsivity and inattention. However, based on the hyperactivity alone, the NK1R$^{-/-}$ mice have face validity as a model of ADHD. Both amphetamine and methylphenidate reduce the hyperactivity of NK1R$^{-/-}$ mice (Yan et al. 2009), demonstrating predictive validity. Furthermore, one study found an association between polymorphisms in or near the *Tacr1* gene and ADHD in humans (Yan et al. 2009). This initial study suggests etiologic validity, although additional studies are needed to confirm the association.

6.2 Monoaminergic Regulation

6.2.1 Dopamine

Studies examining dopaminergic regulation in NK1R$^{-/-}$ mice using conventional microdialysis demonstrated reduced dopamine efflux in prefrontal cortex of NK1R$^{-/-}$ mice compared to wild-type mice but no difference in striatal dopamine efflux (Yan et al. 2009). Additional experiments examining both presynaptic and postsynaptic mechanisms will provide insight into overall dopaminergic tone.

6.2.2 Norepinephrine

An increase in extracellular norepinephrine is observed in the prefrontal cortex of anesthetized NK1R$^{-/-}$ mice compared to control mice (Herpfer et al. 2005; Fisher et al. 2007). Because the reuptake of norepinephrine is normal, it is likely that an increase in release accounts for the increase in extracellular norepinephrine. Indeed, it appears that somodendritic α_{2A}-adrenoceptors, which inhibit the firing rate of locus ceruleus neurons, are desensitized in NK1R$^{-/-}$ mice, consistent with the idea that the excess norepinephrine overflow is caused by an increase in release. It is not yet clear whether the increase in extracellular norepinephrine contributes to the hyperactivity of NK1R$^{-/-}$ mice.

6.2.3 Serotonin

An increase in the firing rate of serotonergic neurons in the dorsal raphe nucleus is observed in anesthetized NK1R$^{-/-}$ mice (Santarelli et al. 2001). Additionally, 5-HT$_{1A}$ serotonin receptor mRNA, receptor density, and the ability of the 5-HT$_{1A}$ serotonin receptor agonist ipsapirone to inhibit dorsal raphe neuronal firing rates are reduced in NKR1$^{-/-}$ mice (Froger et al. 2001). 5-HT$_{1A}$ serotonin receptor desensitization may contribute to the increase in dorsal raphe neuronal firing rates because these receptors normally function to inhibit firing rates. Despite the increase in activity, steady-state extracellular serotonin concentrations are unaffected. However, blockade of serotonin reuptake causes a greater increase in extracellular serotonin in NK1R$^{-/-}$ mice compared to control mice, suggesting that an increase in serotonin release in NK1R$^{-/-}$ mice is neutralized by a concomitant increase in serotonin reuptake (Froger et al. 2001).

6.3 Therapeutic Mechanisms

Although amphetamine reduces locomotor activity of NK1$^{-/-}$ mice, surprisingly the usual increase in striatal extracellular dopamine in response to amphetamine is abrogated. This appears to be a direct effect of the null mutation because pretreatment of normal mice with an NK1 receptor antagonist also abolishes the amphetamine-induced increase in dopamine efflux (Yan et al. 2010). Because there is no dopaminergic response to amphetamine, the mechanism underlying the amphetamine-induced decrease in locomotor activity is unknown, although it is likely that this drug is acting through other monoaminergic systems.

7 What Have We Learned from Animal Models of ADHD?

7.1 Animal Models of ADHD Exhibit Hyperdopaminergic Neurotransmission

Because psychostimulants, which increase catecholamine neurotransmission, have been the primary ADHD treatment for decades, clinicians and researchers

Table 1 Dopamine transmission in rodent models of ADHD

	6-OHDA-lesioned rat	Spontaneously hypertensive rat	*Coloboma* mouse	DAT-KO mouse
Tissue dopamine concentration	↓	↑ [Fc, Mb, Pons] / [Nac, St]	/	↓
Electrically or K^+-evoked DA overflow/release	↓	↓	↑	↓
Uptake/clearance	↓	↓ [2]	ND	↓
Extracellular DA in vivo	↓	↓ [Nac, Vst, Cn] / [Rst, St] ↑ [St, Nac.s]	↑	↑
Psychostimulant-induced DA efflux	↑	↑	↑	↑
D_1 dopamine receptor density	↑ [Cx] ↓ [Rns]	↑	ND	↓ (mRNA)
D_2 dopamine receptor density	↑	↑	↑ (mRNA)	↓
D_1 or D_2 dopamine receptor agonist-induced behavioral response	↑ D_1 ↑ D_2	↑ D_1 ↑ D_2	↑ D_2	↑ D_2 ↓ D_2 (autoreceptor)
D_1 or D_2 dopamine receptor antagonist-induced immobility	↓ D_1 ↓ D_2	↓ D_1 ↓ D_2	↓ D_2	↓ D_2

Cn caudate nucleus, *Cx* cerebral cortex, *Fc* frontal cortex, *Mb* middle brain, *Nac.s* shell of nucleus accumbens, *Nac* nucleus accumbens, *Rns* rostral neostriatum, *St* striatum, *Rst* rostral striatum, *Vst* ventral striatum, ↑ increase, ↓ decrease, / unchanged, *ND* no data
For details, see text

conjectured that hypodopaminergic function is the neurobiological mechanism underlying ADHD. However, a common finding from studies of animal models of ADHD is *hyper*dopaminergic function (Table 1).

A neurochemical sign of hyperdopaminergic function is elevated dopamine overflow. An increase in extracellular dopamine is observed in the dorsal striatum and/or nucleus accumbens of *coloboma* mice, DAT-KO mice, and young SHRs (Gainetdinov et al. 1998; Carboni et al. 2003; Fan and Hess 2007; Heal et al. 2008). Although an overall decrease in dopamine overflow is observed in neonatal 6-OHDA-lesioned rats (33–85% of control levels), the dopamine content in residual dopaminergic terminals is likely to be higher than in control rats, considering that more than 95% of dopamine terminals have degenerated (Castaneda et al. 1990; Herrera-Marschitz et al. 1994; Loupe et al. 2002; Nowak et al. 2007). This disproportionate decrease in dopamine efflux compared to terminal degeneration not only reveals a profound compensation but also suggests hyperdopaminergic transmission in attempt to maintain homeostasis in the residual dopamine terminals.

Changes in dopamine receptor expression are another indication of hyperdopaminergic function in animal models of ADHD. Dopamine receptor expression, particularly the D_2 dopamine receptor subtype, is increased in the 6-OHDA model, SHR, and *coloboma* mouse model, but not in the DAT-KO model, which exhibits extremely high dopamine efflux (Chiu et al. 1982, 1984; Le Fur et al. 1983;

Lim et al. 1989; Broaddus and Bennett 1990; Dewar et al. 1990, 1997; Kirouac and Ganguly 1993; Radja et al. 1993; Sadile 2000; Jones et al. 2001b).

Behavioral studies of animal models of ADHD are consistent with hyperdopaminergic function. For example, 6-OHDA-lesioned rats (Duncan et al. 1987; Johnson and Bruno 1990), *coloboma* mice (Fan and Hess 2007), SHRs (van den Buuse and de Jong 1989; Van den Buuse et al. 1992), and DAT-KO mice (Spielewoy et al. 2000) exhibit a decreased response to dopamine antagonist-induced immobility, suggesting increased dopaminergic neurotransmission in these models (see comments above). In addition, the 6-OHDA, *coloboma*, DAT-KO/KD, and SHR models all show enhanced behavioral responses to dopamine agonists, indicating dopamine receptor supersensitivity (Fuller et al. 1983; Breese et al. 1985; Hynes et al. 1985; Giros et al. 1996; Archer and Fredriksson 2007; Fan and Hess 2007). Considering that increasing dopamine neurotransmission in control animals increases locomotor activity, it makes sense that persistent hyperdopaminergic function in animal models of ADHD underlies hyperactivity.

7.2 Norepinephrine Transmission Plays a Dual Role in ADHD

While both dopamine and norepinephrine are known to regulate motor activity, attention, learning, and cognition, dopamine has been the focus of ADHD research. In the neonatal 6-OHDA-lesioned rat model, selective dopamine depletion is achieved by pretreatment with desipramine, which protects noradrenergic nerves. This model illustrates that dopamine depletion alone is sufficient to produce ADHD-like behaviors such as hyperactivity and inattention (Shaywitz et al. 1976a, b; Miller et al. 1981; Stachowiak et al. 1984).

Behavioral studies using noradrenergic drugs on animal models indicate that norepinephrine transmission does, indeed, affect ADHD symptoms, but the outcomes are mixed. Enhancing norepinephrine transmission by blocking the NET improves hyperactivity in neonatal 6-OHDA-lesioned rats and learning deficits in SHRs (Davids et al. 2002; Moran-Gates et al. 2005; Liu et al. 2008). In contrast, an increase in noradrenergic transmission is observed in both *coloboma* and $NK1R^{-/-}$ mice (Herpfer et al. 2005; Fisher et al. 2007; Jones and Hess 2003). Guanfacine, an α_2-adrenoceptor agonist that decreases norepinephrine transmission through the inhibitory feedback of autoreceptors, reduces hyperactivity in *coloboma* mice and alleviates hyperactivity, inattention, and impulsivity in SHRs (Bruno and Hess 2006; Sagvolden 2006), similar to its effects on ADHD patients. Furthermore, reducing noradrenergic hyperinnervation in *coloboma* mice or blocking α_{2C}-adrenoceptors also reduces the hyperactivity of *coloboma* mice (Jones and Hess 2003; Bruno et al. 2007).

From these studies, one cannot infer a causal relationship between increasing/decreasing norepinephrine neurotransmission and severity of ADHD symptoms. Instead, these studies suggest that norepinephrine has dual effects on ADHD-like behaviors. Depending on the adrenoceptor subtype and subsequent signal transduction pathway that is activated, norepinephrine signaling can either enhance or ameliorate

ADHD symptoms. The norepinephrine uptake inhibitor, atomoxetine, is effective in treating some – but not all – ADHD patients (Garnock-Jones and Keating 2010). The variable response of ADHD patients to atomoxetine may be due to increased activation of both symptom-enhancing and symptom-alleviating adrenoceptor subtypes.

7.3 Serotonin Transmission Plays an Inhibitory Role in ADHD

Currently no serotonergic medications are prescribed in the treatment of ADHD. However, due to "cross-talk" between monoamines, involvement of serotonin in ADHD cannot be excluded. Although the role of serotonin in impulsivity and inattention has not been explored in animal models of ADHD, many studies have examined the role of serotonin in hyperactivity. Some studies using animal models of ADHD implicate serotonin in producing hyperactivity – either directly through 5-HT$_{2A}$ and 5-HT$_{2C}$ serotonin receptor subtypes, or indirectly by modulating dopamine receptor-mediated hyperactivity (Kostrzewa et al. 1993; Barr et al. 2004; Bishop et al. 2004, 2005). In contrast, most studies using animal models of ADHD suggest that serotonin acts to compensate for aberrant dopamine and/or norepinephrine signaling. 6-OHDA-lesioned rats exhibit serotonergic hyperinnervation. In this model, elimination of serotonergic hyperinnervation by administration of the selective serotonergic toxin 5,7-DHT greatly potentiates hyperactivity (Kostrzewa et al. 1994). In both the 6-OHDA-lesioned rat and DAT-KO mouse models, an increase in serotonergic transmission via serotonin agonist m-CPP or SERT blocker, fluoxetine, greatly reduces hyperactivity (Kostrzewa et al. 1994; Gainetdinov et al. 1999; Brus et al. 2004). These studies indicate that serotonin signaling acts to inhibit ADHD symptoms.

7.4 Dopamine and Serotonin Receptors Contribute to the Therapeutic Effects of Psychostimulants

Animal models of ADHD have revealed that changes in all three monoaminergic systems contribute to ADHD. How do psychostimulants modify these systems to produce the improvements in hyperactivity, inattention, and impulsivity?

Studies of animal models indicate that activation of dopamine and serotonin receptors mediates the therapeutic effects of stimulants on ADHD symptoms. Blocking D$_2$ dopamine receptors eliminates the calming effect of amphetamine on *coloboma* mice (Fan et al. 2010), and blocking 5-HT$_2$ serotonin receptors similarly eliminates the therapeutic effects of amphetamine on 6-OHDA-lesioned rats (Heffner and Seiden 1982). On the other hand, potentiating serotonin transmission mimics the effects of psychostimulants in DAT-KO mice (Gainetdinov et al. 1999).

From these studies, it is clear that animal models of ADHD hold great promise for the identification of novel drug targets and therapeutics for the treatment of ADHD.

8 Summary

The neonatal 6-OHDA-lesioned rat, *coloboma* mouse, DAT-KO mouse, and SHR models are at the forefront of ADHD research. These models all bear a phenotypic resemblance to ADHD in that they express hyperactivity, inattention, and/or impulsivity. Compared to ADHD patients, these models also respond similarly to drug challenge and/or have similar genetic polymorphisms. The models illustrate the heterogeneity of ADHD in that they all express ADHD-like behaviors, yet the initial cause (chemical depletion or genetic abnormality) of the behaviors is different for each one.

Neurochemical and behavioral studies of the models indicate aberrations in monoaminergic neurotransmission compared to control animals. Together, they implicate hyperdopaminergic neurotransmission in ADHD symptoms, suggest that norepinephrine has a dual enhancing/inhibitory regulatory role in ADHD symptoms, and indicate that serotonin acts to inhibit abnormal dopamine and norepinephrine signaling. The models have also helped to determine that psychostimulants act through specific dopamine and serotonin receptors, which mediate the therapeutic effects in ADHD. It is unlikely that symptoms arise from a single neurotransmitter dysfunction. Rather, studies of animal models of ADHD suggest that symptoms develop through the complex interactions of monoaminergic neurotransmitter systems.

References

Akiyama K, Yabe K, Sutoo D (1992) Quantitative immunohistochemical distributions of tyrosine hydroxylase and calmodulin in the brains of spontaneously hypertensive rats. Kitasato Arch Exp Med 65:199–208

Archer T, Fredriksson A (2007) Behavioural supersensitivity following neonatal 6-hydroxydopamine: attenuation by MK-801. Neurotox Res 12:113–124

Archer T, Danysz W, Fredriksson A, Jonsson G, Luthman J, Sundstrom E, Teiling A (1988) Neonatal 6-hydroxydopamine-induced dopamine depletions: motor activity and performance in maze learning. Pharmacol Biochem Behav 31:357–364

Archer T, Palomo T, Fredriksson A (2002) Neonatal 6-hydroxydopamine-induced hypo/hyperactivity: blockade by dopamine reuptake inhibitors and effect of acute D-amphetamine. Neurotox Res 4:247–266

Avale ME, Falzone TL, Gelman DM, Low MJ, Grandy DK, Rubinstein M (2004a) The dopamine D4 receptor is essential for hyperactivity and impaired behavioral inhibition in a mouse model of attention deficit/hyperactivity disorder. Mol Psychiatry 9:718–726

Avale ME, Nemirovsky SI, Raisman-Vozari R, Rubinstein M (2004b) Elevated serotonin is involved in hyperactivity but not in the paradoxical effect of amphetamine in mice neonatally lesioned with 6-hydroxydopamine. J Neurosci Res 78:289–296

Barr CL, Feng Y, Wigg K, Bloom S, Roberts W, Malone M, Schachar R, Tannock R, Kennedy JL (2000) Identification of DNA variants in the SNAP-25 gene and linkage study of these polymorphisms and attention-deficit hyperactivity disorder. Mol Psychiatry 5:405–409

Barr AM, Lehmann-Masten V, Paulus M, Gainetdinov RR, Caron MG, Geyer MA (2004) The selective serotonin-2A receptor antagonist M100907 reverses behavioral deficits in dopamine transporter knockout mice. Neuropsychopharmacology 29:221–228

Basura GJ, Walker PD (1999) Serotonin 2A receptor mRNA levels in the neonatal dopamine-depleted rat striatum remain upregulated following suppression of serotonin hyperinnervation. Brain Res Dev Brain Res 116:111–117

Bishop C, Tessmer JL, Ullrich T, Rice KC, Walker PD (2004) Serotonin 5-HT2A receptors underlie increased motor behaviors induced in dopamine-depleted rats by intrastriatal 5-HT2A/2C agonism. J Pharmacol Exp Ther 310:687–694

Bishop C, Daut GS, Walker PD (2005) Serotonin 5-HT2A but not 5-HT2C receptor antagonism reduces hyperlocomotor activity induced in dopamine-depleted rats by striatal administration of the D1 agonist SKF 82958. Neuropharmacology 49:350–358

Bizot JC, Chenault N, Houze B, Herpin A, David S, Pothion S, Trovero F (2007) Methylphenidate reduces impulsive behaviour in juvenile Wistar rats, but not in adult Wistar, SHR and WKY rats. Psychopharmacology (Berl) 193:215–223

Breese GR, Napier TC, Mueller RA (1985) Dopamine agonist-induced locomotor activity in rats treated with 6-hydroxydopamine at differing ages: functional supersensitivity of D-1 dopamine receptors in neonatally lesioned rats. J Pharmacol Exp Ther 234:447–455

Broaddus WC, Bennett JP Jr (1990) Postnatal development of striatal dopamine function. II. Effects of neonatal 6-hydroxydopamine treatments on D1 and D2 receptors, adenylate cyclase activity and presynaptic dopamine function. Brain Res Dev Brain Res 52:273–277

Bruno KJ, Hess EJ (2006) The alpha(2C)-adrenergic receptor mediates hyperactivity of coloboma mice, a model of attention deficit hyperactivity disorder. Neurobiol Dis 23:679–688

Bruno JP, Jackson D, Zigmond MJ, Stricker EM (1987) Effect of dopamine-depleting brain lesions in rat pups: role of striatal serotonergic neurons in behavior. Behav Neurosci 101:806–811

Bruno KJ, Freet CS, Twining RC, Egami K, Grigson PS, Hess EJ (2007) Abnormal latent inhibition and impulsivity in coloboma mice, a model of ADHD. Neurobiol Dis 25:206–216

Brus R, Kostrzewa RM, Perry KW, Fuller RW (1994) Supersensitization of the oral response to SKF 38393 in neonatal 6-hydroxydopamine-lesioned rats is eliminated by neonatal 5,7-dihydroxytryptamine treatment. J Pharmacol Exp Ther 268:231–237

Brus R, Plech A, Kostrzewa RM (1995) Enhanced quinpirole response in rats lesioned neonatally with 5,7-dihydroxytryptamine. Pharmacol Biochem Behav 50:649–653

Brus R, Nowak P, Szkilnik R, Mikolajun U, Kostrzewa RM (2004) Serotoninergics attenuate hyperlocomotor activity in rats. Potential new therapeutic strategy for hyperactivity. Neurotox Res 6:317–325

Budygin EA, Brodie MS, Sotnikova TD, Mateo Y, John CE, Cyr M, Gainetdinov RR, Jones SR (2004) Dissociation of rewarding and dopamine transporter-mediated properties of amphetamine. Proc Natl Acad Sci U S A 101:7781–7786

Butcher SP, Liptrot J, Aburthnott GW (1991) Characterisation of methylphenidate and nomifensine induced dopamine release in rat striatum using in vivo brain microdialysis. Neurosci Lett 122:245–248

Carboni E, Spielewoy C, Vacca C, Nosten-Bertrand M, Giros B, Di Chiara G (2001) Cocaine and amphetamine increase extracellular dopamine in the nucleus accumbens of mice lacking the dopamine transporter gene. J Neurosci 21: RC141: 141–144

Carboni E, Silvagni A, Valentini V, Di Chiara G (2003) Effect of amphetamine, cocaine and depolarization by high potassium on extracellular dopamine in the nucleus accumbens shell of SHR rats. An in vivo microdyalisis study. Neurosci Biobehav Rev 27:653–659

Castaneda E, Whishaw IQ, Lermer L, Robinson TE (1990) Dopamine depletion in neonatal rats: effects on behavior and striatal dopamine release assessed by intracerebral microdialysis during adulthood. Brain Res 508:30–39

Chiu P, Rajakumar G, Chiu S, Kwan CY, Mishra RK (1982) Enhanced [3H]spiroperidol binding in striatum of spontaneously hypertensive rat (SHR). Eur J Pharmacol 82:243–244

Chiu P, Rajakumar G, Chiu S, Kwan CY, Mishra RK (1984) Differential changes in central serotonin and dopamine receptors in spontaneous hypertensive rats. Prog Neuropsychopharmacol Biol Psychiatry 8:665–668

Dal Bo G, Berube-Carriere N, Mendez JA, Leo D, Riad M, Descarries L, Levesque D, Trudeau LE (2008) Enhanced glutamatergic phenotype of mesencephalic dopamine neurons after neonatal 6-hydroxydopamine lesion. Neuroscience 156:59–70

Davids E, Zhang K, Kula NS, Tarazi FI, Baldessarini RJ (2002) Effects of norepinephrine and serotonin transporter inhibitors on hyperactivity induced by neonatal 6-hydroxydopamine lesioning in rats. J Pharmacol Exp Ther 301:1097–1102

Dawson R Jr, Nagahama S, Oparil S (1987) Yohimbine-induced alterations of monoamine metabolism in the spontaneously hypertensive rat of the Okamoto strain (SHR). II. The central nervous system (CNS). Brain Res Bull 19:525–534

De Bruin NM, Kiliaan AJ, De Wilde MC, Broersen LM (2003) Combined uridine and choline administration improves cognitive deficits in spontaneously hypertensive rats. Neurobiol Learn Mem 80:63–79

De Felipe C, Herrero JF, O'Brien JA, Palmer JA, Doyle CA, Smith AJH, Laird JMA, Belmonte C, Cervero F, Hunt SP (1998) Altered nociception, analgesia and aggression in mice lacking the receptor for substance P. Nature 392:394–397

de Villiers AS, Russell VA, Sagvolden T, Searson A, Jaffer A, Taljaard JJ (1995) Alpha 2-adrenoceptor mediated inhibition of [3H]dopamine release from nucleus accumbens slices and monoamine levels in a rat model for attention-deficit hyperactivity disorder. Neurochem Res 20:427–433

Dewar KM, Soghomonian JJ, Bruno JP, Descarries L, Reader TA (1990) Elevation of dopamine D2 but not D1 receptors in adult rat neostriatum after neonatal 6-hydroxydopamine denervation. Brain Res 536:287–296

Dewar KM, Paquet M, Reader TA (1997) Alterations in the turnover rate of dopamine D1 but not D2 receptors in the adult rat neostriatum after a neonatal dopamine denervation. Neurochem Int 30:613–621

Duncan GE, Criswell HE, McCown TJ, Paul IA, Mueller RA, Breese GR (1987) Behavioral and neurochemical responses to haloperidol and SCH-23390 in rats treated neonatally or as adults with 6-hydroxydopamine. J Pharmacol Exp Ther 243:1027–1034

Efron D, Jarman F, Barker M (1997) Methylphenidate versus dexamphetamine in children with attention deficit hyperactivity disorder: a double-blind, crossover trial. Pediatrics 100:E6

Elia J, Borcherding BG, Rapoport JL, Keysor CS (1991) Methylphenidate and dextroamphetamine treatments of hyperactivity: are there true nonresponders? Psychiatry Res 36:141–155

Erinoff L, MacPhail RC, Heller A, Seiden LS (1979) Age-dependent effects of 6-hydroxydopamine on locomotor activity in the rat. Brain Res 164:195–205

Evenden J, Meyerson B (1999) The behavior of spontaneously hypertensive and Wistar Kyoto rats under a paced fixed consecutive number schedule of reinforcement. Pharmacol Biochem Behav 63:71–82

Fan X, Hess EJ (2007) D2-like dopamine receptors mediate the response to amphetamine in a mouse model of ADHD. Neurobiol Dis 26:201–211

Fan X, Xu M, Hess EJ (2010) D2 dopamine receptor subtype-mediated hyperactivity and amphetamine responses in a model of ADHD. Neurobiol Dis 37:228–236

Fauchey V, Jaber M, Caron MG, Bloch B, Le Moine C (2000) Differential regulation of the dopamine D1, D2 and D3 receptor gene expression and changes in the phenotype of the striatal neurons in mice lacking the dopamine transporter. Eur J Neurosci 12:19–26

Ferguson SA, Gough BJ, Cada AM (2003) In vivo basal and amphetamine-induced striatal dopamine and metabolite levels are similar in the spontaneously hypertensive, Wistar-Kyoto and Sprague-Dawley male rats. Physiol Behav 80:109–114

Fisher AS, Stewart RJ, Yan TC, Hunt SP, Stanford SC (2007) Disruption of noradrenergic transmission and the behavioural response to a novel environment in NK1R$^{-/-}$ mice. Eur J Neurosci 25:1195–1204

Fox AT, Hand DJ, Reilly MP (2008) Impulsive choice in a rodent model of attention-deficit/hyperactivity disorder. Behav Brain Res 187:146–152

Froger N, Gardier AM, Moratalla R, Alberti I, Lena I, Boni C, De Felipe C, Rupniak NM, Hunt SP, Jacquot C, Hamon M, Lanfumey L (2001) 5-hydroxytryptamine (5-HT)1A autoreceptor adaptive changes in substance P (neurokinin 1) receptor knock-out mice mimic antidepressant-induced desensitization. J Neurosci 21:8188–8197

Fujita S, Okutsu H, Yamaguchi H, Nakamura S, Adachi K, Saigusa T, Koshikawa N (2003) Altered pre- and postsynaptic dopamine receptor functions in spontaneously hypertensive rat: an animal model of attention-deficit hyperactivity disorder. J Oral Sci 45:75–83

Fuller RW, Hemrick-Luecke SK, Wong DT, Pearson D, Threlkeld PG, Hynes MD III (1983) Altered behavioral response to a D2 agonist, LY141865, in spontaneously hypertensive rats exhibiting biochemical and endocrine responses similar to those in normotensive rats. J Pharmacol Exp Ther 227:354–359

Gainetdinov RR, Jones SR, Fumagalli F, Wightman RM, Caron MG (1998) Re-evaluation of the role of the dopamine transporter in dopamine system homeostasis. Brain Res Brain Res Rev 26:148–153

Gainetdinov RR, Wetsel WC, Jones SR, Levin ED, Jaber M, Caron MG (1999) Role of serotonin in the paradoxical calming effect of psychostimulants on hyperactivity. Science 283:397–401

Garnock-Jones KP, Keating GM (2010) Spotlight on atomoxetine in attention-deficit hyperactivity disorder in children and adolescents. CNS Drugs 24:85–88

Giros B, Jaber M, Jones SR, Wightman RM, Caron MG (1996) Hyperlocomotion and indifference to cocaine and amphetamine in mice lacking the dopamine transporter. Nature 379:606–612

Gong L, Kostrzewa RM, Fuller RW, Perry KW (1992) Supersensitization of the oral response to SKF 38393 in neonatal 6-OHDA-lesioned rats is mediated through a serotonin system. J Pharmacol Exp Ther 261:1000–1007

Gong L, Kostrzewa RM, Perry KW, Fuller RW (1993) Dose-related effects of a neonatal 6-OHDA lesion on SKF 38393- and m-chlorophenylpiperazine-induced oral activity responses of rats. Brain Res Dev Brain Res 76:233–238

Heal DJ, Smith SL, Kulkarni RS, Rowley HL (2008) New perspectives from microdialysis studies in freely-moving, spontaneously hypertensive rats on the pharmacology of drugs for the treatment of ADHD. Pharmacol Biochem Behav 90:184–197

Heffner TG, Seiden LS (1982) Possible involvement of serotonergic neurons in the reduction of locomotor hyperactivity caused by amphetamine in neonatal rats depleted of brain dopamine. Brain Res 244:81–90

Herpfer I, Hunt SP, Stanford SC (2005) A comparison of neurokinin 1 receptor knock-out (NK1$^{-/-}$) and wildtype mice: exploratory behaviour and extracellular noradrenaline concentration in the cerebral cortex of anaesthetized subjects. Neuropharmacology 48:706–719

Herrera-Marschitz M, Luthman J, Ferre S (1994) Unilateral neonatal intracerebroventricular 6-hydroxydopamine administration in rats: II. Effects on extracellular monoamine, acetylcholine and adenosine levels monitored with in vivo microdialysis. Psychopharmacology (Berl) 116:451–456

Hess EJ, Collins KA, Copeland NG, Jenkins NA, Wilson MC (1994) Deletion map of the coloboma (Cm) locus on mouse chromosome 2. Genomics 21:257–261

Hess EJ, Collins KA, Wilson MC (1996) Mouse model of hyperkinesis implicates SNAP-25 in behavioral regulation. J Neurosci 16:3104–3111

Howes LG, Rowe PR, Summers RJ, Louis WJ (1984) Age related changes of catecholamines and their metabolites in central nervous system regions of spontaneously hypertensive (SHR) and normotensive Wistar-Kyoto (WKY) rats. Clin Exp Hypertens A 6:2263–2277

Hynes MD, Langer DH, Hymson DL, Pearson DV, Fuller RW (1985) Differential effects of selected dopaminergic agents on locomotor activity in normotensive and spontaneously hypertensive rats. Pharmacol Biochem Behav 23:445–448

Ikegami M, Ichitani Y, Takahashi T, Iwasaki T (2006) Compensatory increase in extracellular dopamine in the nucleus accumbens of adult rats with neonatal 6-hydroxydopamine treatment. Nihon Shinkei Seishin Yakurigaku Zasshi 26:111–117

Jaber M, Dumartin B, Sagne C, Haycock JW, Roubert C, Giros B, Bloch B, Caron MG (1999) Differential regulation of tyrosine hydroxylase in the basal ganglia of mice lacking the dopamine transporter. Eur J Neurosci 11:3499–3511

Jackson D, Abercrombie ED (1992) In vivo neurochemical evaluation of striatal serotonergic hyperinnervation in rats depleted of dopamine at infancy. J Neurochem 58:890–897

Jentsch JD (2005) Impaired visuospatial divided attention in the spontaneously hypertensive rat. Behav Brain Res 157:323–330

Johnson BJ, Bruno JP (1990) D1 and D2 receptor contributions to ingestive and locomotor behavior are altered after dopamine depletions in neonatal rats. Neurosci Lett 118:120–123

Jones MD, Hess EJ (2003) Norepinephrine regulates locomotor hyperactivity in the mouse mutant coloboma. Pharmacol Biochem Behav 75:209–216

Jones CR, Palacios JM, Hoyer D, Buhler FR (1990) Receptor modification in the brains of spontaneously hypertensive and Wistar-Kyoto rats: regionally specific and selective increase in cerebellar beta 2-adrenoceptors. Br J Clin Pharmacol 30(Suppl 1):174S–177S

Jones SR, Gainetdinov RR, Wightman RM, Caron MG (1998) Mechanisms of amphetamine action revealed in mice lacking the dopamine transporter. J Neurosci 18:1979–1986

Jones SR, Gainetdinov RR, Hu XT, Cooper DC, Wightman RM, White FJ, Caron MG (1999) Loss of autoreceptor functions in mice lacking the dopamine transporter. Nat Neurosci 2:649–655

Jones MD, Williams ME, Hess EJ (2001a) Abnormal presynaptic catecholamine regulation in a hyperactive SNAP-25-deficient mouse mutant. Pharmacol Biochem Behav 68:669–676

Jones MD, Williams ME, Hess EJ (2001b) Expression of catecholaminergic mRNAs in the hyperactive mouse mutant coloboma. Brain Res Mol Brain Res 96:114–121

Kirouac GJ, Ganguly PK (1993) Up-regulation of dopamine receptors in the brain of the spontaneously hypertensive rat: an autoradiographic analysis. Neuroscience 52:135–141

Kirouac GJ, Ganguly PK (1995) Cholecystokinin-induced release of dopamine in the nucleus accumbens of the spontaneously hypertensive rat. Brain Res 689:245–253

Kostrzewa RM, Gong L, Brus R (1993) Serotonin (5-HT) systems mediate dopamine (DA) receptor supersensitivity. Acta Neurobiol Exp (Wars) 53:31–41

Kostrzewa RM, Brus R, Kalbfleisch JH, Perry KW, Fuller RW (1994) Proposed animal model of attention deficit hyperactivity disorder. Brain Res Bull 34:161–167

Kustanovich V, Merriman B, McGough J, McCracken JT, Smalley SL, Nelson SF (2003) Biased paternal transmission of SNAP-25 risk alleles in attention-deficit hyperactivity disorder. Mol Psychiatry 8:309–315

Le Fur G, Guilloux F, Uzan A (1983) Evidence for an increase in [3H]spiperone binding in hypothalamic nuclei during the development of spontaneous hypertension in the rat. Clin Exp Hypertens A 5:1537–1542

Li B, Arime Y, Hall FS, Uhl GR, Sora I (2010) Impaired spatial working memory and decreased frontal cortex BDNF protein level in dopamine transporter knockout mice. Eur J Pharmacol 628:104–107

Lim DK, Ito Y, Hoskins B, Rockhold RW, Ho IK (1989) Comparative studies of muscarinic and dopamine receptors in three strains of rat. Eur J Pharmacol 165:279–287

Linthorst AC, Van den Buuse M, De Jong W, Versteeg DH (1990) Electrically stimulated [3H] dopamine and [14C]acetylcholine release from nucleus caudatus slices: differences between spontaneously hypertensive rats and Wistar-Kyoto rats. Brain Res 509:266–272

Linthorst AC, De Lang H, De Jong W, Versteeg DH (1991) Effect of the dopamine D2 receptor agonist quinpirole on the in vivo release of dopamine in the caudate nucleus of hypertensive rats. Eur J Pharmacol 201:125–133

Linthorst AC, Broekhoven MH, De Jong W, Van Wimersma Greidanus TB, Versteeg DH (1992) Effect of SCH 23390 and quinpirole on novelty-induced grooming behaviour in spontaneously hypertensive rats and Wistar-Kyoto rats. Eur J Pharmacol 219:23–28

Linthorst AC, De Jong W, De Boer T, Versteeg DH (1993) Dopamine D1 and D2 receptors in the caudate nucleus of spontaneously hypertensive rats and normotensive Wistar-Kyoto rats. Brain Res 602:119–125

Liu LL, Yang J, Lei GF, Wang GJ, Wang YW, Sun RP (2008) Atomoxetine increases histamine release and improves learning deficits in an animal model of attention-deficit hyperactivity disorder: the spontaneously hypertensive rat. Basic Clin Pharmacol Toxicol 102:527–532

Loupe PS, Zhou X, Davies MI, Schroeder SR, Tessel RE, Lunte SM (2002) Fixed ratio discrimination training increases in vivo striatal dopamine in neonatal 6-OHDA-lesioned rats. Pharmacol Biochem Behav 74:61–71

Lubow RE, Josman ZE (1993) Latent inhibition deficits in hyperactive children. J Child Psychol Psychiatry 34:959–973

Luque JM, Guillamon A, Hwang BH (1991) Quantitative autoradiographic study on tyrosine hydroxylase mRNA with in situ hybridization and alpha 2 adrenergic receptor binding in the locus coeruleus of the spontaneously hypertensive rat. Neurosci Lett 131:163–166

Luthman J, Fredriksson A, Lewander T, Jonsson G, Archer T (1989a) Effects of d-amphetamine and methylphenidate on hyperactivity produced by neonatal 6-hydroxydopamine treatment. Psychopharmacology (Berl) 99:550–557

Luthman J, Fredriksson A, Sundstrom E, Jonsson G, Archer T (1989b) Selective lesion of central dopamine or noradrenaline neuron systems in the neonatal rat: motor behavior and monoamine alterations at adult stage. Behav Brain Res 33:267–277

Luthman J, Brodin E, Sundstrom E, Wiehager B (1990a) Studies on brain monoamine and neuropeptide systems after neonatal intracerebroventricular 6-hydroxydopamine treatment. Int J Dev Neurosci 8:549–560

Luthman J, Lindqvist E, Young D, Cowburn R (1990b) Neonatal dopamine lesion in the rat results in enhanced adenylate cyclase activity without altering dopamine receptor binding or dopamine- and adenosine 3':5'-monophosphate-regulated phosphoprotein (DARPP-32) immunoreactivity. Exp Brain Res 83:85–95

Luthman J, Friedemann M, Bickford P, Olson L, Hoffer BJ, Gerhardt GA (1993a) In vivo electrochemical measurements and electrophysiological studies of rat striatum following neonatal 6-hydroxydopamine treatment. Neuroscience 52:677–687

Luthman J, Friedemann MN, Hoffer BJ, Gerhardt GA (1993b) In vivo electrochemical measurements of exogenous dopamine clearance in normal and neonatal 6-hydroxydopamine-treated rat striatum. Exp Neurol 122:273–282

Luthman J, Lindqvist E, Ogren SO (1995) Hyperactivity in neonatally dopamine-lesioned rats requires residual activity in mesolimbic dopamine neurons. Pharmacol Biochem Behav 51:159–163

Luthman J, Friedemann MN, Hoffer BJ, Gerhardt GA (1997) In vivo electrochemical measurements of serotonin clearance in rat striatum: effects of neonatal 6-hydroxydopamine-induced serotonin hyperinnervation and serotonin uptake inhibitors. J Neural Transm 104:379–397

Masuo Y, Ishido M, Morita M, Oka S, Niki E (2004) Motor activity and gene expression in rats with neonatal 6-hydroxydopamine lesions. J Neurochem 91:9–19

Mill J, Curran S, Kent L, Gould A, Huckett L, Richards S, Taylor E, Asherson P (2002) Association study of a SNAP-25 microsatellite and attention deficit hyperactivity disorder. Am J Med Genet 114:269–271

Miller FE, Heffner TG, Kotake C, Seiden LS (1981) Magnitude and duration of hyperactivity following neonatal 6-hydroxydopamine is related to the extent of brain dopamine depletion. Brain Res 229:123–132

Molina-Holgado E, Dewar KM, Descarries L, Reader TA (1994) Altered dopamine and serotonin metabolism in the dopamine-denervated and serotonin-hyperinnervated neostriatum of adult rat after neonatal 6-hydroxydopamine. J Pharmacol Exp Ther 270:713–721

Moran-Gates T, Zhang K, Baldessarini RJ, Tarazi FI (2005) Atomoxetine blocks motor hyperactivity in neonatal 6-hydroxydopamine-lesioned rats: implications for treatment of attention-deficit hyperactivity disorder. Int J Neuropsychopharmacol 8:439–444

Myers MM, Whittemore SR, Hendley ED (1981) Changes in catecholamine neuronal uptake and receptor binding in the brains of spontaneously hypertensive rats (SHR). Brain Res 220:325–338

Myers MM, Musty RE, Hendley ED (1982) Attenuation of hyperactivity in the spontaneously hypertensive rat by amphetamine. Behav Neural Biol 34:42–54

Nowak P, Bortel A, Dabrowska J, Oswiecimska J, Drosik M, Kwiecinski A, Opara J, Kostrzewa RM, Brus R (2007) Amphetamine and mCPP effects on dopamine and serotonin striatal in vivo microdialysates in an animal model of hyperactivity. Neurotox Res 11:131–144

Oke AF, Adams RN (1978) Selective attention dysfunctions in adult rats neonatally treated with 6-hydoxydopamine. Pharmacol Biochem Behav 9:429–432

Papa M, Diewald L, Carey MP, Esposito FJ, Gironi Carnevale UA, Sadile AG (2002) A rostro-caudal dissociation in the dorsal and ventral striatum of the juvenile SHR suggests an anterior hypo- and a posterior hyperfunctioning mesocorticolimbic system. Behav Brain Res 130:171–179

Raber J, Mehta PP, Kreifeldt M, Parsons LH, Weiss F, Bloom FE, Wilson MC (1997) Coloboma hyperactive mutant mice exhibit regional and transmitter-specific deficits in neurotransmission. J Neurochem 68:176–186

Radja F, Descarries L, Dewar KM, Reader TA (1993) Serotonin 5-HT1 and 5-HT2 receptors in adult rat brain after neonatal destruction of nigrostriatal dopamine neurons: a quantitative autoradiographic study. Brain Res 606:273–285

Raison S, Weissmann D, Rousset C, Pujol JF, Descarries L (1995) Changes in steady-state levels of tryptophan hydroxylase protein in adult rat brain after neonatal 6-hydroxydopamine lesion. Neuroscience 67:463–475

Russell VA (2000) The nucleus accumbens motor-limbic interface of the spontaneously hypertensive rat as studied in vitro by the superfusion slice technique. Neurosci Biobehav Rev 24:133–136

Russell VA (2002) Hypodopaminergic and hypernoradrenergic activity in prefrontal cortex slices of an animal model for attention-deficit hyperactivity disorder – the spontaneously hypertensive rat. Behav Brain Res 130:191–196

Russell V, de Villiers A, Sagvolden T, Lamm M, Taljaard J (1998) Differences between electrically-, ritalin- and D-amphetamine-stimulated release of [3H]dopamine from brain slices suggest impaired vesicular storage of dopamine in an animal model of Attention-Deficit Hyperactivity Disorder. Behav Brain Res 94:163–171

Russell V, Allie S, Wiggins T (2000) Increased noradrenergic activity in prefrontal cortex slices of an animal model for attention-deficit hyperactivity disorder – the spontaneously hypertensive rat. Behav Brain Res 117:69–74

Sadile AG (2000) Multiple evidence of a segmental defect in the anterior forebrain of an animal model of hyperactivity and attention deficit. Neurosci Biobehav Rev 24:161–169

Sagvolden T (2006) The alpha-2A adrenoceptor agonist guanfacine improves sustained attention and reduces overactivity and impulsiveness in an animal model of Attention-Deficit/Hyperactivity Disorder (ADHD). Behav Brain Funct 2:41

Sagvolden T, Metzger MA, Schiorbeck HK, Rugland AL, Spinnangr I, Sagvolden G (1992) The spontaneously hypertensive rat (SHR) as an animal model of childhood hyperactivity (ADHD): changed reactivity to reinforcers and to psychomotor stimulants. Behav Neural Biol 58:103–112

Sagvolden T, Aase H, Zeiner P, Berger D (1998) Altered reinforcement mechanisms in attention-deficit/hyperactivity disorder. Behav Brain Res 94:61–71

Santarelli L, Gobbi G, Debs PC, Sibille ET, Blier P, Hen R, Heath MJ (2001) Genetic and pharmacological disruption of neurokinin 1 receptor function decreases anxiety-related behaviors and increases serotonergic function. Proc Natl Acad Sci USA 98:1912–1917

Shaywitz BA, Klopper JH, Yager RD, Gordon JW (1976a) Paradoxical response to amphetamine in developing rats treated with 6-hydroxydopamine. Nature 261:153–155

Shaywitz BA, Yager RD, Klopper JH (1976b) Selective brain dopamine depletion in developing rats: an experimental model of minimal brain dysfunction. Science 191:305–308

Shaywitz BA, Klopper JH, Gordon JW (1978) Methylphenidate in 6-hydroxydopamine-treated developing rat pups. Effects on activity and maze performance. Arch Neurol 35:463–469

Shen HW, Hagino Y, Kobayashi H, Shinohara-Tanaka K, Ikeda K, Yamamoto H, Yamamoto T, Lesch KP, Murphy DL, Hall FS, Uhl GR, Sora I (2004) Regional differences in extracellular dopamine and serotonin assessed by in vivo microdialysis in mice lacking dopamine and/or serotonin transporters. Neuropsychopharmacology 29:1790–1799

Snyder AM, Zigmond MJ, Lund RD (1986) Sprouting of serotoninergic afferents into striatum after dopamine-depleting lesions in infant rats: a retrograde transport and immunocytochemical study. J Comp Neurol 245:274–281

Sora I, Wichems C, Takahashi N, Li XF, Zeng Z, Revay R, Lesch KP, Murphy DL, Uhl GR (1998) Cocaine reward models: conditioned place preference can be established in dopamine- and in serotonin-transporter knockout mice. Proc Natl Acad Sci U S A 95:7699–7704

Spielewoy C, Roubert C, Hamon M, Nosten-Bertrand M, Betancur C, Giros B (2000) Behavioural disturbances associated with hyperdopaminergia in dopamine-transporter knockout mice. Behav Pharmacol 11:279–290

Stachowiak MK, Bruno JP, Snyder AM, Stricker EM, Zigmond MJ (1984) Apparent sprouting of striatal serotonergic terminals after dopamine-depleting brain lesions in neonatal rats. Brain Res 291:164–167

Sulzer D, Chen TK, Lau YY, Kristensen H, Rayport S, Ewing A (1995) Amphetamine redistributes dopamine from synaptic vesicles to the cytosol and promotes reverse transport. J Neurosci 15:4102–4108

Towle AC, Criswell HE, Maynard EH, Lauder JM, Joh TH, Mueller RA, Breese GR (1989) Serotonergic innervation of the rat caudate following a neonatal 6-hydroxydopamine lesion: an anatomical, biochemical and pharmacological study. Pharmacol Biochem Behav 34:367–374

Tsuda K, Tsuda S, Masuyama Y, Goldstein M (1990) Norepinephrine release and neuropeptide Y in medulla oblongata of spontaneously hypertensive rats. Hypertension 15:784–790

van den Buuse M, de Jong W (1989) Differential effects of dopaminergic drugs on open-field behavior of spontaneously hypertensive rats and normotensive Wistar-Kyoto rats. J Pharmacol Exp Ther 248:1189–1196

van den Buuse M, Linthorst AC, Versteeg DH, de Jong W (1991) Role of brain dopamine systems in the development of hypertension in the spontaneously hypertensive rat. Clin Exp Hypertens A 13:653–659

van den Buuse M, Jones CR, Wagner J (1992) Brain dopamine D-2 receptor mechanisms in spontaneously hypertensive rats. Brain Res Bull 28:289–297

Wultz B, Sagvolden T, Moser EI, Moser MB (1990) The spontaneously hypertensive rat as an animal model of attention-deficit hyperactivity disorder: effects of methylphenidate on exploratory behavior. Behav Neural Biol 53:88–102

Yamada S, Ashizawa N, Nakayama K, Tomita T, Hayashi E (1989) Decreased density of alpha 2-adrenoceptors in medulla oblongata of spontaneously hypertensive rats. J Cardiovasc Pharmacol 13:440–446

Yan TC, Hunt SP, Stanford SC (2009) Behavioural and neurochemical abnormalities in mice lacking functional tachykinin-1 (NK1) receptors: a model of attention deficit hyperactivity disorder. Neuropharmacology 57:627–635

Yan TC, McQuillin A, Thapar A, Asherson P, Hunt SP, Stanford SC, Gurling H (2010) NK1 (TACR1) receptor gene 'knockout' mouse phenotype predicts genetic association with ADHD. J Psychopharmacol 24:27–38

Yousfi-Alaoui MA, Hospital S, Garcia-Sanz A, Badia A, Clos MV (2001) Presynaptic modulation of K+-evoked [3H]dopamine release in striatal and frontal cortical synaptosomes of normotensive and spontaneous-hypertensive rats. Neurochem Res 26:1271–1275

Zhang K, Davids E, Tarazi FI, Baldessarini RJ (2002a) Serotonin transporter binding increases in caudate-putamen and nucleus accumbens after neonatal 6-hydroxydopamine lesions in rats: implications for motor hyperactivity. Brain Res Dev Brain Res 137:135–138

Zhang K, Tarazi FI, Davids E, Baldessarini RJ (2002b) Plasticity of dopamine D4 receptors in rat forebrain: temporal association with motor hyperactivity following neonatal 6-hydroxydopamine lesioning. Neuropsychopharmacology 26:625–633

Zhuang X, Oosting RS, Jones SR, Gainetdinov RR, Miller GW, Caron MG, Hen R (2001) Hyperactivity and impaired response habituation in hyperdopaminergic mice. Proc Natl Acad Sci U S A 98:1982–1987

Rat Models of ADHD

Terje Sagvolden and Espen Borgå Johansen

Contents

1 Introduction 303
2 Criteria for a Valid Animal Model of ADHD 304
   2.1 Behavioral Differences Among Strains of Rats 304
   2.2 Genetic Differences Among Strains 305
3 Applying Validity Criteria to Animal Research 307
   3.1 WKY Heterogeneity: SHR/NCrl and WKY/NCrl
         Versus WKY/NHsd Controls 307
   3.2 ADHD: Defining Features and Situational Factors 308
   3.3 Age and Development 309
4 Implications for Understanding ADHD 310
5 Conclusions 310
References 311

Abstract Showing that an animal is hyperactive is not sufficient for it to be accepted as a model of ADHD. Based on behavioral, genetic, and neurobiological data, the spontaneously hypertensive rat (SHR) obtained from Charles River, Germany, (SHR/NCrl) is at present the best-validated animal model of ADHD. One Wistar Kyoto substrain (WKY/NHsd), obtained from Harlan, UK, is its most appropriate control. Another WKY substrain (WKY/NCrl) obtained from Charles River, Germany, is inattentive, has distinctly different genetics and neurobiology, and provides a promising model for the predominantly inattentive subtype of ADHD (ADHD-I) if one wants to investigate categorical ADHD subtypes. In this case, also, the WKY/NHsd substrain should be used as control. Although

---

Note. Strain nomenclature is based on the Rat Genome Database (Twigger et al. 2007; Rat Genome Database 2008).

T. Sagvolden
Department of Physiology, Institute of Basic Medical Sciences, University of Oslo, NO-0317 Oslo, Norway
e-mail: terje.sagvolden@medisin.uio.no

E.B. Johansen (✉)
Akershus University College, PO Box 423, 2001 Lillestrøm, Norway
e-mail: EspenBorga.Johansen@hiak.no

other rat strains may behave like WKY/NHsd rats, neurobiological results indicate significant differences when compared to the WKY/NHsd substrain, making them less suitable as controls for the SHR/NCrl. Thus, there are no obvious behavioral differences among the various SHRs, but there are behavioral and neurobiological differences among the WKY strains. The use of WKY/NCrl, outbred Wistar, Sprague Dawley, or other rat strains as controls for SHR/NCrl may produce spurious neurobiological effects and erroneous conclusions. Finally, model data yield support to independent hyperactivity and inattention dimensions in ADHD behavior.

Keywords Animal models · Attention-deficit/hyperactivity disorder · Genetics · Neuroanatomy · Neurophysiology · Validation

Abbreviations

ADHD	Attention-deficit/hyperactivity disorder
ADHD-C	Attention-deficit/hyperactivity disorder combined subtype
ADHD-H	Attention-deficit/hyperactivity disorder predominantly hyperactive-impulsive subtype
ADHD-I	Attention-deficit/hyperactivity disorder predominantly inattentive subtype
DA/OlaHsd	Inbred rats from Harlan, UK
IMAGE	International multi-center ADHD gene (project)
LEW/NHsd	Lewis rats from Harlan, UK
PVG/Mol	Inbred hooded rats from Møllegaard Breeding Centre, Denmark
RT-PCR	Real-time polymerase chain reaction
SD/MolTac	Outbred Sprague Dawley rats from Møllegaard Breeding Centre, Denmark
SD/NTac (NTac:SD)	Taconic Sprague Dawley rats
SHR	Spontaneously hypertensive rat
SHR/N	Inbred SHR from NIH
SHR/NCrl	Inbred SHR from Charles River, Germany
SHR/NMol	Inbred SHR from Møllegaard Breeding Centre, Denmark
SNP	Single nucleotide polymorphism
SSLP	Simple sequence length polymorphisms
WH/HanTac (also known as: HanTac:WH)	Outbred Wistar Hannover GALAS rats from Taconic Europe
WHHA/Edh (now WKHA/N)	Inbred rat from a cross between SHR and WKY with selection for high spontaneous activity and low systolic blood pressure at the University of Vermont College of Medicine, USA

WHHT/Edh (now WKHT/N)	Inbred rat from a cross between SHR and WKY with selection for normal spontaneous activity and high systolic blood pressure at the University of Vermont College of Medicine, USA
Wistar/Mol	Outbred from Møllegaard Breeding Centre, Denmark
WKY/N	Inbred WKY from NIH, USA
WKY/NHsd	Inbred WKY from Harlan Europe, UK
WKY/NicoCrlf	Inbred WKY from Charles River, France
WKY/NMolTac (also known as: WKY/NMol)	WKY from Møllegaard Breeding Centre, Denmark

1 Introduction

Attention-deficit/hyperactivity disorder (ADHD) is a developmental disorder where all clinical criteria are behavioral. It is a heterogeneous disorder affecting about 5% of children (Faraone and Mick 2010), and its prevalence is similar in different cultures (Dwivedi and Banhatti 2005; Meyer et al. 2004; Rohde et al. 2005). The heterogeneity may be sorted along two independent behavioral dimensions: inattention and hyperactivity impulsiveness (Lahey and Willcutt 2010). DSM-IV (American Psychiatric Association 2000) attempts to reduce the heterogeneity by subdividing ADHD into three subtypes: the predominantly inattentive subtype (ADHD-I); the predominantly hyperactive-impulsive subtype (ADHD-H); and the combined subtype (ADHD-C). ADHD places the child at increased risk of school failure, juvenile delinquency, criminality, substance abuse, and HIV/AIDS as a consequence of sexual promiscuity and disregard for preventative measures (Barkley et al. 2004; Molina et al. 2002; Kahn et al. 2002).

There have been many attempts to explain the origins of ADHD symptoms. A learning-theory perspective is gaining ground for the case of ADHD-C. The dynamic developmental theory of ADHD (Johansen et al. 2002, 2009; Sagvolden et al. 2005a; Johnson et al. 2009; Sagvolden and Archer 1989) suggests that less efficient dopamine-mediated reinforcement processes and deficient extinction of previously reinforced behavior may explain behavioral changes that are often described as either poor "executive functions" (Tannock 1998) or "response disinhibition" (Barkley 1997). This learning-theory perspective predicts specific neuronal changes related to synaptic plasticity and long-term potentiation (LTP) (Sagvolden et al. 2005a).

A reinforcer is not defined in terms of previous events, but defined in terms of the behavioral changes that follow the reinforcer. For a reinforcer to alter behavior, events need to occur within a limited time-frame, but the duration of this time-frame also depends on attentional and memory variables. This is important both in basic laboratory research, where it is often overlooked, and in analysis of ADHD, which is associated with poor attention and memory (Martinussen et al. 2005; Willcutt et al. 2005).

Animal models are helpful in medical research (Sagvolden et al. 2009). There are many putative animal models of ADHD (Roessner et al. 2010; Pardey et al. 2009; Sagvolden et al. 2009; Vendruscolo et al. 2009; Sanabria and Killeen 2008; DasBanerjee et al. 2008; Heal et al. 2008; Kostrzewa et al. 2008). However, it is important to emphasize that the DSM-IV definition of ADHD does not say "always hyperactive." Thus, although several molecular and genetic manipulations may produce hyperactive animals (Vendruscolo et al. 2009; Ruocco et al. 2009; Yan et al. 2009; Dalley et al. 2009; Kostrzewa et al. 2008), hyperactivity alone is insufficient for the animal to qualify as a model of ADHD. It is important to consider whether children with ADHD would be hyperactive in a similar test or situation (Johansen et al. 2009).

This review concentrates on the best-validated animal model of ADHD: the spontaneously hypertensive rat (SHR) obtained from Charles River, Germany (SHR/NCrl) (Rat Genome Database 2008) (see the Abbreviations section) with the Wistar Kyoto rat, obtained from Harlan, UK (WKY/NHsd), as the reference strain in an animal model for ADHD-C. However, WKY rats obtained from Charles River, Germany (WKY/NCrl), are a promising model for the predominantly inattentive subtype of ADHD (ADHD-I) when the WKY/NHsd STRAIN is used as control. Use of both substrains as models of ADHD is potentially interesting even if ADHD is not regarded as separate subtypes, but as one disorder with the severity of symptoms varying along two independent dimensions: inattentiveness and hyperactivity impulsiveness.

2 Criteria for a Valid Animal Model of ADHD

Because the diagnosis of ADHD is based on behavior, the validation of animal models must also be based on behavior. If valid animal models were to be found, one would expect many of the same fundamental genetic and neurobiological alterations to be common in the human and the animal case. Thus, an ADHD animal model should mimic the fundamental behavioral characteristics of ADHD (face validity), conform to a theoretical rationale (construct validity), and predict correlates of ADHD in humans as regards behavior, genetics, and neuronal functions not shown previously in clinical settings (predictive validity) (Sagvolden 2000; Sagvolden et al. 2009). Although a variety of rat and mouse strains exhibit hyperactivity (Russell et al. 2005), few meet the complete set of criteria for model validation.

2.1 Behavioral Differences Among Strains of Rats

The SHR displays the major symptoms of ADHD (inattention, hyperactivity, and impulsivity) that, like ADHD, develop over time when reinforcers are infrequent

(Li et al. 2007; van den Bergh et al. 2006; Sagvolden 2000; Johansen et al. 2005b; Sagvolden et al. 1998, 2005b). As in children with ADHD (Sonuga-Barke et al. 1992), SHRs are more sensitive to delayed reinforcement (Johansen and Sagvolden 2005; Johansen et al. 2005b), consistent with a steepened delay-of-reinforcement gradient found in SHR relative to controls (Johansen et al. 2007). This means that a reinforcer has to be given immediately following the correct behavior to be efficient in the SHR, while reinforcers could be delayed somewhat in controls and still affect behavior. In addition, as in children with ADHD (Castellanos et al. 2005; Aase et al. 2006), there is increased intraindividual variability and variability in the individual SHR's behavior within the task, relative to controls (Perry et al. 2010a, b).

There is systematic overactivity, impulsiveness, and sustained attention deficit in the SHRs obtained from: NIH (SHR/N), the Møllegaard Breeding Centre, Denmark (SHR/NMol); Charles River, Italy (SHR/CrlIco); and Charles River, Germany (SHR/NCrl). By contrast (to these SHRs) neither the hypertensive WHHT/Edh nor the hyperactive WHHA/Edh substrains showed any systematic overactivity, impulsiveness, or sustained attention deficit, although the WHHA/Edh does appear to be overactive in fear-provoking open-field tests (Sagvolden et al. 2009).

The development of overactivity, impulsiveness, and sustained attention deficit in the SHRs appear to be poorly correlated [see Fig. 2 in (Sagvolden et al. 2005b)]. Medication affects these behaviors differently in the SHR (Sagvolden 2006; Sagvolden and Xu 2008). Thus, it may appear that inattention and overactivity-impulsiveness are two independent behavioral dimensions in the SHR just as they may be in children with ADHD (Lahey and Willcutt 2010).

Behaviorally, the WKY/NHsd, WKY/N, and the WKY/NMolTac are all normal in which these WKY substrains may not differ behaviorally from either WH/HanTac Wistar rats; SD/MolTac; SD/NTac Sprague Dawley rats; hooded PVG/Mol rats; outbred Wistar/Mol rats; or the offspring of DA/OlaHsd females, time-mated with LEW/NHsd Lewis males (Harlan, UK) (Sagvolden 2000; Sagvolden et al. 2009). However, the WKY/NHsd substrain is the preferred control on the basis of genetic and neurobiological considerations (see below).

2.2 Genetic Differences Among Strains

To investigate whether SHR/NCrl rats show changes in expression in systems relevant to ADHD, we (DasBanerjee et al. 2008) have analyzed ADHD candidate genes identified as a part of the International Multi-center ADHD Gene project (IMAGE), and their biological neighbors (collectively referred to as IMAGE genes) (Kuntsi et al. 2006). The IMAGE gene biological neighbors are defined as any gene that was part of the same gene or protein family as an IMAGE gene, or has a well-established direct relationship with an IMAGE gene.

The SHR/NCrl rats showed significant changes in a set of IMAGE genes: a number of these genes are relevant for a learning-theory perspective of ADHD-C. The dynamic developmental theory of ADHD (Johansen et al. 2009; Sagvolden et al. 2005a; Johnson et al. 2009; Sagvolden and Archer 1989) suggests that defective interactions between dopamine and glutamate alter synaptic plasticity and LTP. On a behavioral level, such a faulty interaction may give rise to less efficient dopamine-mediated reinforcement processes and deficient extinction of previously reinforced behavior, and these differences could explain both inattention and overactivity-impulsiveness associated with ADHD (Sagvolden et al. 2005a).

Some of these genes showed *decreased* expression across tissues in ~65-day-old SHR/NCrl rats compared with WKY/NHsd rats: these included the ionotropic glutamate NMDA-binding protein *(Grina)*, the NMDA-like 1A complex *(Grinl1a)*; the NR2D subunit *(Grin2d)*; the AMPA receptor subunit GluR-3 *(Gria3)*; the alpha stimulating, olfactory-type guanine nucleotide-binding protein *(Gnal/Golf)*; the norepinephrine transporter NET *(Slc6a2)*; calmodulin 3 *(Calm3)*; calcium/calmodulin-dependent protein kinases *Camk1*, *Camk2a*, and *Camk2g)*; synaptotagmin III *(Syt3)*; and syntaxin-binding protein 1 *(Stxbp1)*. Gnal *(Golf)* is coupled to the dopamine receptor, DRD1, and plays a major role in excitatory dopamine transmission in the striatum. Significant relationships have been observed between certain SNPs in *Gnal* and symptoms of inattention and hyperactivity/impulsivity in ADHD children (Laurin et al. 2008).

In contrast, other genes showed *increased* expression (mRNA) in the SHR/NCrl rats compared to WKY/NHsd rats: these included the AMPA receptor subunit Glu-R2 subunit *(Gria2)*; the NMDA subunits NR1 and NR2C *(Grin1* and *Grin2c)*; calcium/calmodulin-dependent protein kinase kinase 1 *(Camkk1)*; catechol-*O*-methyltransferase *(Comt)*; the dopamine transporter DAT1 *(Slc6a3)*; the dopamine receptor D1 interacting protein *(DRD1ip)*; the 5-hydroxytryptamine (serotonin) receptor *(Htr3b)*; the calmodulin-binding protein striatin *(Strn)*; syntaxin 11 *(Stx11)*; syntaxin 17 *(Stx17)*; nicotinic cholinergic alpha polypeptide 9 receptor *(Chrna9)*; mu opioid receptor 1 *(Oprm1)*; hairy and enhancer of split 6 *(Hes6)*; and aquaporin 3 *(Aqp3)*. A complete list of significantly altered genes is available in DasBanerjee et al. (2008).

Based on blood samples, no between-strain differences in DNA were observed for either the DRD2 or the DRD4 genes, suggesting that neither gene is likely to mediate the behavioral differences between the WKY and SHR strains. In contrast, WKY/SHR differences were observed in the third exon of DAT1. While these mutations do not result in direct amino acid changes to the DAT protein, it is possible that they mediate some other process that explains the differences in DAT expression and function in the two strains (Mill et al. 2005).

The dopamine receptor (DRD1)-interacting protein *(DRD1ip)*, calcyon, represents a brain-specific protein involved in DRD1/DRD5 receptor-mediated calcium signaling. In our data, the SHR/NCrl had a twofold increase in expression of calcyon mRNA compared with WKY/NHsd rats. This is in agreement with a recent study that examined calcyon mRNA expression in the frontal-striatal circuitry of 3-, 5-, and 10-week-old SHR and WKY rats (Heijtz et al. 2007). Such a changed expression of *DRD1ip* may indicate an underlying disruption of reinforcement processes mediated by dopamine (Schultz 2010).

A major function of dopaminergic transmission is to modulate fast, ionotropic synaptic transmission mediated by the neurotransmitter glutamate. Thus, the observed changes in gene expression for subunits of both AMPA and NMDA glutamatergic receptors may profoundly affect neuronal function. Electrophysiological studies revealed two potential consequences of such changes (Jensen et al. 2009). First, in male SHR/NCrl and WKY/NHsd rats, at postnatal day 28, the AMPA receptor-mediated transmission at the CA3-to-CA1 synapses was reduced in the stratum radiatum of the hippocampus. Second, the NMDAR containing *Grin2b* (aka *GluN2B*) subunits contributed substantially to induction of LTP in SHR/NCrl, but not in WKY/NHsd. In human ADHD, there is evidence for genetic polymorphism of both *Grin2a* and *Grin2b* subunits of the NMDA receptor (Turic et al. 2004; Dorval et al. 2007), which might mean that synaptic plasticity associated with learning, reinforcement, and extinction may be altered in ADHD individuals as well (Sagvolden et al. 2005a).

Human and animal data indicate that the mu opioid receptor 1 (*Oprm1*) is associated with substance abuse disorders (Berrendero et al. 2002; Zhang et al. 2006). Individuals with ADHD show strong substance dependence (Faraone et al. 2007). Thus, it is possible that substance dependence in ADHD may be modulated by *Oprm1*.

3 Applying Validity Criteria to Animal Research

A large number of studies support the use of SHR as the best animal model of ADHD. However, there are also researchers who question the validity of the SHR/NCrl model (Ferguson and Cada 2003; van den Bergh et al. 2006). This section highlights a few important factors that may have contributed to some of the inconsistencies in the literature regarding the value of SHR as an animal model of ADHD.

3.1 WKY Heterogeneity: SHR/NCrl and WKY/NCrl Versus WKY/NHsd Controls

From a genetic point of view, the best candidate for a control strain is the progenitor strain of SHR/NCrl: i.e., the WKY. However, the various WKY substrains are not equally suited to serve as controls due to genetic and behavioral differences. For instance, genome-wide analyses show that the WKY/NCrl rats are more similar to the SHR/NCrl than to the WKY/NHsd rats (Sagvolden et al. 2008). Behaviorally, WKY/NCrl rats are more similar to the WKY/NHsd strain in some tasks, but are more similar to SHR/NCrl in others. We will argue that the SHR/NCrl strain, with the WKY/NHsd substrain acting as controls, is the best animal model of ADHD-C if this subtype really exists, or ADHD with individually highly variable dimensions of inattention and overactivity (Perry et al. 2010a, b) in a dimensional view of ADHD (Lahey and Willcutt 2010).

The newly described genetic and behavioral changes in the WKY/NCrl make this a promising model of ADHD-I (Sagvolden et al. 2008) if subtypes of ADHD exist. Both the WKY/NCrl and SHR/NCrl strains are inattentive relative to Sprague Dawley and Wistar/HanTac controls strains. However, WKY/NCrl rats are neither hyperactive nor impulsive, like the SHR/NCrl rat (Sagvolden et al. 2008). It is conceivable; however, that inattention is a phenomenon by itself and not necessarily associated with ADHD. Then, the WKY/NCrl might not be a model of ADHD, but of some other disorder mainly associated with inattention.

Independent of whether or not the WKY/NCrl is a model of ADHD, the heterogeneity between the WKY substrains makes it imperative that researchers provide information about the substrain and breeder used in their studies to enable empirical findings to be adequately evaluated by others.

3.2 ADHD: Defining Features and Situational Factors

One issue that might lead to disagreement regarding the validity of SHR/NCrl as an animal model of ADHD is how findings are interpreted and extrapolated. A defining feature of ADHD-C and of ADHD-H is hyperactivity. However, the DSM-IV definition of ADHD does not say "always hyperactive," but includes statements like "have persisted for at least 6 months to a degree that is maladaptive and inconsistent with the developmental level" or "present in more than two or more settings." Some animal researchers seem to assume that ADHD implies persistent hyperactivity. Thus, if hyperactivity is not found in the animal model (in the specific test used in the present study), it is not a valid model of ADHD. These researchers fail to ask an additional, central question: "Are children with ADHD always hyperactive?" The answer to that question is "no" based on findings reported in the research literature, clinical experience, and reports from parents and teachers.

As in people with ADHD, the degree of behavioral problems in SHR depends on the task. Thus, the conclusion that a particular animal model is not valid for studies of ADHD, based on results from one test, only, may simply be incorrect. This point emphasizes the importance of good, reliable, translational tests that can be used in the animal model as well as in children with ADHD to test the correspondence between ADHD hyperactivity and hyperactivity in the animal model.

A second, related issue is the uncritical reliance on ADHD research literature when designing animal model studies. Such studies may refer to findings that report the presence of a particular behavioral change or cognitive deficit, which is then investigated in the animal model. Researchers may sometimes conclude that the results do not support continued use of an ADHD model, because a behavioral change or cognitive deficit that has been reported in the ADHD literature is absent in the animal model. However, many behavioral measures and cognitive concepts studied in ADHD, e.g., many aspects of "executive functions," are not defining features of the disorder. The literature on children diagnosed with ADHD is inconsistent regarding most of these cognitive or behavioral measures. Furthermore, if a clinician observes a child with all the symptoms of ADHD, but without the behavioral change or specific cognitive deficit in question, she or he would not automatically conclude that this child does not have ADHD. Thus, categorical conclusions on the validity of animal models based solely on one such measure may be erroneous.

3.3 Age and Development

The lack of a positive response to medication is a final issue that sometimes is used as an argument against the SHR/NCrl model of ADHD. As the greater majority of patients with ADHD *do* respond positively, an animal model of ADHD should do the same. However, a positive response to medication is not a defining feature of ADHD: up to one in five children diagnosed with ADHD will similarly not respond positively (Faraone and Buitelaar 2010).

Several studies find that psychostimulants improve symptoms of inattention, hyperactivity, and impulsivity in SHR/NCrl (Sagvolden et al. 1992; Wultz et al. 1990; Myers et al. 1982; Sagvolden and Xu 2008). When some researchers do not find ameliorating effects of medication in SHR/NCrl, it is important to consider whether the behavioral measures are improved by medication in children with ADHD. Furthermore, we may need to adopt a developmental perspective. The effect of psychostimulant treatment in young and adolescent individuals may not be the same as in adults; medication may interact with brain development and neuronal pruning to produce its effects (Shaw et al. 2009; Bizot et al. 2007).

In this developmental perspective, we examined the expression of genes involved in dopamine signaling and metabolism in the dorsal striatum and ventral mesencephalon of SHR/NCrl and WKY/NCrl, as well as three reference control strains (WKY/NHsd, WK/HanTac, and SD/NTac) using quantitative real-time RT-PCR. In addition, we determined striatal dopamine transporter (DAT) density, by ligand-binding assay, in the two ADHD-like strains at different developmental stages and after methylphenidate treatment. In adult rats, the mRNA expression of DAT and tyrosine hydroxylase was elevated in SHR/NCrl and WKY/NCrl rats compared to control strains: differences in DAT and tyrosine hydroxylation expression between SHR/NCrl and WKY/NCrl rats were also evident. During normal development, changes in striatal DAT densities occurred in both strains, with lower densities in WKY/NCrl than SHR/NCrl after postnatal day 25. Two weeks of methylphenidate treatment, during different developmental stages, was associated with decreased striatal DAT density in both rat strains compared to the non-treated rats with more pronounced effects followed by prepubertal treatment (Roessner et al. 2010).

Thus, use of old, hypertensive SHRs may potentially produce misleading results when studying SHR/NCrl as an animal model of ADHD. Hypertension can have deleterious effects on the brain function and produce spurious results. Studies of the SHR/NCrl model should preferably use young, prehypertensive animals to avoid this possible confound, although young adults with ADHD may be hypertensive as well as obese (Fuemmeler et al. 2010).

4 Implications for Understanding ADHD

The dynamic developmental theory of ADHD (Johansen et al. 2005a; Sagvolden et al. 2005a) suggests that reduced dopaminergic transmission changes fundamental behavioral selection mechanisms. This arises from deficient reinforcement of successful behavior, combined with deficient extinction (elimination) of unsuccessful behavior. In SHR/NCrl, neurobiological evidence for such factors is found both in the reduced dopamine efficacy (Sagvolden et al. 2009; Roessner et al. 2010) and in altered LTP in hippocampal slices (Jensen et al. 2009).

Such deficient selection mechanisms will slow the association ("chunking") of simple response units into longer, more elaborate chains of adaptive behavioral elements that function as higher-order behavioral units (Miller 1956; Aase and Sagvolden 2005; Aase et al. 2006; Perry et al. 2010a, b). Whenever behavioral units are chunked together into a chain of responses that is emitted in this context, each behavioral unit reliably precedes the next with high predictability. Consequently, deficient or slowed chunking of behavior will increase intraindividual variability. This is observed in children with ADHD and in the SHR (Aase and Sagvolden 2006; Johansen et al. 2009; Perry et al. 2010a, b).

5 Conclusions

There are no obvious behavioral differences among the various SHRs, but there are behavioral and neurobiological differences among the WKY strains. Several strains of rats may behave like WKY/NHsd rats; genetic studies indicate significant differences between various "normal" strains. Thus, Sprague Dawley rats may be a poor control for the SHR/NCrl, particularly in neurobiological studies. Given that the Wistar WH/HanTac rats and WKY/NCrl deviate both genetically and behaviorally from the WKY/NHsd, the use of these strains as controls for SHRs may produce spurious neurobiological differences. Thus, WKY/NHsd is the most appropriate control for SHR/NCrl. As a consequence, data may be misinterpreted if researchers or readers do not pay attention to the strain or substrain that was used in a study.

It is likely that lack of attention to such factors has led to erroneous conclusions in studies involving the SHR, WKY, and other comparison strains, in model studies of ADHD. The SHR/NCrl is the best-validated animal model of ADHD. Genetic and neurobiological data strengthen such a conclusion. Recent data suggest that the WKY/NCrl is inattentive, but it is unclear whether this substrain can be used as a model of ADHD.

The availability of validated ADHD animal models has substantial implications for research. Unlike some disorders, such as schizophrenia or bipolar disorder (for which there exist brain tissue resource centers), brain tissue is not available for ADHD patients. Animal models provide a source of such tissue for studies of gene expression, epigenetics, neuroanatomy, cellular neurophysiology, and other methods. Animal models of ADHD can also be used to search for ADHD genes using linkage or association analysis and to search for gene–environment interactions by exposing susceptible animals to environmental toxins (e.g., polychlorinated biphenyls) suspected to be risk factors for ADHD (DasBanerjee et al. 2008; Holene et al. 1998; Kuehn 2010). The SHR/NCrl is clearly useful for these.

Acknowledgment Financial support for the work described herein was mainly from the Research Council of Norway.

References

Aase H, Sagvolden T (2005) Moment-to-moment dynamics of ADHD behaviour. Behav Brain Funct 1:12

Aase H, Sagvolden T (2006) Infrequent, but not frequent, reinforcers produce more variable responding and deficient sustained attention in young children with attention-deficit/hyperactivity disorder (ADHD). J Child Psychol Psychiat 47:457–471

Aase H, Meyer A, Sagvolden T (2006) Moment-to-moment dynamics of ADHD behaviour in South African children. Behav Brain Funct 2:11

American Psychiatric Association (2000) Diagnostic and statistical manual of mental disorders: DSM-IV-TR. American Psychiatric Association, Washington, DC

Barkley RA (1997) Attention-deficit/hyperactivity disorder, self-regulation, and time: toward a more comprehensive theory. J Dev Behav Pediatr 18:271–279

Barkley RA, Fischer M, Smallish L, Fletcher K (2004) Young adult follow-up of hyperactive children: antisocial activities and drug use. J Child Psychol Psychiatry 45:195–211

Berrendero F, Kieffer BL, Maldonado R (2002) Attenuation of nicotine-induced antinociception, rewarding effects, and dependence in mu-opioid receptor knock-out mice. J Neurosci 22:10935–10940

Bizot JC, Chenault N, Houze B, Herpin A, David S, Pothion S, Trovero F (2007) Methylphenidate reduces impulsive behaviour in juvenile Wistar rats, but not in adult Wistar, SHR and WKY rats. Psychopharmacology 193:215–223

Castellanos FX, Sonuga-Barke EJ, Scheres A, Di Martino A, Hyde C, Walters JR (2005) Varieties of attention-deficit/hyperactivity disorder-related intra-individual variability. Biol Psychiatry 57:1416–1423

Dalley JW, Fryer TD, Aigbirhio FI, Brichard L, Richards HK, Hong YT, Baron JC, Everitt BJ, Robbins TW (2009) Modelling human drug abuse and addiction with dedicated small animal positron emission tomography. Neuropharmacol 56(Suppl 1):9–17

DasBanerjee T, Middleton FA, Berger DF, Lombardo JP, Sagvolden T, Faraone SV (2008) A comparison of molecular alterations in environmental and genetic rat models of ADHD: a pilot study. Am J Medical Genet B Neuropsychiatr Genet 147B:1554–1563

Dorval KM, Wigg KG, Crosbie J, Tannock R, Kennedy JL, Ickowicz A, Pathare T, Malone M, Schachar R, Barr CL (2007) Association of the glutamate receptor subunit gene GRIN2B with attention-deficit/hyperactivity disorder. Genes Brain Behav 6:444–452

Dwivedi KN, Banhatti RG (2005) Attention deficit/hyperactivity disorder and ethnicity. Arch Dis Child 90(Suppl 1):i10–i12

Faraone SV, Buitelaar J (2010) Comparing the efficacy of stimulants for ADHD in children and adolescents using meta-analysis. Eur Child Adolesc Psychiatry 19:353–364

Faraone SV, Mick E (2010) Molecular genetics of attention deficit hyperactivity disorder. Psychiatr Clin North Am 33:159–180

Faraone SV, Biederman J, Wilens TE, Adamson J (2007) A naturalistic study of the effects of pharmacotherapy on substance use disorders among ADHD adults. Psychol Med 37:1743–1752

Ferguson SA, Cada AM (2003) A longitudinal study of short- and long-term activity levels in male and female spontaneously hypertensive, Wistar-Kyoto, and Sprague-Dawley rats. Behav Neurosci 117:271–282

Fuemmeler BF, Ostbye T, Yang C, McClernon FJ, Kollins SH (2010) Association between attention-deficit/hyperactivity disorder symptoms and obesity and hypertension in early adulthood: a population-based study. Int J Obes. Oct 26. [Epub ahead of print]

Heal DJ, Smith SL, Kulkarni RS, Rowley HL (2008) New perspectives from microdialysis studies in freely-moving, spontaneously hypertensive rats on the pharmacology of drugs for the treatment of ADHD. Pharmacol Biochem Behav 90:184–197

Heijtz RD, Alexeyenko A, Castellanos FX (2007) Calcyon mRNA expression in the frontal-striatal circuitry and its relationship to vesicular processes and ADHD. Behav Brain Funct 3:33

Holene E, Nafstad I, Skaare JU, Sagvolden T (1998) Behavioural hyperactivity in rats following postnatal exposure to sub- toxic doses of polychlorinated biphenyl congeners 153 and 126. Behav Brain Res 94:213–224

Jensen V, Rinholm JE, Johansen TJ, Medin T, Storm-Mathisen J, Sagvolden T, Hvalby O, Bergersen LH (2009) N-methyl-d-aspartate receptor subunit dysfunction at hippocampal glutamatergic synapses in an animal model of attention-deficit/hyperactivity disorder. Neuroscience 158:353–364

Johansen EB, Sagvolden T (2005) Behavioral effects of intra-cranial self-stimulation in an animal model of attention-deficit/hyperactivity disorder (ADHD). Behav Brain Res 162:32–46

Johansen EB, Aase H, Meyer A, Sagvolden T (2002) Attention-deficit/hyperactivity disorder (ADHD) behaviour explained by dysfunctioning reinforcement and extinction processes. Behav Brain Res 130:37–45

Johansen EB, Sagvolden T, Aase H, Russell VA (2005a) Authors' response: the dynamic developmental theory of attention-deficit/hyperactivity disorder (ADHD): present status and future perspectives. Behav Brain Sci 28:451–468

Johansen EB, Sagvolden T, Kvande G (2005b) Effects of delayed reinforcers on the behavior of an animal model of attention-deficit/hyperactivity disorder (ADHD). Behav Brain Res 162:47–61

Johansen EB, Killeen PR, Sagvolden T (2007) Behavioral variability, elimination of responses, and delay-of-reinforcement gradients in SHR and WKY rats. Behav Brain Funct 3:60

Johansen EB, Killeen PR, Russell VA, Tripp G, Wickens JR, Tannock R, Williams J, Sagvolden T (2009) Origins of altered reinforcement effects in ADHD. Behav Brain Funct 5:7

Johnson KA, Wiersema JR, Kuntsi J (2009) What would Karl Popper say? Are current psychological theories of ADHD falsifiable? Behav Brain Funct 5:15

Kahn JA, Kaplowitz RA, Goodman E, Emans SJ (2002) The association between impulsiveness and sexual risk behaviors in adolescent and young adult women. J Adolesc Health 30:229–232

Kostrzewa RM, Kostrzewa JP, Kostrzewa RA, Nowak P, Brus R (2008) Pharmacological models of ADHD. J Neural Transm 115:287–298

Kuehn BM (2010) Increased risk of ADHD associated with early exposure to pesticides, PCBs. J Am Med Ass 304:27–28

Kuntsi J, Neale BM, Chen W, Faraone SV, Asherson P (2006) The IMAGE project: methodological issues for the molecular genetic analysis of ADHD. Behav Brain Funct 2:27

Lahey BB, Willcutt EG (2010) Predictive validity of a continuous alternative to nominal subtypes of attention-deficit/hyperactivity disorder for DSM-V. J Clin Child Adolesc Psychol 39:761–775

Laurin N, Ickowicz A, Pathare T, Malone M, Tannock R, Schachar R, Kennedy JL, Barr CL (2008) Investigation of the G protein subunit Galphaolf gene (GNAL) in attention deficit/hyperactivity disorder. J Psychiatr Res 42:117–124

Li Q, Lu G, Antonio GE, Mak YT, Rudd JA, Fan M, Yew DT (2007) The usefulness of the spontaneously hypertensive rat to model attention-deficit/hyperactivity disorder (ADHD) may be explained by the differential expression of dopamine-related genes in the brain. Neurochem Int 50:848–857

Martinussen R, Hayden J, Hogg-Johnson S, Tannock R (2005) A meta-analysis of working memory impairments in children with attention-deficit/hyperactivity disorder. J Am Acad Child Adolesc Psychiatry 44:377–384

Meyer A, Eilertsen DE, Sundet JM, Tshifularo JG, Sagvolden T (2004) Cross-cultural similarities in ADHD-like behaviour amongst South African primary school children. S Afr J Psychol 34:123–139

Mill J, Sagvolden T, Asherson P (2005) Sequence analysis of Drd2, Drd4, and Dat1 in SHR and WKY rat strains. Behav Brain Funct 1:24

Miller GA (1956) The magical number seven plus or minus two: some limits on our capacity for processing information. Psychol Rev 63:81–97

Molina BS, Bukstein OG, Lynch KG (2002) Attention-deficit/hyperactivity disorder and conduct disorder symptomatology in adolescents with alcohol use disorder. Psychol Addict Behav 16:161–164

Myers MM, Musty RE, Hendley ED (1982) Attenuation of hyperactivity in the spontaneously hypertensive rat by amphetamine. Behav Neural Biol 34:42–54

Pardey MC, Homewood J, Taylor A, Cornish JL (2009) Re-evaluation of an animal model for ADHD using a free-operant choice task. J Neurosci Methods 176:166–171

Perry GM, Sagvolden T, Faraone SV (2010a) Intra-individual variability in genetic and environmental models of attention-deficit/hyperactivity disorder. Am J Med Genet B Neuropsychiatr Genet 153B:1094–1101

Perry GM, Sagvolden T, Faraone SV (2010b) Intraindividual variability (IIV) in an animal model of ADHD – the Spontaneously Hypertensive Rat. Behav Brain Funct 6:56

Rat Genome Database (2008) http://rgd.mcw.edu

Roessner V, Sagvolden T, DasBanerjee T, Middleton FA, Faraone SV, Walaas SI, Becker A, Rothenberger A, Bock N (2010) Methylphenidate normalizes elevated dopamine transporter densities in an animal model of the attention-deficit/hyperactivity disorder combined type, but not to the same extent in one of the attention-deficit/hyperactivity disorder inattentive type. Neurosci 167:1183–1191

Rohde LA, Szobot C, Polanczyk G, Schmitz M, Martins S, Tramontina S (2005) Attention-deficit/hyperactivity disorder in a diverse culture: do research and clinical findings support the notion of a cultural construct for the disorder? Biol Psychiatr 57:1436–1441

Ruocco LA, Carnevale UA, Sadile AG, Sica A, Arra C, Di MA, Topo E, D'Aniello A (2009) Elevated forebrain excitatory l-glutamate, l-aspartate and d-aspartate in the Naples high-excitability rats. Behav Brain Res 198:24–28

Russell VA, Sagvolden T, Johansen EB (2005) Animal models of attention-deficit hyperactivity disorder. Behav Brain Funct 1:9

Sagvolden T (2000) Behavioral validation of the spontaneously hypertensive rat (SHR) as an animal model of attention-deficit/hyperactivity disorder (AD/HD). Neurosci Biobehav Rev 24:31–39

Sagvolden T (2006) The alpha-2A adrenoceptor agonist guanfacine improves sustained attention and reduces overactivity and impulsiveness in an animal model of Attention-Deficit/Hyperactivity Disorder (ADHD). Behav Brain Funct 2:41

Sagvolden T, Archer T (1989) Future perspectives on ADD research – an irresistible challenge. In: Sagvolden T, Archer T (eds) Attention deficit disorder: clinical and basic research. Lawrence Erlbaum Associates, Hillsdale, NJ, pp 369–389

Sagvolden T, Xu T (2008) l-Amphetamine improves poor sustained attention while d-amphetamine reduces overactivity and impulsiveness as well as improves sustained attention in an animal model of Attention-Deficit/Hyperactivity Disorder (ADHD). Behav Brain Funct 4:3

Sagvolden T, Metzger MA, Schiørbeck HK, Rugland AL, Spinnangr I, Sagvolden G (1992) The spontaneously hypertensive rat (SHR) as an animal model of childhood hyperactivity (ADHD): changed reactivity to reinforcers and to psychomotor stimulants. Behav Neural Biol 58:103–112

Sagvolden T, Aase H, Zeiner P, Berger DF (1998) Altered reinforcement mechanisms in Attention-Deficit/Hyperactivity Disorder. Behav Brain Res 94:61–71

Sagvolden T, Johansen EB, Aase H, Russell VA (2005a) A dynamic developmental theory of Attention-Deficit/Hyperactivity Disorder (ADHD) predominantly hyperactive/impulsive and combined subtypes. Behav Brain Sci 28:397–468

Sagvolden T, Russell VA, Aase H, Johansen EB, Farshbaf M (2005b) Rodent models of attention-deficit/hyperactivity disorder. Biol Psychiatr 57:1239–1247

Sagvolden T, DasBanerjee T, Zhang-James Y, Middleton F, Faraone S (2008) Behavioral and genetic evidence for a novel animal model of Attention-Deficit/Hyperactivity Disorder Predominantly Inattentive Subtype. Behav Brain Funct 4:56

Sagvolden T, Johansen EB, Woien G, Walaas SI, Storm-Mathisen J, Bergersen LH, Hvalby O, Jensen V, Aase H, Russell VA, Killeen PR, DasBanerjee T, Middleton FA, Faraone SV (2009) The spontaneously hypertensive rat model of ADHD–the importance of selecting the appropriate reference strain. Neuropharmacology 57:619–626

Sanabria F, Killeen PR (2008) Evidence for impulsivity in the Spontaneously Hypertensive Rat drawn from complementary response-withholding tasks. Behav Brain Funct 4:7

Schultz W (2010) Dopamine signals for reward value and risk: basic and recent data. Behav Brain Funct 6:24

Shaw P, Sharp WS, Morrison M, Eckstrand K, Greenstein DK, Clasen LS, Evans AC, Rapoport JL (2009) Psychostimulant treatment and the developing cortex in attention deficit hyperactivity disorder. Am J Psychiatr 166:58–63

Sonuga-Barke EJ, Taylor E, Sembi S, Smith J (1992) Hyperactivity and delay aversion–I. The effect of delay on choice. J Child Psychol Psychiatr 33:387–398

Tannock R (1998) Attention deficit hyperactivity disorder: advances in cognitive, neurobiological, and genetic research. J Child Psychol Psychiatr 39:65–99

Turic D, Langley K, Mills S, Stephens M, Lawson D, Govan C, Williams N, Van den BM, Craddock N, Kent L, Owen M, O'Donovan M, Thapar A (2004) Follow-up of genetic linkage findings on chromosome 16p13: evidence of association of N-methyl-D aspartate glutamate receptor 2A gene polymorphism with ADHD. Mol Psychiatr 9:169–173

Twigger SN, Shimoyama M, Bromberg S, Kwitek AE, Jacob HJ (2007) The rat genome database, update 2007–easing the path from disease to data and back again. Nucleic Acids Res 35: D658–D662

van den Bergh FS, Bloemarts E, Chan JS, Groenink L, Olivier B, Oosting RS (2006) Spontaneously hypertensive rats do not predict symptoms of attention-deficit hyperactivity disorder. Pharmacol Biochem Behav 83:380–390

Vendruscolo LF, Izidio GS, Takahashi RN (2009) Drug reinforcement in a rat model of attention deficit/hyperactivity disorder–the Spontaneously Hypertensive Rat (SHR). Curr Drug Abuse Rev 2:177–183

Willcutt EG, Doyle AE, Nigg JT, Faraone SV, Pennington BF (2005) Validity of the executive function theory of attention-deficit/hyperactivity disorder: a meta-analytic review. Biol Psychiatr 57:1336–1346

Wultz B, Sagvolden T, Moser EI, Moser MB (1990) The spontaneously hypertensive rat as an animal model of attention-deficit hyperactivity disorder: effects of methylphenidate on exploratory behavior. Behav Neural Biol 53:88–102

Yan TC, Hunt SP, Stanford SC (2009) Behavioural and neurochemical abnormalities in mice lacking functional tachykinin-1 (NK1) receptors: a model of attention deficit hyperactivity disorder. Neuropharmacology 57:627–635

Zhang L, Kendler KS, Chen X (2006) The mu-opioid receptor gene and smoking initiation and nicotine dependence. Behav Brain Funct 2:28

Epigenetics: Genetics Versus Life Experiences

Josephine Elia, Seth Laracy, Jeremy Allen, Jenelle Nissley-Tsiopinis, and Karin Borgmann-Winter

Contents

1 Introduction	318
2 General Processes Involved in Epigenetics	319
2.1 Chromosome Organization	319
2.2 DNA Methylation	320
2.3 Transcription Factors	321
3 Are Epigenetic Effects Relevant in ADHD?	321
4 ADHD and Epigenetic Processes	322
4.1 Imprinting	323
4.2 Gene–Environmental Interactions: Two-Hit Hypothesis	323
5 ADHD Medications and Gene Expression	326
5.1 ADHD Meds and Gene/Protein Expression	326
5.2 ADHD Medications and Immediate Early Genes	330
5.3 ADHD Medications, Gene Expression, and Adverse Effects	331
5.4 Gene Expression and Potential Novel Pathways in ADHD	331
6 Summary	331
References	332

Abstract Epigenetics is the field of research that examines alterations in gene expression caused by mechanisms other than changes in DNA sequence. ADHD is highly heritable; however, epigenetics are considered relevant in potentially explaining the variance not accounted for by genetic influence. In this chapter, some of the well-known processes of epigenetics, such as chromosome organization, DNA methylation, and effects of transcriptional factors are reviewed along with studies examining the role of these processes in the pathophysiology of ADHD. Potential epigenetic factors conferring risk for ADHD at various

J. Elia (✉), S. Laracy, J. Allen, J. Nissley-Tsiopinis, and K. Borgmann-Winter
The Children's Hospital of Philadelphia, Science Center, 3440 Market St, Philadelphia, PA 19104, USA
e-mail: elia@email.chop.edu

C. Stanford and R. Tannock (eds.), *Behavioral Neuroscience of Attention Deficit Hyperactivity Disorder and Its Treatment*, Current Topics in Behavioral Neurosciences 9, DOI 10.1007/7854_2011_144, © Springer-Verlag Berlin Heidelberg 2011, published online 5 July 2011

developmental stages, such as alcohol, tobacco, toxins, medications, and psychosocial stressor are discussed. Animal studies investigating ADHD medications and changes in CNS Gene/Protein Expression are also explored since they provide insight into the neuronal pathways involved in ADHD pathophysiology. The current limited data suggest that identification of the epigenetic processes involved in ADHD is extremely important and may lead to potential interventions that may be applied to modify the expression of deleterious, as well as protective, genes.

Keywords ADHD · DNA methylation · Epigenetics · Gene transcription · Gene–environment

Abbreviations

ADHD	Attention deficit hyperactivity disorder
alpha2A-AR	α_{2A}-Adrenoceptor
ApT	Adenine-thymine base pair
CD	Conduct disorder
CpG	Cytosine-guanine base pair
CREB	Cyclic nucleotide response element binding protein
DAT	Dopamine transporter
DES	Diethylstilbestrol
DZ	Dizygotic
GATA-1	Erythroid transcription factor
GC	Guanine-cytosine
hERG	Human Ether-a-go-go
IEG	Immediate early gene
IMAGE	International Multicentre ADHD Genetics Project
MAOB	Monoamine oxidase-B
MPH	Methylphenidate
MSN-D1'	Medium sized spiny neurons expressing dopamine D1 receptors
MZ	Monozygotic
ODD	Oppositional defiant disorder
VNTR	Variable number tandem repeat

1 Introduction

Epigenetics is a rapidly growing field of research that examines alterations in gene expression caused by mechanisms other than changes in DNA sequence. The molecular basis of epigenetics involves activation or repression of certain genes without changing the basic structure of the DNA. The most common example is the transformation of a pluripotent cell line in the embryo that leads to differentiated cells such as muscle, bone, and neurons. These epigenetic changes are preserved over cell division so that muscle cells continue to produce muscle cells and not bone cells (reviewed by: (Li 2002; Robertson 2005; Tsankova et al. 2007; Feinberg 2008). Most epigenetic changes are limited to one individual over the course of his/her lifetime, but if a mutation in the DNA has been caused, some epigenetic changes are inherited from one generation to the next (Chandler 2007).

Although DNA serves as a library of genes, determining which genes are expressed or silenced during different developmental stages falls under the control of various epigenetic processes. Some of the processes that determine whether genes are expressed or silenced are well known, such as DNA methylation (Robertson 2005) and transcription factors (Latchman 1997). Other, less well-defined, epigenetic processes in humans include paramutation (one allele can cause a heritable change in the expression of the homologous allele) (Hollick et al. 1997) and bookmarking (the phenotype of a lineage of cells is maintained such that a specific cell type continues to divide into the same cell type) (Sarge and Park-Sarge 2005). The effects of environmental factors, including teratogens, such as metals and air pollutants, and endocrine-disrupting chemicals, such as diethylstilbestrol (DES), on the epigenetic processes of DNA methylation, histone modifications, and microRNA are also being determined (Jirtle and Skinner 2007). Whether such modifications are transmitted transgenerationally in humans remains to be determined (Baccarelli and Bollati 2009). In this chapter, we review the general processes involved in epigenetics and explore their relevance to ADHD.

2 General Processes Involved in Epigenetics

2.1 Chromosome Organization

Chromatin consists of DNA, histone proteins, and non-histone proteins. As depicted in Figure 1, DNA is coiled around histone proteins that are organized in sets of eight (nucleosome). When chromatin is condensed, it is inactive (heterochromatin) and does not allow transcription of genes. Small groups of nucleosomes can open to an active state (euchromatin) as a result of various mechanisms including methylation, acetylation, phosphorylation, ubiquitylation, and SUMOylation (Berger 2007; Feinberg 2008). Some chromatin, where histones and DNA are methylated and bound to repressor proteins, are never accessible to transcription. Other portions of chromatin are in "repressed" or "permissive states" to facilitate critical processes, such as cell division, or to modulate the regulation of neural developmental maturation.

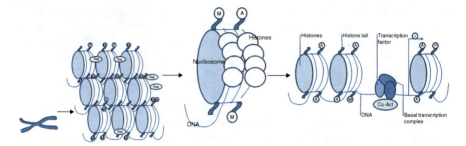

Fig. 1 DNA is tightly wrapped around the nucleosome, composed of 8 histone proteins. Methylation of histone proteins results in repression while acetylation results in a permissive state that facilitates gene expression. (Figure adapted from Tsankova et al. (2007))

2.2 DNA Methylation

The DNA library is composed of over 3 billion base pairs, adenine-thymine (ApT) and cytosine-guanine (CpT) bound by a phosphate. In about 2–7% of the CpG dinucloetide pairs, the cytosine is methylated (mCpG) and this usually, but not always (Wu et al. 2010), results in repression of transcription. In the past, it was thought that CpG dinucleotides in the unmethylated state were clustered (referred to as CpG islands) near transcription start sites (promoters) thus promoting gene transcription. However, unbiased genome-wide analyses have shown that DNA methylation is widespread across the euchromatic portion of mammalian genomes and predominantly takes place in regions outside proximal promoters, including intergenic regions and gene bodies. Wu and colleagues have shown that although methylation of DNA sequences in promoters tends to be repressive, methylation of DNA sequences beyond the promoters can actually promote gene expression (Wu et al. 2010).

The most dramatic example of transcriptional repression is seen in females where one of the X chromosomes is extensively methylated and rendered inactive (Goto and Monk 1998). DNA methylation can also occur in the parental germ line. A gene that is methylated in the female germ line, but not in the male germ line, results in an organism where the paternal gene is expressed in the somatic cells, whereas the maternal gene is silenced. The term used to describe a gene that is marked in some way, so that it remembers which parent it came from, is known as "imprinting" (Wood and Oakey 2006). During the development of the gametes, this methylation imprint is erased and re-established during oogenesis but not during spermatogenesis. The methylated genes inherited from one sex can be unmethylated when it is passed onto the offspring of the opposite sex so that these methylation imprints are re-set at each generation depending on the sex of the organism (Reik et al. 2001).

Cancer was the first human disease shown to result from epigenetic activation of tumor suppression and/or silencing of tumor-suppressor genes (Feinberg and

Tycko 2004). Some neuropsychiatric disorders that are examples of DNA methylation include Fragile X syndrome. This is caused by an abnormal expansion of CGG repeats in the FMR1 gene that make it unstable. This results in methylation of DNA and silencing of the gene that codes for the FMR protein leading to the most common form of mental retardation, autistic behaviors, macrocephaly, long and narrow face with large ears, macroorchidism, hypotonia (Levenson and Sweatt 2005; Lim et al. 2005) and ADHD (Baumgardner et al. 1995). In Rett Syndrome, the loss of developmental milestones has been attributed to abnormal gene expression in the brain caused by a lack of a normal MeCP2 protein that recognizes methylated DNA and helps to repress gene expression (Amir et al. 1999).

DNA methylation depends on dietary methionine and folate. Diets deficient in methionine have been reported to lead to DNA hypomethylation and higher rates of liver cancer in rats (Wilson et al. 1984). Hypomethylation of the agouti gene has also been shown to affect coat color in mice (Waterland and Jirtle 2003).

2.3 Transcription Factors

As reviewed by Latchman, for a gene to be expressed, DNA needs to be transcribed to RNA (Latchman 1997). The process, occurring in the nucleus of eukaryotic cells, is mediated by positive and negative regulatory proteins that bind to specific regions of the DNA and stimulate or inhibit transcription (Karin 1990). These groups of proteins are called transcription factors. Basal transcription factors are proteins that facilitate the proper alignment of RNA polymerase on the DNA template strand, whereas special transcription factors, such as those involved in the regulation of heat, light, and hormone inducible genes, bind to enhancers located in the vicinity of a gene (Thomas and Chiang 2006).

3 Are Epigenetic Effects Relevant in ADHD?

Twin studies attempting to distinguish between genetic and environmental influences in Attention Deficit Hyperactivity Disorder (ADHD) have reported heritability ranging between 30 and 90% (Kuntsi et al. 2005; Price et al. 2005; Saudino et al. 2005; Hay et al. 2007; Polderman et al. 2007; Derks et al. 2008; Mick and Faraone 2008; Ouellet-Morin et al. 2008; Wood et al. 2008). Epigenetics, therefore, would be relevant since the rest of the variance may be explained by nonadditive genetic influences, such as dominant effects (Derks et al. 2008), as well as shared and nonshared environmental influences (Hay et al. 2004, 2007; Larsson et al. 2004; Kuntsi et al. 2005) that could result from epigenetic factors.

ADHD persists throughout the lifespan (Scahill and Schwab-Stone 2000; Barkley et al. 2006). However, some of the clinical symptoms vary throughout development. Twin studies indicate that hyperactive symptoms are more stable in early and middle childhood, whereas attention problems are more stable in late childhood and adolescence (van der Valk et al. 1998, Larsson et al. 2004, Rietveld et al. 2004, Hay et al. 2007). Both inattention and hyperactivity subscales have been shown to be highly heritable and to substantially share genetic effects (McLoughlin et al. 2007). Thus, the decrease in hyperactivity noted with aging could potentially be due to epigenetic factors decreasing the expression of genes involved in activity levels.

ADHD medications are known to affect neurotransmitter function in humans. However, animal studies are pointing to epigenetic effects as well. In one study, over 700 genes involved in the formation, maturation, and stability of neural connections were elevated in the striatum of rats treated with methylphenidate (Adriani et al. 2006b). ADHD medications also change the expression of immediate early genes (IEGs) and define brain areas potentially involved in ADHD (Lin et al. 1996; Yano and Steiner 2005b).

4 ADHD and Epigenetic Processes

The dopamine transporter gene (*DAT; SLC6A3*), the most extensively studied candidate gene in ADHD, confers a small amount of risk for ADHD (Mick and Faraone 2008). This may be because factors involved in its expression contribute to the phenotypic variability. A recent report highlights features of this gene that render it sensitive to epigenetic factors (Shumay et al. 2010). Some of these features, summarized in Table 1, include the large number of variable number of tandem repeats (VNTRs) that indicate a tendency for chromatin structure to remain highly accessible to modifiers. High CpG density throughout the gene, especially in its promoter sequence, as well as the several transcription factors, such as Sp1 which mediates hormone dependent gene activation, further increase this gene's vulnerability to epigenetic influences.

Furthermore, *DAT* mRNA expression has been reported to decrease with age (Bannon and Whitty 1997) and in response to medication (Volkow et al. 2005), environmental factors (Volkow et al. 2005), and pathogens (Wang et al. 2004).

Table 1 Major epigenetic processes and their potential role in SLC6A3 (ADHD risk gene) expression

Chromosome organization	Over 90 VNTRs in SLC6A3 gene body indicate tendency to open chromatin structure and increase accessibility to modifiers (Shumay et al. 2010)
DNA methylation	SLC6A3 is Cytosine guanine base pair dense especially in its promoter (Shumay et al. 2010)
Transcription regulation	SLC6a3's flanking sequences contain binding sites for GATA-1, CREB, C-Myc, cis-acting regulatory elements. (Shumay et al. 2010)

4.1 Imprinting

Whereas paternal and maternal genomic contributions may be equal, they are not functionally equivalent. In the earliest studies of imprinting, maternal deficiency and paternal duplication for the distal region of chromosome 2 resulted in hyperkinetic mice whereas the opposite, maternal duplication/paternal deficiency, resulted in hypokinetic mice (Cattanach and Kirk 1985).

In human studies, parental gender and parent-of-origin effects may play a role in ADHD, conferring different quantitative or qualitative phenotypic manifestations. Several family-based studies have reported paternal overtransmission to ADHD-affected individuals of alleles at ADHD candidate genes, including *HTR1B* (Hawi et al. 2002; Quist et al. 2003; Smoller et al. 2006), *SNAP25* (Brophy et al. 2002; Kustanovich et al. 2003; Mill et al. 2004), *BDNF* (Kent et al. 2005), *DDC* (Hawi et al. 2001), *SLC6A4* (Hawi et al. 2005; Banerjee et al. 2006) *DRD4, DRD5, SLC6A3, TPH2* (Hawi et al. 2005). Maternal overtransmission was reported for *GNAL* (Laurin et al. 2008). Other studies, however, reported negative findings (Kim et al. 2007; Laurin et al. 2007) including the recent International Multicentre ADHD Genetics Project (IMAGE) study that tested 47 autosomal genes for overall association as well as a gene-specific effect of the parent of origin (Anney et al. 2008).

4.2 Gene–Environmental Interactions: Two-Hit Hypothesis

Animal studies have shown that malnutrition, maternal stress, infection, and toxic compounds (lead, bisphenol) influence prenatal brain development in circuits relevant to ADHD. For example, dopaminergic and serotonergic deficiencies were reported in young adult rats exposed prenatally to bacterial lipopolysaccharide (Wang et al. 2009). Low-dose prenatal and neonatal bisphenol exposure resulted in deficits in development of synaptic plasticity in rat dorsal striatum (Zhou et al. 2009). Prenatal protein deprivation increased postpubertal behavioral response to dopamine agonists (e.g., amphetamine) (Palmer et al. 2008). Hyperactivity and alteration of the midbrain dopaminergic system were reported in the male offspring of maternally stressed mice (Son et al. 2007). Serotonin depletion in pregnant rats has been shown to result in decreased whole brain tissue 5-hydroxyindoleactic acid and 5-hydroxytryptamine concentrations and increased locomotor activity in adult offspring (Vataeva et al. 2007). In children, exposure to polychlorinated bisphenyls (PCBs) (Eubig et al. 2010), lead (Cho et al. 2010; Eubig et al. 2010) and tobacco (Cho et al. 2010) have been associated with ADHD symptoms.

Also of clinical importance is prenatal exposure to synthetic glucocorticoids frequently used in the management of preterm labor. As reviewed by Kappor and colleagues (Kapoor et al. 2008), synthetic glucocorticoids enter the fetal circulation, especially in late gestation, and exert epigenetic effects by attenuating the fetal hypothalamic-pituitary-axis. Whereas these medications are effective in improving lung maturation in the developing fetus, repeated antenatal exposure to synthetic glucocorticoids has been associated with ADHD in children (French et al. 2004).

Overexpression of dopamine-related genes, including cyclin-dependent kinase inhibitor 1C (*Cdkn1c*), known to be critical for dopaminergic neuronal development, has also been reported in the offspring of mouse dams, which had been given a diet deficient in proteins throughout pregnancy and lactation, *versus* a control group that received adequate protein. In these animals, methylation of the promoter region of *Cdkn1c* was decreased by half and Cdkn1c mRNA expression was increased across brain regions in those animals that were hyperactive and had altered reward processing (Vucetic et al. 2010). Embryonic growth retardation and low birth weight have been shown in transgenic mice where Cdkn1c was overexpressed (Andrews et al. 2007). This, or other epigenetic factors, may explain the risk of lower birth weight associated with ADHD in monozygotic (MZ) birth weight-discordant twin pairs (Sharp et al. 2003; Leff et al. 2004; Asbury et al. 2006; Lehn et al. 2007) where the lighter twin from both the monozygotic (MZ) and dizygotic (DZ) birth weight-discordant twins showed higher ADHD ratings (Hultman et al. 2007).

In humans, maternal mutations in genes coding for enzymes involved in serotonin synthesis have been reported to potentially confer risk for ADHD in their offspring (Halmoy et al. 2010). In rat pups exposed to glucocuroticoids in the perinatal period, the organization of midbrain dopaminergic populations was permanently altered. Furthermore in males, the cytoarchitecture resembled that of females (McArthur et al. 2007). These *changes* may result from *changes* in neuroanatomy or neuronal or glial functions arising from epigenetic *changes*.

Maternal smoking during pregnancy has also been associated with ADHD in some studies (Naeye and Peters 1984; Fried and Makin 1987; McIntosh et al. 1995; Milberger et al. 1996, 1997; O'Callaghan et al. 1997; Mick et al. 2002; O'Connor et al. 2002; Kotimaa et al. 2003; Thapar et al. 2003; Langley et al. 2005; Linnet et al. 2005; Braun et al. 2006; Huizink and Mulder 2006; Neuman et al. 2007), although to a lesser degree when genetic risk is controlled for (Knopik et al. 2006). Comorbid conditions may play a role here since prenatal nicotine exposure has also been associated with Oppositional Defiant Disorder (ODD) and Conduct Disorder (CD: (Orlebeke et al. 1999; Day et al. 2000; Wakschlag et al. 2002; Maughan et al. 2004; Huizink and Mulder 2006). A longitudinal study that considered potential overlapping factors and included low birth weight and normal birth weight children found that prenatal maternal smoking was strongly linked to ODD and CD, independent of birth weight, but not to ADHD. Only low birth weight was found to be associated with ADHD in this study. Maternal smoking was also confounded by maternal drug abuse and educational level (Nigg and Breslau 2007).

In a 2006 study (Brookes et al. 2006), in which 28.6% of mothers smoked cigarettes (approximately 3 months during gestation) and 57.8% drank alcohol at some time during pregnancy, a significant interaction was found for the 10/3 haplotype encompassing the *SLC6A3* gene and maternal alcohol use, whereas no interaction with genotype was observed for maternal smoking. Kahn and colleagues reported that children homozygous for the *SLC6A3* VNTR 480/480 genotype and also exposed to nicotine *in utero* had higher parental ratings on measures of hyperactivity-impulsivity and oppositional behaviors (Kahn et al. 2003). In a study of twin pairs, for the 24% of mothers who reported smoking during pregnancy their offspring had higher incidence of ADHD symptoms than those not exposed to prenatal nicotine. Risk for ADHD was greater in twins with the *SLC6A3* 440 allele who were exposed to *in utero* nicotine than twins who had neither risk factor, while no significant interaction was found for *SLC6A3* 480 allele carriers. The *DRD4* 7-repeat (Neuman et al. 2007) and *CHRNA4* variants (Todd and Neuman 2007) were also associated with higher ADHD risk for individuals with prenatal exposure, although gestational age and birth weights were not accounted for in these studies. No gene by environment interactions were detected for *DRD4* 7-repeat and prenatal nicotine exposure in a large cohort of 539 ADHD-affected children and 407 nonaffected siblings who had participated in the IMAGE study (Altink et al. 2008).

Although gestational alcohol exposure has also been associated with ADHD (Bhatara et al. 2006; Fryer et al. 2007), twin studies do not confirm this association (Neuman et al. 2006). Offspring of female MZ and DZ twins, concordant or discordant for alcohol use during pregnancy, had a higher risk of having ADHD than controls (Knopik et al. 2006). A pilot study, reporting higher risk of ADHD in children of parents with substance use disorder (alcohol and other substances), did not control for CD or other comorbidities (Wilens et al. 2005). Studies have also not taken into account maternal levels of alcohol dehydrogenase activity, which may be important given that maternal absence of the alcohol dehydrogenase allele (*ADH1B*$_*$3*) and prenatal alcohol exposure resulted in higher ADHD symptoms in offspring (Jacobson et al. 2006).

Psychosocial adversity (measured as a composite of family adversity that included overcrowding, marital discord, lack of social support) in adolescents homozygous for the *SLC6A3* 480 allele resulted in greater ADHD symptoms than in adolescents with other genotypes or with more favorable environments (Laucht et al. 2007).

Epigenetic factors may also be relevant in ADHD comorbidity. By attenuating the expression of α_{2A}-adrenoceptors in neonatal rat brainstem, through antisense technology and RNA interference, a decrease in anxiety-related behaviors and increases in α_{2A}-adrenoceptor densities in the hypothalamus of adult animals have been reported (Shishkina et al. 2004).

Investigating gene–environment interactions is complicated by numerous confounding effects. For example, use of tobacco by pregnant women is independently linked to other factors that could potentially confer risk, such as lower birth weight (Secker-Walker et al. 1997), stress (Rodriguez and Bohlin 2005), premature rupture of membranes and placental abruption (Andres and Day 2000). Regular smoking is also more prevalent in women alcoholics, who are also more likely to smoke during pregnancy (Knopik et al. 2005, 2006).

5 ADHD Medications and Gene Expression

The mechanism of action of ADHD medications is not well-understood (Heal et al. 2011). Most studies have focused on synaptic reuptake or release of neurotransmitters (Swanson and Volkow 2002) but this may be only one mechanism. Administration of methylphenidate in adolescent rats has been reported to upregulate over 700 genes in the striatum involved in neural and synaptic plasticity. These included three main groups: (1) genes involved in migration of immature neural/glial cells and/or growth of new axons; (2) genes involved in axonal myelination and stabilization of myelinating glia–axon contacts; and (3) genes involved in mature processes such as intercellular junctions, neurotransmitter receptors and proteins responsible for transport and anchoring (Adriani et al. 2006a, b).

5.1 ADHD Meds and Gene/Protein Expression

As summarized in Table 2, methylphenidate has also been reported to increase cortical gene expressions of *Egr1* (*Zif 268*), a mammalian transcription factor whose induction is associated with neuronal activity (Yano and Steiner 2005a), and regulation of synaptobrevin II, a protein important for vesicular exocytosis (Petersohn and Thiel 1996). Chronic methylphenidate administration increased both the density of dendritic spines in MSN-D1 (medium sized spiny neurons expressing dopamine D1 receptors) from the core and shell of the nucleus accumbens and MSN-D2 (MSN expressing dopamine D2 receptors from the shell of the nucleus accumbens). Expression of *ΔFosB*, which is implicated in the control of the reward system in the brain and may play a role in drug addiction (Hope 1998), was increased by chronic methylphenidate administration only in MSN-D1 from all areas of the striatum.

Chronic *d*-amphetamine administration *has been reported* to increase synaptic protein expression (spinophilin) in the striatum as well as in septum, hippocampus, amygdala, and cingulate cortex (Boikess and Marshall 2008). Rats given an escalating dose of *d*-amphetamine (1–8 mg/kg) for five weeks were found to have upregulation of spinophilin and the vesicular glutamate transporter gene, *VGLUT1*, in the thalamus, lateral hypothalamus, and habenula (Boikess et al. 2010). *d*-Amphetamine treatment (>1 month) increased the length of dendrites, the density of dendritic spines, and the number of branched spines in the prefrontal cortex and the nucleus accumbens (Robinson and Kolb 1997). Increased spine density has *also been reported* in the striatum (Li et al. 2003). Acute administration of *d*-amphetamine increased expression of *Ras-GRF1* but not *GRF2* in the striatum and prefrontal cortex. No changes were seen in the hippocampus (Parelkar et al. 2009). *D*-Amphetamine administration also downregulates a GABA synthesizing enzyme (GAD_{67} expression in the olfactory bulb of mouse deficient in monoamine oxidase B (MAOB) (Yin et al. 2010) while in a separate study, the neurons

Table 2 ADHD medications and changes in CNS gene/protein expression in animal models

	Acute/sub-chronic MPH	Chronic MPH	Acute AMPH	Chronic AMPH	Atomoxetine	α-Adrenergic	Modafinil
Striatum	*c-fos* (Lin et al. 1996)	*Homer 1a* (Adriani et al. 2006a)	*c-fos* (Graybiel et al. 1990; Snyder-Keller 1991; Johansson et al. 1994; Dalia and Wallace 1995; Konradi et al. 1996; Badiani et al. 1998; Rotllant et al. 2010)				
	Homer 1a (Yano and Steiner 2005a)	ΔFosB (Kim et al. 2009)					
	EGR1 (Zif 268) (Yano and Steiner 2005a)		*Ras-GFR1* (Parelkar et al. 2009)				
	Homer 1, Shank 2, MPP3, Grik2, Gabrγ1, Gabrβ3, Htr7, Adrα1b (Adriani et al. 2006b)		Spinophilin (Boikess et al. 2010)				
	c-fos (Lin et al. 1996; Trinh et al. 2003; Koda et al. 2010)		*c-fos* (Lillrank et al. 1996; Badiani et al. 1998; Carr and Kutchukhidze 2000; Day et al. 2001; Ostrander et al. 2003a; Miyamoto et al. 2004; Colussi-Mas et al. 2007; Rotllant et al. 2010)	Spinophilin (Boikess and Marshall 2008)			
	Homer 1a (Yano and Steiner 2005a)		Ras-GFR1				
	EGR1 (Zif 268) (Yano and Steiner 2005a)						
Pre-Frontal Cortex					*c-fos* (Koda et al. 2010)		
Cingulate Cortex							

(continued)

Table 2 (continued)

	Acute/sub-chronic MPH	Chronic MPH	Acute AMPH	Chronic AMPH	Atomoxetine	α-Adrenergic	Modafinil
Amygdala			c-fos (Lillrank et al. 1996; Badiani et al. 1998; Carr and Kutchukhidze 2000; Day et al. 2001; Ostrander et al. 2003a; Miyamoto et al. 2004; Colussi-Mas et al. 2007; Rotllant et al. 2010)	Spinophilin (Boikess and Marshall 2008)			
Habenula			c-fos (Lillrank et al. 1996; Badiani et al. 1998; Carr and Kutchukhidze 2000; Day et al. 2001; Ostrander et al. 2003a; Miyamoto et al. 2004; Colussi-Mas et al. 2007; Rotllant et al. 2010)	Spinophilin, VGLUT1 (Boikess et al. 2010)			
Hypothalamus	c-fos (Lin et al. 1996)		c-fos (Lillrank et al. 1996; Badiani et al. 1998; Carr and Kutchukhidze 2000; Day et al. 2001; Ostrander et al. 2003a; Miyamoto et al. 2004; Colussi-Mas et al. 2007; Rotllant et al. 2010)	Spinophilin, VGLUT1 (Boikess et al. 2010)			c-fos (Lin et al. 1996)

Ventral tegmental area		c-fos (Lillrank et al. 1996; Badiani et al. 1998; Carr and Kutchukhidze 2000; Day et al. 2001; Ostrander et al. 2003a; Miyamoto et al. 2004; Colussi-Mas et al. 2007; Rotllant et al. 2010)
Raphe nuclei		GABA (Rotllant et al. 2010) c-fos (Lillrank et al. 1996; Badiani et al. 1998; Carr and Kutchukhidze 2000; Day et al. 2001; Ostrander et al. 2003a; Miyamoto et al. 2004; Colussi-Mas et al. 2007; Rotllant et al. 2010)
Thalamus	Spinophilin, *VGLUT1* (Boikess et al. 2010)	
Brainstem		Caspace-3 mRNA (Dygalo et al. 2004)

activated by *d*-amphetamine in the ventral tegmental area were predominantly GABAergic (Rotllant et al. 2010).

It is also important to note that acute and chronic administration as well as the time of exposure during development may be important. As shown with methylphenidate, exposure during adolescence resulted in upregulation of Grik2 (gene encoding for a subunit of a kainite glutamate receptor known to mediate excitatory neurotransmission and synaptic plasticity in brain) and Htr7 (cell receptor involved in serotonergic neurotransmission) into adulthood (Adriani et al. 2006a). Clonidine, an α2-adrenoceptor agonist, has been reported to increase the level of apoptotic enzyme caspace-3MRNA expression in the brainstem of rats during fetal development and after birth, but not in older pups, suggesting that it facilitates cell death in the developing brain (Dygalo et al. 2004).

5.2 ADHD Medications and Immediate Early Genes

Animal studies investigating IEGs that are expressed in the brain, such as *c-Fos* and homer 1a, are providing a window into brain areas activated by ADHD medications. These IEGs are activated transiently and rapidly in response to a wide variety of cellular stimuli including medications. As summarized in Table 2, acute administration of methylphenidate and *d*-amphetamine has been shown to increase expression of *c-Fos* the striatum (caudate-putamen), cortex (mediofrontal), amygdala, lateral habenula, paraventricular nuclei of the hypothalamus, ventral tegmental area, and raphe nuclei (Graybiel et al. 1990; Johansson et al. 1994; Dalia and Wallace 1995; Konradi et al. 1996; Lin et al. 1996; Badiani et al. 1998; Carr and Kutchukhidze 2000; Day et al. 2001; Penner et al. 2002; Brandon and Steiner 2003; Ostrander et al. 2003a, b; Trinh et al. 2003; Miyamoto et al. 2004; Yano and Steiner 2005b). These findings indicate that methylphenidate and *d*-amphetamine affect transcription and synaptic plasticity of regulatory proteins in specific corticostriatal circuits, such as those implicated in attentional functions. Methylphenidate has also been reported to increase cortical gene expressions of *Homer 1a* (Yano and Steiner 2005a) and *c-fos* (Lin et al. 1996; Trinh et al. 2003). They also support dopaminergic-enhancing activity for methylphenidate and *d*-amphetamine since the caudate nucleus receives projections from mesencephalic dopaminergic neurons, with their neurotransmission mediated by dopamine D1 and D2 receptors (Graybiel et al. 1990). Furthermore, the mediofrontal cortex is the target of the mesoneocorital system originating from the dopaminergic ventral tegmental area of Tsai (VTA-A10) (Swanson 1982). Treatment with methylphenidate resulted in a greater number of cortical cells expressing *c-fos* than did treatment with a bioequivalent dose of *d*-amphetamine, suggesting that the dopaminergic mesoneocortical system may be more sensitive to the former compound (Lin et al. 1996). In contrast, modafanil induced *c-Fos* expression in neurons of the anterior hypothalamic nucleus and adjacent suprachiasmatic borders suggesting that many areas may be involved in maintaining alertness (Lin et al. 1996).

5.3 ADHD Medications, Gene Expression, and Adverse Effects

Investigating gene expression is also being used as a tool better to understand adverse effects of ADHD medications. An example is atomoxetine, which was associated with mild but significant increases in heart rate and blood pressure and mild QT prolongation (Wernicke et al. 2003). However, further studies have shown that atomoxetine blocked hERG channels in a dose-dependent manner (Scherer et al. 2009): *hERG* (human Ether-a-go-go related gene) encodes for the α subunit of a potassium ion channel protein (I_{kr}) that mediates the repolarization current of the cardiac action potential (Kiehn et al. 1999). Atomoxetine induced a prolonged action potential in guinea pig cardiomyocytes – the prolongation was in the plateau phase with only minor effects on phase 3 repolarization, suggesting that the proarrhythmic potential is low. High atomoxetine concentrations also resulted in a reduction of *hERG* expression but this did not occur at atomoxetine concentrations within the therapeutic range (Scherer et al. 2009). This suggests that there may be greater risk in subjects with overdoses, in those with slower metabolism, due to *2D6* variants and impaired hepatic and renal clearance (Scherer et al. 2009), or who are taking a combination of medications, as has been reported in two case reports (Barker et al. 2004; Sawant and Daviss 2004).

5.4 Gene Expression and Potential Novel Pathways in ADHD

Gene expression studies with ADHD medications may also point to mechanisms of action of ADHD medications not yet considered in ADHD. For example, in the CNS, noradrenaline inhibits the production of inflammatory factors. Treatment with atomoxetine may limit the CNS inflammatory activity, not directly, but indirectly by increasing noradrenaline availability, which in turn suppresses expression of inflammatory cytokines (O'Sullivan et al. 2009).

6 Summary

The role of epigenetics in normal development, health, and neuropsychiatric diseases is clearly extremely important. Identifying the processes involved in ADHD will allow the development of interventions that may be applied to modify the expression of deleterious genes as well as protective genes. Some of the tools being used to identify areas of methylation throughout the genome include high-throughput array-based methylation analysis (Irizarry et al. 2008) as well as methylated DNA immunoprecipitation (Weber et al. 2005). In the future, these tools, as well as newer ones, applied to ADHD samples, may reveal areas of the genome that have not yet been considered important in ADHD.

Gene expression enhanced by ADHD medications are also providing a window into the brain areas involved in ADHD, as seen by studies investigating IEGs. Animal studies are also showing the importance of the length of time of exposure to medications, as well as the stage of development. As yet, we do not have the tools needed to identify neurons inhibited by these medications. Finally, gene expression studies with ADHD medications may also point to neuronal pathways or mechanisms not yet considered in ADHD.

References

Adriani W, Leo D, Greco D, Rea M, di Porzio U, Laviola G, Perrone-Capano C (2006a) Methylphenidate administration to adolescent rats determines plastic changes on reward-related behavior and striatal gene expression. Neuropsychopharmacology 31:1946–1956

Adriani W, Leo D, Guarino M, Natoli A, Di Consiglio E, De Angelis G, Traina E, Testai E, Perrone-Capano C, Laviola G (2006b) Short-term effects of adolescent methylphenidate exposure on brain striatal gene expression and sexual/endocrine parameters in male rats. Ann N Y Acad Sci 1074:52–73

Altink ME, Arias-Vasquez A, Franke B, Slaats-Willemse DI, Buschgens CJ, Rommelse NN, Fliers EA, Anney R, Brookes KJ, Chen W, Gill M, Mulligan A, Sonuga-Barke E, Thompson M, Sergeant JA, Faraone SV, Asherson P, Buitelaar JK (2008) The dopamine receptor D4 7-repeat allele and prenatal smoking in ADHD-affected children and their unaffected siblings: no gene-environment interaction. J Child Psychol Psychiatry 49:1053–1060

Amir RE, Van den Veyver IB, Wan M, Tran CQ, Francke U, Zoghbi HY (1999) Rett syndrome is caused by mutations in X-linked MECP2, encoding methyl-CpG-binding protein 2. Nat Genet 23:185–188

Andres RL, Day MC (2000) Perinatal complications associated with maternal tobacco use. Semin Neonatol 5:231–241

Andrews SC, Wood MD, Tunster SJ, Barton SC, Surani MA, John RM (2007) Cdkn1c (p57Kip2) is the major regulator of embryonic growth within its imprinted domain on mouse distal chromosome 7. BMC Dev Biol 7:53

Anney RJ, Hawi Z, Sheehan K, Mulligan A, Pinto C, Brookes KJ, Xu X, Zhou K, Franke B, Buitelaar J, Vermeulen SH, Banaschewski T, Sonuga-Barke E, Ebstein R, Manor I, Miranda A, Mulas F, Oades RD, Roeyers H, Rommelse N, Rothenberger A, Sergeant J, Steinhausen HC, Taylor E, Thompson M, Asherson P, Faraone SV, Gill M (2008) Parent of origin effects in attention-deficit hyperactivity disorder (ADHD): analysis of data from the international multi-center ADHD genetics (IMAGE) program. Am J Med Genet B Neuropsychiatr Genet 147B:1495–1500

Asbury K, Dunn JF, Plomin R (2006) Birthweight-discordance and differences in early parenting relate to monozygotic twin differences in behaviour problems and academic achievement at age 7. Dev Sci 9:F22–F31

Baccarelli A, Bollati V (2009) Epigenetics and environmental chemicals. Curr Opin Pediatr 21:243–251

Badiani A, Oates MM, Day HE, Watson SJ, Akil H, Robinson TE (1998) Amphetamine-induced behavior, dopamine release, and c-fos mRNA expression: modulation by environmental novelty. J Neurosci 18:10579–10593

Banerjee E, Sinha S, Chatterjee A, Gangopadhyay PK, Singh M, Nandagopal K (2006) A family-based study of Indian subjects from Kolkata reveals allelic association of the serotonin transporter intron-2 (STin2) polymorphism and attention-deficit-hyperactivity disorder (ADHD). Am J Med Genet B Neuropsychiatr Genet 141B:361–366

Bannon MJ, Whitty CJ (1997) Age-related and regional differences in dopamine transporter mRNA expression in human midbrain. Neurology 48:969–977

Barker MJ, Benitez JG, Ternullo S, Juhl GA (2004) Acute oxcarbazepine and atomoxetine overdose with quetiapine. Vet Hum Toxicol 46:130–132

Barkley RA, Fischer M, Smallish L, Fletcher K (2006) Young adult outcome of hyperactive children: adaptive functioning in major life activities. J Am Acad Child Adolesc Psychiatry 45:192–202

Baumgardner TL, Reiss AL, Freund LS, Abrams MT (1995) Specification of the neurobehavioral phenotype in males with fragile X syndrome. Pediatrics 95:744–752

Berger SL (2007) The complex language of chromatin regulation during transcription. Nature 447:407–412

Bhatara V, Loudenberg R, Ellis R (2006) Association of attention deficit hyperactivity disorder and gestational alcohol exposure: an exploratory study. J Atten Disord 9:515–522

Boikess SR, Marshall JF (2008) A sensitizing d-amphetamine regimen induces long-lasting spinophilin protein upregulation in the rat striatum and limbic forebrain. Eur J Neurosci 28:2099–2107

Boikess SR, O'Dell SJ, Marshall JF (2010) A sensitizing d-amphetamine dose regimen induces long-lasting spinophilin and VGLUT1 protein upregulation in the rat diencephalon. Neurosci Lett 469:49–54

Brandon CL, Steiner H (2003) Repeated methylphenidate treatment in adolescent rats alters gene regulation in the striatum. Eur J Neurosci 18:1584–1592

Braun JM, Kahn RS, Froehlich T, Auinger P, Lanphear BP (2006) Exposures to environmental toxicants and attention deficit hyperactivity disorder in U.S. children. Environ Health Perspect 114:1904–1909

Brookes KJ, Mill J, Guindalini C, Curran S, Xu X, Knight J, Chen CK, Huang YS, Sethna V, Taylor E, Chen W, Breen G, Asherson P (2006) A common haplotype of the dopamine transporter gene associated with attention-deficit/hyperactivity disorder and interacting with maternal use of alcohol during pregnancy. Arch Gen Psychiatry 63:74–81

Brophy K, Hawi Z, Kirley A, Fitzgerald M, Gill M (2002) Synaptosomal-associated protein 25 (SNAP-25) and attention deficit hyperactivity disorder (ADHD): evidence of linkage and association in the Irish population. Mol Psychiatry 7:913–917

Carr KD, Kutchukhidze N (2000) Chronic food restriction increases fos-like immunoreactivity (FLI) induced in rat forebrain by intraventricular amphetamine. Brain Res 861:88–96

Cattanach BM, Kirk M (1985) Differential activity of maternally and paternally derived chromosome regions in mice. Nature 315:496–498

Chandler VL (2007) Paramutation: from maize to mice. Cell 128:641–645

Cho SC, Kim BN, Hong YC, Shin MS, Yoo HJ, Kim JW, Bhang SY, Cho IH, Kim HW (2010) Effect of environmental exposure to lead and tobacco smoke on inattentive and hyperactive symptoms and neurocognitive performance in children. J Child Psychol Psychiatry 51:1050–1057

Colussi-Mas J, Geisler S et al (2007) Activation of afferents to the ventral tegmental area in response to acute amphetamine: a double-labelling study. Eur J Neurosci 26(4):1011–1025

Dalia A, Wallace LJ (1995) Amphetamine induction of c-fos in the nucleus accumbens is not inhibited by glutamate antagonists. Brain Res 694:299–307

Day NL, Richardson GA, Goldschmidt L, Cornelius MD (2000) Effects of prenatal tobacco exposure on preschoolers' behavior. J Dev Behav Pediatr 21:180–188

Day HE, Badiani A, Uslaner JM, Oates MM, Vittoz NM, Robinson TE, Watson SJ Jr, Akil H (2001) Environmental novelty differentially affects c-fos mRNA expression induced by amphetamine or cocaine in subregions of the bed nucleus of the stria terminalis and amygdala. J Neurosci 21:732–740

Derks EM, Hudziak JJ, Dolan CV, van Beijsterveldt TC, Verhulst FC, Boomsma DI (2008) Genetic and environmental influences on the relation between attention problems and attention deficit hyperactivity disorder. Behav Genet 38:11–23

Dygalo NN, Bannova AV, Kalinina TS, Shishkina GT (2004) Clonidine increases caspase-3 mRNA level and DNA fragmentation in the developing rat brainstem. Brain Res Dev Brain Res 152:225–231

Eubig PA, Aguiar A, Schantz SL (2010) Lead and PCBs as risk factors for attention deficit/hyperactivity disorder. Environ Health Perspect 118:1654–1667

Feinberg AP (2008) Epigenetics at the epicenter of modern medicine. JAMA 299:1345–1350

Feinberg AP, Tycko B (2004) The history of cancer epigenetics. Nat Rev Cancer 4:143–153

French NP, Hagan R, Evans SF, Mullan A, Newnham JP (2004) Repeated antenatal corticosteroids: effects on cerebral palsy and childhood behavior. Am J Obstet Gynecol 190:588–595

Fried PA, Makin JE (1987) Neonatal behavioural correlates of prenatal exposure to marihuana, cigarettes and alcohol in a low risk population. Neurotoxicol Teratol 9:1–7

Fryer SL, McGee CL, Matt GE, Riley EP, Mattson SN (2007) Evaluation of psychopathological conditions in children with heavy prenatal alcohol exposure. Pediatrics 119:e733–e741

Goto T, Monk M (1998) Regulation of X-chromosome inactivation in development in mice and humans. Microbiol Mol Biol Rev 62:362–378

Graybiel AM, Moratalla R, Robertson HA (1990) Amphetamine and cocaine induce drug-specific activation of the c-fos gene in striosome-matrix compartments and limbic subdivisions of the striatum. Proc Natl Acad Sci U S A 87:6912–6916

Halmoy A, Johansson S, Winge I, McKinney JA, Knappskog PM, Haavik J (2010) Attention-deficit/hyperactivity disorder symptoms in offspring of mothers with impaired serotonin production. Arch Gen Psychiatry 67:1033–1043

Hawi Z, Foley D, Kirley A, McCarron M, Fitzgerald M, Gill M (2001) Dopa decarboxylase gene polymorphisms and attention deficit hyperactivity disorder (ADHD): no evidence for association in the Irish population. Mol Psychiatry 6:420–424

Hawi Z, Dring M, Kirley A, Foley D, Kent L, Craddock N, Asherson P, Curran S, Gould A, Richards S, Lawson D, Pay H, Turic D, Langley K, Owen M, O'Donovan M, Thapar A, Fitzgerald M, Gill M (2002) Serotonergic system and attention deficit hyperactivity disorder (ADHD): a potential susceptibility locus at the 5-HT(1B) receptor gene in 273 nuclear families from a multi-centre sample. Mol Psychiatry 7:718–725

Hawi Z, Segurado R, Conroy J, Sheehan K, Lowe N, Kirley A, Shields D, Fitzgerald M, Gallagher L, Gill M (2005) Preferential transmission of paternal alleles at risk genes in attention-deficit/hyperactivity disorder. Am J Hum Genet 77:958–965

Hay D, Bennett KS, McStephen M, Rooney R, Levy F (2004) Attention deficit-hyperactivity disorder in twins: a developmental genetic analysis. Aust J Psychol 56:99–107

Hay DA, Bennett KS, Levy F, Sergeant J, Swanson J (2007) A twin study of attention-deficit/hyperactivity disorder dimensions rated by the strengths and weaknesses of ADHD-symptoms and normal-behavior (SWAN) scale. Biol Psychiatry 61:700–705

Heal DJ, Smith SL, Findling RL (2011) ADHD: Current and future therapeutics. Curr Topics Behav Neurosci DOI 10.1007/7854_2011_125

Hollick JB, Dorweiler JE, Chandler VL (1997) Paramutation and related allelic interactions. Trends Genet 13:302–308

Hope BT (1998) Cocaine and the AP-1 transcription factor complex. Ann N Y Acad Sci 844:1–6

Huizink AC, Mulder EJ (2006) Maternal smoking, drinking or cannabis use during pregnancy and neurobehavioral and cognitive functioning in human offspring. Neurosci Biobehav Rev 30:24–41

Hultman CM, Torrang A, Tuvblad C, Cnattingius S, Larsson JO, Lichtenstein P (2007) Birth weight and attention-deficit/hyperactivity symptoms in childhood and early adolescence: a prospective Swedish twin study. J Am Acad Child Adolesc Psychiatry 46:370–377

Irizarry RA, Ladd-Acosta C, Carvalho B, Wu H, Brandenburg SA, Jeddeloh JA, Wen B, Feinberg AP (2008) Comprehensive high-throughput arrays for relative methylation (CHARM). Genome Res 18:780–790

Jacobson SW, Carr LG, Croxford J, Sokol RJ, Li TK, Jacobson JL (2006) Protective effects of the alcohol dehydrogenase-ADH1B allele in children exposed to alcohol during pregnancy. J Pediatr 148:30–37

Jirtle RL, Skinner MK (2007) Environmental epigenomics and disease susceptibility. Nat Rev Genet 8:253–262

Johansson B, Lindstrom K, Fredholm BB (1994) Differences in the regional and cellular localization of c-fos messenger RNA induced by amphetamine, cocaine and caffeine in the rat. Neuroscience 59:837–849

Kahn RS, Khoury J, Nichols WC, Lanphear BP (2003) Role of dopamine transporter genotype and maternal prenatal smoking in childhood hyperactive-impulsive, inattentive, and oppositional behaviors. J Pediatr 143:104–110

Kapoor A, Petropoulos S, Matthews SG (2008) Fetal programming of hypothalamic-pituitary-adrenal (HPA) axis function and behavior by synthetic glucocorticoids. Brain Res Rev 57:586–595

Karin M (1990) Too many transcription factors: positive and negative interactions. New Biol 2:126–131

Kent L, Green E, Hawi Z, Kirley A, Dudbridge F, Lowe N, Raybould R, Langley K, Bray N, Fitzgerald M, Owen MJ, O'Donovan MC, Gill M, Thapar A, Craddock N (2005) Association of the paternally transmitted copy of common Valine allele of the Val66Met polymorphism of the brain-derived neurotrophic factor (BDNF) gene with susceptibility to ADHD. Mol Psychiatry 10:939–943

Kiehn J, Lacerda AE, Brown AM (1999) Pathways of HERG inactivation. Am J Physiol 277: H199–H210

Kim JW, Waldman ID, Faraone SV, Biederman J, Doyle AE, Purcell S, Arbeitman L, Fagerness J, Sklar P, Smoller JW (2007) Investigation of parent-of-origin effects in ADHD candidate genes. Am J Med Genet B Neuropsychiatr Genet 144B:776–780

Kim Y, Teylan MA, Baron M, Sands A, Nairn AC, Greengard P (2009) Methylphenidate-induced dendritic spine formation and DeltaFosB expression in nucleus accumbens. Proc Natl Acad Sci U S A 106:2915–2920

Knopik VS, Sparrow EP, Madden PA, Bucholz KK, Hudziak JJ, Reich W, Slutske WS, Grant JD, McLaughlin TL, Todorov A, Todd RD, Heath AC (2005) Contributions of parental alcoholism, prenatal substance exposure, and genetic transmission to child ADHD risk: a female twin study. Psychol Med 35:625–635

Knopik VS, Heath AC, Jacob T, Slutske WS, Bucholz KK, Madden PA, Waldron M, Martin NG (2006) Maternal alcohol use disorder and offspring ADHD: disentangling genetic and environmental effects using a children-of-twins design. Psychol Med 36:1461–1471

Koda K, Ago Y, Cong Y, Kita Y, Takuma K, Matsuda T (2010) Effects of acute and chronic administration of atomoxetine and methylphenidate on extracellular levels of noradrenaline, dopamine and serotonin in the prefrontal cortex and striatum of mice. J Neurochem 114:259–270

Konradi C, Leveque JC, Hyman SE (1996) Amphetamine and dopamine-induced immediate early gene expression in striatal neurons depends on postsynaptic NMDA receptors and calcium. J Neurosci 16:4231–4239

Kotimaa AJ, Moilanen I, Taanila A, Ebeling H, Smalley SL, McGough JJ, Hartikainen AL, Jarvelin MR (2003) Maternal smoking and hyperactivity in 8-year-old children. J Am Acad Child Adolesc Psychiatry 42:826–833

Kuntsi J, Rijsdijk F, Ronald A, Asherson P, Plomin R (2005) Genetic influences on the stability of attention-deficit/hyperactivity disorder symptoms from early to middle childhood. Biol Psychiatry 57:647–654

Kustanovich V, Merriman B, McGough J, McCracken JT, Smalley SL, Nelson SF (2003) Biased paternal transmission of SNAP-25 risk alleles in attention-deficit hyperactivity disorder. Mol Psychiatry 8:309–315

Langley K, Rice F, van den Bree MB, Thapar A (2005) Maternal smoking during pregnancy as an environmental risk factor for attention deficit hyperactivity disorder behaviour. A review. Minerva Pediatr 57:359–371

Larsson JO, Larsson H, Lichtenstein P (2004) Genetic and environmental contributions to stability and change of ADHD symptoms between 8 and 13 years of age: a longitudinal twin study. J Am Acad Child Adolesc Psychiatry 43:1267–1275

Latchman DS (1997) Transcription factors: an overview. Int J Biochem Cell Biol 29:1305–1312

Laucht M, Skowronek MH, Becker K, Schmidt MH, Esser G, Schulze TG, Rietschel M (2007) Interacting effects of the dopamine transporter gene and psychosocial adversity on attention-deficit/hyperactivity disorder symptoms among 15-year-olds from a high-risk community sample. Arch Gen Psychiatry 64:585–590

Laurin N, Feng Y, Ickowicz A, Pathare T, Malone M, Tannock R, Schachar R, Kennedy JL, Barr CL (2007) No preferential transmission of paternal alleles at risk genes in attention-deficit hyperactivity disorder. Mol Psychiatry 12:226–229

Laurin N, Ickowicz A, Pathare T, Malone M, Tannock R, Schachar R, Kennedy JL, Barr CL (2008) Investigation of the G protein subunit Galphaolf gene (GNAL) in attention deficit/hyperactivity disorder. J Psychiatr Res 42:117–124

Leff S, Costigan TE, Power TJ (2004) Using participatory action research to develop a playground-based prevention program. J Sch Psychol 42:3–21

Lehn H, Derks EM, Hudziak JJ, Heutink P, van Beijsterveldt TC, Boomsma DI (2007) Attention problems and attention-deficit/hyperactivity disorder in discordant and concordant monozygotic twins: evidence of environmental mediators. J Am Acad Child Adolesc Psychiatry 46:83–91

Levenson JM, Sweatt JD (2005) Epigenetic mechanisms in memory formation. Nat Rev Neurosci 6:108–118

Li E (2002) Chromatin modification and epigenetic reprogramming in mammalian development. Nat Rev Genet 3:662–673

Li Y, Kolb B, Robinson TE (2003) The location of persistent amphetamine-induced changes in the density of dendritic spines on medium spiny neurons in the nucleus accumbens and caudate-putamen. Neuropsychopharmacology 28:1082–1085

Lillrank SM, Lipska BK, Bachus SE, Wood GK, Weinberger DR (1996) Amphetamine-induced c-fos mRNA expression is altered in rats with neonatal ventral hippocampal damage. Synapse 23:292–301

Lim JH, Booker AB, Fallon JR (2005) Regulating fragile X gene transcription in the brain and beyond. J Cell Physiol 205:170–175

Lin JS, Hou Y, Jouvet M (1996) Potential brain neuronal targets for amphetamine-, methylphenidate-, and modafinil-induced wakefulness, evidenced by c-fos immunocytochemistry in the cat. Proc Natl Acad Sci U S A 93:14128–14133

Linnet KM, Wisborg K, Obel C, Secher NJ, Thomsen PH, Agerbo E, Henriksen TB (2005) Smoking during pregnancy and the risk for hyperkinetic disorder in offspring. Pediatrics 116:462–467

Maughan B, Taylor A, Caspi A, Moffitt TE (2004) Prenatal smoking and early childhood conduct problems: testing genetic and environmental explanations of the association. Arch Gen Psychiatry 61:836–843

McArthur S, McHale E, Gillies GE (2007) The size and distribution of midbrain dopaminergic populations are permanently altered by perinatal glucocorticoid exposure in a sex- region- and time-specific manner. Neuropsychopharmacology 32:1462–1476

McIntosh DE, Mulkins RS, Dean RS (1995) Utilization of maternal perinatal risk indicators in the differential diagnosis of ADHD and UADD children. Int J Neurosci 81:35–46

McLoughlin G, Ronald A, Kuntsi J, Asherson P, Plomin R (2007) Genetic support for the dual nature of attention deficit hyperactivity disorder: substantial genetic overlap between the inattentive and hyperactive-impulsive components. J Abnorm Child Psychol 35:999–1008

Mick E, Faraone SV (2008) Genetics of attention deficit hyperactivity disorder. Child Adolesc Psychiatr Clin N Am 17:261–284, vii-viii

Mick E, Biederman J, Faraone SV, Sayer J, Kleinman S (2002) Case-control study of attention-deficit hyperactivity disorder and maternal smoking, alcohol use, and drug use during pregnancy. J Am Acad Child Adolesc Psychiatry 41:378–385

Milberger S, Biederman J, Faraone SV, Chen L, Jones J (1996) Is maternal smoking during pregnancy a risk factor for attention deficit hyperactivity disorder in children? Am J Psychiatry 153:1138–1142

Milberger S, Biederman J, Faraone SV, Chen L, Jones J (1997) Further evidence of an association between attention-deficit/hyperactivity disorder and cigarette smoking. Findings from a high-risk sample of siblings. Am J Addict 6:205–217

Mill J, Richards S, Knight J, Curran S, Taylor E, Asherson P (2004) Haplotype analysis of SNAP-25 suggests a role in the aetiology of ADHD. Mol Psychiatry 9:801–810

Miyamoto S, Snouwaert JN, Koller BH, Moy SS, Lieberman JA, Duncan GE (2004) Amphetamine-induced Fos is reduced in limbic cortical regions but not in the caudate or accumbens in a genetic model of NMDA receptor hypofunction. Neuropsychopharmacology 29:2180–2188

Naeye RL, Peters EC (1984) Mental development of children whose mothers smoked during pregnancy. Obstet Gynecol 64:601–607

Neuman RJ, Lobos E et al (2006) Prenatal smoking exposure and dopaminergic genotypes interact to cause a severe ADHD subtype. Biol Psychiatry

Neuman RJ, Lobos E, Reich W, Henderson CA, Sun LW, Todd RD (2007) Prenatal smoking exposure and dopaminergic genotypes interact to cause a severe ADHD subtype. Biol Psychiatry 61:1320–1328

Nigg JT, Breslau N (2007) Prenatal smoking exposure, low birth weight, and disruptive behavior disorders. J Am Acad Child Adolesc Psychiatry 46:362–369

O'Callaghan MJ, Williams GM, Andersen MJ, Bor W, Najman JM (1997) Obstetric and perinatal factors as predictors of child behaviour at 5 years. J Paediatr Child Health 33:497–503

O'Connor TG, Heron J, Golding J, Beveridge M, Glover V (2002) Maternal antenatal anxiety and children's behavioural/emotional problems at 4 years. Report from the Avon Longitudinal Study of Parents and Children. Br J Psychiatry 180:502–508

Orlebeke JF, Knol DL, Verhulst FC (1999) Child behavior problems increased by maternal smoking during pregnancy. Arch Environ Health 54:15–19

Ostrander MM, Richtand NM, Hermann JP (2003a) Stress and amphetamine induce Fos expression in medial prefrontal cortex neurons containing glucocorticoid receptors. Brain Res 990:209–214

Ostrander MM, Badiani A, Day HE, Norton CS, Watson SJ, Akil H, Robinson TE (2003b) Environmental context and drug history modulate amphetamine-induced c-fos mRNA expression in the basal ganglia, central extended amygdala, and associated limbic forebrain. Neuroscience 120:551–571

O'Sullivan JB, Ryan KM, Curtin NM, Harkin A, Connor TJ (2009) Noradrenaline reuptake inhibitors limit neuroinflammation in rat cortex following a systemic inflammatory challenge: implications for depression and neurodegeneration. Int J Neuropsychopharmacol 12:687–699

Ouellet-Morin I, Wigg KG, Feng Y, Dionne G, Robaey P, Brendgen M, Vitaro F, Simard L, Schachar R, Tremblay RE, Perusse D, Boivin M, Barr CL (2008) Association of the dopamine transporter gene and ADHD symptoms in a Canadian population-based sample of same-age twins. Am J Med Genet B Neuropsychiatr Genet 147B:1442–1449

Palmer AA, Brown AS, Keegan D, Siska LD, Susser E, Rotrosen J, Butler PD (2008) Prenatal protein deprivation alters dopamine-mediated behaviors and dopaminergic and glutamatergic receptor binding. Brain Res 1237:62–74

Parelkar NK, Jiang Q, Chu XP, Guo ML, Mao LM, Wang JQ (2009) Amphetamine alters Ras-guanine nucleotide-releasing factor expression in the rat striatum in vivo. Eur J Pharmacol 619:50–56

Penner MR, McFadyen MP, Pinaud R, Carrey N, Robertson HA, Brown RE (2002) Age-related distribution of c-fos expression in the striatum of CD-1 mice after acute methylphenidate administration. Brain Res Dev Brain Res 135:71–77

Petersohn D, Thiel G (1996) Role of zinc-finger proteins Sp1 and zif268/egr-1 in transcriptional regulation of the human synaptobrevin II gene. Eur J Biochem 239:827–834

Polderman TJ, Derks EM, Hudziak JJ, Verhulst FC, Posthuma D, Boomsma DI (2007) Across the continuum of attention skills: a twin study of the SWAN ADHD rating scale. J Child Psychol Psychiatry 48:1080–1087

Price TS, Simonoff E, Asherson P, Curran S, Kuntsi J, Waldman I, Plomin R (2005) Continuity and change in preschool ADHD symptoms: longitudinal genetic analysis with contrast effects. Behav Genet 35:121–132

Quist JF, Barr CL, Schachar R, Roberts W, Malone M, Tannock R, Basile VS, Beitchman J, Kennedy JL (2003) The serotonin 5-HT1B receptor gene and attention deficit hyperactivity disorder. Mol Psychiatry 8:98–102

Reik W, Dean W, Walter J (2001) Epigenetic reprogramming in mammalian development. Science 293:1089–1093

Rietveld MJ, Hudziak JJ, Bartels M, van Beijsterveldt CE, Boomsma DI (2004) Heritability of attention problems in children: longitudinal results from a study of twins, age 3 to 12. J Child Psychol Psychiatry 45:577–588

Robertson KD (2005) DNA methylation and human disease. Nat Rev Genet 6:597–610

Robinson TE, Kolb B (1997) Persistent structural modifications in nucleus accumbens and prefrontal cortex neurons produced by previous experience with amphetamine. J Neurosci 17:8491–8497

Rodriguez A, Bohlin G (2005) Are maternal smoking and stress during pregnancy related to ADHD symptoms in children? J Child Psychol Psychiatry 46:246–254

Rotllant D, Marquez C, Nadal R, Armario A (2010) The brain pattern of c-fos induction by two doses of amphetamine suggests different brain processing pathways and minor contribution of behavioural traits. Neuroscience 168:691–705

Sarge KD, Park-Sarge OK (2005) Gene bookmarking: keeping the pages open. Trends Biochem Sci 30:605–610

Saudino KJ, Ronald A, Plomin R (2005) The etiology of behavior problems in 7-year-old twins: substantial genetic influence and negligible shared environmental influence for parent ratings and ratings by same and different teachers. J Abnorm Child Psychol 33:113–130

Sawant S, Daviss SR (2004) Seizures and prolonged QTc with atomoxetine overdose. Am J Psychiatry 161:757

Scahill L, Schwab-Stone M (2000) Epidemiology of ADHD in school-age children. Child Adolesc Psychiatr Clin N Am 9:541–555, vii

Scherer D, Hassel D, Bloehs R, Zitron E, von Lowenstern K, Seyler C, Thomas D, Konrad F, Burgers HF, Seemann G, Rottbauer W, Katus HA, Karle CA, Scholz EP (2009) Selective noradrenaline reuptake inhibitor atomoxetine directly blocks hERG currents. Br J Pharmacol 156:226–236

Secker-Walker RH, Vacek PM, Flynn BS, Mead PB (1997) Smoking in pregnancy, exhaled carbon monoxide, and birth weight. Obstet Gynecol 89:648–653

Sharp WS, Gottesman RF, Greenstein DK, Ebens CL, Rapoport JL, Castellanos FX (2003) Monozygotic twins discordant for attention-deficit/hyperactivity disorder: ascertainment and clinical characteristics. J Am Acad Child Adolesc Psychiatry 42:93–97

Shishkina GT, Kalinina TS, Dygalo NN (2004) Attenuation of alpha2A-adrenergic receptor expression in neonatal rat brain by RNA interference or antisense oligonucleotide reduced anxiety in adulthood. Neuroscience 129:521–528

Shumay E, Fowler JS, Volkow ND (2010) Genomic features of the human dopamine transporter gene and its potential epigenetic States: implications for phenotypic diversity. PLoS One 5: e11067

Smoller JW, Biederman J, Arbeitman L, Doyle AE, Fagerness J, Perlis RH, Sklar P, Faraone SV (2006) Association between the 5HT1B receptor gene (HTR1B) and the inattentive subtype of ADHD. Biol Psychiatry 59:460–467

Snyder-Keller AM (1991) Striatal c-fos induction by drugs and stress in neonatally dopamine-depleted rats given nigral transplants: importance of NMDA activation and relevance to sensitization phenomena. Exp Neurol 113:155–165

Son GH, Chung S, Geum D, Kang SS, Choi WS, Kim K, Choi S (2007) Hyperactivity and alteration of the midbrain dopaminergic system in maternally stressed male mice offspring. Biochem Biophys Res Commun 352:823–829

Swanson LW (1982) The projections of the ventral tegmental area and adjacent regions: a combined fluorescent retrograde tracer and immunofluorescence study in the rat. Brain Res Bull 9:321–353

Swanson JM, Volkow ND (2002) Pharmacokinetic and pharmacodynamic properties of stimulants: implications for the design of new treatments for ADHD. Behav Brain Res 130:73–78

Thapar A, Fowler T, Rice F, Scourfield J, van den Bree M, Thomas H, Harold G, Hay D (2003) Maternal smoking during pregnancy and attention deficit hyperactivity disorder symptoms in offspring. Am J Psychiatry 160:1985–1989

Thomas MC, Chiang CM (2006) The general transcription machinery and general cofactors. Crit Rev Biochem Mol Biol 41:105–178

Todd RD, Neuman RJ (2007) Gene-environment interactions in the development of combined type ADHD: evidence for a synapse-based model. Am J Med Genet B Neuropsychiatr Genet 144B:971–975

Trinh JV, Nehrenberg DL, Jacobsen JP, Caron MG, Wetsel WC (2003) Differential psychostimulant-induced activation of neural circuits in dopamine transporter knockout and wild type mice. Neuroscience 118:297–310

Tsankova N, Renthal W, Kumar A, Nestler EJ (2007) Epigenetic regulation in psychiatric disorders. Nat Rev Neurosci 8:355–367

van der Valk JC, Verhulst FC, Neale MC, Boomsma DI (1998) Longitudinal genetic analysis of problem behaviors in biologically related and unrelated adoptees. Behav Genet 28:365–380

Vataeva LA, Kudrin VS, Vershinina EA, Mosin VM, Tiul'kova EI, Otellin VA (2007) Behavioral alteration in the adult rats prenatally exposed to para-chlorophenylalanine. Brain Res 1169:9–16

Volkow ND, Wang GJ, Fowler JS, Ding YS (2005) Imaging the effects of methylphenidate on brain dopamine: new model on its therapeutic actions for attention-deficit/hyperactivity disorder. Biol Psychiatry 57:1410–1415

Vucetic Z, Totoki K, Schoch H, Whitaker KW, Hill-Smith T, Lucki I, Reyes TM (2010) Early life protein restriction alters dopamine circuitry. Neuroscience 168:359–370

Wakschlag LS, Pickett KE, Cook E Jr, Benowitz NL, Leventhal BL (2002) Maternal smoking during pregnancy and severe antisocial behavior in offspring: a review. Am J Public Health 92:966–974

Wang GJ, Chang L, Volkow ND, Telang F, Logan J, Ernst T, Fowler JS (2004) Decreased brain dopaminergic transporters in HIV-associated dementia patients. Brain 127:2452–2458

Wang S, Yan JY, Lo YK, Carvey PM, Ling Z (2009) Dopaminergic and serotoninergic deficiencies in young adult rats prenatally exposed to the bacterial lipopolysaccharide. Brain Res 1265:196–204

Waterland RA, Jirtle RL (2003) Transposable elements: targets for early nutritional effects on epigenetic gene regulation. Mol Cell Biol 23:5293–5300

Weber M, Davies JJ, Wittig D, Oakeley EJ, Haase M, Lam WL, Schubeler D (2005) Chromosome-wide and promoter-specific analyses identify sites of differential DNA methylation in normal and transformed human cells. Nat Genet 37:853–862

Wernicke JF, Faries D, Girod D, Brown J, Gao H, Kelsey D, Quintana H, Lipetz R, Michelson D, Heiligenstein J (2003) Cardiovascular effects of atomoxetine in children, adolescents, and adults. Drug Saf 26:729–740

Wilens TE, Hahesy AL, Biederman J, Bredin E, Tanguay S, Kwon A, Faraone SV (2005) Influence of parental SUD and ADHD on ADHD in their offspring: preliminary results from a pilot-controlled family study. Am J Addict 14:179–187

Wilson MJ, Shivapurkar N, Poirier LA (1984) Hypomethylation of hepatic nuclear DNA in rats fed with a carcinogenic methyl-deficient diet. Biochem J 218:987–990

Wood AJ, Oakey RJ (2006) Genomic imprinting in mammals: emerging themes and established theories. PLoS Genet 2:e147

Wood AC, Rijsdijk F, Saudino KJ, Asherson P, Kuntsi J (2008) High heritability for a composite index of children's activity level measures. Behav Genet 38:266–276

Wu H, Coskun V, Tao J, Xie W, Ge W, Yoshikawa K, Li E, Zhang Y, Sun YE (2010) Dnmt3a-dependent nonpromoter DNA methylation facilitates transcription of neurogenic genes. Science 329:444–448

Yano M, Steiner H (2005a) Methylphenidate (Ritalin) induces Homer 1a and zif 268 expression in specific corticostriatal circuits. Neuroscience 132:855–865

Yano M, Steiner H (2005b) Topography of methylphenidate (ritalin)-induced gene regulation in the striatum: differential effects on c-fos, substance P and opioid peptides. Neuropsychopharmacology 30:901–915

Yin HS, Chen K, Shih JC, Tien TW (2010) Down-regulated GABAergic expression in the olfactory bulb layers of the mouse deficient in monoamine oxidase B and administered with amphetamine. Cell Mol Neurobiol 30:511–519

Zhou R, Zhang Z, Zhu Y, Chen L, Sokabe M (2009) Deficits in development of synaptic plasticity in rat dorsal striatum following prenatal and neonatal exposure to low-dose bisphenol A. Neuroscience 159:161–171

Sexual Differentiation of the Brain and ADHD: What Is a Sex Difference in Prevalence Telling Us?

Jaylyn Waddell and Margaret M. McCarthy

Contents

1 Hormone-Defined Critical Periods of Development and Pathology 342
2 The Hypothalamus: A Model of Many Mechanisms 346
3 Hormones, Neurogenesis, and Apoptosis 348
4 Hormonal Modulation of Excitation Via Effects on GABA 350
5 Hormonal Modulation of Excitation Via Effects on Dopamine 352
6 Conclusion 353
References 354

Abstract Sexual differentiation of the brain is a function of various processes that prepare the organism for successful reproduction in adulthood. Release of gonadal steroids during both the perinatal and the pubertal stages of development organizes many sex differences, producing changes in brain excitability and morphology that endure across the lifespan. To achieve these sexual dimorphisms, gonadal steroids capitalize on a number of distinct mechanisms across brain regions. Comparison of the developing male and female brain provides insight into the mechanisms through which synaptic connections are made, and circuits are organized that mediate sexually dimorphic behaviors. The prevalence of most psychiatric and neurological disorders differ in males versus females, including disorders of attention, activity and impulse control. While there is a strong male bias in incidence of attention deficit and hyperactivity disorders, the source of that bias remains controversial. By elucidating the biological underpinnings of male versus female brain development, we gain a greater understanding of how hormones and genes do and do not contribute to the differential vulnerability in one sex versus the other.

Keywords Development · Estradiol · Gonadal steroids · Sensitive periods · Sexual differentiation

J. Waddell (✉) and M.M. McCarthy
Department of Physiology, School of Medicine, University of Maryland, Baltimore, MD 21201, USA
e-mail: jwadd001@umaryland.edu

Abbreviations

ARC	Arcuate nucleus (of the hypothalamus)
AVPV	Anteroventral periventricular nucleus of the hypothalamus
BNST	Bed nucleus of the stria terminalis
COX-2	Cycloogenase-2
FSH	Follicle-stimulating hormone
KCC2	K^+–Cl^- cotransporter
LH	Luteinizing hormone
MAP	Mitogen-activated protein (kinase)
NKCC1	Na^+–K^+–$2Cl^-$ cotransporter
POA	Preoptic area
SDN	Sexually dimorphic nucleus
SN	Substantia nigra
SNB	Spinal nucleus of the bulbocavernosus
VTA	Ventral tegmental area

1 Hormone-Defined Critical Periods of Development and Pathology

Three basic subtypes of Attention-Deficit/Hyperactivity Disorder (ADHD) describe the two basic dimensions of the disorder: inattention and hyperactivity/impulsivity (American Psychiatric Association 2000). The predominantly inattentive type of ADHD is associated with academic impairment, and is the most common (American Psychiatric Association 2000). The predominantly hyperactive/impulsive and combined subtypes are strongly predictive of externalizing behaviors, characterized by aggression and disruptive behavior (Graetz et al. 2001, 2005). In clinically referred populations, males are estimated to suffer from ADHD 2:1 to 9:1 compared to females (Anderson et al. 1987; Bird et al. 1988). However, sampling of nonreferred populations suggests the incidence may not differ, but rather reflects lower clinical referrals in females (Rucklidge 2010). More generally, pathologies including symptoms of hypermobility, impulsivity, aggression, and disruptive behaviors are more common in boys, thus increasing comorbidity of disruptive behavior disorders in males significantly above that in females (Abikoff et al. 2002; Gaub and Carlson 1997; Rucklidge 2010). Overall, males are more likely to suffer from disorders that manifest early in development, such as ADHD and learning disabilities (American Psychiatric Association 2000), whereas females are more likely to develop mood disorders with a later onset, such as depression (American Psychiatric Association 2000; see Martel et al. 2009 for review). Critical periods of gonadal steroid release correspond to this divergence at the onset of psychiatric disorders in males and females. It is not possible to effectively separate biological variables from societal and cultural influences on human brain development and behavior. Conversely, discerning the impact of experience and external influence on brain development in animal models is modestly successful, at best, but biological variables are readily explored and provide insight into the cellular basis for both variability and vulnerability.

Comparison of the developing male and female brain presents an opportunity to assess potential mechanisms of developmental differences between the sexes. As the brain prepares itself for successful reproduction, gonadal steroids can contribute to pathology by increasing vulnerability, and exacerbating symptoms in males earlier in development. ADHD is a complex developmental disorder characterized by a high degree of heritability (Faraone et al. 2005). Despite the reported sex bias in incidence, none of the genetic loci identified to date as potential contributors to the disorder are located on the X or Y chromosome (Sharp et al. 2009). The genetic contribution to ADHD is multifactorial, with each gene making small but significant contributions to overall risk. The sex bias in prevalence suggests a point of convergence of genetic and hormonal influences. Thus, high circulating concentrations of gonadal steroids in the developing male may be an additional risk factor, interacting with genetic predisposition.

Although controversial, Geschwind proposed that males are at a higher risk for learning disabilities and hyperactivity because testosterone slows development of the brain in early development, rendering males vulnerable to insult for longer periods of time, allowing a wider range of behavioral outcomes (Geschwind and Galaburda 1985; Morris et al. 2004). Consistent with this view, high levels of testosterone in the perinatal period leads to increased neural lateralization, by promoting cell death in the right hemisphere of the brain, and slowing development of the left hemisphere (Geschwind and Galaburda 1985; Goodman 1991). Functional brain imaging of ADHD-diagnosed children reveals reduced volume of the right hemisphere as well as abnormalities of the corpus callosum (reviewed in Seidman et al. 2005). Smaller brain volumes appear to be fixed in ADHD patients across childhood and adolescence suggesting that genetic and early life experiences that contribute to ADHD induce seemingly permanent changes in the brain (Castellanos et al. 2002). Thus, enhanced lateralization of the brain may render cognitive abilities less flexible and permit some of the paramount symptoms of ADHD, such as impulsivity. Indeed, adult females exhibit flexibility in learning strategy in response to fluctuations in estrogens that is not evident in males (Korol and Kolo 2002). This flexibility relies on the absence of perinatal gonadal steroids. In humans, the menstrual cycle drives performance differences in sexually dimorphic tasks, particularly those related to motor and spatial ability (Hampson and Kimura 1998). Gender-specific patterns of lateralized activity have been demonstrated in language processing tasks as well, with females exhibiting greater hemispheric connectivity (Bitan et al. 2010). Although the relationship between hemispheric connectivity and cognitive ability is not completely clear, these results highlight sex differences in nonreproductive areas involved in cognition (Bitan et al. 2010).

Gonadal steroids define developmental sensitive periods of plasticity that shape reproductive success in adulthood (e.g., Phoenix et al. 1959; Rhees et al. 1990a, b). Organization of the male brain is achieved, in part, by two critical periods of elevated testosterone. In rodents, testosterone surges during the last few days of gestation in male fetuses and remains elevated until shortly after birth (Fig. 1; Corbier et al. 1992; Weisz and Ward 1980). In humans, testosterone is elevated in the fetus during the latter portion of the second trimester and remains high for several months after birth but subsequently declines to undetectable levels. During this hormonally defined sensitive window, testosterone, and its metabolites, estradiol and dihydrotestosterone (DHT), evoke permanent changes in brain structure and function (Maclusky and Naftolin 1981; Phoenix et al. 1959). A second postnatal critical period for sexual differentiation of the brain occurs in both sexes during the peripubertal phase. This is characterized again by a surge in testosterone in the male and the onset of the estrous (animal) or menstrual (human) cycle in the female (Schulz and Sisk 2006; Schulz et al. 2004). At this time, social and mating behaviors are shaped by both hormones and experience, refining further the neural circuitry involved in reproductive success.

Most of our current understanding of sexual differentiation of the brain comes from the study of animal models, especially laboratory rats and mice. Sex differences in the brain are most robust and reliable in brain regions directly involved in sex behavior, such as specific nuclei of the hypothalamus. These differences can be volumetric, meaning a region is larger in one sex. Alternatively, or in addition to volume differences, sex difference may be connective, such that the type or amount

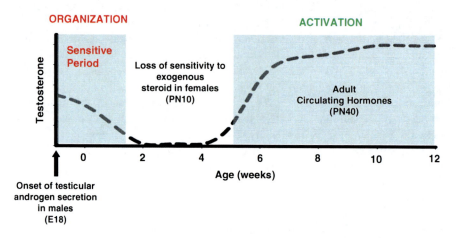

Fig. 1 Sexual differentiation of the brain is a consequence of sex differences in gonadal steroid synthesis during both perinatal and peripubertal sensitive periods. In male rats, the production of testicular androgens begins prenatally, around embryonic day 18, and defines the beginning of the perinatal sensitive period. During this time, the female ovary is quiescent. Sex differences in brain and behavior in adulthood are largely determined by the actions of steroids during the perinatal period, with an additional organizational effect at puberty

of synapses, or size of a particular projection, differs between males and females. Sex differences are also found in the amount of neurochemicals, or neurotransmitters, or the intrinsic excitability of particular classes of neurons (e.g., Nunez and McCarthy 2007; Perrot-Sinal et al. 2003; Davis et al. 1999). In rodents, estradiol is synthesized, by the enzyme aromatase cytochrome P450, from testicularly derived androgens, and mediates many of the sex-specific brain differences that endure through the lifespan (for review, see McCarthy 2008). Male rats castrated shortly after birth exhibit adult sex behaviors characteristic of females following estrogen and progesterone administration. If castration occurs later, after the first few days of life, males do not exhibit female sex behaviors (Fig. 2). In parallel, females exposed to high levels of testosterone or estrogen any time between late gestation and the first 10 days or so of life will not exhibit typical female sex behavior as adults. Instead, the hormonally masculinized female exhibits male sex behaviors when treated with testosterone in adulthood. Testosterone or estradiol exposure after this sensitive developmental window does not change adult female sex behavior (Gerall et al. 1992; Weisz and Ward 1980). Thus, as outlined by Phoenix et al. (1959), gonadal steroids act within critical windows of development to change the brain permanently to support male or female sex behavior in adulthood.

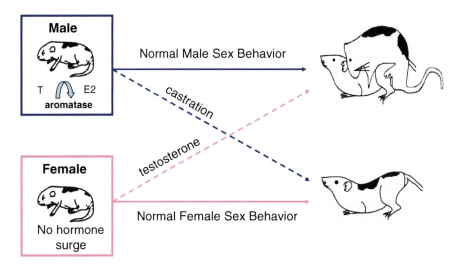

Fig. 2 Testosterone and its aromatized product, estradiol, in the perinatal period are necessary for expression of masculine sex behavior in the adult. Testicularly derived steroids organize the hypothalamus and other brain areas to support mounting of the female, intromission and ejaculation in adulthood. The absence of androgens and estradiol is necessary for normal development of the female brain. This absence of steroids is necessary for the sexually receptive posture in the adult female, termed lordosis. Administration of exogenous testosterone to the neonatal female induces masculinization of adult brain and behavior. Likewise, removal of male gonadal steroids by castration in the neonatal male induces feminized sex behavior in the adult male

2 The Hypothalamus: A Model of Many Mechanisms

The mechanisms by which the brain achieves both the male and the female patterns of sex behavior are diverse. Many of these mechanisms are exemplified in the developing hypothalamus (McCarthy 2008; Schwarz and McCarthy 2008). Within the hypothalamus there are dissociable zones with distinct patterns of connectivity. The periventricular zone of the hypothalamus is situated along the medial ventricular wall, where sensitivity to circulating peptides and gonadal steroids is maximal and most direct (Simerly 2002). From this vantage point, the periventricular zone regulates steroid secretion from the anterior pituitary and is characterized by dense, internally reciprocal connections (Simerly and Swanson 1988; Simerly 2002). This region is also connected to extrahypothalamic structures involved in olfaction, pheromone processing, and social behavior (see Swanson and Petrovich 1998 for review). The medial zone of the hypothalamus innervates the periventricular zone and is reciprocally connected to limbic pathways (Canteras and Swanson 1992; Risold and Swanson 1987). Through these diverse connections, the hypothalamus integrates neuroendocrine responses to environmental cues and helps to guide motivated behavior. The hypothalamus is critically involved in circuits that regulate the response to stress and reward and is a critical mediator of sex behavior. It is not surprising that this brain region is characterized by sexual dimorphisms in morphology and connectivity.

The preoptic area (POA) lies just rostral to, and is closely associated with, the hypothalamus. Neurons located here mediate both male and female specific behaviors that are critical for successful reproduction and rearing of offspring (Numan and Numan 1995; Numan and Callahan 1980; Simerly 2002). In the male, the POA is critical for copulatory behavior, including mounting, intromission, and ejaculation (Christensen et al. 1977). In the adult female, this area supports maternal behaviors, such as nest-building, pup retrieval and protection of offspring (for review, see: Numan 2006). To support these sex-specific behaviors, the POA and other subregions of the hypothalamus undergo critical periods of plasticity, driven largely by exposure to gonadal steroids in the same manner as seen for the hypothalamus and both regions are nexus points for the integration of social behaviors that involve motivation and reward.

In rodents, aromatization of testosterone into estradiol initiates complex interactions between cells to build the circuitry necessary for reproduction. The arcuate nucleus (ARC) lies within the periventricular zone of the hypothalamus, where it regulates release of the gonadotropin hormones, follicle-stimulating hormone (FSH) and luteinizing hormone (LH), from the pituitary (Ojeda and Urbanski 1994; Simerly 2002). Astrocytes in the ARC of neonatal rats exhibit sexually dimorphic morphology. Astrocytes provide metabolic substrates to neighboring neurons and maintain homeostasis in the extrasynaptic space through regulation of glutamate recycling (Brown and Ransom 2007; He and Sun 2007; Nave and Trapp 2008). Beyond these roles, accruing evidence suggests that astrocytes secrete trophic factors as well as neurotransmitters that shape the number and strength of neuronal synapses or spines (Barker and Ullian 2010; Christopherson et al. 2005; Hamilton and Attwell 2010; Kozlov et al. 2006; Ullian et al. 2001). Dendritic spines are the direct source of excitatory signals to neurons; differences in spine number and shape determine electrical and chemical activity in the neuron (Hayashi and Majewski 2005; Bourne and Harris 2008; Sorra and Harris 2000). The ARC of neonatal male rats has more fully differentiated astrocytes, characterized by long, thin processes and many branches (Mong et al. 1996). This sex difference is the direct result of higher testosterone in developing males (Mong et al. 1996) and is coupled with decreased neuronal spine density within the ARC (Mong et al. 1999, 2001). These data suggest that astrocytes participate in synaptic patterning by guiding or blocking the path of neurites. Thus, modulation of astrocyte differentiation affects circuit structure and activity. Sexual differentiation of such mechanisms may contribute to sex differences in motivated behavior (e.g., Becker 2009).

Astrocyte morphology in the male POA is also more complex than in females. Male astrocytes have more branches and longer primary processes and like other subregions of the hypothalamus, astrocyte complexity can be masculinized by estradiol, as can the higher density of dendritic spine synapses (Amateau and McCarthy 2002). Estradiol enhances astrocyte complexity in the POA but it is not known whether this is necessary for the action of estradiol on neuronal spines. Estradiol increases neuronal spine density in the neonatal male POA through upregulation of cyclooxygenase-2 (COX-2; Amateau and McCarthy 2002, 2004), the rate-liming enzyme that converts arachidonic acid to prostaglandins (Brock et al. 1999; Yang and Chen 2008). Prostaglandins mediate inflammation and fever, but are also emerging as regulators of synaptic plasticity in both the developing and the adult brain (Vidensky et al. 2003; Yang and Chen 2008).

A current working model proposes that estradiol acts first in neurons to increase COX-2 concentration and activity, which increases PGE2 release from neurons to act on neighboring astrocytes. This, in turn, increases glutamate release from astrocytes and increases neuronal spine density (Bezzi et al. 1998; McCarthy 2008). Estradiol administration to neonatal females increases COX-2 and PGE2 levels to that of males in the developing POA, and is sufficient to enhance astrocyte complexity and neuronal spine density (Amateau and McCarthy 2002, 2004). Further, female pups, treated with PGE2 within the first few postnatal days, exhibit male-typic sex behavior in adulthood. Likewise, inhibition of COX-2 in the neonatal male brain results in a profound disruption of sex behavior (Amateau and McCarthy 2004). These results predict that males are more vulnerable to insult from commonly used COX inhibitors (nonsteroidal anti-inflammatory drugs), such as aspirin. In brain areas not directly involved in reproduction, such as the hippocampus, COX-2 also modulates excitatory glutamatergic transmission, although estradiol does not appear to change COX-2 activity in the neonatal hippocampus (Amateau and McCarthy 2004). Nevertheless, the action of COX-2 is modulated by a number of stimuli, including seizures, brain injury and NMDA receptor activation (Yang and Chen 2008), but the role of this ubiquitous signaling system in normal brain development remains poorly understood.

Estradiol establishes higher excitatory synapse density in the male mediobasal hypothalamus through mechanisms that differ from those of the POA. Estradiol increases dendrite length and the number of excitatory synapses in the mediobasal hypothalamus via protein synthesis and protein synthesis-independent pathways (Schwarz et al. 2008). In presynaptic neurons, estradiol increases glutamate release which, in turn, activates MAP kinase pathways through NMDA and AMPA receptors in the postsynaptic neuron. This sequence of pre- and postsynaptic modulation of excitation induces a protein synthesis-dependent increase in spine number (Schwarz et al. 2008). Direct pharmacological enhancement of glutamate transmission in the medial basal hypothalamus is sufficient to masculinize spine density. Thus, estradiol establishes masculine levels of spines by modulating glutamate release between neurons (Schwarz et al. 2008).

Administration of testosterone in the neonate also changes astrocyte and neuronal sensitivity to gonadal steroids in the adult. The adult female hypothalamus undergoes synaptic remodeling in response to changing estrogen levels across the phases of the estrus cycle (Garcia-Segura et al. 1994a). The intact male or masculinized female brain does not exhibit this plasticity in adulthood (Garcia-Segura et al. 1994b). Thus, neonatal differentiation of ARC astrocytes by testosterone or estradiol may suppress sensitivity to circulating hormones in the adult. Similar changes in hormone sensitivity in the adult brain induced by neonatal testosterone administration have been demonstrated in areas not directly involved in sex behavior, such as the hippocampus (Garcia-Segura et al. 1998; Gould et al. 1990; Leranth et al. 2003; MacLusky et al. 2006; Woolley and McEwen 1992). Administration of estrogens does not change neuronal morphology in the male hippocampus, but can increase neuronal spine density in the female hippocampus (MacLusky et al. 2006). Thus, the perinatal period is a sensitive period during which gonadal steroids determine sensitivity to the adult hormonal milieu demonstrating the impact gonadal steroids have on the brain throughout the lifespan.

3 Hormones, Neurogenesis, and Apoptosis

Hormones can promote cell survival or cell death, depending on the brain region and sex of the animal (Forger 2006). Testosterone and its metabolites inhibit cell death in the perinatal POA, the bed nucleus of the stria terminalis (BNST) and the spinal nucleus of the bulbocavernosus (SNB), all of which are larger in the gonadally intact male (Arai et al. 1996; Breedlove and Arnold 1983; Gotsiridze et al. 2007; Guillamon et al. 1988; Nordeen et al. 1985; Murakami and Arai 1989). In the rat, sex differences in cell death contribute to sexual dimorphisms in brain circuitry (Forger et al. 2004). The most striking example is the sexually dimorphic nucleus (SDN) of the POA (Gorski et al. 1978). The male SDN is 3–5 times larger than that of the female: this difference is driven by higher rates of cell death in females, postnatally which are due to a lack of testosterone and its metabolite, estradiol (Davis et al. 1996). In males, testosterone decreases cell death in the SDN, as well as other brain regions involved in sex behavior (Dodson and Gorski 1992; Gotsiridze et al. 2007; Nordeen et al. 1985; Murakami and Arai 1989; Davis et al. 1996). Castration of male rats at birth causes a reduction in SDN size. Conversely, administration of testosterone or estradiol increases SDN size in the newborn female rat (Dohler et al. 1982; Gorski 1984).

In most areas that have been studied, gonadal steroids modulate cell death through two proteins, Bcl2 and Bax (Forger et al. 2004; Holmes et al. 2009; Zup and Forger 2002; Zup et al. 2003). The Bcl2 protein family regulates apoptotic programs in many cell types (Merry and Korsmeyer 1997). Testosterone and its metabolites can increase Bcl2, which inhibits the apoptotic action of Bax, which is a second class of proteins in this family (Garcia-Segura et al. 1998; Pike 1999; Zup and Forger 2002). Specific subpopulations of cells are vulnerable to cell death during specific critical phases and the impact of exogenous hormone administration depends on the brain region and sex of the animal. For instance, the anteroventral periventricular nucleus of the hypothalamus (AVPV) contains a greater number of dopaminergic neurons in females, which can be reduced to that in males by administration of testosterone or estradiol (Simerly et al. 1985; Waters and Simerly 2009; Krishnan et al. 2009). Deletion of Bax eliminates volumetric sex differences in the BNST and the AVPV (Forger et al. 2004).

Puberty is a second developmental period of brain organization that refines circuits that have sexually differentiated during early neural development (Schulz et al. 2009). During puberty, gonadal steroids profoundly shape social behaviors in both sexes (Schulz and Sisk 2006). At this time, both testicular and ovarian steroids mediate the addition of new cells in sexually dimorphic brain regions. The sexually dimorphic pattern of cell genesis parallels that observed perinatally: new cells are added in the brain regions of the sex that exhibits the larger volume in adulthood. Thus, more proliferating cells are detected in the pubertal female AVPV and, likewise, more proliferating cells are detected in the male SDN (Ahmed et al. 2006). The addition of new cells during the prepubertal period has not been directly tied to behavior or hormones but hormones during the pubertal transition further differentiate sexual behavior, aggression, and territoriality (Schulz et al. 2004; Schulz and Sisk 2006). Thus, gonadal steroids also define, in part, this sensitive period during which social behavior is shaped to support reproductive success in adulthood.

In rats, a distinct period of rough and tumble play begins to emerge around 19 days of age and continues to develop and peak between 25 and 40 days of age (Panksepp 1981). This play occurs at higher frequencies in male rats (Olioff and Stewart 1978; Panksepp 1981; Pellis et al. 1994). A complex repertoire of defensive and aggressive behaviors is shaped through social interaction. Deprivation of play fighting produces long-lasting deficits in social behavior, and increases displays of anxiety and aggression (Bell et al. 2010; Bock et al. 2008; Panksepp and Beatty 1980; van den Berg et al. 1999). This reflects a second critical period in plasticity, during which overabundant synaptic connections are pruned as development of the brain progresses (Zehr et al. 2006). The lack of permanence of some neural connections is highlighted by experience-evoked changes, through which stabilization of synapses and dendritic arborizations is refined (e.g., Rakic et al. 1986;

Bock et al. 2005, 2008). In prefrontal cortical regions, which are extensively connected to the limbic system, neuronal morphology is shaped by stress and social isolation in the neonatal and adolescent rat (Helmeke et al. 2001; Ovtscharoff and Braun 2001; Poeggel et al. 2003; Bock et al. 2005). Stressors experienced in early development can be reversed or ameliorated through socialization during puberty (Bock et al. 2008). Interestingly, many children diagnosed with ADHD experience an amelioration of symptoms at puberty (Gittelman et al. 1985). This may relate to the maturation of the frontal cortex which modulates response inhibition, impulsivity, and dopaminergic reward circuits (Bell et al. 2010; Bock et al. 2008).

4 Hormonal Modulation of Excitation Via Effects on GABA

Males are more vulnerable than females to disorders of excitability and movement, such as epilepsy, Tourette's syndrome, ADHD, and Parkinson's disease (Haaxma et al. 2007; McHugh and Delanty 2008; Shulman 2007). An additional contributing variable to insult in males could be the enhanced neuronal excitation induced by steroids and manifest through multiple systems as highlighted above. Processes that lead to higher cellular excitation are inherently fraught with greater risk for disruption, either by inadvertently triggering cell death programs, or over-exuberant innervation and inappropriate pruning of extraneous synapses. Hormone-mediated sex differences in the developmental maturation of neurons in brain areas involved in cognition, movement, and motivation produce sex differences in the timing of sensitive periods such that events can differentially alter excitation between the sexes, in both direction and magnitude (Auger et al. 2001; Nunez and McCarthy 2007; Galanopoulou 2006, 2008a, b; Perrot-Sinal et al. 2003).

Excitation in the developing brain is mediated through both GABAergic and glutamatergic systems. In early cell development, GABA induces cell membrane depolarization and is the primary excitatory neurotransmitter in immature neurons. GABA also functions as a trophic factor, regulating neurite outgrowth and spine formation (Cancedda et al. 2007; Cherubini et al. 1991; Owens and Kriegstein 2002; Pfeffer et al. 2009; Plotkin et al. 1997; Sipila et al. 2006; Spritzer 2006). As cells mature, GABA gradually becomes inhibitory and glutamate emerges as the primary excitatory neurotransmitter (e.g., Ben Ari et al. 1997; Tyzio et al. 1999). Gonadal steroids play a complex role in this developmental progression. Here, we discuss the mechanisms of this "developmental switch" and differences in its developmental progression between the sexes. Across seemingly disparate brain regions, males switch from depolarizing GABA to hyperpolarizing GABA more slowly than females (Nunez and McCarthy 2007, 2008; Galanopoulou 2006, 2008a; Kyrozis et al. 2006; Perrot-Sinal et al. 2003, 2007).

The developmental progression, from excitatory to inhibitory GABA, is driven by the balance between two cation-chloride cotransporters: the $Na^+-K^+-2Cl^-$ cotransporter, NKCC1, and the K^+-Cl^- cotransporter, KCC2 (Riviera et al. 1999). Males have higher levels of NKCC1 protein in the first week of life relative to females in both

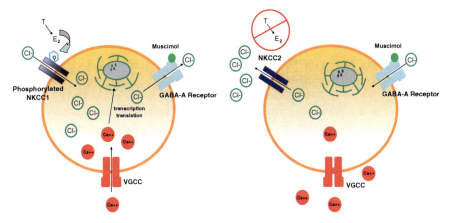

Fig. 3 GABA is largely excitatory in the developing brain. This is due to a shift in the reversal potential for Cl$^-$. Administration of the GABA-A agonist, muscimol, opens Cl$^-$ channels, resulting in Cl$^-$ efflux and depolarization of the cell. This depolariziation is sufficient to activate voltage-gated Ca^{2+} channels and allow influx of Ca^{2+}. Gonadal steroids enhance the excitatory actions of GABA in immature cells manifesting as increased Ca^{2+} entry, increased numbers of cells exhibiting excitation in response to GABA-A agonists and extension of the developmental expression of depolarizing GABA. Gonadal steroids increase the amount and activity of the Na$^+$–K$^+$–Cl$^-$ cotransporter, NKCC1, possibly through its phosphorylation. Depolarizing action of GABA is more protracted in the developing male brain relative to females, suggesting that the absence of gonadal steroids allows this earlier switch in females

the hypothalamus and the hippocampus (Nunez and McCarthy 2007; Perrot-Sinal et al. 2007). The sex difference in magnitude and duration of NKCC1 expression is robust in early postnatal life, but dissipates by 2 weeks of age (Galanopoulou 2008a; Nunez and McCarthy 2007; Perrot-Sinal et al. 2007). Sex differences in chloride transporter expression correspond to robust GABA-mediated sex differences in measures of Ca^{2+}-dependent cell excitability (Auger et al. 2001; Nunez and McCarthy 2008; Perrot-Sinal et al. 2003). Female hippocampal neurons exhibit desensitization to GABA stimulation and so do not respond to a second exposure to the GABA-A receptor agonist muscimol (Nunez and McCarthy 2008). This sex difference represents not only a difference in the magnitude of excitation, but also cellular adaption to GABAergic excitation. Application of gonadal steroids can reverse this, suggesting that androgens promote excitability in response to GABA by attenuating desensitization (Nunez and McCarthy 2007, 2008). Estradiol similarly promotes depolarizing responses to GABA by extending the duration of expression of NKCC1 in the developing hippocampus (Fig. 3; Nunez and McCarthy 2007).

In the developing hippocampus, GABA-mediated excitation is critically involved in the incorporation of new neurons into circuits, followed by promotion of neurite outgrowth and synaptogenesis (Estrada et al. 2006; Ge et al. 2006). This influence of GABA on cell maturation and incorporation into mature circuitry is also important in adult hippocampal neurogenesis (Esposito et al. 2005; Tozuka et al. 2005). The sequential contribution of GABA and glutamate corresponds to the development of the dendrite and increase in soma size (Tyzio et al. 1999; Demarque et al. 2002). Perturbation of the temporal progression from GABA to glutamate-mediated excitation can induce profound changes in cell morphology (Barbin et al. 1993; Cancedda et al. 2007). Depolarizing GABA may promote cell development in synaptically silent, immature cells, through paracrine modulation of cell morphology (Barbin et al. 1993; Demarque et al. 2002). Excitability in the developing substantia nigra (SN) is similarly modulated by gonadal steroids (Galanopoulou and Moshe 2003; Galanopoulou 2006). The SN plays an established role in seizure control and movement (Giorgi et al. 2007; Iadarola and Gale 1992). Males are more likely to suffer from both seizure and movement disorders, suggesting a sexual dimorphism in this brain region (McHugh and Delanty 2008).

Developmental GABA responsivity in the SN is sexually dimorphic (Galanopoulou 2006; Kyrozis et al. 2006). For instance, 2 weeks after birth, the GABA-A agonist, muscimol, has a proconvulsive effect when injected into the male rat SN, and increases Ca^{2+} influx (Galanopoulou et al. 2003; Galanopoulou 2006; Sperber et al. 1987). The same treatment does not induce convulsions (i.e., seizures) or Ca^{2+} entry in female rat pups of the same age (Galanopoulou 2006; Veliskova and Moshe 2001). This difference is dependent upon testicularly derived steroids (Veliskova and Moshe 2001). As in the hippocampus, expression of KCC2 is slower to develop in males than females (Galanopoulou et al. 2003). Both androgen and estrogen receptor expression differ between the sexes, and also differ across development within the SN (Ravizza et al. 2003). At birth, the male SN contains more androgen and estrogen receptors, and this pattern reverses by the first day of life (Ravizza et al. 2002, 2003). Administration of testosterone to females abolishes this sex difference (Ravizza et al. 2003). Testosterone or DHT, a potent androgen receptor agonist, upregulates KCC2 mRNA in both sexes, whereas estradiol downregulates KCC2 mRNA in males, but not females (Galanopoulou and Moshe 2003; Galanopoulou 2006). This opposing effect of androgen relative to estradiol suggests a complex steroid sensitivity that is dependent on the sex of the animal.

5 Hormonal Modulation of Excitation Via Effects on Dopamine

Gonadal steroids contribute to sex differences in dopaminergic circuits and transmission. These differences may be particularly relevant to sex differences in prevalence of ADHD as dopamine-modulating drugs are most widely used to treat ADHD. Because of their established role in reward, motivation, decision making and cognition, dopaminergic pathways and the prefrontal cortex have been the subject of intense study in regard to the neurobiology of ADHD (e.g., Seidman et al. 2005). Ascending dopaminergic projections from the ventral tegmental area (VTA), which have a key role in reward pathways, guide motivated behaviors that differ between the sexes, such as rearing of offspring (Becker 2009). These circuits are modulated by gonadal steroids during both perinatal development and puberty. Exposure to steroids at these critical phases of development induces sex differences in the effects of hormones in the adult (Kritzer 1997, 1998, 2003; Stewart and Rajabi 1994). In the neonate, testicularly derived steroids are necessary for masculine neuronal architecture and neuronal survival in both the VTA and the SN (Becker 2009). Projections from these dopaminergic midbrain structures to prefrontal cortex and other association cortices are sexually dimorphic (Kritzer and Cruetz 2008). Castration at birth reduces catecholamine activity in the prefrontal cortex and cortical regions regulating motor control and attention (Kritzer 1998; Stewart and Rajabi 1994). Furthermore, removal of perinatal steroids through castration also produces hemispheric differences in dopamine projections to cortical circuits, suggesting that these pathways are lateralized in a sex-specific manner (Kritzer 1998). The BNST, a sexually differentiated area of the limbic system, sends excitatory projections to the VTA (Hines et al. 1992; Jalabert et al. 2009), which have been directly tied to impulsive behavior in animal models of drug-seeking behavior (Aston-Jones and Harris 2004). Precisely how gonadal steroids might contribute to sex differences in this projection remains largely unexplored.

Sex differences have also been noted in the distribution and density of dopamine receptors in the striatum, the nucleus accumbens, and the prefrontal cortex during juvenile development in rats, with males exhibiting a much higher increase in receptor expression during puberty than females (Andersen and Teicher 2000). Symptoms of ADHD often attenuate during and after puberty, suggesting that gonadal steroids may shape the disorder across development (Gittelman et al. 1985). As noted above, social experience during puberty can attenuate some deleterious effects of early-life stress (Bock et al. 2008). It is possible that those children receiving treatment, whether psychotherapeutic or pharmacological, achieve healthy social interactions during this second critical window.

6 Conclusion

Although it is clear that perinatal exposure to testosterone and its metabolites influence neural development through a myriad of mechanisms, the contribution of steroids to sex differences in prevalence of pathology is not well understood. ADHD is characterized by inattentiveness, impulsivity, and hyperactivity (American Psychiatric Association 2000). If sex is a predictor of pathologies such as ADHD, then we predict that aberrant developmental processes will occur during periods of dynamic changes in exposure to gonadal steroids. In males, there is a perinatal sensitive period of elevated androgens and estrogens that females do not experience, and this is a dynamic period for cell birth, death, differentiation, and synaptogenesis. Puberty is a second period of dynamic steroid hormone profiles, with females experiencing dramatic increases in estrogens and progesterone, which are themselves dynamic due to the cyclical nature of female reproduction, while males have a resurgence in androgen production which remains elevated throughout adulthood. The onset of puberty, and its attendant hormonal changes, is considerably earlier in females: as much as 5 years is normal in humans. This, combined with different maturation rates for distinct portions of the brain in males and females, creates a complex and variable developing brain that is further influenced by experience, environment, and behavioral expression itself. Understanding the relative contribution of each variable to normal brain development in males versus females, and to then translate this into clinically meaningful predictors or indicators of pathology, is a goal best achieved by studying each in isolation as well as an integrated whole.

References

Abikoff HB, Jensen PS, Arnold LLE et al (2002) Observed classroom behavior of children with ADHD: relationship to gender and comorbidity. J Abnorm Child Psychol 30:349–359

Ahmed EI, Zehr JL, Schulz KM, Lorenz BH, DonCarlos LL, Sisk CL (2006) Pubertal hormones modulate the addition of new cells to sexually dimorphic brain regions. Nat Neurosci 11:995–997

Amateau SK, McCarthy MM (2002) A novel mechanism of dendritic spine plasticity involving estradiol induction of protaglandin E2. J Neurosci 22:8586–8596

Amateau SK, McCarthy MM (2004) Induction of PGE2 by estradiol mediates developmental masculinization of sex behavior. Nat Neurosci 7:643–650

American Psychiatric Association (2000) Diagnostic and statistical manual of mental disorders, 4th edn. APA, Washington DC

Andersen SL, Teicher MH (2000) Sex differences in dopamine receptors and their relevance to ADHD. Neurosci Biobehav Rev 24:137–141

Anderson JC, Williams S, McGee R, Silva PA (1987) DSM-III disorders in preadolescent children: prevalence in a large sample from the general population. Arch Gen Psychiatry 44:69–76

Arai Y, Sekine Y, Murakami S (1996) Estrogen and apoptosis in the developing sexually dimorphic preoptic area in female rats. Neurosci Res 25:403–407

Aston-Jones G, Harris GC (2004) Brain substrates for increased drug seeking during protracted withdrawal. Neuropharmacology 47:167–179

Auger AP, Perrot-Sinal TS, McCarthy MM (2001) Excitatory versus inhibitory GABA as a divergence point in steroid-mediated sexual differentiation of the brain. Proc Natl Acad Sci 98:8059–8064

Barbin G, Pollard H, Gaïarsa JL, Ben-Ari Y (1993) Involvement of GABAA receptors in the outgrowth of cultured hippocampal neurons. Neurosci Lett 152:150–154

Barker AJ, Ullian EM (2010) Astrocytes and synaptic plasticity. Neuroscientist 16:40–50

Becker JB (2009) Sexual differentiation of motivation: a novel mechanism? Horm Behav 55:646–654

Bell HC, Pellis SM, Kolb B (2010) Juvenile peer play experience and the development of the orbitofrontal cortices. Behav Brain Res 207:7–13

Ben Ari Y, Khazipov R, Leinekugel X, Caillard O, Gaiarsa JL (1997) GABAA, NMDA and AMPA receptors: a developmentally regulated 'menage a trois'. Trends Neurosci 20:523–529

Bezzi P, Camingnoto G, Pasti L, Vesce S, Rossi D, Rizzini BL, Pozzan T, Volterra A (1998) Prostaglandins stimulate calcium-dependent glutmate release in astrocytes. Nat 391:281–285

Bird HR, Canino G, Rubio-Stipec M, Gould MS, Ribera J, Sesman M, Woodbury M, Huertas-Goldman S, Pagan A et al (1988) Estimates of the prevalence of childhood maladjustment in a community survey in Puerto Rico. Arch Gen Psychiatry 45:1120–1126

Bitan T, Lifshitz A, Breznitz Z, Booth JR (2010) Bidirectional connectivity between hemispheres occurs at multiple levels in language processing but depends on sex. J Neurosci 30:11576–11585

Bock J, Gruss M, Braun K (2005) Experience-induced changes of dendritic spine densities in the prefrontal and sensory cortex: correlation with developmental time windows. Cereb Cortex 15:802–808

Bock J, Murmu RP, Ferdman N, Leshem M, Braun K (2008) Refinement of dendritic and synaptic networks in the rodent anterior cingulate and orbitofrontal cortex: critical impact of early and late social experience. Dev Neurobiol 68:685–695

Bourne JN, Harris KM (2008) Balancing structure and function at hippocampal dendritic spines. Annu Rev Neurosci 31:47–67

Breedlove SM, Arnold AP (1983) Hormonal control of a developing neuromuscular system. I. Complete demasculinization of the male rat spinal nucelus of the bulbocavernosus using the anti-androgen flutamide. J Neurosci 3:424–432

Brock TG, McNish RW, Peters-Golden M (1999) Arachidonic acid is preferentially metabolized by cyclooxygenase-2 to prostacyclin and prostaglandin E2. J Biol Chem 274:11660–1166

Brown AM, Ransom BR (2007) Astrocyte glycogen and brain energy metabolism. Glia 55:1263–1267

Cancedda L, Fiumelli H, Chen K, Poo M-M (2007) Excitatory GABA action is essential for morphological maturation of cortical neurons in vivo. J Neurosci 27:5224–5239

Canteras NS, Swanson LW (1992) Projections of the ventral subiculum to the amygdala, septum, and hypothalamus: a PHAL anterograde tract-tracing study in the rat. J Comp Neurol 324:180–194

Castellanos FX, Lee PP, Sharp W, Jeffries NO, Greenstein DK, Clasen LS, Blumenthal JD, James RS, Ebens CL, Walter JM, Zijdenbos A, Evans AC, Giedd JN, Rapoport JL (2002) Developmental trajectories of brain volume abnormalities in children and adolescents with attention-deficit/hyperactivity disorder. JAMA 288:1740–1748

Cherubini E, Gaiarsa JL, Ben-Ari Y (1991) GABA: an excitatory transmitter in early postnatal life. Trends Neurosci 14:515–519

Christensen LW, Nance DM, Gorski RA (1977) Effects of hypothalamic and preoptic lesions on reproductive behavior in male rats. Brain Res Bull 2:137–141

Christopherson KS, Ullian EM, Stokes CC, Mullownet CE, Hell JW, Agah A et al (2005) Thrombopondins are astrocyte-secreted proteins that promote CNS synaptogenesis. Cell 120:421–433

Corbier P, Edwards DA, Roffi J (1992) The neonatal testosterone surge: a comparative study. Arch Int Physiol Biochim Biophys 100:127–131

Davis EC, Shryne JE, Gorski RA (1996) Structural sexual dimorphisms in the anteroventral periventricular nucleus of the rat hypothalamus are sensitive to gonadal steroids perinatally, but develop peripubertally. Neuroendocrinology 63:142–148

Davis AM, Ward SC, Selmanoff M, Herbison AE, McCarthy MM (1999) Developmental sex differences in amino acid neurotransmitter levels in hypothalamic and limbic areas of rat brain. Neuroscience 90:1471–1482

Demarque M, Represa A, Becq H, Khalilov I, Ben-Ari Y, Aniksztejn L (2002) Paracrine intercellular communication by a Ca2+ and SNARE-dependent release of GABA and glutamate prior to synapse formation. Neuron 36:1051–1061

Dodson RE, Gorski RA (1992) Testosterone propionate administration prevents the loss of neurons within the central part of the medial preopic nucleus. J Neurobiol 24:80–88

Dohler KD, Coquelin A, Davis F, Hines M, Shryne JE, Gorski RA (1982) Differentiation of the sexually dimorphic nucleus in the preoptic area of the rat brain is determined by the perinatal hormone environment. Neurosci Lett 33:295–298

Esposito MS, Piatti VC, Laplagne DA, Morgenstern NA, Ferrari CC, Pitossi FJ, Schinder AF (2005) Neuronal differentiation in the adult hippocampus recapitulates embryonic development. J Neurosci 25:10074–10086

Estrada M, Uhlen P, Ehrlich BE (2006) Ca2+ oscillations induced by testosterone enhance neurite outgrowth. J Cell Sci 119:733–743

Faraone SV, Perlis RH, Doyle AE, Smoller JW, Goralnick JJ, Holmgren MA, Sklar P (2005) Molecular genetics of attention-deficit/hyperactivity disorder. Biol Psychiatry 57:1313–1323

Forger NG (2006) Cell death and sexual differentiation of the nervous system. Neuroscience 138:929–938

Forger NG, Rosen GJ, Waters EM, Jacob D, Simerly RB, de Vries GJ (2004) Deletion of Bax eliminates sex differences in the mouse forebrain. Proc Natl Acad Sci 101:13666–13671

Galanopoulou AS (2006) Sex- and cell-type-specific patterns of GABAA receptor and estradiol-mediated signaling in the immature rat substantia nigra. Eur J Neurosci 23:2423–2430

Galanopoulou AS (2008a) Dissociated gender-specific effects of recurrent seizures on GABA signaling in CA1 pyramidal neurons: role of GABA-A receptors. J Neurosci 28:1557–1567

Galanopoulou AS (2008b) Sexually dimorphic expression of KCC2 and GABA function. Epilepsy Res 80:99–113

Galanopoulou AS, Moshe SL (2003) Role of sex hormones in the sexually dimorphic expression of KCC2 in rat substantia nigra. Exp Neurol 184:1003–1009

Galanopoulou AS, Kyrozis A, Claudio OI, Stanton PK, Moshe SL (2003) Sex-specific KCC2 expression and GABA-A receptor function in rat substantia nigra. Exp Neurol 183:628–637

Garcia-Segura LM, Luquin S, Parduca A, Naftolin F (1994a) Gonadal hormone regulation of glial fibrillary acidic protein immunoreactivity and glial ultrastructure in the rat neuroendocrine hypothalamus. Glia 10:59–69

Garcia-Segura LM, Chowen JA, Duenanas M, Torres-Aleman I, Naftolin F (1994b) Gonadal steroids as promotoers of neuro-glial plasticity. Psychoneuroendocrinology 19:317–345

Garcia-Segura LM, Cardona-Gomez P, Naftolin F, Chowen JA (1998) Estradiol upregulates Bcl-2 expression in adult brain neurons. Neuroreport 9:593–597

Gaub M, Carlson CL (1997) Gender differences in ADHD: a meta-analysis and critical review. J Am Acad Child Adolesc Psychiatry 36:1036–1045

Ge S, Goh ELK, Sailor KA, Kitabatake Y, G-l M, Song H (2006) GABA regulates synaptic integration of newly generated neurons in the adult brain. Nature 439:589–593

Gerall AA, Moltz H, Ward IL (1992) Sexual differentiation: handbook of behavioral neurobiology. Plenum, New York

Geschwind N, Galaburda AM (1985) Cerebral lateralization. Biological mechanisms, associations, and pathology: I. A hypothesis and a program for research. Arch Neurol 42:428–459

Giorgi FS, Veliskova J, Chudomel O, Kyrozis A, Moshe SL (2007) The role of substantia nigra pars reticulata in modulating clonic seizures is determined by testosterone levels during the immediate postnatal period. Neurobiol Dis 25:73–79

Gittelman R, Mannuzza S, Shenker R, Bonagura N (1985) Hyperactive boys almost grown up; I: psychiatric status. Arch Gen Psychiatry 42:937–947

Goodman R (1991) Developmental disorders and structural brain development. In: Rutter M, Casaer P (eds) Biological risk factors for psychosocial disorders. Cambridge University Press, Cambridge, pp 260–310

Gorski RA (1984) Critical role for the medial preoptic area in the sexual differentiation of the brain. Prog Brain Res 61:129–146

Gorski RA, Gordon JH, Shryne JE, Southam AM (1978) Evidence for a morphological sex difference within the medial preoptic area of the rat brain. Brain Res 148:333–346

Gotsiridze T, Kang N, Jacob D, Forger NG (2007) Development of sex differences in the principal nucleus of the bed nucleus of the stria terminalis of mice: role of Bax-dependent cell death. Dev Neurobiol 67:355–362

Gould E, Woolley CS, Frankfurt M, McEwen BS (1990) Gonadal steroids regulate spine synapse density in hippocampal pyramidal cells in adulthood. J Neurosci 10:1286–1291

Graetz BW, Sawyer MG, Hazell PL, Arney F, Baghurst P (2001) Validity of DSV-IV subtypes in a nationally representative sample of Australian children and adolesents. J Am Acad Child Adol Psychiatry 40:1410–1417

Graetz BW, Sawyer MG, Baghurst P (2005) Gender differences among children with DSM-IV ADHD in Australia. J Am Acad Child Adol Psychiatry 44:159–168

Guillamon A, Segovia S, del Abril A (1988) Early effects of gonadal steroids on the neuron number in the medial posterior region and the lateral division of the bed nucleus of the stria terminalis in the rat. Brain Res 44:281–290

Haaxma C, Bloem BR, Borm GF et al (2007) Gender differences in Parkinson's disease. J Neurol Neurosurg Psychiatry 78:819–824

Hamilton NB, Attwell D (2010) Do astrocytes really exocytose neurotransmitters? Nat Rev 11:227–238

Hampson E, Kimura D (1998) Reciprocal effects of hormonal fluctuations on human motor and perceptural-spatial skills. Behav Neurosci 102:356–359

Hayashi Y, Majewski A (2005) Dendritic spine geometry: functional implication and regulation. Neuron 46:529–532

He F, Sun YE (2007) Glial cells more than support cells? Int J Cell Biol 39:661–665

Helmeke C, Ovtscharoff W Jr, Poeggel G, Braun K (2001) Juvenile emotional experience alters synaptic inputs on pyramidal neurons in the anterior cingulate cortex. Cereb Cortex 11:717–727

Hines M, Allen LS, Gorski RA (1992) Sex differences in subregions of the medial nucleus of the amygdala and the bed nucleus of the stria terminalis of the rat. Brain Res 579:321–326

Holmes MM, McCutcheon J, Forger NG (2009) Sex differences in NeuN and androgen receptor-positive cells in the bed nucleus of the stria terminalis are due to BAX-dependent cell death. Neuroscience 158:1251–1256

Iadarola MJ, Gale K (1992) Substantia nigra: site of anticonvulsant activity mediated by g-aminobutyric acid. Science 218:1237–1240

Jalabert M, Aston-Jones G, Herzog E, Manzoni O, Georges F (2009) Role of the bed nucleus of the stria terminalis in the control of the ventral tegmental area dopamine neurons. Prog Neuropsychopharmacol Biol Psychiatr 33:1336–1346

Korol DL, Kolo LL (2002) Estrogen-induced changes in place and response learning in young adult female rats. Behav Neurosci 116:411–420

Kozlov AS, Angulo MC, Audinat E, Charpak S (2006) Target cell-specific modulation of neuronal activity by astrocytes. Proc Natl Acad Sci 103:10058–10063

Krishnan S, Intlekofer KA, Aggison LK, Petersen SL (2009) Central role of TRAF-interacting protein in a new model of brain sexual differentiation. Proc Natl Acad Sci 106:16692–16697

Kritzer MF (1997) Selective colocalization of immunoreactivity for intracellular gonadal hormone receptors and tyrosine hydroxylase in the ventral tegmental area, substantia nigra, and retrorubral fields in the rat. J Comp Neurol 379:247–260

Kritzer MF (1998) Perinatal gonadectomy exerts regionally selective, lateralized effects on the density of axons immunoreactive for tyrosine hydroxylase in the cerebral cortex of adult male rats. J Neurosci 18:10735–10748

Kritzer MF (2003) Long-term gonadectomy affects the density of tyrosine hydroxylase- but not dopamine-beta-hydroxylase-, choline acetyltransferase-or serotonin-immunoreactive axons in the medial prefrontal cortices of adult male rats. Cereb Cortex 13:282–296

Kritzer MF, Cruetz LM (2008) Region and sex differences in constituent dopamine neurons and immunoreactivity for intracellular estrogen and androgen receptors in mesocortical projections in rats. J Neurosci 28:9525–9535

Kyrozis A, Chudomel O, Moshe SL, Galanopoulou AS (2006) Sex-dependent maturation of GABAA receptor-mediated synaptic events in rat substantia nigra reticulata. Neurosci Lett 398:1–5

Leranth C, Petnehazy O, MacLusky NJ (2003) Gonadal hormones affect spine synaptic density in the CA1 hippocampal subfield of male rats. J Neurosci 23:1588–1592

MacLusky NJ, Naftolin F (1981) Sexual differentiation of the central nervous system. Science 211:1294–1302

MacLusky NJ, Hajszan T, Prange-Kiel J, Leranth C (2006) Androgen modulation of hippocampal synaptic plasticity. Neuroscience 138:957–965

Martel MM, Klump K, Nigg JT, Breedlove SM, Sisk CL (2009) Potential hormonal mechanisms of attention-deficit/hyperactivity disorder and major depressive disorder: a new perspective. Horm Behav 55:465–479

McCarthy MM (2008) Estradiol and the developing brain. Physiol Rev 88:91–134

McHugh JC, Delanty N (2008) Epidemiology and classification of epilepsy: gender comparisons. Int Rev Neurobiol 83:11–26

Merry DE, Korsmeyer SJ (1997) Bcl-2 gene family in the nervous system. Annu Rev Neurosci 20:245–267

Mong JA, Kurzweil RL, Davis AM, Rocca MS, McCarthy MM (1996) Evidence for sexual differentiation of glia in rat brain. Horm Behav 30:553–562

Mong JA, Glaser E, McCarthy MM (1999) Gonadal steroid promote glial differentiation and alter neuronal morphology in the developing hypothalamus in a regionally specific manner. J Neurosci 19:1464–1472

Mong JA, Roberts RC, Kelly JJ, McCarthy MM (2001) Gonadal steroids reduce the density of axospinous synapses in the developing rat arcuate nucleus: an electron microscopy study. J Comp Neurol 432:259–267

Morris JA, Jordan CL, Breedlove SM (2004) Sexual differentiation of the vertebrate nervous system. Nat Neurosci 7:1034–1039

Murakami S, Arai Y (1989) Neuronal death in the developing sexually dimorphic periventricular nucleus of the preoptic area in the female rat: effect of neonatal androgen treatment. Neurosci Lett 102:185–190

Nave KA, Trapp BD (2008) Axon-glial signaling and the glial supportof axon function. Annu Rev Neurosci 31:535–561

Nordeen EJ, Nordeen KW, Sengelaub DR, Arnold AP (1985) Androgens prevent normally occurring cell death in a sexually dimorphic spinal nucleus. Science 229:671–673

Numan M (2006) Hypothalamic neural circuits regulating maternal responsivenesstoward infants. Behav Cogn Neurosci Rev 5:163–190

Numan M, Callahan EC (1980) The connections of the medial preoptic region and maternal behavior in the rat. Physiol Behav 25:653–665

Numan M, Numan MJ (1995) Importance of pup-related sensory inputs and maternal performance for the expression of Fos-like immunoreactivity in the preoptic area and ventral bed nucleus of the stria terminalis of postpartum rats. Behav Neurosci 109:135–149

Nunez JL, McCarthy MM (2007) Evidence for an extended duration of GABA-mediate excitation in the developing male versus female hippocampus. Dev Neurobiol 67:1879–1890

Nunez JL, McCarthy MM (2008) Androgens predispose males to GABA-A-mediated excitotoxicity in the developing hippocampus. Exp Neurol 210:699–708

Ojeda SR, Urbanski HF (1994) Puberty in the rat. In: Knobil E, Neill JD (eds) Physiology of reproduction. Raven, New York, pp 453–486

Olioff M, Stewart J (1978) Sex differences in the play behavior of prepubescent rats. Physiol Behav 20:113–115

Ovtscharoff W Jr, Braun K (2001) Maternal separation and social isolation modulate the postnatal development of synaptic composition in the infralimbic cortex of Octodon degus. Neuroscience 104:33–40

Owens DF, Kriegstein AR (2002) Is there more to GABA than synaptic inhibition? Nat Rev Neurosci 3:725–727

Panksepp J (1981) The ontogeny of play in rats. Dev Psychobiol 14:327–332

Panksepp J, Beatty WW (1980) Social deprivation and play in rats. Behav Neural Biol 30:197–206

Pellis SM, Pellis VC, McKenna MM (1994) A feminine dimension in the play fighting of rats (Rattus norvegicus) and its defeminization neonatally by androgens. J Comp Psychol 108:68–73

Perrot-Sinal TS, Auger AP, McCarthy MM (2003) Excitatory actions of GABA in developing brain are mediated by L-type Ca2+ channels and dependent on age, sex and brain region. Neuroscience 116:995–1003

Perrot-Sinal TS, Sinal CJ, Reader JC, Speert DB, McCarthy MM (2007) Sex differences in the chloride cotransporters, NKCC1 and KCC2, in the developinghypothalamus. J Neuroendocrinol 19:302–308

Pfeffer CK, Stein V, Keating DJ, Maier H, Rinke I, Rudhard Y, Hentschke M, Rune GM, Jentsch TJ, Hubner CA (2009) NKCC1-dependent GABAergic excitation drives synaptic network maturation during early hippocampal development. J Neurosci 29:3419–3430

Phoenix CH, Goy RW, Gerall AA, Young WC (1959) Organizing action of prenatally administered testosterone proprionate on the tissues mediating mating behavior in the female guinea pig. Endocrinology 65:369–382

Pike CJ (1999) Estrogen modulates neuronal Bcl-xL expression and beta-amyloid-induced apoptosis: relevance to Alzheimer's disease. J Neurochem 72:1552–1563

Plotkin MD, Snyder EY, Hebert SC, Delpire E (1997) Expression of the Na-K-2Cl- cotransporter is developmentally regulated in postnatal rat brains: a possible mechanism underlying GABA's excitatory role in immature brain. J Neurobiol 33:781–795

Poeggel G, Helmeke C, Abraham A, Schwabe T, Friedrich P, Braun K (2003) Juvenile emotional experience alters synaptic composition in the rodent cortex, hippocampus, and lateral amygdala. Proc Natl Acad Sci 100:16137–16142

Rakic P, Bourgeois JP, Eckenhoff MF, Zecevic N, Goldman-Rakic PS (1986) Concurrent overproduction of synapses in diverse regions of the primate cerebral cortex. Science 11:232–235

Ravizza T, Galanopoulou AS, Moshe JL (2002) Sex differences in androgen and estrogen receptor expression in rat substantia nigra during development: an immunohistochemical study. Neuroscience 115:685–696

Ravizza T, Veliskova J, Moshe JL (2003) Testosterone regulates androgen and estrogen receptor immunoreactivity in rat substantia nigra pars reticulata. Neurosci Lett 338:57–61

Rhees RW, Shryne JE, Gorski RA (1990a) Onset of the hormone-sensitive perinatal period for sexual differentiation of the sexually dimorphic nucleus of the preoptic area in female rats. J Neurobiol 21:781–786

Rhees RW, Shryne JE, Gorski RA (1990b) Termination of the hormone-sensitive period for differentiation of the sexually dimorphic nucleus of the preoptic area in male and female rats. Brain Res Dev Brain Res 52:17–23

Risold PY, Swanson LW (1987) Connections of the rat lateral septal complex. Brain Res Rev 24:115–195

Riviera C, Voipio J, Payne JA, Ruusuvuori E, Lahtinen H, Lamsa K, Pirvola U, Saarma M, Kaila K (1999) The $K^+.Cl^-$ co-transporter KCC2 renders GABA hyperpolarizing during neuronal maturation. Nature 397:251–255

Rucklidge JJ (2010) Gender differences in attention-deficit/hyperactivity disorder. Psychiatr Clin N Am 33:357–373

Schulz KM, Sisk CL (2006) Pubertal hormones, the adolescent brain, and the maturation of social behaviors: lessons from the Syrian hamster. Mol Cell Endocrinol 254–255:120–126

Schulz KM, Richardson HN, Zehr JL, Osetek AJ, Menard TA, Sisk CL (2004) Gonadal hormones masculinize and defeminize reproductive behaviors during puberty in the male Syrian hamster. Horm Behav 45:242–249

Schulz KM, Molenda-Figueira HA, Sisk CL (2009) Back to the future: the organizational-activational hypothesis adapted to puberty and adolescence. Horm Behav 55:597–604

Schwarz JM, McCarthy MM (2008) Steroid induced sexual differentiation of the brain: multiple pathways, one goal. J Neurochem 105:1561–1572

Schwarz JM, Liang S-L, Thompson SM, McCarthy MM (2008) Estradiol induces hypothalamic dendritic spines by enhancing glutamate release: a mechanism for organizational sex differences. Neuron 58:584–598

Seidman LJ, Valera EM, Makris N (2005) Structural brain imaging of attention-deficit/hyperactivity disorder. Biol Psychiatry 57:1263–1272

Sharp SI, McQuillin A, Gurling HMD (2009) Genetics of attention- deficit hyperactivity disorder (ADHD). Neuropharmacology 57:590–600

Shulman LM (2007) Gender differences in Parkinson's disease. Gend Med 4:8–18

Simerly RB (2002) Wired for reproduction: organization and development of sexually dimorphic circuits in the mammalian forebrain. Annu Rev Neurosci 25:507–536

Simerly RB, Swanson LW (1988) Projections of the medial preoptic nucleus: a Phaseolus vulgaris leucoagglutinin anterograde tract tracing study in the rat. J Comp Neurol 270:209–242

Simerly RB, Swanson LW, Handa RJ, Gorski RA (1985) Influence of perinatal androgen on the sexually dimorphic distribution of tryrosine hydorxylase-immunoreactive cells and fibers in the anterovertral periventricular nucleus of the rat. Neuroendocrinology 40:501–510

Sipila ST, Schuchmann S, Voipio J, Yamada J, Kaila K (2006) The cation-chloride cotransporter NKCC1 promotes sharp waves in the neonatal rat hippocampus. J Physiol 573:765–773

Sorra KE, Harris KM (2000) Overview on the structure, composition, function, development, and plasticity of hippocampal dendritic spines. Hippocampus 10:501–511

Sperber EF, Wurpel WBY, JN MSL (1987) Nigral infusions of muscimol or bicuculline facilitate seizures in developing rats. Brain Res 465:243–250

Spritzer NC (2006) Electrical activity in early neuronal development. Nature 444:707–712

Stewart J, Rajabi H (1994) Estradiol derived from testosterone in prenatal life affects the development of catecholamine systems in the frontal cortex in the male rat. Brain Res 646:157–160

Swanson LW, Petrovich GD (1998) What is the amygdala? Trends Neurosci 21:323–331

Tozuka Y, Fukuda S, Namba T, Seki T, Hisatsune T (2005) GABAergic excitation promotes neuronal differentiation in adult hippocampal progenitor cells. Neuron 47:803–815

Tyzio R, Represa A, Jorquera I, Ben-Ari Y, Gozlan H, Aniksztejn L (1999) The establishment of GABAergic and glutamatergic synapses on CA1 pyramidal neurons is sequential and correlates with the development of the apical dendrite. J Neurosci 19:10372–10382

Ullian EM, Sapperstein SK, Christopherson KS, Barres BA (2001) Control of synapse number by glia. Science 291:657–661

van den Berg CL, Hol T, van Ree JM, Spruijt BM, Everts H, Koolhaas JM (1999) Play is indispensable for an adequate development of coping with social challenges in the rat. Dev Psychobiol 34:129–138

Veliskova J, Moshe SL (2001) Sexual dimorphism and developmental regulation of substantia nigra function. Ann Neurol 50:506–601

Vidensky S, Zhang Y, hand T, Goellner J, Shaffer A, Isakson P, Andreasson K (2003) Neuronal overexpression of COX-2 results in dominant production of PGE2 and altered fever response. Neuromol Med 3:15–28

Waters EM, Simerly RB (2009) Estrogen induces caspase-dependent cell death during hypothalamic development. J Neurosci 29:9714–9718

Weisz J, Ward IL (1980) Plasma testosterone and progesterone titers of pregnant rats, their male and female fetuses and neonatal offspring. Endocrinology 106:306–313

Woolley CS, McEwen BS (1992) Estradiol mediates fluctuation in hippocampal synapse density during the estrous cycle in the adult rat. J Neurosci 12:2549–2554

Yang H, Chen C (2008) Cyclooxygenase-2 (COX-2) in synaptic signaling. Curr Pharm Des 14:1443–1451

Zehr JL, Todd BJ, Schulz KM, McCarthy MM, Sisk CL (2006) Dendritic pruning of the medial amygdala during pubertal development of the male Syrian hamster. J Neurobiol 66:578–590

Zup SL, Forger NG (2002) Testosterone regulates BCL-2 immunoreactivity in a sexually dimorphic motor pool of adult rats. Brain Res 950:312–316

Zup SL, Carrier H, Waters E, Tabor A, Bengston L, Rosen G, Simerly R, Forger NG (2003) Overexpression of Bcl-2 reduces sex differences in neuron number in the brain and spinal cord. J Neurosci 23:2357–2362

ADHD: Current and Future Therapeutics

David J. Heal, Sharon L. Smith and Robert L. Findling

Contents

1 Introduction 362
2 ADHD Drugs: A Translational Pharmacology Evaluation 363
    2.1 The Amphetamines 365
    2.2 Methylphenidate 369
    2.3 Bupropion 375
    2.4 Guanfacine and Other α_2-Adrenoceptor Agonists 376
    2.5 Modafinil 377
3 Clinical Experience with Drugs Used to Treat ADHD 379
4 New Developments in the Field of Research into Drugs to Treat ADHD and Future Directions 381
5 A Summary of the Current Status and Future Prospects of Drug Therapy for ADHD 384
References 385

Abstract The stimulants, amphetamine and methylphenidate, have long been the mainstay of attention-deficit hyperactivity disorder (ADHD) therapy. They are rapidly effective and are generally the first medications selected by physicians. In the development of alternative pharmacological approaches, drug candidates have been evaluated with a wide diversity of mechanisms. All of these developments have contributed real progress in the field, but there is still much room for improvement and unmet clinical need in ADHD pharmacotherapy. The availability of a wide range of compounds with a high degree of specificity for individual monoamines (dopamine and noradrenaline) and/or different pharmacological mechanisms has refined our understanding of the essential elements for optimum pharmacological effect in managing ADHD. In this chapter, we review the pharmacology of the different classes of drug used to treat ADHD and provide a

D.J. Heal (✉) and S.L. Smith
RenaSci Consultancy Ltd, BioCity, Nottingham, NG1 1GF, UK
e-mail: david.heal@renasci.co.uk

R.L. Findling
University Hospitals Case Medical Center, Case Western Reserve University School of Medicine, Cleveland, Ohio 44106-2205, USA

neurochemical rationale, predominantly from the use of in vivo microdialysis experiments, to explain their relative efficacy and potential to elicit side effects. In addition, we will consider how predictions based on results from animal models translate into clinical outcomes. The treatment of ADHD is also described from the perspective of the physician. Finally, the new research development for drugs to treat ADHD is discussed.

Keywords α_2-Adrenoceptor agonists · ADHD · Clinical experience · Microdialysis · New treatments · Releasing agents · Reuptake inhibitors · Stimulants

Abbreviations

CNS	Central nervous system
COMT	Catechol-O-methyltransferase
DAT	Dopamine reuptake transporter
HVA	Homovanillic acid
MAO	Monoamine oxidase
NET	Norepinephrine reuptake transporter
PET	Positron emission tomography
PFC	Prefrontal cortex
SERT	Serotonin reuptake transporter
SHR	Spontaneously hypertensive rat
SSRI	Selective serotonin reuptake inhibitor
XR	Extended release

1 Introduction

It is more than 60 years since Bradley (1937) first used DL-amphetamine to treat the psychiatric, behavioural and cognitive disorder that we now call attention-deficit hyperactivity disorder (ADHD). This discovery was made purely by accident as the original intention of the psychiatrist, Bradley, was to treat headaches in children with learning and behavioural problems. These headaches resulted from the children receiving a pneumoencephalogram that was used as part of the clinical diagnosis. Since then, the stimulants, amphetamine and methylphenidate, have become the pre-eminent class of drugs used to treat this disorder. The pharmaceutical industry has created ever more sophisticated approaches to maximise their therapeutic efficacy and minimise side effects. Much effort has also been expended on research into new pharmacological approaches for ADHD treatment, particularly with non-stimulant drugs: e.g. atomoxetine and guanfacine. A list of drugs currently and formerly used to treat ADHD is shown in Table 1. From a comparison of drugs approved in Europe versus the USA, it is apparent that many more agents

Table 1 Current and former drugs used to treat ADHD

Generic drug name	Trade names	Approved for use USA	Approved for use Europe
DL-Amphetamine[a]	Benzedrine®	×	×
D-Amphetamine	Dexedrine®, Dexedrine Spansules®[b]	✓	✓[c]
L-Amphetamine[a]	Cydril	×	×
"Mixed-enantiomers/ mixed-salts" amphetamine (3:1 mixture of D- and L-isomers)	Adderall®, Adderall XR®[b]	✓	×
Lisdexamfetamine (D-amphetamine prodrug)	Vyvanse®	✓	×
DL-Methylphenidate[a] (erythro + threo isomers)	Centedrine®	×	×
DL-*threo*-Methylphenidate	Ritalin®, Ritalin SR®[b], Metadate CD®[b], Concerta®[b], Concerta XL®, Equasym XL®, Medikinet XL®, Daytrana®[b]	✓	✓
D-*threo*-Methylphenidate	Focalin®, Focalin XR®[b]	✓	×
Atomoxetine	Strattera®	✓	✓
Guanfacine	Intuniv®[b]	✓	×
Bupropion[d]	Wellbutrin®, Wellbutrin SR®[b]	×	×

[a]Product withdrawn
[b]Extended release formulation
[c]Refractory ADHD only
[d]Not approved as an ADHD treatment

are available in US formularies. In Europe, ADHD is predominantly managed using DL-methylphenidate products or atomoxetine, with D-amphetamine restricted to the treatment of refractory ADHD. In the USA, D-amphetamine-based products comprise a major portion of the market, along with various methylphenidate formulations and more recently, guanfacine.

In this chapter, we will review the pharmacology of the different classes of drugs used in ADHD and provide a neurochemical rationale to explain their relative efficacy and potential to elicit side effects. In addition, we will consider how predictions based on results from animal models translate into clinical outcomes. The treatment of ADHD is also described from the perspective of the physician.

2 ADHD Drugs: A Translational Pharmacology Evaluation

Clinically effective ADHD drugs have highly defined pharmacological characteristics. ADHD drugs can be sub-classified (Fig. 1) according to their mechanism of action or neurotransmitter target(s). With the exception of the α_2-adrenoceptor

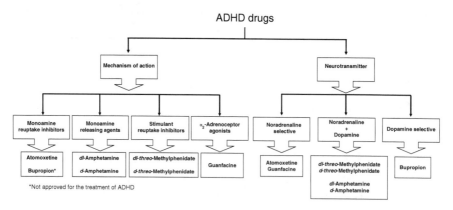

Fig. 1 The classification of ADHD drugs by mechanism of action and neurotransmitter target

agonists, clonidine and guanfacine, all drug treatments for ADHD act indirectly to potentiate and prolong the action of the catecholamines. They achieve this by stimulating catecholamine release from presynaptic nerve terminals, inhibition of monoamine (dopamine and noradrenaline) reuptake and/or inhibiting catabolism by monoamine oxidase (MAO). Although it was originally considered that there was little to differentiate between the pharmacodynamics of monoamine reuptake inhibitors (e.g. atomoxetine) and releasing agents (e.g. amphetamine), the use of in vivo techniques, such as intracerebral microdialysis, has revealed profound differences in the functional consequences of these two mechanisms. Another factor that contributed to a blurring of the distinction between monoamine reuptake inhibitors and releasing agents in the ADHD field is the unique and enigmatic pharmacology of the so-called "stimulant reuptake inhibitor", methylphenidate. Thus, although the stimulants are similar in terms of efficacy and side effect profiles in patients, they are not necessarily interchangeable.

In the following sections, we describe results that have been predominantly obtained using *in vivo* microdialysis. These illustrate how the pharmacological effects of the different classes of drugs provide insights into their relative efficacy and side effect profile in ADHD treatment. When describing the actions of drugs in microdialysis experiments, the term "efflux" is often used to describe the net result of two opposing active processes: i.e. increased extracellular neurotransmitter concentration following impulse-dependent release, minus the decrease in concentration mediated by transmitter reuptake and/or catabolism by MAO and catechol-*O*-methyltransferase. The reuptake transporters that regulate synaptic monoamine concentrations in the CNS are NET, DAT and SERT (for noradrenaline, dopamine and serotonin, respectively). It is important to remember that neurotransmitter concentrations in microdialysates are surrogate markers for changes in extracellular neurotransmitter concentration. The recovery of monoamines using microdialysis is only 10–20%, and as the sampling area of the microdialysis membrane (mm) is huge compared with the size of monoaminergic neurones (μm), the technique is incapable of estimating synaptic neurotransmitter concentrations.

2.1 The Amphetamines

Of the current generation of amphetamine formulations, it is the more potent D-isomer that is used for ADHD treatment (Table 1); L-amphetamine (Cydril®) disappeared from the formulary many years ago. Although each isomer of amphetamine enhances release of both catecholamines, D-amphetamine has a preferential effect on dopamine efflux, whereas L-amphetamine has a more balanced action on both dopamine and noradrenaline (Fig. 2). The difference in the catecholaminergic profiles of D- and L-amphetamine has been exploited in the mixed-salts/mixed-enantiomers amphetamine products, Adderall® and Adderall Extended-Release (Adderall XR®).

The pharmacological actions of D- and L-amphetamine are complex with two or three additive or synergistic mechanisms affecting synaptic monoamine concentrations (Table 2). The primary therapeutic action of amphetamines is monoamine release. Amphetamine is chemically and structurally related to dopamine and noradrenaline and is a competitive substrate for NET and DAT. It is actively transported into monoaminergic nerve terminals (Hurd and Ungerstedt 1989) and induces impulse-independent release of cytosolic catecholamines (Sulzer and Rayport 1990). At moderate to high doses of amphetamine, displaced catecholamines are expelled by reverse-transport into the synaptic cleft by a mechanism independent of neuronal firing (Carboni et al. 1989). Since amphetamine's isomers compete with endogenous catecholamines for transport into presynaptic terminals via NET and DAT, they delay synaptic neurotransmitter clearance. Metabolism of these neurotransmitters is also impeded by amphetamine's isomers as they weakly inhibit MAO by acting as competitive substrates (Mantle et al. 1976).

The combination of these pharmacological effects on the efflux of noradrenaline and dopamine in the brain of freely moving spontaneously hypertensive rats (SHR) is shown in Fig. 2 (Cheetham et al. 2007; Heal et al. 2009). There is an initial steep rise in catecholamine outflow, peaking within 30 min. Increased dopamine and noradrenaline efflux exceed 500%, even with low doses of D-amphetamine. Since the releasing mechanism of high doses of D-amphetamine cannot be attenuated by reducing neuronal firing rate or by autoreceptor control, the only determinant of its magnitude of effect is the rate of entry into catecholaminergic terminals via NET and DAT. That, in turn depends on the amount of amphetamine available for transport. Thus, this lack of a dose-ceiling differentiates the amphetamines from classical reuptake inhibitors such as atomoxetine and bupropion.

L-Amphetamine's effects on noradrenaline and dopamine efflux in the brains of SHR are powerful (Fig. 2), with peak increases of ~500% within 45 min of dosing. Despite similarities with the pharmacodynamic profile of D-amphetamine, there are important differences that are believed to be of clinical significance. First, L-amphetamine is approximately threefold less potent than the D-isomer (Fig. 2) and, second, at low doses (with greater clinical relevance), L-amphetamine increases efflux of noradrenaline and dopamine to the same extent. The SHR has been proposed as a model of ADHD (Sagvolden et al. 1992). However, qualitatively

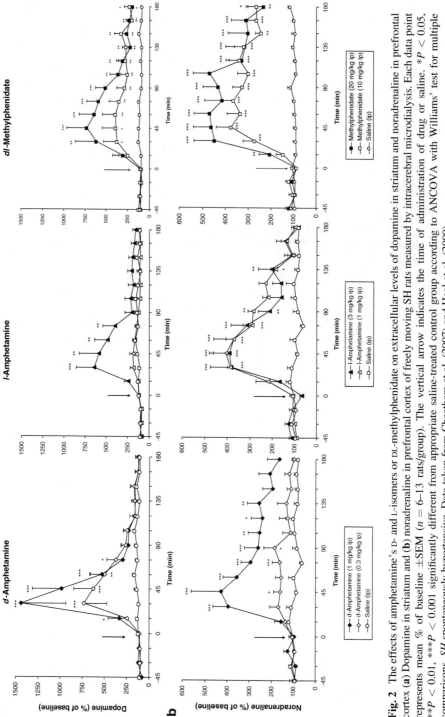

Fig. 2 The effects of amphetamine's D- and L-isomers or DL-methylphenidate on extracellular levels of dopamine in striatum and noradrenaline in prefrontal cortex (**a**) Dopamine in striatum and (**b**) noradrenaline in prefrontal cortex of freely moving SH rats measured by intracerebral microdialysis. Each data point represents mean % of baseline ±SEM ($n = 6$–13 rats/group). The vertical arrow indicates the time of administration of drug or saline. $*P < 0.05$, $**P < 0.01$, $***P < 0.001$ significantly different from appropriate saline-treated control group according to ANCOVA with Williams' test for multiple comparisons. *SH* spontaneously hypertensive. Data taken from Cheetham et al. (2007) and Heal et al. (2009)

Table 2 A summary of the mode of action of various drugs to treat ADHD

ADHD drug	Pharmacological actions	Catecholaminergic profile
D-Amphetamine	NA + DA release NA + DA reuptake inhibition MAO inhibition	DA ≥ NA
L-Amphetamine	NA + DA release NA + DA reuptake inhibition MAO inhibition	NA = DA
DL-Methylphenidate	NA + DA reuptake inhibition	NA = DA
D-Methylphenidate	NA + DA reuptake inhibition	NA = DA
Modafinil	Enigmatic DA reuptake inhibition	NA = DA
Atomoxetine	NA reuptake inhibition	NA = (DA)
Bupropion	DA reuptake inhibition	DA = NA
Guanfacine	α_{2A}-Adrenoceptor agonist	NA

Effects in parentheses are indirect actions
NA noradrenaline, *DA* dopamine, *MAO* monoamine oxidase

comparable pharmacodynamic effects have been demonstrated in normal rats: e.g. for striatal dopamine efflux including D- and L-amphetamine (Kuczenski et al. 1995) and for release of dopamine and noradrenaline in the prefrontal cortex (PFC; Géranton et al. 2003; Pum et al. 2007). When D- and L-amphetamine doses are increased beyond those relevant to therapeutic action (the experimental dose that can reproduce clinically achievable/tolerable exposure), their pharmacological profiles change radically. They become profoundly dopaminergic in character with peak increases of dopamine efflux of 4,000–5,000% (D-amphetamine) and 3,000–4,000% (L-amphetamine) (Cheetham et al. 2007; Heal et al. 2009). This aspect is unlikely to be relevant to amphetamine's therapeutic effects, but is almost certainly the reason why it is abused as a psychostimulant.

The monoamine reuptake transporters do not display good substrate specificity. As a consequence, amphetamine's isomers can be sequestered by SERT and evoke the release of serotonin (Heal et al. 1998). In terms of magnitude, its effect is probably greater than those of the serotonin selective reuptake inhibitors (SSRIs) and equivalent to the serotonin releasing agent, D-fenfluramine (Heal et al. 1998).

2.1.1 Clinical Profile of the Amphetamines and Accuracy of Predictions Based on Preclinical Results

Based on their pharmacological profiles, D- and L-amphetamine were predicted to be highly efficacious in the treatment of ADHD. This prediction has been confirmed (Table 3) for D-amphetamine, L-amphetamine (Arnold et al. 1972), racemic amphetamine (Gross 1976), Adderall (Pelham et al. 1999) and the D-amphetamine prodrug, lisdexamfetamine (Vyvanse®) (Findling et al. 2008b).

It is generally accepted that there is no difference in the efficacy of the current generation of amphetamine-based drugs and methylphenidate for ADHD treatment

Table 3 A summary of the pharmacological characteristics of various ADHD drugs

Drug	Pharmacological class	Neurotransmitter targets[a]	Pharmacological characteristics — Site of action	Time of peak effect	Effect size[b] Dopamine	Effect size[b] Noradrenaline	Effect[b] Ceiling	Effect[b] Duration[b]	Psychostimulant/ euphoriant	Clinical efficacy (response rate)
D-Amphetamine	Releasing agent	Noradrenaline Dopamine 5-HT	Intraneuronal Extraneuronal	<1.0 h	>1000%	>500%	No	2.0–3.0 h	Yes	~70%
L-Amphetamine	Releasing agent	Noradrenaline Dopamine 5-HT	Intraneuronal Extraneuronal	<1.0 h	<500%	<500%	No	1.0–2.0 h	Yes	ND
DL-Methylphenidate D-Methylphenidate	Reuptake inhibitor	Noradrenaline Dopamine (Histamine)	Extraneuronal	<1.0 h	>500%	>500%	No	3.0–4.0 h	Yes	~70%
Atomoxetine	Reuptake inhibitor	Noradrenaline (Dopamine in PFC) (Histamine)	Extraneuronal	≥1.0 h	<500%	<500%	Yes	>4.0 h	No	50–60%
Bupropion	Reuptake inhibitor	(Noradrenaline) Dopamine	Extraneuronal	≥1.0 h (PFC) <1.0 h (STR)	<500%	<500%	Yes	1.0–2.0 h	No	Weak/equivocal
Guanfacine	α$_{2A}$-Agonist	Noradrenaline	Extraneuronal	<1.0 h	Decrease (PFC)[c]	Decrease (PFC)[c]	Yes	ND	No	50–60%[d]
Modafinil	Enigmatic	Noradrenaline Dopamine (Histamine)	Extraneuronal	>1.0 h	<500% (PFC)	<500% (PFC)	ND	2.0–3.0 h	Equivocal	50–60%

PFC prefrontal cortex, *STR* striatum, *ND* not determined

[a]Neurotransmitter targets. Monoamines in parentheses are indirect effects
[b]Parameters determined by microdialysis experiments in freely moving rats
[c]Effect predicted on the basis of microdialysis experiments previously performed with other α$_2$-adrenoceptor agonists (Ihalainen and Tanila 2002)
[d]Estimated from previous studies (Scahill et al. 2001)

(Pelham et al. 1990). Head-to-head clinical trials comparing amphetamines and non-stimulant drugs in the treatment of ADHD are relatively scarce. Wigal et al. (2005) compared the efficacy of Adderall XR® against atomoxetine (Strattera®). The former was maximally efficacious 2 h after dosing, while the peak effect for atomoxetine was not observed for 7 h. When time-courses of peak efficacy were compared over the dosing period, there was no difference between Adderall XR® and atomoxetine. More recent clinical trials have described the superior efficacy of Adderall XR® versus atomoxetine (Faraone et al. 2007). No trials were conducted to compare the efficacy and safety of either bupropion or guanfacine versus amphetamine-based drugs. However, responder rates of about 50–60% suggest that they may be less efficacious than stimulant medications (Conners et al. 1996; Scahill et al. 2001). In a long-term trial in which guanfacine XR was given alone and in combination with stimulants, drop-outs through lack of efficacy were considerably higher for the monotherapy (Sallee et al. 2009).

Since D-amphetamine was used as an appetite suppressant and is still used for the treatment of narcolepsy, it is not surprising that anorexia, weight loss and insomnia are common adverse events for all amphetamine-based medications (Findling et al. 2008b). Other adverse events include: nausea, vomiting, abdominal cramps, increased blood pressure and heart rate, and possibly exacerbation of motor tics (Findling et al. 2008a). The central actions of the amphetamines that produce these side effects are also responsible for their efficacy in treating ADHD. Optimising the dose is, therefore, the key to risk/benefit. In short, the greater efficacy of the amphetamines in treating ADHD, compared with either classical reuptake inhibitors (atomoxetine and bupropion) or the α_2-adrenoceptor agonist (guanfacine), is consistent with their faster and more powerful increase in extraneuronal catecholamines (Table 3).

2.2 Methylphenidate

Methylphenidate has a piperidine chemical structure with two chiral centres that give rise to four stereo-isomers: D-*threo*(2R-2'R)-methylphenidate, L-*threo*(2S:2'S)-methylphenidate, D-*erythro*(2R:2'S)-methylphenidate and L-*erythro*(2S:2'R)-methylphenidate. It was initially marketed in mixed form (Centedrin™) until it was discovered that the *erythro*-isomers were devoid of CNS activity. All currently approved methylphenidate products are composed of either a 50:50 mixture of D- and L-*threo*-methylphenidate or D-*threo*-methylphenidate. We refer to these drugs as racemic, or DL-methylphenidate, and D-methylphenidate, respectively (Table 1).

In vitro, racemic methylphenidate is a moderately potent inhibitor of noradrenaline and dopamine reuptake (Ki values = 100–250 ηM; Andersen 1989), but not of serotonin (Richelson and Pfenning 1984). The D-enantiomer is approximately tenfold more potent as a catecholamine reuptake inhibitor in vitro than the corresponding L-enantiomer (Ferris et al. 1972), indicating that the majority of the pharmacological effect of the racemate is delivered by the D-isomer.

At pharmacologically relevant concentrations, DL-methylphenidate does not inhibit MAO (Sharman 1966).

Since methylphenidate has only a single pharmacological mechanism of action and moderate potency as an inhibitor of noradrenaline or dopamine reuptake, it would be predicted to produce only a modest and gradual enhancement of catecholaminergic function in vivo. However, microdialysis experiments demonstrate that this prediction is incorrect. The effects of DL-methylphenidate on extraneuronal concentrations of dopamine and noradrenaline in rat brain are shown in Fig. 2 (Cheetham et al. 2007; Heal et al. 2009). In the SHR brain, DL-methylphenidate dose-dependently increased (>500%) the efflux of noradrenaline and dopamine (Fig. 2). Its pharmacodynamics are similar to those of the amphetamines, with peak effects occurring 30–45 min after dosing: larger doses of DL-methylphenidate produce greater increases in catecholamine efflux (Fig. 2). In contrast to the action of the amphetamines, where increased extraneuronal dopamine returns to basal levels more rapidly than noradrenaline, sustained efflux of both catecholamines occurs more than 3 h after administration of DL-methylphenidate. In terms of magnitude, speed of onset and the lack of a dose-effect ceiling, the actions of DL-methylphenidate are comparable to those of the amphetamines (Fig. 2). Studies performed in normal, outbred rats have observed similar effects (Kuczenski and Segal 1997). In microdialysis experiments performed in our laboratory, the D-enantiomer was >10-fold more potent than the L-form and 10 mg/kg of D-methylphenidate produced the same increased efflux of noradrenaline in PFC and dopamine in striatum as a 20 mg/kg dose of the racemate (Heal et al. 2008).

The lack of SERT affinity for DL-methylphenidate in vitro (Richelson and Pfenning 1984) is mirrored by a lack of effect on serotonin efflux in vivo (Kuczenski and Segal 1997).

The pharmacodynamics of DL-methylphenidate on catecholamine efflux in the PFC differentiate it from atomoxetine (Fig. 4). Not only is the atomoxetine-induced increase in noradrenaline efflux considerably smaller, but there is also a ceiling for its effect (Figs. 2 and 4). The profound difference between the pharmacological profiles of methylphenidate and classical reuptake inhibitors (e.g. GBR 12909) is emphasised when these drugs are investigated using reverse dialysis. Nomikos et al. (1990) determined the effect of reverse dialysis of methylphenidate, GBR 12909 and D-amphetamine on the extracellular concentration of dopamine in striatum of freely moving rats. DL-Methylphenidate and D-amphetamine evoked rapid, maximal increases in dopamine efflux: the incremental increases were more exaggerated as the drug concentrations increased. In contrast, GBR 12909 induced a plateau of dopamine efflux more slowly and the amplitude of the increments diminished as the drug concentration increased. Since reverse dialysis provides rapid access to the site of pharmacological action, sub-optimal pharmacokinetics cannot explain these moderate and gradual increases in monoamine efflux. An absolute requirement for intact neuronal firing in the pharmacological actions of methylphenidate and the classical reuptake inhibitors has been demonstrated in many studies (Nomikos et al. 1990).

Overall, the ability of DL-methylphenidate to evoke rapid and substantial increases in dopamine efflux in striatum (Heal et al. 2008) and nucleus accumbens

(Gerasimov et al. 2000) is not typical of the action of a classical reuptake inhibitor. Heal (2008) postulated that methylphenidate and cocaine are allosteric modulators of DAT and induce a firing-dependent reversal of dopamine transport. Hence, they appear to act as "inverse agonists", reversing the usual direction of dopamine transport by DAT. This would explain the similarity between the profiles of ADHD medications like methylphenidate and the amphetamines. Moreover, this hypothesis for the mode of action of methylphenidate (Heal 2008) could explain why it has powerful psychostimulant and reinforcing effects in humans (Smith and Davis 1977), whereas classical dopamine reuptake inhibitors, such as bupropion, do not (Miller and Griffiths 1983).

2.2.1 Clinical Profile of Methylphenidate and Accuracy of Predictions Based on Preclinical Results

Clinical trials have shown no substantial difference between the efficacy of methylphenidate and amphetamine-based medications (Efron et al. 1997a). Various formulations of DL-methylphenidate were effective in both short- and long-term clinical trials of ADHD (Barkley et al. 1991; Findling et al. 2008a). Consistent with a rapid increase in catecholamine efflux in striatum and PFC, DL-methylphenidate is maximally efficacious within 2 h of oral administration (Quinn et al. 2004). Substantial benefit is often seen within 1 week of treatment onset (Wolraich et al. 2001).

Since DL-methylphenidate and amphetamine-based drugs are equally efficacious, they are likely to be more effective than non-stimulant ADHD drugs. Thus, osmotically released DL-methylphenidate (Concerta®) was more effective than atomoxetine in several clinical trials (Newcorn et al. 2008). When bupropion was compared with immediate-release, DL-methylphenidate, the stimulant was more efficacious (Barrickman et al. 1995). No comparisons have yet been performed on the efficacy and safety of DL-methylphenidate versus guanfacine.

The most common side effects of DL-methylphenidate are insomnia, anorexia, emotional lability, changes in blood pressure/heart rate and gastrointestinal effects. These occur to a similar extent for all formulations (Efron et al. 1997b). In a study comparing DL-methylphenidate versus D-amphetamine in children, there were no differences in frequency or severity of adverse events (Efron et al. 1997a, b).

The more potent enantiomer of racemic methylphenidate (D-methylphenidate) has been developed as an ADHD medication (Focalin® and Focalin XR®). Part of the rationale for this approach was to expand the therapeutic window of methylphenidate in ADHD therapy. This argument is somewhat flawed by the fact that ~90% of the efficacy and side effects of DL-methylphenidate are mediated by the more active D-enantiomer. When D-methylphenidate and the racemate have been compared in clinical trials, their efficacy and adverse event profiles are identical (Quinn et al. 2004).

When the clinical outcome results for D- and DL-methylphenidate are compared with data obtained using in vivo microdialysis experiments, their relative efficacy and the spectrum and severity of their side effects are consistent with their predicted actions.

2.2.2 Atomoxetine

Atomoxetine is a potent and selective noradrenaline reuptake inhibitor approved in Europe and USA as a non-stimulant drug for the treatment of ADHD. Since dopamine is often considered to be the more important mediator in the treatment of ADHD, it is essential to understand how the unusual catecholaminergic neuroanatomy in the PFC contributes to the pharmacological actions of atomoxetine.

Compared with the nigrostriatal or mesolimbic systems, dopaminergic innervation in the PFC is sparce. The major difference lies in the low density of DAT sites on PFC dopaminergic neurones (Hitri et al. 1991). Clearance of dopamine via DAT in the PFC is slow, allowing it to diffuse away from its site of release (Cass and Gerhardt 1995). Catecholamine reuptake transporters have relatively little substrate selectivity, and a substantial proportion of dopamine is sequestered into noradrenergic terminals via NET (Morón et al. 2002). The mechanism for dopamine reuptake and catabolism in the PFC is illustrated in Fig. 3. These unusual, if not unique, characteristics of dopamine regulation in the PFC play a major role in determining the catecholaminergic profile of atomoxetine. Atomoxetine induces moderate and sustained increases in extracellular concentrations of both noradrenaline and dopamine in the PFC of rats

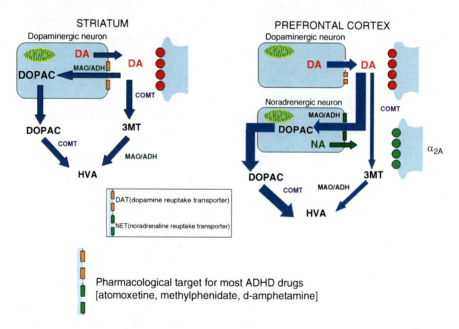

Fig. 3 Differences in dopamine reuptake and catabolism between the prefrontal cortex and striatum. In the PFC, the number of DAT sites is low compared with other dopaminergic terminal fields like the striatum. Most of the dopamine that is neuronally released in the PFC is sequestered into noradrenergic neurones via NET sites. *DOPAC* dihydroxyphenylacetic acid, *3MT* 3-methoxytyramine, *HVA* homovanillic acid, *MAO/ADH* monoamine oxidase/aldehyde dehydrogenase, *COMT* catechol-*O*-methyltransferase

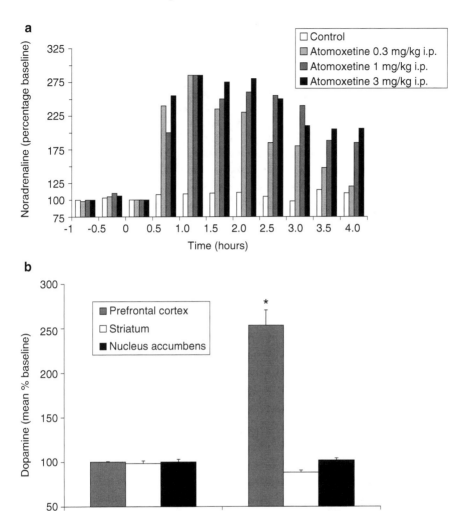

Fig. 4 The effects of atomoxetine on catecholamine efflux in the brains of freely-moving rats measured by intracerebral microdialysis. (**a**) The effect of atomoxetine on extracellular noradrenaline in PFC of Sprague-Dawley rats. All doses of atomoxetine significantly different ($P < 0.025$) from saline-treated controls throughout the 4-h experimental period according to Duncan's post hoc test ($n = 5$–6 rats/group). Administration of atomoxetine or saline was at 0 hours. (**b**) The effect of atomoxetine on extracellular dopamine levels in PFC (3 mg/kg i.p.), nucleus accumbens (3 mg/kg i.p.) and striatum (10 mg/kg i.p.) of Sprague-Dawley rats ($n = 5$–6). Atomoxetine significantly increased extracellular dopamine concentrations throughout the 4-h period only in the PFC. *$P < 0.05$ significantly different from saline treated controls according to Duncan's post hoc test. Data taken from Bymaster et al. (2002)

(Fig. 4). However, atomoxetine does not enhance dopamine efflux in the striatal or mesolimbic terminal fields: regions where dopamine reuptake is by DAT. Compared with the effects of methylphenidate or amphetamine (Fig. 2), the increases in noradrenaline and dopamine efflux evoked by atomoxetine are smaller, slower to plateau and there is a dose ceiling (Fig. 4; Bymaster et al. 2002). This limit on atomoxetine's ability to increase extraneuronal noradrenaline and dopamine is typical of classical monoamine reuptake inhibitors. It is caused by progressive reduction in neuronal firing and autoreceptor feedback inhibition that gradually counterbalances the reduced clearance of neurotransmitter.

Enhancement of catecholaminergic neurotransmission in the PFC is thought to be fundamental for ADHD treatment. Due to the unusual neuroanatomy of PFC catecholaminergic neurones, atomoxetine increases synaptic concentrations of noradrenaline and dopamine. However, it is a highly selective noradrenaline reuptake inhibitor and does not increase dopaminergic neurotransmission elsewhere (Fig. 4). Due to this regional selectivity, atomoxetine is predicted to lack psychostimulant side effects. On the negative side, dysregulation of the striato-cortical catecholamine system might have a role in the aetiology of ADHD (Durston 2003). Therefore, this regional selectivity may partly account for its reduced efficacy compared with the stimulants.

2.2.3 Clinical Profile of Atomoxetine and Accuracy of Predictions Based on Preclinical Results

The efficacy of atomoxetine has been shown in a number of clinical trials (Kelsey et al. 2004; Buitelaar et al. 2007). Both the inattentive and impulsive/hyperactive aspects of the disorder respond to atomoxetine (Michelson et al. 2001). Its efficacy was not influenced by co-morbid generalised anxiety disorder, depression, oppositional defiant disorder, tics or Tourette's syndrome. Although comparator trials generally report that atomoxetine is less effective in treating ADHD than the stimulants, there have been exceptions (Wang et al. 2007). No clinical trials have compared atomoxetine against guanfacine or bupropion.

Adverse events with atomoxetine treatment include decreased appetite, abdominal pain, nausea/vomiting, somnolence, fatigue, dry mouth, loss of libido and erectile dysfunction (Michelson et al. 2003). Overall, the incidence rates and severity of atomoxetine's adverse events are similar to those of methylphenidate and the amphetamines (Wigal et al. 2005). Drug-experienced volunteers reported some similarities between the subjective effects of atomoxetine and the stimulants, but its effects were mild in comparison (Jasinski et al. 2008), indicating that it has little potential for recreational abuse. The reported increases of blood pressure and heart rate (Michelson et al. 2001) are, in general, similar in magnitude to those observed with amphetamine (Wigal et al. 2005) or methylphenidate (Newcorn et al. 2008).

Adjunctive therapy with DL-methylphenidate in atomoxetine partial responders has been successful (Wilens et al. 2009), but this also increases the rates of insomnia, irritability and loss of appetite (Hammerness et al. 2009). This combination therapy has not included amphetamine because blockade of NET by atomoxetine prevents entry of amphetamine into presynaptic noradrenergic terminals (Sofuoglu et al. 2009).

The clinical profile of atomoxetine, therefore, fits well with predictions from intracerebral microdialysis experiments in terms of its relative efficacy as an ADHD therapy, its side effect profile and lack of abuse potential (Table 3).

2.3 Bupropion

Bupropion is a moderately selective, but weak, dopamine reuptake inhibitor developed as an atypical antidepressant (Wellbutrin®) and as an aid to smoking cessation (Zyban®). It has been postulated that bupropion produces its effects by blocking both dopamine and noradrenaline reuptake (Cooper et al. 1994). The potential benefit of bupropion in the treatment of ADHD has been explored and it is used off-label. Moreover, the clinical lessons that have been learned, together with extensive knowledge of its pharmacology, provide useful insights for development of new drugs.

In microdialysis experiments (Li et al. 2002), bupropion produced gradual increases in extracellular noradrenaline (~300%) and dopamine (~250%) in PFC of freely moving rats, 60–90 min post-dosing. It is highly probable that noradrenaline reuptake inhibition plays a role in this outcome. First, because highly selective dopamine reuptake inhibitors, e.g. GBR 12909, produce little increase in dopamine and noradrenaline efflux in PFC (Pozzi et al. 1994). Second, bupropion increases noradrenaline and dopamine efflux in brain areas that lack the unusual catecholaminergic neuronal architecture of the PFC (Hasegawa et al. 2005), indicating that this drug inhibits DAT and NET in vivo. Bupropion also enhances extraneuronal concentration of dopamine in striatum and nucleus accumbens (Sidhpura et al. 2007), although its maximum effect is small compared with D-amphetamine (Nomikos et al. 1990).

These preclinical results indicate that bupropion has the appropriate pharmacology for ADHD treatment, but is predicted to be less efficacious than the stimulants. On the other hand, these results also indicate that bupropion would have a much lower potential for abuse than the psychostimulants.

2.3.1 Clinical Profile of Bupropion and Accuracy of Predictions Based on Preclinical Results

In children, significant improvement in ADHD symptoms was found with bupropion (Casat et al. 1987), although smaller than that seen with the stimulants. In a more recent trial in adolescents, significant therapeutic benefit was not observed (Daviss et al. 2001). In adults, Wilens et al. (2001, 2005) observed a moderate improvement in subjects receiving bupropion. The adverse events that are frequently reported with bupropion are typical of catecholamine reuptake inhibitors: dry mouth, insomnia, chest pain, nausea, dizziness, constipation and irritability (Conners et al. 1996).

Clinical trials with drug-experienced and drug-naïve volunteers investigated whether bupropion has psychostimulant-like abuse potential. In Table 3, bupropion produced no abuse signal across a wide range of doses, including those above the therapeutic range (Miller and Griffiths 1983). The data demonstrate that bupropion lacks psychostimulant and/or euphoriant potential, which is the predicted outcome, based on the speed and magnitude of its action to increase synaptic dopamine concentrations.

When viewed overall, the clinical outcomes for efficacy, side effects and abuse liability fit well with predictions based on the results from the microdialysis experiments (Table 3). The exception is the failure of bupropion to show only low efficacy in children and none in adolescents with ADHD.

2.4 Guanfacine and Other α_2-Adrenoceptor Agonists

In view of their sedative actions, the α_2-adrenoceptor agonists were a logical choice for the potential treatment of ADHD. Clonidine was the first drug in this class to be used off-label, but more recently, interest has focused on guanfacine.

In humans, guanfacine has moderate affinity ($K_i = 50$–146 ηM) for α_{2A}-adrenoceptor subtypes and has a 10- to 24-fold selectivity for α_{2A}- versus α_{2B}-adrenoceptors/α_{2C}-adrenoceptors (Uhlén et al. 1994). Based on the results mainly from primate experiments, it was thought that therapeutic effects of the α_2-adrenoceptor agonists in ADHD were mediated by activation of postsynaptic α_{2A}-adrenoceptors in PFC (Arnsten 2006). Although evidence to support this hypothesis is compelling, microdialysis experiments revealed that effects of the α_2-adrenoceptor agonists on catecholaminergic neuronal transmission in PFC extend beyond postsynaptic α_{2A}-adrenoceptor activation.

The unusual neuronal architecture of the PFC (Fig. 3) has profound implications for the actions of α_2-adrenoceptor agonists on catecholaminergic function. Several research groups have reported that noradrenaline and dopamine efflux in this region is modulated by changes in α_2-adrenergic function. Thus, nonselective α_2-adrenoceptor agonists, like clonidine, reduce the extracellular concentrations of noradrenaline and dopamine (Devoto et al. 2003). It is likely that the reductions in neurotransmitter efflux result from the actions of these agonists at α_{2A}-adrenergic auto- and hetero-receptors on the catecholaminergic nerve cell bodies and terminals. The overall effect of an α_2-adrenoceptor agonist in the PFC will be to enhance noradrenergic neurotransmission via postsynaptic α_{2A}-adrenoceptors, to attenuate signalling via other postsynaptic adrenergic receptor subtypes, and to reduce dopaminergic neurotransmission. Thus, the pharmacological profile of the α_2-adrenoceptor agonists differs from other ADHD drugs, which evoke broad spectrum increases in noradrenergic and dopaminergic neurotransmission in the PFC.

2.4.1 Clinical Profile of Guanfacine and Other α_2-Adrenoceptor Agonists and Accuracy of Predictions Based on Preclinical Results

In randomised, double-blind, placebo-controlled clinical trials, guanfacine was an effective ADHD treatment (Scahill et al. 2001; Biederman et al. 2008a). These trials were performed mainly in the combined hyperactive/impulsive-inattentive ADHD subgroup. However, guanfacine was less efficacious in the minority of subjects with the inattentive subtype of ADHD: response rates of 50–60% placed it alongside other non-stimulant drugs in terms of relative efficacy.

Open-label trials to assess the long-term efficacy and safety of guanfacine XR (Intuniv®) revealed drop-out rates >80% (Biederman et al. 2008b) with discontinuations for lack of efficacy reported as 12.5% (Sallee et al. 2009). In the population who completed 1 or 2 years in these trials, guanfacine maintained its efficacy.

Frequently reported adverse events with guanfacine include somnolence, fatigue, sedation, upper abdominal pain, dry mouth, nausea and dizziness. These are all consistent with the role of α_{2A}-adrenoceptors in the induction of sedation, or are similar to those observed with other noradrenergic drugs (Biederman et al. 2008a,b). Reductions in systolic and diastolic blood pressure and heart rate were also found. Most changes were not clinically relevant, but serious episodes of orthostatic hypotension and syncope have been reported in ~2% of subjects in long-term trials. Weight gain has also been observed in a small number of subjects (Biederman et al. 2008b).

The preclinical pharmacology of α_{2A}-adrenoceptor agonists, including their effects on cognitive function, sedation, regulation of blood pressure, and their actions to reduce the efflux of catecholamines in the PFC, accurately predicts the relative efficacy of these drugs in the treatment of ADHD and their adverse events.

2.5 *Modafinil*

Modafinil is approved as a treatment for narcolepsy and it also increases vigilance and cognitive function. Modafinil did not complete clinical development as a potential medication for ADHD. Nonetheless, it is used off-label and is an interesting drug to explore in terms of its efficacy and side effect profile, and as a mechanistic approach for the development of novel treatments for this disorder.

Modafinil's pharmacology is poorly understood, but several putative mechanisms have been postulated to account for its therapeutic actions. This drug has low affinity for DAT (~3–6 μm; Zolkowska et al. 2009) and no measurable affinity for NET or SERT (Madras et al. 2006). However, in positron emission tomography (PET) experiments, modafinil occupied a significant proportion of DAT and NET sites in monkey striatum and thalamus (Madras et al. 2006; Andersen et al. 2010). Moreover, its pharmacological effects were abolished or attenuated by agents that decrease catecholaminergic, particularly α_1-adrenoceptor-mediated, function (Minzenberg and Carter 2008).

Modafinil induces behavioural effects indicative of enhanced dopaminergic neurotransmission: it increases motor activity in normal and Parkinsonian monkeys (Jenner et al. 2000); partially generalises to a cocaine-like discriminative stimulus in rodents and, at high doses, is reinforcing in monkeys and humans (Gold and Balster 1996). Modafinil is an agonist of D_2 receptors. The wake-promoting effects of this drug are abolished in mice lacking DAT sites and D_1/D_2 receptors (Wisor et al. 2001).

In microdialysis studies, modafinil increases extracellular dopamine in the nucleus accumbens of rats (Murillo-Rodríguez et al. 2007), and striata of dogs (Wisor et al. 2001) and monkeys (Andersen et al. 2010). In the PFC of rats, a high dose increased extracellular dopamine, noradrenaline and serotonin that was slow in onset (peak increase 60–120 min post-injection) and modest in amplitude (de Saint Hilaire et al. 2001). In human PET studies, therapeutic doses of modafinil occupied DAT sites and modestly increased dopamine efflux in striatum (Volkow et al. 2009). In terms of its effects on CNS monoamines, the most powerful action of modafinil was to increase extracellular serotonin (Ferraro et al. 2002). However, this action is not predicted to contribute to modafinil's potential benefit in ADHD.

In summary, modafinil's effects on cognition and vigilance, together with evidence that it enhances central catecholaminergic function, provide a reasonable rationale to suggest that this drug may be of benefit in the treatment of ADHD.

2.5.1 Clinical Profile of Modafinil and Accuracy of Predictions Based on Preclinical Results

In clinical trials, modafinil improved ADHD symptoms (Biederman et al. 2005). The efficacy of modafinil was robust, but relatively modest in magnitude. The response rate was 40–50% with no clear difference between efficacy in patients previously responsive to stimulants versus those were not, or who were treatment-naive (Wigal et al. 2006).

Since no head-to-head comparison has been performed between modafinil and any established stimulant ADHD medication, there is no firm information on its relative efficacy. It is likely that modafinil would be less effective than the stimulants. Modafinil improved ADHD symptoms within the first week of treatment, but its maximum effect took 7–9 weeks. Reported side effects (insomnia, decreased appetite, headache and weight loss; Wigal et al. 2006) are consistent with enhanced dopaminergic and noradrenergic neurotransmission in the CNS.

Unlike the stimulants, modafinil increases vigilance without associated tolerance or sensitisation (Bastuji and Jouvet 1988). In the UK, modafinil is not a Controlled Drug as in other countries within Europe and the USA. Clinical studies in drug-experienced volunteers have produced contradictory results. Jasinski (2000) observed that modafinil produced a dose-dependent reporting of "drug liking" and "high" in polydrug users that was not observed by Rush et al. (2002) in cocaine-experienced subjects. Recently, Volkow et al. (2009) reported that normal subjects receiving similar doses of modafinil did not report any abuse- or drug-related subjective effects. In these trials, modafinil lacked the profound psychostimulant properties of methylphenidate, amphetamine and cocaine (Jasinski 2000). Although modafinil has weak reinforcing effects, it can nonetheless give rise to psychoactive effects typical of psychostimulant drugs. Taking all these factors into account, modafinil appeared to be a promising approach in the search for an ADHD drug with low potential for abuse. However, the FDA issued a non-approvable letter for the use of modafinil to treat ADHD in 2006. This decision was based on a single adverse reaction in clinical trials of a patient developing a rash suggestive of Stevens–Johnson syndrome (a serious hypersensitivity reaction affecting the skin and mucous membranes). Since then, the FDA has excluded the use of modafinil in any paediatric indication and the development of this drug for the treatment of ADHD has been discontinued, although it is still used off-label.

The relative efficacy of modafinil in ADHD, its side effect profile and its low level of abuse liability are generally consistent with its reported effects on catecholamine efflux in microdialysis experiments. However, the increases of dopamine and noradrenaline efflux are less profound and slower in onset than those evoked by amphetamines or methylphenidate (Table 3; Fig. 2) and are consistent with the prediction of low efficacy in reducing ADHD symptoms compared with the stimulants. Increases in extraneuronal dopamine induced by modafanil are modest and relatively slow in onset, which is consistent with modafinil's reported low level of abuse liability.

3 Clinical Experience with Drugs Used to Treat ADHD

When faced with a patient for whom a diagnosis of ADHD is being considered, a key first step is a careful and thorough assessment. The differential diagnosis of the symptoms by which ADHD is characterised is quite extensive. Many psychiatric and general medical conditions, as well as non-pathological life conditions, can cause a patient to express symptoms of ADHD. Presuming that the patient does indeed have ADHD, options include non-pharmacological and pharmacological interventions to address the impairments. Considering the potential benefits associated with both forms of treatment, as well as their combination, discussions between the clinician and patient, and, if appropriate, the patient's family, are pivotal. ADHD is a chronic condition and developing an effective therapeutic relationship between the patient and provider can begin during the treatment planning process. Experience suggests that patients and their families often do not want the clinician to make treatment decisions for them. Rather, they want to be educated about treatment options so that they can make the decision that they feel meets their needs. However, there are times when patients and their guardians ask what the clinician would do. In this section, the treatment options and their rationale will be discussed from the physician's perspective.

Physicians recommend combined interventions for ADHD that include pharmacological, environmental (educational/vocational) and behavioural components. This is done in the hope not only of reducing symptoms, but also to improve functioning and mastery of previously challenging undertakings. A patient may ask for medication alone, wish to eschew medication or may not wish to involve a patient's school in drug administration. As a general rule, it seems reasonable to accommodate a patient's/guardian's request, while also monitoring treatment outcomes, and considering adding or removing forms of intervention with time. Presuming that pharmacotherapy is chosen as a treatment option, the first consideration is what medication(s) should be considered "first line". Despite the multiple treatment options, physicians generally recommend a stimulant as the initial intervention. Although they are imperfect agents associated with abuse liability and well-documented side effects, they are "first line" pharmacotherapy. This is partly due to the volume of information available on their efficacy and tolerability that has accumulated over many decades.

There are patients for whom stimulants might be problematic and inappropriate: e.g. those with a history of substance abuse or Tourette's disorder. Before choosing between the stimulants in these patients, non-stimulant treatments will be considered: e.g. atomoxetine or guanfacine XR. Although there are no empirical head-to-head data, physicians generally choose guanfacine for youths with tics, or with particular difficulty with impulsivity and hyperactivity, but atomoxetine if inattention or anxiety predominate. However, if pharmacotherapy is started with a stimulant, the next question is whether or not this should be a methylphenidate- or amphetamine-based preparation. Since the efficacy of these compounds is equivalent across populations, the approach to the individual patient needs to be based on other parameters. This might include eschewing a formulation that shares the same active ingredient as a medication that the patient/family does not wish to administer. As the lay public seems to be familiar with stimulant formulations, people may have pre-existing concerns. These pre-conceived ideas may be due to, e.g., positive or negative responses, or side effects that occurred in family or acquaintances. In some instances, treatment availability and cost are key parameters. Once the class of drug and its active ingredient is chosen, the next question is whether or not the patient should start on a short- or longer-acting preparation. This decision is one that is frequently driven by patient and family concerns. Although it generally ameliorates with time, at the initiation of pharmacotherapy, caution about medication-related side effects may be substantial and there may be reluctance to take longer-acting formulations. In order to minimise duration of medication exposure, some patients may wish to start on short-acting preparations.

Medical advances over the past decade in the pharmacotherapy of ADHD have made it easier for patients and prescribers to individualise care. However, all available treatments are imperfect. Stimulants are still associated with periods of time where their benefits "wear off". Furthermore, some patients cannot tolerate, or do not benefit from, pharmacotherapy with this class of drugs. Therefore, there is ample scope for drug development in ADHD. Some of the characteristics that would be possessed by the "ideal" drug would include the following:

1. Safety, both acutely and with chronic use across the lifecycle
2. No abuse liability
3. Once-daily dosing, while providing 24 h of benefit
4. Optimal symptom reduction within 1 h of administration
5. Equally effective for inattentive as well as hyperactivity/impulsivity symptoms
6. Will not require sustained treatment over days/weeks in order for benefits to be maximised

4 New Developments in the Field of Research into Drugs to Treat ADHD and Future Directions

A review of drug candidates in the late-stage development for ADHD treatment is dominated by the application of formulation technologies to existing drugs or by "me too" pharmacology. Pharmaceutical companies have applied a range of formulation technologies (bead formulations, slow-release matrices and transdermal patches) to methylphenidate or amphetamine to produce novel, once-daily medications.

Drug candidates acting at various novel molecular targets have been evaluated to determine their efficacy and safety as potential ADHD medications (Table 4).

Table 4 Compounds in development

Compound number	Generic name	Company	Mode of action	Development status
Reuptake inhibitors				
LY2216684	–	Eli Lilly	NA reuptake inhibitor	PIII
SEP228432	–	Sepracor	Triple uptake inhibitor	PI
DOV102677	–	Dov Pharmaceuticals	Triple uptake inhibitor	D
GSK372475 (NS2359)	–	GSK/NeuroSearch	Triple uptake inhibitor	D
–	(R)-Sibutramine	Sepracor	Triple uptake inhibitor	D
SPD473	–	Shire	Triple uptake inhibitor	D
Nicotinic agents				
ABT089	Pozanicline	Abbott/NeuroSearch	α_4/β_2 partial agonist	PII
ABT894	Sofinicline	Abbott/NeuroSearch	α_4/β_2 agonist	PII
AZD3480	–	AstraZeneca/ Targacept	α_4/β_2 partial agonist	PII

(continued)

Table 4 (continued)

Compound number	Generic name	Company	Mode of action	Development status
AZD1446	–	AstraZeneca/ Targacept	α_4/β_2 agonist	PII
GTS21	–	CoMentis	α_7 agonist	PII
TC5619	–	Targacept	α_7 agonist	PII
–	Lobeline	Yaupon	α_7 agonist	PII
Histaminergic agents				
JNJ31001074	–	J & J	H_3 antagonist	PII
SAR110894	–	Sanofi-Aventis	H_3 antagonist	D
AMPA modulators				
Org26576	–	Merck	AMPA modulator	PII
CX717	–	Cortex pharmaceuticals	AMPA agonist	D
SPD420	–	Shire	AMPA modulator	D
Miscellaneous targets				
NWP06	–	NextWave pharmaceuticals	Not disclosed	PII
SPN811	–	Sepracor	Not disclosed	PI
SPN812	–	Sepracor	Not disclosed	PC
NSD867	–	NeuroSearch	Not disclosed	PI
KP106	–	KemPharm	α_{1A}/α_{2A} agonist	PI

PI phase I, *PII* phase II, *PIII* phase III, *PC* preclinical, *D* discontinued

When predicting the likely efficacy and safety of new therapeutic approaches in ADHD, the knowledge gained from existing drugs can be helpful. The pharmacological characteristics of the most effective drugs for treating ADHD, the stimulants, are summarised below and in Table 3:

1. These drugs produce large and rapid increases in the synaptic concentration of catecholamines in the PFC.
2. There is no obvious ceiling on the magnitude of their effect on catecholamine efflux.
3. The most efficacious ADHD drugs also enhance dopaminergic neurotransmission in sub-cortical brain regions.

However, some caveats have to be taken into consideration. For example, lack of information in the public domain indicates that drugs that are selective dopamine releasing agents, or noradrenaline reuptake inhibitors with the pharmacological characteristics of methylphenidate, have not been evaluated as potential ADHD therapies. Hence, it is impossible to know whether sub-cortical dopamine efflux is a critical component of maximal efficacy in an ADHD medication, or alternatively, whether a drug with a selective noradrenergic mechanism that is as powerful as methylphenidate or amphetamine could rival the efficacy of the stimulants. In addition, it is important to remember that effective drugs have actions in multiple symptom domains; reduction of hyperactivity, impulsivity and concomitant social misconduct, while increasing vigilance, attention and cognitive performance.

Any drug candidate that fails to provide benefit in all these areas may be valuable only as an adjunctive therapy.

Several triple reuptake inhibitors have been taken into clinical development in ADHD and some have potent dopamine reuptake inhibitor properties. Development of four of them has been terminated; mostly for lack of sufficient efficacy indicating that a non-selective monoamine reuptake inhibition profile is not appropriate for the treatment of ADHD.

Once-daily formulations of other α_2-adrenoceptor agonists apart from guanfacine, e.g. clonidine, are now in late-stage development as alternative non-stimulant treatments for ADHD.

Development of nicotinic receptor agonists is the approach that has received the most attention in the search for new ADHD drugs. Partial and full agonists of the α_4/β_2 and α_7 receptor subtypes are undergoing clinical evaluation. The α_4/β_2 receptor agonists, ABT418, ABT089 and ABT594, reduce distractibility and increase vigilance and cognitive performance in rodents and primates (Buccafusco et al. 1995, 2007). Compounds of this type also increase dopamine, noradrenaline and 5-HT turnover in whole-brain (Tani et al. 1997) and increase release of [^{3}H]dopamine from striatal synaptosomes (Sharples et al. 2000). In a placebo-controlled, crossover, pilot trial in adult ADHD, Wilens et al. (1999) observed that the α_4/β_2 receptor agonist, ABT418, produced moderate, but statistically significant, reductions in several ADHD severity scales. ABT418 was more effective in the inattentive than the hyperactive/impulsive subgroup, but of greater importance was the observation that the efficacy of this compound diminished as the severity of the disorder increased. More recently, Wilens et al. (2006) reported results for the α_4/β_2 receptor partial agonist, ABT089, in adult ADHD. This compound produced statistically significant reductions in ADHD severity scales, but the effects were small and not dose-dependent. Therefore, based on the available data, nicotinic α_4/β_2 receptor agonists are unlikely to provide a significant advance in ADHD pharmacotherapy. Nicotinic α_7 receptor agonists have also been shown to reduce distractibility and improve cognitive function in animals (Buccafusco et al. 2007) and to modulate catecholaminergic neurotransmission in vitro and in vivo (Kaiser and Wonnacott 2000). Various α_7 receptor agonists are currently being evaluated in ADHD, but there is no information in the public domain on their efficacy or safety.

Over the past few years, preclinical studies have shown that histamine H_3 antagonists improve cognitive function (e.g. see review by Stocking and Letavic 2008). JNJ31001074 (J&J) is a histamine H_3 receptor antagonist that is currently in clinical development for cognitive deficits in ADHD. With the *caveat* that no firm conclusions can be based on a single discontinuation, the termination of another H_3 receptor antagonist, SAR110894 (Sanofi-Aventis), suggests that this pharmacological approach may not be suitable for use in ADHD. This hypothesis is supported by preclinical investigations indicating that H_3 receptor antagonists are being targeted mainly at cognitive dysfunction in schizophrenia and Alzheimer's disease (Medhurst et al. 2007).

In the SHR, disturbances in glutamate, as well as dopamine and noradrenaline function suggest a defect in brain areas required for reward associated learning and memory (e.g. see review by Russell 2003). Lehola et al. (2004) reported an abnormality in the NMDA receptor of SHRs, which they suggested could contribute to their ADHD-related symptoms. These findings parallel a recent clinical study. In adolescents with ADHD and normal comparators, proton magnetic resonance spectroscopy was used to scan their brains before and after treatment with methylphenidate XR. The preliminary findings indicated glutamatergic abnormalities in adolescents with ADHD, which could be rectified with MPH treatment (Hammerness et al. 2010). The AMPA receptor modulator, Org26576 (Merck), is currently in Phase II clinical evaluation in ADHD. However, although no information is available on this particular drug candidate, the termination of two other AMPA receptor modulators (Table 4) suggests that compounds with this pharmacology may not be effective as ADHD treatments.

While some of the new approaches may be clinically efficacious, thereby offering more choice to physicians, none appears to be equal to the stimulants. The preclinical pharmacology of these drugs shows that their ability to enhance catecholaminergic neurotransmission in vivo is at best moderate. These observations reinforce the view that the relative efficacy of ADHD drugs is strongly associated with their ability to directly or indirectly enhance catecholaminergic neurotransmission.

5 A Summary of the Current Status and Future Prospects of Drug Therapy for ADHD

The introduction of the stimulants revolutionised the treatment of ADHD for a number of reasons. These drugs are remarkably effective, and the percentage of non-responders is low compared to drug therapy in many other psychiatric indications. Moreover, the rapid appearance of their benefits in ADHD treatment provides an important early indicator of success or failure. For this reason, amphetamine or methylphenidate formulations are generally the first selected by physicians. Another advantage is that these drugs manipulate the function of the catecholamines and serotonin, probably the most widely studied and best understood of the CNS neurotransmitters. The availability of a wide range of compounds with a high degree of specificity for individual monoamines and/or different pharmacological mechanisms has refined our understanding of the essential elements for optimum pharmacological effect in managing the disorder. It has also provided valuable information from which to construct hypotheses about the neuroanatomy and neurotransmitter dysregulation that are causal in ADHD. The validity of these hypotheses has been tested in clinical trials and confirmed by the success of some new drugs and the failure of others.

The development of sustained-release preparation stimulants has expanded treatment options for young patients with ADHD. Since ADHD is a psychiatric and cognitive disorder characterised by inattentiveness, poor concentration, distractibility and impulsivity, subjects are unsuited to self-medication, particularly with short-acting drugs that need to be administered two or three times daily at reasonably precise intervals. In circumventing the need for multiple-daily dosing, formulation and prodrug strategies have played a critical role in improving the pharmacotherapy of ADHD, and now there are multiple treatment options for every type of medication. All of these developments have contributed real progress in the field, but there is still much room for improvement and unmet clinical need in ADHD pharmacotherapy.

In the search for alternative pharmacological approaches, drug candidates have been evaluated with a wide diversity of mechanisms – many of which act as modulators of catecholaminergic function in the CNS. Preliminary clinical data indicate that some of these compounds are likely to be effective as ADHD drugs, but none looks as if they will deliver comparable efficacy equal to the stimulants or provide the breakthrough in efficacy and safety that is required. It may be too early to be judgemental on the future prospects for the introduction of advantaged new drugs for the treatment of ADHD. They may add significantly to the list of drugs available to treat ADHD but, equally, it is also clear that the stimulants and other catecholaminergic drugs are likely to be the mainstay of pharmacotherapy in this disorder for many years to come.

References

Andersen PH (1989) The dopamine uptake inhibitor GBR 12909: selectivity and molecular mechanism of action. Eur J Pharmacol 166:493–504

Andersen ML, Kessler E, Murnane KS et al (2010) Dopamine transporter-related effects of modafinil in rhesus monkeys. Psychopharmacol 210:439–448

Arnold LE, Wender PH, McCloskey K et al (1972) Levoamphetmaine and dextroamphetamine: comparative efficacy in the hyperkinetic syndrome. Assessment of target symptoms. Arch Gen Psychiatry 27:819–822

Arnsten AFT (2006) Fundamentals of attention-deficit/hyperactivity disorder: circuits and pathways. J Clin Psychiatry 67:7–12

Barkley RA, DuPaul GJ, McMurray MB (1991) Attention deficit disorder with and without hyperactivity: clinical response to three dose levels of methylphenidate. Pediatrics 87:519–531

Barrickman LL, Perry PJ, Allen AJ et al (1995) Bupropion versus methylphenidate in the treatment of attention-deficit hyperactivity disorder. J Am Acad Child Adolesc Psychiatry 34:649–657

Bastuji H, Jouvet M (1988) Successful treatment of idiopathic hypersomnia and narcolepsy with modafinil. Prog Neuropsychopharmacol Biol Psychiatry 12:695–700

Biederman J, Swanson JM, Wigal SB et al (2005) Efficacy and safety of modafinil film-coated tablets in children and adolescents with attention-deficit/hyperactivity disorder: results of a randomized, double-blind, placebo-controlled, flexible-dose study. Pediatrics 116: e777–e784

Biederman J, Melmed RD, Patel A et al (2008a) A randomized, double-blind, placebo-controlled study of guanfacine extended release in children and adolescents with attention-deficit/hyperactivity disorder. Pediatrics 121:e73–e84

Biederman J, Melmed RD, Patel A et al (2008b) Long-term, open-label extension study of guanfacine extended release in children and adolescents with ADHD. CNS Spectr 13:1047–1055

Bradley C (1937) Behaviour of children receiving benzedrine. Am J Psychiatry 94:577–585

Buccafusco JJ, Jackson WJ, Terry AV Jr et al (1995) Improvement in performance of a delayed matching-to-sample task by monkeys following ABT-418: a novel cholinergic channel activator for memory enhancement. Psychopharmacol 120:256–266

Buccafusco JJ, Terry AV Jr, Decker MW et al (2007) Profile of nicotinic acetylcholine receptor agonists ABT-594 and A-582941, with differential subtype selectivity, on delayed matching accuracy by young monkeys. Biochem Pharmacol 15:1202–1211

Buitelaar JK, Michelson D, Danckaerts M et al (2007) A randomized, double-blind study of continuation treatment for attention-deficit/hyperactivity disorder after 1 year. Biol Psychiatry 61:694–699

Bymaster FP, Katner JS, Nelson DL et al (2002) Atomoxetine increases extracellular levels of norepinephrine and dopamine in prefrontal cortex of rat: a potential mechanism for efficacy in attention deficit/hyperactivity disorder. Neuropsychopharmacol 27:699–711

Carboni E, Imperato A, Perezzani L et al (1989) Amphetamine, cocaine, phencyclidine and nomifensine increase extracellular dopamine concentrations preferentially in the nucleus accumbens of freely moving rats. Neurosci 28:653–661

Casat CD, Pleasants DZ, Van Wyck FJ (1987) A double-blind trial of bupropion in children with attention deficit disorder. Psychopharmacol Bull 23:120–122

Cass WA, Gerhardt GA (1995) *In vivo* assessment of dopamine uptake in rat medial prefrontal cortex: comparison with dorsal striatum and nucleus accumbens. J Neurochem 65:201–207

Cheetham SC, Kulkarni RS, Rowley HL et al (2007) The SH rat model of ADHD has profoundly different catecholaminergic responses to amphetamine's enantiomers compared with Sprague-Dawleys. Society for Neurosciences. Abstract 386.14. Available online at www.sfn.org

Conners CK, Casat CD, Gualtieri CT et al (1996) Bupropion hydrochloride in attention deficit disorder with hyperactivity. J Am Acad Child Adolesc Psychiatry 35:1314–1321

Cooper BR, Wang CM, Cox RF et al (1994) Evidence that the acute behavioural and electrophysiological effects of bupropion (Wellbutrin) are mediated by a noradrenergic mechanism. Neuropsychopharmacol 11:133–141

Daviss WB, Bentivoglio P, Racusin R et al (2001) Bupropion sustained release in adolescents with comorbid attention-deficit/hyperactivity disorder and depression. J Am Acad Child Psychiatry 40:307–314

de Saint Hilaire Z, Orosco M, Rouch C et al (2001) Variations in extracellular monoamines in the prefrontal cortex and medial hypothalamus after modafinil administration: a microdialysis study in rats. Neuroreport 12:3533–3537

Devoto P, Flore G, Longu G et al (2003) Origin of extracellular dopamine from dopamine and noradrenaline neurons in the medial prefrontal and occipital cortex. Synapse 50:200–205

Durston S (2003) A review of the biological bases of ADHD: what have we learned from imaging studies? Ment Retard Dev Disabil Res Rev 9:184–195

Efron D, Jarman F, Barker M (1997a) Methylphenidate versus dexamphetamine in children with attention deficit hyperactivity disorder: a double-blind, crossover trial. Pediatrics 100:E6

Efron D, Jarman F, Barker M (1997b) Side effects of methylphenidate and dexamphetamine in children with attention deficit hyperactivity disorder: a double-blind, crossover trial. Pediatrics 100:662–666

Faraone SV, Wigal SB, Hodgkins P (2007) Forecasting three-month outcomes in a laboratory school comparison of mixed amphetamine salts extended release (Adderall XR) and atomoxetine (Strattera) in school-aged children with ADHD. J Atten Disord 11:74–82

Ferraro L, Fuxe K, Tanganelli S et al (2002) Differential enhancement of dialysate serotonin levels in distinct brain regions of the awake rat by modafinil: possible relevance for wakefulness and depression. J Neurosci Res 68:107–112

Ferris RM, Tang FLM, Maxwell RA (1972) A comparison of the capacities of isomers of amphetamine, deoxypiprarol and methylphenidate to inhibit the uptake of tritiated catecholamine into rat cerebral cortex slices, synaptosomal preparations of rat cerebral cortex, hypothalamus and striatum and into adrenergic nerves of rabbit aorta. J Pharmacol Exp Ther 34:447–449

Findling RL, Bukstein OG, Melmed RD (2008a) A randomized, double-blind, placebo-controlled, parallel-group study of methylphenidate transdermal system in pediatric patients with attention-deficit/hyperactivity disorder. J Clin Psychiatry 69:149–159

Findling RL, Childress AC, Krishnan S et al (2008b) Long-term effectiveness and safety of lisdexamfetamine dimesylate in school-aged children with attention-deficit/hyperactivity disorder. CNS Spectr 13:614–620

Géranton SM, Heal DJ, Stanford SC (2003) Differences in the mechanisms that increase noradrenaline efflux after administration of d-amphetamine: a dual-probe microdialysis study in rat frontal cortex and hypothalamus. Br J Pharmacol 139:1441–1448

Gerasimov MR, Franceschi M, Volkow N et al (2000) Comparison between intraperitoneal and oral methylphenidate administration: a microdialysis and locomotor activity study. J Pharm Exp Ther 295:51–57

Gold LH, Balster RL (1996) Evaluation of the cocaine-like discriminative stimulus effects and reinforcing effects of modafinil. Psychopharmacol 126:286–292

Gross MD (1976) A comparison of dextro-amphetamine and racemic-amphetamine in the treatment of the hyperkinetic syndrome or minimal brain dysfunction. Dis Nerv Syst 37:14–16

Hammerness P, Georgiopoulos A, Doyle RL et al (2009) An open study of adjunct OROS-methylphenidate in children who are atomoxetine partial responders: II. Tolerability and pharmacokinetics. J Child Adolesc Psychopharmacol 19:493–499

Hammerness P, Biederman J, Petty C, Henin A, Moore CM (2010) Brain biochemical effect of methylphenidate treatment using proton magnetic spectroscopy in youth with attention-deficit hyperactivity disorder: a controlled pilot study. CNS Neurosci Ther. Dec 8 [Epub ahead of print]

Hasegawa H, Meeusen R, Sarre S et al (2005) Acute dopamine/norepinephrine reuptake inhibition increases brain and core temperature in rats. J Appl Physiol 99:1397–1401

Heal DJ (2008) A method for identifying a compound for treating a disorder or condition associated with dysfunction of monoamine neurotransmission. UK Patent Application GB 2447 949

Heal DJ, Cheetham SC, Prow MR et al (1998) A comparison of the effects on central 5-HT function of sibutramine hydrochloride and other weight-modifying agents. Br J Pharmacol 125:301–308

Heal DJ, Smith SL, Kulkarni RS et al (2008) New perspectives from microdialysis studies in freely-moving, spontaneously hypertensive rats on the pharmacology of drugs for the treatment of ADHD. Pharmacol Biochem Behav 90:184–197

Heal DJ, Cheetham SC, Smith SL (2009) The neuropharmacology of ADHD drugs *in vivo*: insights into efficacy and safety. Neuropharmacol 57:608–618

Hitri A, Venable D, Nguyen HQ et al (1991) Characteristics of [^{3}H]GBR 12935 binding in the human and rat frontal cortex. J Neurochem 56:1663–1672

Hurd YL, Ungerstedt U (1989) Ca^{2+} dependence of the amphetamine, nomifensine, and Lu 19-005 effect on *in vivo* dopamine transmission. Eur J Pharmacol 166:261–269

Ihalainen JA, Tanila H (2002) *In vivo* regulation of dopamine and noradrenaline release by α_{2A}-adrenoceptors in the mouse prefrontal cortex. Eur J Neurosci 15:1789–1794

Jasinski DR (2000) An evaluation of the abuse potential of modafinil using methylphenidate as a reference. J Psychopharmacol 14:53–60

Jasinski DR, Faries DE, Moore RJ et al (2008) Abuse liability assessment of atomoxetine in a drug-abusing population. Drug Alcohol Depend 95:140–146

Jenner P, Zeng BY, Smith LA et al (2000) Antiparkinsonian and neuroprotective effects of modafinil in the mptp-treated common marmoset. Exp Brain Res 133:178–188

Kaiser S, Wonnacott S (2000) α-Bungarotoxin-sensitive nicotinic receptors indirectly modulate [^{3}H]dopamine release in rat striatal slices via glutamate release. Mol Pharmacol 58:312–318

Kelsey DK, Sumner CR, Casat CD et al (2004) Once-daily atomoxetine treatment for children with attention-deficit/hyperactivity disorder, including an assessment of evening and morning behavior: a double-blind, placebo-controlled trial. Pediatrics 114:e1–e8

Kuczenski R, Segal DS (1997) Effects of methylphenidate on extracellular dopamine, serotonin, and norepinephrine: comparison with amphetamine. J Neurochem 68:2032–2037

Kuczenski R, Segal DS, Cho AK et al (1995) Hippocampus norepinephrine, caudate dopamine and serotonin, and behavioral responses to the stereoisomers of amphetamine and methamphetamine. J Neurosci 15:1308–1317

Lehola M, Kellaway L, Russell VA (2004) NMDA receptor function in the prefrontal cortex of a rat model for attention-deficit hyperactivity disorder. Metab Brain Dis 19:35–42

Li SX, Perry KW, Wong DT (2002) Influence of fluoxetine on the ability of bupropion to modulate extracellular dopamine and norepinephrine concentration in three mesocorticolimbic areas of rats. Neuropharmacol 42:181–190

Madras BK, Xie Z, Lin Z et al (2006) Modafinil occupies dopamine and norepinephrine transporters *in vivo* and modulates the transporters and trace amine activity *in vitro*. J Pharmacol Exp Ther 319:561–569

Mantle TJ, Tipton KF, Garrett NJ (1976) Inhibition of monoamine oxidase by amphetamine and related compounds. Biochem Pharmacol 25:2073–2077

Medhurst AD, Atkins AR, Beresford IJ et al (2007) GSK189254, a novel H_3 receptor antagonist that binds to histamine H_3 receptors in Alzheimer's disease brain and improves cognitive performance in preclinical models. J Pharmacol Exp Ther 321:1032–1045

Michelson D, Faries D, Wernicke J et al (2001) Atomoxetine in the treatment of children and adolescents with attention-deficit/hyperactivity disorder: a randomized, placebo-controlled, dose-response study. Pediatrics 108:E83–E92

Michelson D, Adler L, Spencer S (2003) Atomoxetine in adults with ADHD: two randomized placebo-controlled studies. Biol Psychiatry 53:112–120

Miller L, Griffith J (1983) A comparison of bupropion, dextroamphetamine, and placebo in mixed-substance abusers. Psychopharmacol 80:199–205

Minzenberg MJ, Carter CS (2008) Modafinil: a review of neurochemical actions and effects on cognition. Neuropsychopharmacol 33:1477–1502

Morón JA, Brockington A, Wise RA et al (2002) Dopamine uptake through the norepinephrine transporter in brain regions with low levels of the dopamine transporter: evidence from knock-out mouse lines. J Neurosci 22:389–395

Murillo-Rodríguez E, Haro R, Palomero-Rivero M et al (2007) Modafinil enhances extracellular levels of dopamine in the nucleus accumbens and increases wakefulness in rats. Behav Brain Res 176:353–357

Newcorn JH, Kratochvil CJ, Allen AJ et al (2008) Atomoxetine and osmotically released methylphenidate for the treatment of attention deficit hyperactivity disorder: acute comparison and differential response. Am J Psychiatry 165:721–730

Nomikos GG, Damsma G, Wenkstern D et al (1990) *In vivo* characterization of locally applied dopamine uptake inhibitors by striatal microdialysis. Synapse 6:106–112

Pelham WE, Greenslade KE, Vodde-Hamilton M et al (1990) Relative efficacy of long-acting stimulants on children with attention deficit-hyperactivity disorder: a comparison of standard methylphenidate, sustained-release methylphenidate, sustained-release dextroamphetamine, and pemoline. Pediatrics 86:226–237

Pelham WE, Aronoff HR, Midlam JK et al (1999) A comparison of Ritalin and Adderall: efficacy and time-course in children with attention-deficit/hyperactivity disorder. Pediatrics 103:e43

Pozzi L, Invernizzi R, Cervo L et al (1994) Evidence that extracellular concentration of dopamine are regulated by noradrenergic neurons in the frontal cortex of rats. J Neurochem 63:195–200

Pum M, Carey RJ, Huston JP et al (2007) Dissociating effects of cocaine and *d*-amphetamine on dopamine and serotonin in the perirhinal, entorhinal, and prefrontal cortex of freely moving rats. Psychopharmacol 193:375–390

Quinn D, Wigal S, Swanson J et al (2004) Comparative pharmacodynamics and plasma concentrations of d-threo-methylphenidate hydrochloride after single doses of d-threo-methylphenidate hydrochloride and d, l-threo-methylphenidate hydrochloride in a double-blind, placebo-controlled, crossover laboratory school study in children with attention-deficit/hyperactivity disorder. J Am Acad Child Adolesc Psychiatry 43:1422–1429

Richelson E, Pfenning M (1984) Blockade by antidepressants and related compounds of biogenic amine uptake into rat brain synaptosomes: most antidepressants selectively block norepinephrine uptake. Eur J Pharmacol 104:277–286

Rush CR, Kelly TH, Hays LR et al (2002) Acute behavioral and physiological effects of modafinil in drug abusers. Behav Pharmacol 13:105–115

Russell VA (2003) Dopamine hypofunction possibly results from a defect in glutamate stimulated release of dopamine in the nucleus accumbens shell of a rat model for attention deficit hyperactivity disorder – the spontaneously hypertensive rat. Neurosci Biobehav Rev 27:671–682

Sagvolden T, Metzger MA, Schiørbeck HK et al (1992) The spontaneously hypertensive rat (SHR) as an animal model of childhood hyperactivity (ADHD): changed reactivity to reinforcers and to psychomotor stimulants. Behav Neural Biol 58:103–112

Sallee FR, Lyne A, Wigal T et al (2009) Long-term safety and efficacy of guanfacine extended release in children and adolescents with attention-deficit/hyperactivity disorder. J Child Adolesc Psychopharmacol 19:215–226

Scahill L, Chappell PB, Kim YS (2001) A placebo-controlled study of guanfacine in the treatment of children with tic disorders and attention deficit hyperactivity disorder. Am J Psychiatry 158:1067–1074

Sharman DF (1966) Changes in the metabolism of 3, 4-dihydroxyphenylethylamine (dopamine) in the striatum of the mouse induced by drugs. Br J Pharmacol Chemother 28:158–163

Sharples CGV, Kaiser S, Soliakov L et al (2000) UB-165: a novel nicotinic agonist with subtype selectivity implicates the α_4/β_2* subtype in the modulation of dopamine release from rat striatal synaptosomes. J Neurosci 20:2783–2791

Sidhpura N, Redfern P, Rowley H et al (2007) Comparison of the effects of bupropion and nicotine on locomotor activation and dopamine release *in vivo*. Biochem Pharmacol 74:1292–1298

Smith RC, Davis JM (1977) Comparative effects of *d*-amphetamine, *l*-amphetamine, and methylphenidate on mood in man. Psychopharmacol 53:1–12

Sofuoglu M, Hill K, Kosten T et al (2009) Atomoxetine attenuates dextroamphetamine effects in humans. Am J Drug Alcohol Abuse 35:412–416

Stocking EM, Letavic MA (2008) Histamine H3 antagonists as wake-promoting and pro-cognitive agents. Curr Top Med Chem 8:988–1002

Sulzer D, Rayport S (1990) Amphetamine and other psychostimulants reduce pH gradients in midbrain dopaminergic neurons and chromaffin granules: a mechanism of action. Neuron 5:797–808

Tani Y, Saito K, Tsuneyoshi A et al (1997) Nicotinic acetylcholine receptor (nACh-R) agonist-induced changes in brain monoamine turnover in mice. Psychopharmacol 129:225–232

Uhlén S, Porter AC, Neubig RR (1994) The novel alpha-2 adrenergic radioligand [^{3}H]-MK912 is alpha-2C selective among human alpha-2A, alpha-2B and alpha-2C adrenoceptors. J Pharmacol Exp Ther 271:1558–1565

Volkow ND, Fowler JS, Logan J et al (2009) Effects of modafinil on dopamine and dopamine transporters in the male human brain: clinical implications. J Am Med Assoc 301:1148–1154

Wang Y, Zheng Y, Du Y et al (2007) Atomoxetine versus methylphenidate in paediatric outpatients with attention deficit hyperactivity disorder: a randomized, double-blind comparison trial. Aust NZ J Psychiatry 41:222–230

Wigal SB, McGough JJ, McCracken JT et al (2005) A laboratory school comparison of mixed amphetamine salts extended release (Adderal XR®) and atomoxetine (Strattera®) in school-aged children with attention deficit/hyperactivity disorder. J Atten Disord 9:275–289

Wigal SB, Biederman J, Swanson JM et al (2006) Efficacy and safety of modafinil film-coated tablets in children and adolescents with or without prior stimulant treatment for attention-deficit/hyperactivity disorder: pooled analysis of 3 randomized, double-blind, placebo-controlled studies. Prim Care Companion J Clin Psychiatry 8:352–360

Wilens TE, Biederman J, Spencer TJ et al (1999) A pilot controlled clinical trial of ABT-418, a cholinergic agonist, in the treatment of adults with attention deficit hyperactivity disorder. Am J Psychiatry 156:1931–1937

Wilens TE, Spencer TJ, Biederman J et al (2001) A controlled clinical trial of bupropion for attention deficit hyperactivity disorder in adults. Am J Psychiatry 158:282–288

Wilens TE, Haight BR, Horrigan JP et al (2005) Bupropion XL in adults with attention-deficit/hyperactivity disorder: a randomized, placebo-controlled study. Biol Psychiatry 57:793–801

Wilens TE, Verlinden MH, Adler LA et al (2006) ABT-089, a neuronal nicotinic receptor partial agonist, for the treatment of attention-deficit/hyperactivity disorder in adults: results of a pilot study. Biol Psychiatry 59:1065–1070

Wilens TE, Hammerness P, Utzinger L et al (2009) An open study of adjunct OROS-methylphenidate in children and adolescents who are atomoxetine partial responders: I. Effectiveness. J Child Adolesc Psychopharmacol 19:485–492

Wisor JP, Nishino S, Sora I et al (2001) Dopaminergic role in stimulant-induced wakefulness. J Neurosci 21:1787–1794

Wolraich ML, Greenhill LL, Pelham W et al (2001) Randomized, controlled trial of OROS methylphenidate once a day in children with attention-deficit/hyperactivity disorder. Pediatrics 108:883–892

Zolkowska D, Jain R, Rothman RB et al (2009) Evidence for the involvement of dopamine transporters in behavioral stimulant effects of modafinil. J Pharmacol Exp Ther 329:738–746

The Four Causes of ADHD: A Framework

Peter R. Killeen, Rosemary Tannock, and Terje Sagvolden

Contents

1 Introduction: The Causal Framework 392
   1.1 Formal Causes 393
   1.2 Efficient Causes 394
   1.3 Material Causes 394
   1.4 Final Causes 395
   1.5 Recurrent Causes 395
2 The Formal Causes of ADHD 396
   2.1 The Proximate Formal Causes of ADHD: What It Is 396
   2.2 The Ultimate Formal Causes of ADHD: Theories About It 402
3 The Efficient Causes of ADHD: Triggers of Syndrome and Symptoms 405
   3.1 Ultimate Efficient Causes of ADHD: From Genotype to Phenotype 405
   3.2 Proximate Efficient Causes of Episodes of ADHD Behavior 406
4 The Material Causes of ADHD: Its Substrates 407
   4.1 Ultimate Material Cause: Genetics and Epigenetics 407
   4.2 Proximate Material Causes: Neurophysiology 407
5 Final Causes; What Is ADHD Good for? 409
   5.1 Ultimate Function: Evolutionary Utility 411
   5.2 Proximate Function: Immediate Reinforcement 412
6 Recurrent Causes: How Feedback Exacerbates
or Meliorates the Syndrome and Symptoms 413
   6.1 Ultimate Causes: the Baldwin Effect 413
   6.2 Proximate Causes: Reinforcement 413

P.R. Killeen (✉)
Psychology Department, Arizona State University, Tempe, AZ 85287, USA
e-mail: Killeen@asu.edu

R. Tannock
Neuroscience & Mental Health Research Program, The Hospital for Sick Children, 555 University Avenue, Toronto, ON, Canada M5G 1X8

T. Sagvolden
Department of Physiology, Institute of Basic Medical Sciences, University of Oslo, P.O. Box 1103 Blindern, 0317 Oslo, Norway

7	Therapy: Controlling Causes	414
7.1	Formal Causes: Redefinition	414
7.2	Efficient Causes: Context Modification	415
7.3	Material Causes: Brain Modification	415
7.4	Final Causes: The Uses of ADHD	416
References		416

Abstract In addition to the symptoms singled out by the diagnostic criteria for Attention-Deficit Hyperactivity Disorder (ADHD), a comprehensive definition should inform us of the events that trigger ADHD in both its acute and chronic manifestations; the neurobiology that underlies it; and the evolutionary forces that have kept it in the germ line of our species. These factors are organized in terms of Aristotle's four kinds of "causes," or explanations: formal, efficient, material, and final. This framework systematizes the nosology, biology, psychology, and evolutionary pressures that cause ADHD.

Keywords Causal framework · Decision theory · Function · Substrate · Triggers

Abbreviations

AUC	Area under the curve
DBD	Disruptive Behavior Disorders Scale
DSM	Diagnostic and Statistical Manual
ICD	International Classification of Diseases
MAO	Monoamine oxidase
MPH	Methylphenidate
ROC	Relative operating characteristic

1 Introduction: The Causal Framework

Attention-Deficit Hyperactivity Disorder is a complex multidimensional syndrome, characterized by its diagnosis, causes, biological substrates, and effects on individuals and society. These aspects are variously ordered and emphasized by theories of the condition; but that literature is large and diffuse. This chapter is a response to the call by Coghill and associates (Coghill et al. 2005) for an analysis that crosses multiple levels, "integrating environmental and social processes of genetic and neurobiological influence" (p. 105). It is an attempt to carry the torch passed forward by Sagvolden and associates in their dynamic developmental theory of ADHD (Sagvolden et al. 2005).

Millennia ago, Aristotle provided a generic framework for what constitutes a full explanation; it remains a useful candidate structure for situating our knowledge about ADHD. Aristotle noted four aspects of explanation (Ross 1936), which he called reasons or "becauses" (Hocutt 1974). (Modern appreciation of his framework was undermined by the mistranslation of the keyword as "causes"; Santayana 1957). The thing to be understood – here, ADHD – is first defined; this is the lowest layer of a formal description. Such a sketch gets us in the ballpark and facilitates communication with nonexperts, but by itself is insufficient to understand a phenomenon. The other [be]causes tell us about the action on the field: Its origins (efficient causes) and its purposes (final causes); they characterize the material machinery (balls and bats and bases) and its formal rules. No one kind of causal analysis is sufficient for understanding a phenomenon, or even for a complete definition. Events that are alike in some aspects – say, an individual who satisfies DSM criteria for a disorder – may differ in etiology or underlying mechanism; not all children who satisfy diagnostic criteria for ADHD have the same condition.

In this section, the four causes are reviewed in thumbnail and in the following sections applied more explicitly to ADHD. Each of the four causes has a more general, remote, *ultimate* layer, and a more detailed, immediate, *proximate* layer. Formal causes are forms; things of similar structure that explicate the phenomenon. Efficient causes are triggers that exist in the immediate or evolutionary context. Material causes comprise structure and process within the organism. Final causes involve selection by consequences: selection of incidences of behavior or of organisms with certain traits.

1.1 Formal Causes

These are the analogs, metaphors, and models with which we describe phenomena, and which permit us to communicate about them, to characterize, predict, and control them. They can range from simple descriptions using English, to complex ones using tensor calculus. The physicist's favorite formal cause is a differential equation. The chemist's is a molecular model. The behaviorist's is the three-term contingency of stimulus, response, and reinforcer. Formal causes, at lower, proximate, levels of sophistication and detail, serve to define the phenomenon, to provide the basic criteria with which we identify it; at higher, more general levels they are attempts to explain the phenomenon. Facts and theories are parts of the Formal domain, as they attempt at different levels of specificity to characterize the nature of the phenomenon. Theories are subject to frequent revision, correction, or rejection; facts change at a slower rate, as seen in the slower paced change of criteria in the various editions of the DSM.

Models such as these, of fact and theory, do not dictate the other causes of the processes they describe. Simple harmonic motion describes equally well the swing of a pendulum or the vibration of a molecule. Different models capture different aspects of a phenomenon, some at a molar level, and some at a more molecular level – as seen throughout this book. Our sense of familiarity with the structure of the model is transferred to the phenomenon with which it is put in correspondence and constitutes part of the process of "explanation." As our understanding of the subject advances, so also does our formulation of its formal causes.

Models need not be symbolic. Maquettes permit architects to experiment with shapes of buildings, prototypes permit designers to trace streamlines of cars and planes in wind tunnels. Animals are evolved structures that often have useful correspondences with an object of enquiry. One of the first steps in the search for a cure of a disease is to find an animal model of it. Such animal models are alternative formal "becauses" – representations – of select aspects of the condition (Killeen 1999). Some rat strains may be useful models of some, but not all, aspects of ADHD (Fan et al. 2011; Russell et al. 2005; Sagvolden and Johansen 2011).

1.2 Efficient Causes

These are events that occur before a change of state and trigger it (sufficient causes); or do not occur, and by their absence preclude it (necessary causes). They are what most people think of as "causes." Sufficient causes identify the early parts of a stream of events that suffice to bring about the later parts. A *proximate* sufficient cause of ADHD behavior is any one of the events that causes a symptom associated with ADHD, such as the requirement to delay a prepotent response while in turn taking or queuing. It refers to a trigger of the current *state* of the organism; in this case, jumping ahead in line. *Ultimate* efficient causes set up the machinery of the organism to be susceptible to the syndrome – whose symptoms can then be triggered into action by the more proximate efficient causes. It refers to longer-term, more molar *trait* variables (Nigg et al. 2002). Necessary causes are a component of sufficient causes that are required for an outcome, and which may be manipulated in therapeutic interventions. Therapies may address proximate efficient causes with, for example, classroom interventions, and ultimate causes with, for example, prenatal counseling.

1.3 Material Causes

These are the substrates, the underlying mechanisms. Neurobiological mechanisms of ADHD, a pervasive theme of this book, exemplify material causes. Until we can "open the hood" and look inside, we do not fully understand a phenomenon. If modern automobile engines confound the observer, so much more does the most complicated engine in the universe, the human brain (Uttal 2008). Several authors in this volume describe the slow but real progress in characterizing the state of the brain that underlies ADHD (e.g., Mahone et al. 2011; Vaidya 2011). Therapies may address the proximate machinery by, for example, bolstering neurohormonal concentrations where those are insufficient.

1.4 Final Causes

The final cause of an entity or process is its *raison d'etre* – what it does that has brought about or sustained its existence. Not all phenomena have final causes. A biological feature may exist because of its function (i.e., what gave it a selective advantage), or because it was associated with another function that increased fitness. Some features derive from random drift, intensified by restricted breeding populations; thus not all biological features can support genuine inferences to function.

Final causes are not time-reversed efficient causes; they are not teleologies. Explanations in terms of reinforcement or evolutionary fitness or laws of least action are all valid explanations in terms of final causes. Whenever individuals seek to understand a strange machine and ask: "What is it for?" they are requesting a final cause. There may be many final causes for a phenomenon; proximate causes may involve expectations, intentions, or the results of a history of reinforcement; ultimate causes may involve evolutionary pressures. The aspects of ADHD that keep it manifest at a high level in the population constitute an ultimate final cause of the syndrome.

The four becauses have also been called theoretical, causal, reductive, and functional explanations. No single type is definitive. To understand ADHD, we must know something about the immediate stimuli that are necessary and sufficient for its manifestation; the underlying physiological condition, its function, and how best to talk about it – in toto, a comprehensive theory of ADHD. The relation of these parts to the whole is shown in Fig. 1.

1.5 Recurrent Causes

Many phenomena are recurrent – they are parts of a system that they affect, and whose output redounds to affect them: They are *closed-loop* systems. Aristotle noted that a man may exercise for health, then, healthier, he exercises yet more, and continues to strengthen. A thermostat's *function* is to control temperature, it is *triggered* by a change in temperature, its *machinery* may involve the warped expansion of a bimetallic strip, and it is *formally* represented as a negative feedback loop in linear systems theory. Closed-loop systems may stabilize their environment, as above, or amplify instabilities in the case of positive feedback systems. Aristotle's fifth cause is familiar to any parent of ADHD children; the attempt to damp overly rambunctious behavior (negative feedback), to reinforce on-target behavior (positive feedback); or in the escalating stridency of maladaptive control attempts (positive feedback driving interaction in a tragic spiral). It is represented in Fig. 1 by recurrent arrows from final causes to state (reinforcement); and to trait (selective advantage in turbulent environments), with the information transmitted through the proximate and ultimate material substrates.

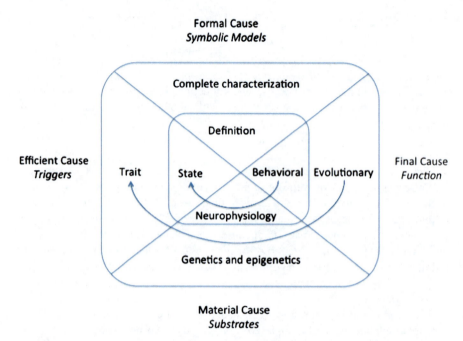

Fig. 1 The causal framework that organizes the contributions of this volume. Each of Aristotle's four causes constitutes issues that must be addressed before a phenomenon is understood. The inner set are proximate, or molecular causes; the outer set are ultimate, or molar causes. Exposition begins with a simple definition or description, such as found in psychiatric manuals. These simple models map criterial behaviors onto vernacular words and numbers. Triggers of ADHD symptoms (states) are proximate efficient causes; triggers of the phenotype (trait) are ultimate efficient causes. Necessary causes at each level are sought that, when removed, will lessen or remove the syndrome or symptom. Material causes comprise the machinery that causes the symptom (proximate: neurophysiology) and syndrome (ultimate: genetic). The events that maintain the behavior, such as immediate gratification, or the syndrome, such as enhanced fitness of the extended phenotype, are the final causes. The recursive *arrows* show that such outcomes can modify the system to change its sensitivity to correlated stimuli or responses; in the short term this is called attention and learning, in the long term it is called Darwinian evolution. A general theory of ADHD, towards which the chapters of this volume strive, constitutes the highest level of formal cause

2 The Formal Causes of ADHD

2.1 The Proximate Formal Causes of ADHD: What It Is

2.1.1 How the DSM Defines It

The DSM-IV uses five lettered criteria to define ADHD. Criterion A1 identifies 9 symptoms of inattention and, A2, 9 symptoms of impulsivity/hyperactivity; evidence of persistent and maladaptive manifestations of at least 6 of the former or 6 of the latter are required for "categorization." Criteria B (age of onset) and C (symptom presence in two or more settings) require temporal and contextual generality in the symptoms; Criterion D (domains of impairment) requires that the symptoms damage progress toward achievements valued by society; Criterion E (exclusionary diagnoses) lists the diagnoses which trump ADHD in accounting for the symptoms. They do so either because those symptoms *ensemble* are more crippling than ADHD, such as major psychoses, or because they are better accounted for by another mental disorder. Because of the substantial comorbidities associated with all psychiatric syndromes, and because they often preclude diagnosing as "ADHD" symptoms that otherwise satisfy the DSM, the prevalence of ADHD is greater than suggested by its nominal prevalence rates of 5–8% of children and adolescents, and 2–4% adults (Polanczyk and Jensen 2008). Criterion B also causes an underestimation of prevalence, and may reduce diagnostic accuracy by increasing false negatives more than true negatives (Kieling et al. 2010). Barkley and associates (Barkley et al. 2007, p. 128) note that half of their well-validated sample of adults who satisfy the criteria for "ADHD" would be lost if the criterion of 7 years were enforced. Whether or not those excluded have ADHD, other than by *de jure* satisfying all the criteria except early onset, awaits a definition of ADHD that is not so dependent on a list of symptoms.

Notably absent from this list, and from the DSM5 criteria, are material causes and nonverbal measures. Also notable is the requirement of impairment in social context; this justifiable "let-well-enough alone" decision biases evaluations of the public cost of ADHD by omitting those individuals who have enough symptoms of inattention or hyperactivity, but whose context or condition enables them to cope with it or turn it to their advantage; or, for the "primarily inattentive type," to simply coast along under the radar. Inclusion of such subsets in the definition would increase the number of individuals diagnosed by 1/3 (Gordon et al. 2006), while decreasing the estimates of per capita cost to the individual and society. Gender, ethnicity, and region add additional variance to purely symptom-based diagnosis (Schneider and Eisenberg 2006).

2.1.2 Scientific Continua, Diagnostic Categories

It is important to distinguish between *adhd* and "ADHD": the former refers to a constellation of dimensional parameters, an as yet poorly identified phenotype (Castellanos et al. 2005; Gottesman and Gould 2003; Kuntsi et al. 2005; Waldman 2005), that envelopes a region of a multidimensional character space. An individual with *adhd* is *a person who sometimes manifests symptoms satisfying the DSM criteria for assigning the label "ADHD."* By contrast, "ADHD," in quotes, is the label for a subset of individuals who have passed the standard criteria – and therefore are located in an extreme region of that space (but not so extreme that comorbidities take diagnostic precedence; Hudziak et al. 1998; Levy et al. 1997). The standard acronym, ADHD, is used when this distinction is not at issue. Individuals with *adhd* are not labeled "ADHD" if they have any of the conditions listed in Section E of the DSM, or if their distribution and timing of symptoms are not just so – five apiece in categories A1 and A2 are not as dispositive as six only in either of the categories. Six of A1 or a half dozen of A2 does not define a threshold at which a disease occurs: Mathematics and reading test scores decrease linearly with the number of ADHD symptoms (Currie and Stabile 2006). Instead, they prescribe a criterion for action at which the utility of psychosocial interventions, special education, or pharmacotherapy crosses a cost/benefit line beyond which the community is willing to expend resources and risk side effects (Matza et al. 2005; Schlander 2010; Secnik et al. 2005). As we learn more about the causes of *adhd*, and in particular the material causes of it, the correlation between the two categories will improve. But there will always be political and economic aspects to both the disorder (e.g., Muir and Zegarac 2001) and to the diagnosis and therapy, both necessary to set rational thresholds for intervention. This distinction between what we know and how we should act exemplifies the classic *tai chi* of science versus policy (Killeen 2003).

2.1.3 Signal Detection Theory Reconciles Decisional Criteria and Evidential Continua

How many symptoms *should* be required before categorization? As more are included, the probability of a false positive (saying that a normal individual has "ADHD") decreases – but so too does the probability of a true positive (saying that a person with *adhd* has "ADHD"). A good way to organize our thinking about this trade-off is with Signal Detectability Theory (e.g., Swets 1973), an instantiation of decision theory. Figure 2 is the foundation of the metaphor: Evidence for a policy is arrayed on the x-axis. It could be the strength of a radar signal, the first principal component of an EEG signal, or a weighted average of scores on inattentive and hyperactivity tests. Above the axis are two hypothetical distributions, the one on the left displaying the distribution of scores due to "noise" – here, the distribution for normal individuals; the one on the right the distribution of evidence due to a "signal" – here, the scores of individuals with *adhd*. A board stipulates when the evidence is strong enough to pronounce "ADHD" by setting a criterion so that all scores to the right qualify. This diagnostic model assumes the existence of two populations, *normal* and *adhd,* yet it remains useful for the present case, where psychiatric problems occur even when those symptoms are subthreshold for classification (e.g., Malmberg et al. 2011).

There are two ways that diagnoses can be correct: (a) a true positive, or *hit*, whose probability is given by the area under the *adhd* distribution to the right of the criterion; (b) a true negative, or *correct rejection*, whose probability is given by the area under the *normal* distribution to the left of the criterion. There are two ways in which decisions can be incorrect: (c) a false positive, or *false alarm*, whose probability is given by the area under the *normal* distribution to the right of the criterion; (d) a false negative, or *miss*, whose probability is given by the area under the *adhd* distribution to the left of the criterion. Improving the power of the

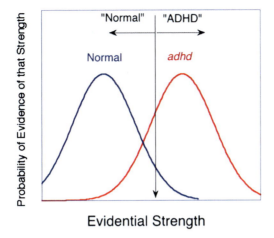

Fig. 2 The Signal Detection Theory model of diagnosis. The *left curve* describes the probability that a normal individual will present evidence of "ADHD" of varying severity. The *right curve* is the probability that an individual who in fact has the trait of *adhd* will present evidence of "ADHD" of varying severity. The diagnostic criteria, working from the best evidence available, parses these individuals into "Normal" and "ADHD." The probability that that diagnosis is correct for normals is the area under the *left curve*, to the left of the criterion. The area under that curve to the right of the criterion gives the probability of a false positive, or false alarm. The probability that the diagnosis is correct for those with *adhd* is the area under the right curve to the right of the criterion. The area under that curve to the left of the criterion is a false negative, or miss. It is the task of scientist to increase the separation of the curves relative to their variance (this is the effect size, d, of a diagnosis), thus improving accuracy. It is the task of professional organizations to set the criterion so as to maximize the benefit/cost ratio of the four outcomes. As the criterion is moved from right to left, the probabilities of hits and false alarms increase, sweeping out a ROC, shown in Figs. 3 and 4. This conventional model treats *adhd* as a taxon – that is, that it represents a distinct population

evidence with better tests can decrease these errors, moving the distributions apart. Thereupon, the criterion should be placed where it maximizes benefits and minimizes costs. If there are cheap effective therapies with few side effects, the criterion should be placed to the left, to minimize misses. Think seat belts for passengers in cars. If the therapy is expensive or may be more harmful than the disease, the criterion should be placed far to the right. Think of Rofecoxib, a nonsteroidal anti-inflammatory drug that was marketed to treat osteoarthritis and which has now been withdrawn over safety concerns. In all decisions, the personal and societal cost of a psychiatric label should be included in the location of the criterion.

To show how the location of the criterion affects accuracy, a graph called a *Relative Operating Characteristic* (ROC; formerly known as Receiver Operating Characteristic; Beck and Shultz 1986; Jensen and Poulsen 1992; Swets et al. 2000) displays the probability of a true positive against the probability of a false positive, as the criterion is moved along the evidence axis. This is exemplified by the research of Faraone et al. (1993), who compared the diagnostic efficiency of four

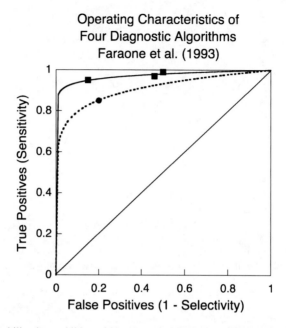

Fig. 3 The probability that a child would be correctly labeled "ADHD" is shown by the ordinate, plotted as a function of the probability that they would be incorrectly categorized as the abscissae (continuous curve; data from Faraone et al. 1993). The area under that curve measures the accuracy of the diagnosis independent of the location of the criterion. Because three algorithms fall along the same ROC, they do not differ in their accuracy, but provide more or less stringent criteria for labeling. A similar ROC may be inferred from the performance of the DSM-III-R (*circle*) in field settings (*dotted line*; Spitzer and Siegel 1990); the smaller area it subtends indicates lesser accuracy in that context

algorithms against psychiatric assessments: (a) Their preferred algorithm, a set of 3 clusters of questions, with a positive answer to any one of the questions in each of the three clusters qualifying the individual; (b) the best 14 of the questions; or (c) all 21 of the questions. Figure 3 shows that these criteria all fall along the same *ROC*, the power function $H = F^b$, where F is the false alarm rate (1 − *specificity*), H the hit rate (*sensitivity*), and $b = 0.028$. That they fell along the same curve indicates that none of the algorithms was more sensitive than the others; rather, they effected more or less stringent criteria for categorization. The area under this curve (*AUC*; $1/(1 + b)$), a criterion-free measure of accuracy, is an impressive 97%. The sensitivity and selectivity of the DSM-III-R in field trials (dotted line Fig. 3; Spitzer and Siegel 1990), falls below that of the other tests, yet delivers an excellent 91% area.

The high accuracy of these algorithms is due to the use of questions that, explicitly or implicitly, form a part of the clinical judgments against which they are validated. It is as much a measure of reliability as of validity. A completely independent predictor of ADHD is provided by Meyer and Sagvolden (2006). They measured the fine motor skills of 528 children diagnosed with the Disruptive

Behavior Disorders (DBD) scale (Pelham et al. 1992; Pillow et al. 1998). With the average scores on their two motor coordination tasks as the evidence axis, varying the criteria gives the ROC in Fig. 4., whose AUC is 0.72, impressive for a nonverbal measure independent of the verbal responses used to categorize. Leth-Steensen et al. (2000) have reported yet more impressive ROCs for nonverbal predictors using characteristics of response latencies as their dependent variable. Others (e.g., Jacobi-Polishook et al. 2009; Kooistra et al. 2009) contribute to this literature.

Note that we are not predicting *adhd* with these data, but rather "ADHD" as characterized by other scales, such as the DBD. Those scales are themselves imperfect, limiting the accuracy of any scale validated against them. Improved diagnosis is thus a bootstrap process, using established criteria to validate new measures, which may eventually displace the traditional measures (Mota and Schachar 2000). As much of the variance in the diagnosis of "ADHD" arises from clinicians' decisions regarding informants, instrumentation, and the method for aggregating information across them as it does from the condition of the child. In a comprehensive evaluation of the SNAP-IV criteria (Bussing et al. 2008) the inter-rater reliability between parent and teacher ratings was less than 50%; Wolraich et al. (2004) report parent–teacher inter-rater agreement kappas around 0.2. These mediocre levels of agreement underscore the potential utility of more objective tests as part of the formal definition of "ADHD" (Ohashi et al. 2010).

Categorical criteria are necessary for treatment (e.g., Chen et al. 1994), but restrict the information that is communicated by "ADHD" to less than 1 bit. Conversely, the square root of the average motor scores of Meyer and Sagvolden's (2006) children, a "dimensional" measure, accounted for 75% of the variance in the DBD scales. This is better than the agreement between different human raters, as

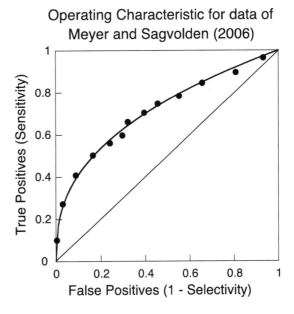

Fig. 4 The probability that a child at the first, second, etc. quintiles of average motor coordination scores (*circles*) would be correctly labeled "*ADHD*" according to the DBD scale, is shown by the ordinate, plotted as a function of the probability that they would be incorrectly so categorized. Data from (Meyer and Sagvolden 2006)

noted above. It is clear then that when studying the neurobiological aspects of any syndrome, investigators must avail themselves of dimensional measures whenever available, as they can make the difference between the discovery of significant relations and mere trends (Bobb et al. 2006).

The area under the ROC curves (AUC) shown in Figs. 3 and 4 provides an index of what we know about the character of the children from the task, independent of where the diagnostic criterion is set. (When these are power functions with exponent b, the area is simply $1/(1 + b)$; given a single point, it is $ln(F)/(ln(F) + ln(H))$, where ln is the natural logarithm). This area is precisely equivalent to the probability of replication – that is, the probability that an equal-powered replication attempt will return supportive evidence (Irwin 2009; Killeen 2006). Such indices, rather than categorical status which confounds criterion with detectability, are the touchstones for evaluating both diagnostic and remedial tools; and for giving policy experts an unbiased index of epidemiological and therapeutic impacts. Decisions of how many symptoms to include, or how to aggregate scores from different observers, will inevitably change the criterion, but may leave the information available from the test – the AUC – relatively invariant (see, e.g., the top ROC in Fig. 3). Using only the criteria that teachers AND parents agree upon, for instance, will move the criterion to a more conservative position (rightward in Fig. 2), shifting the allocation of subtypes and reducing the number diagnosed. Using all symptoms that either parents OR teachers indicate moves the criterion left, increasing the number categorized in general, and the number as "combined types" in particular (Rowland et al. 2008; Valo and Tannock 2010). The child's condition has not changed, but by altering which symptoms, how many, or on who's say-so they count, the criterion has of necessity moved. The use of different criteria in different countries is an important factor in the different national rates of "ADHD" (Singh 2008). Once these issues of numbers and aggregations of symptoms are recognized as criterial decisions, they can be formulated to optimize benefits/costs of diagnoses.

2.2 The Ultimate Formal Causes of ADHD: Theories About It

Formal causes are the description of the form of a phenomenon; at the proximate level its definition; at the ultimate level, everything we have to say about it that characterizes its nature, origins, and effects. There are numerous theories of ADHD that emphasize different aspects of the phenomenon. These may be broadly categorized as formal/cognitive, material/biological, and efficient/environmental. It is impossible to respect all the theories that have been promulgated to explain ADHD in this section, but the following examples will give a sense of some of the dominant approaches.

2.2.1 Theories Emphasizing Cognition

There are many theories of normal psychological function that have been enlisted in the attempt to understand ADHD. The referents or explanatory principles already reside in the formal realm, and ADHD is characterized as a failure of one or more of those hypothetical mechanisms.

One important theory of cognitive functions parses automatic processes from control processes – the higher order, executive functions. Michael Posner is among many who have developed this framework, (e.g., Fodor 1983; Posner and Petersen 1990). The executive of a company is not a specialist, but rather organizes, emphasizes, invests resources, and plans for the future. It is these executive functions – self-monitoring, attentional control, information updating, inhibitory control – that are compromised in ADHD, so it should not be surprising that the most impactful general theory of ADHD cites "executive function" in its title (Barkley 1997). A subsequent meta-analysis (Willcutt et al. 2005) concluded that the population of individuals with ADHD is associated with significant weaknesses in several domains most closely associated with executive functions. But, whereas such weaknesses may characterize the group, as many as half of ADHD children may be free from significant executive dysfunction (Lambek et al. 2010; Loo et al. 2007). Such dysfunction therefore is not definitive of ADHD (Nigg et al. 2005).

The cognitive-energetics theory (e.g., Sergeant 2000, 2005) holds that in ADHD there is insufficient activation to maintain functions on three levels – executive, computational, and energetic state. Delay aversion theory (e.g., Solanto et al. 2001; Sonuga-Barke et al. 2008) emphasizes one particular executive function, the ability to tolerate delays.

The dynamic developmental theory (Sagvolden et al. 2005) argues that in ADHD, a shorter delay-of-reinforcement gradient results in the accumulation of responses that are selected by both scheduled and unscheduled reinforcers, and poor development of extended sequences of responses (Aase and Sagvold'en 2005; Johansen et al. 2009a, b). Dysfunctional extinction mechanisms will fail to prune some inefficient responses occurring early in the chain. The poorer habits and skills under poorer stimulus control will emerge as impulsivity, inattention, and overactivity.

The advantage of face validity and familiarity that benefit these approaches is balanced by the tenuousness of all current psychological theories of the mind on which they are predicated. One of the reasons for the lack of definitive theories of ADHD is that there are many routes to ADHD, making it impossible to accurately characterize this heterogeneous disorder, defined by symptoms shared with other disorders, with a single unified theoretical explanation, such as "inhibition failure" or "executive dysfunction" (Nigg et al. 2005). Constructs such as "impulsivity" and "excessive delay discounting" that we associate with ADHD may, as Williams (2010) argues, be but higher-level summaries of lower-level states and traits, whose various and shifting constellations give rise, in their heterogeneous way, to the various manifestations of the syndrome.

Another kind of formal "theory" of ADHD involves the proximate description, the very definition of ADHD as a dysfunction. Some critics such as Szasz (2003) have argued that ADHD was invented rather than discovered, and reflects a medicalization of normal personality differences. Clarion voices have been raised against this position (Timimi and Taylor 2004). Conversely, one might define ADHD therapeutically, as the disorder that is mitigated by the administration of stimulants. The DSM is primarily used to dictate therapies, and such functional criteria could improve that utility. But such psychopharmacopragmatism would thwart progress toward more fundamental understanding, would discourage the invention of novel therapies, and would represent the worst case scenario of medicalization of individual differences.

2.2.2 Theories Emphasizing Neurophysiology

The above examples of cognitive theories typically also reach down to the material causes, the underlying neurophysiology (e.g., Sonuga-Barke 2003). Conversely, other theories start from the material causes and reach up. Nigg and Casey (2005, p. 785), for instance, emphasized the "joint operations of frontostriatal and frontocerebellar neural loops in detecting and predicting what and when important events ... will occur and their interaction with frontoamygdala loops in assigning emotional significance to these events." They related dysfunctions in those circuits to the dysfunctions in behavior that might follow.

The most frequently prescribed medication for children is methylphenidate (MPH), a dopamine re-uptake inhibitor. It is therefore not surprising that a number of dopaminergic theories of ADHD have arisen (Levy and Swanson 2001; Swanson et al. 2000; Tripp and Wickens 2008), with most emphasizing the role of dopamine in the reinforcement process.

Todd and Botteron (2001), suggested that ADHD may be viewed as cortical energy-deficit syndrome caused by catecholamine-mediated hypofunctionality of astrocyte glucose and glycogen metabolism. This suggestion was fostered by Russell and associates (Russell et al. 2006), who hypothesized that ADHD involves: (1) Impaired lactate production by astrocytes, providing insufficient ATP to maintain ion gradients across neuronal and glial cell membranes; and (2) impaired myelin synthesis making axons less efficient, and slowing responses. The unmyelinated dopamine neurons are intrinsically less efficient, and thus especially susceptible to slowed firing resulting from insufficient energy supply. These energetics hypotheses complement the cognitive energetics theories mentioned above.

2.2.3 Theories Emphasizing Environment

Theories that emphasize efficient or final causes have received less attention in the scientific literature than those based on other theories of cognition, or on neurobiological dysfunction. There are many hypotheses that ADHD results from environmental toxins. Lead, organophosphate pesticides, and flame retardants have been shown to trigger ADHD in susceptible individuals. But ADHD is approximately as common in countries where those toxins are, for now, less prevalent. Presence or absence of particular ingredients in diets cannot account for the majority of cases of ADHD. There have also been theories concerning the evolutionary origins of ADHD, discussed in Sect. 4, and in Adriani et al. (2011).

3 The Efficient Causes of ADHD: Triggers of Syndrome and Symptoms

A comprehensive theory of ADHD must encompass all four causes: its etiology, neurobiology, and functional aspects, along with descriptive and theoretical models. It is important to separate the causes of a particular symptom of ADHD that occurs on a particular occasion from those events that created a child who is liable to those and other symptoms, and which together cause him to be labeled "ADHD." The former are *proximate* efficient causes, and the later *ultimate* efficient causes (see Fig. 1).

3.1 Ultimate Efficient Causes of ADHD: From Genotype to Phenotype

Prenatal causal factors. The large heritability of ADHD (Asherson and Gurling 2011), and the identification of several genetic markers, suggests that ADHD probands had always been at risk. Are there environmental factors that trigger the genes or inhibit their action? Maternal smoking during pregnancy has been shown to be associated with the emergence of ADHD (Langley et al. 2005; Sciberras et al. 2011). But there is a potential risk of too heavy a reliance on epidemiological correlations. The association may be due to the higher prevalence of smoking in mothers who have ADHD (Knopik et al. 2006; Lindblad and Hjern 2010), and the associated lower birth weights of their children (Agrawal et al. 2010; Nigg and Breslau 2007). Nonetheless, individuals with both genetic risk factors (dopamine transporter or receptor polymorphisms) and prenatal exposure to tobacco are 3 to 9 times as likely to be subsequently diagnosed "ADHD" (Neuman et al. 2007). ADHD is also associated with prenatal alcohol consumption (but that type of ADHD may constitute a different endophenotype than nonalcohol-engendered ADHD: Burden et al. 2010).

Maternal exposure to environmental toxins, such as lead (Froehlich et al. 2009), organochlorines (Sagiv et al. 2010), and possibly cannabis (Huizink and Mulder 2006), increase the risk of ADHD. Because the risk factors tend to cluster and interact, these correlations need to be treated with caution. Williams and Ross (2007) provide a useful overview of the prenatal toxins that play a causal role in psychiatric disorder.

Perinatal factors such as birth trauma, anoxia, and prematurity have long been recognized as causal factors for ADHD (Barry and Gill 2007; Doyle 2004). Unfortunately, some of the very interventions aimed to mitigate such risks may themselves cause ADHD (Kapoor et al. 2008). Threats continue through early life, including factors that are situational – familial difficulties, unemployment, single parent family, low SES, low education, poor neighborhoods, parental psychopathologies – or functional – low parental responsivity, high and critical expressed emotion, over or under stimulation, harsh or coercive disciplinary techniques, and conflictual marital relationships. The role of stress is discussed in depth by Fairchild in this volume. The possibility that these, or even more indirect triggers of ADHD operate through epigenetic mechanisms (Mill and Petronis 2008), is considered by Elia in this volume. Kieling and associates provide a recent review of the neurobiology of ADHD (Kieling et al. 2008) that adds important details to this overview.

3.2 Proximate Efficient Causes of Episodes of ADHD Behavior

We may also ask what are the triggers of the *particular behaviors* that characterize ADHD. Many characteristic symptoms of ADHD (e.g., fidgeting, out of seat, difficulty concentrating, distractibility) manifest in the following situations: where there is little demand for active engagement, such as listening to a teacher's multiple or lengthy explanation of a new concept; which impose heavy demands on working memory and other executive functions, such as writing a persuasive argument or performing a series of mental computations (e.g., Kofler et al. 2010; Lauth et al. 2006; Rapport et al. 2009); and which provide little reinforcement (Pelham et al. 1986). Random stimuli, such as hallway noise, can trigger off-task behavior in ADHD students (Carroll et al. 2006). Some symptoms are rooted in poorly established habits under poor situational control. For instance, boys with ADHD rely on inappropriate entry strategies (disruptive attention seeking) when seeking to join in games with unfamiliar peers, which over time "turns off" the peers who respond negatively and reject future interaction with these ADHD peers (Ronk et al. 2011). Reinforcement strengthens preceding behavior regardless of whether the parent or teacher deems the behavior correct or disruptive (Catania 1971, 2005).

It is a truism that it is impossible to do just one thing to a complicated system. Treatments for some disorders often give rise to others (Whitaker 2010). This may be the case also for ADHD, whether it is the mother treated for risk of prematurity, or the child treated for asthma (Saricoban et al. 2011). Even inappropriate behavioral interventions by well-intended but misguided caregivers or teachers may push symptoms over the criteria for categorization.

4 The Material Causes of ADHD: Its Substrates

4.1 Ultimate Material Cause: Genetics and Epigenetics

ADHD is highly heritable (~ 0.7–0.8, just under the values for height and IQ; Bouchard 2004; Faraone and Mick 2010), and is polygenetic, with interactions between gene and environment playing a central role in its manifestation (Kahn et al. 2003). The most frequently investigated genes belong to the dopaminergic neuromodulatory system (Gizer et al. 2009). Genome-wide association studies (GWAS) enable the identification of ADHD genes in a hypothesis-free manner (reviewed by Banaschewski et al. 2010; Franke et al. 2009). A surprising finding of GWAS is the involvement of genes related to cell adhesion and cell migration, such as cadherin, CDH13, in ADHD etiology. Genes involved in regulating synaptic excitability and neuronal plasticity are also implicated. Such findings may explain the apparent immaturity, smaller brains, reduced IQ, and reinforcement and extinction deficits associated with ADHD. This active area of research is reviewed by Asherson and Gurling (2011).

4.2 Proximate Material Causes: Neurophysiology

The ultimate material causes – the genetic and epigenetic factors that shape the brains of children that are later characterized as "ADHD" – set the anatomical stage for the processes that occur later as those children interact with their school and home environments. These neurophysiological substrates are the proximate material causes of ADHD. Some areas of the brain (the right prefrontal cortex, the basal ganglia, and the vermis of the cerebellum) are smaller in ADHD. What is different about their function that cause the symptoms associated with ADHD?

4.2.1 Sex, Cigarettes, and MAO

An unresolved mystery surrounding ADHD is its greater prevalence in boys than girls. Some, but not all, of the difference in incidence can be attributed to differential referral rates (see: Wadell and McCarthy in this volume; Heptinstall and Taylor 2002). A fundamental difference between the sexes is the presence of an extra X-chromosome in females. Although the redundant chromosome is largely inhibited, some genetic expression occurs. One of the genes on the X-chromosome that codes for monoamine oxidase (MAO; Chen et al. 1992; Kochersperger et al. 1986; Levy et al. 1989) is not inhibited on its sister chromosome and could buffer transcription errors for females. In the case of a mutation that destroys MAO type A (MAOA), the female carriers were unaffected but the males showed impulsive aggression and lowered IQ (Brunner et al. 1993). Furthermore, genetic polymorphisms of MAOA are associated with the hyperactive–impulsive type of ADHD, and with greater impulsivity scores, but only in males (Liu et al. 2011).

MAO deactivates monoamines such as norepinephrine and dopamine; it is competitively inhibited by amphetamine (Pliszka 2005), a classic treatment for ADHD. Domschke and associates (Domschke et al. 2005; cf. Rommelse et al. 2008) found a significant association of a more active MAOA allele with ADHD; this would depress vesicular stores and exocytotic release of dopamine and serotonin, but the concentration of synaptic transmitter would be restored by reuptake inhibitors such as MPH (Guimarães et al. 2009). In several small studies, MAO inhibitors were shown to be as effective as amphetamine in reducing ADHD symptoms with comparable or fewer side effects (Mohammadi et al. 2004; Rubinstein et al. 2006; Trott et al. 1992; Zametkin et al. 1985). However, most of these studies used orally administered l-deprenyl (selegiline), which is an irreversible inhibitor of the type B form of MAO. This is curious, given that irregularities in MAOA are most closely associated with ADHD and that dopaminergic and noradrenergic neurons contain primarily MAOA and should be unaffected by B-inhibitors. There are several hypotheses as to why MAOB inhibitors were efficacious: (a) glial cells, also responsible for the recycling of neurotransmitters, are the major repository of MAOB in the brain (Perdan et al. 2009; Tipton 1973); (b) selegiline inhibits the uptake of dopamine and noradrenaline into presynaptic nerves and increases the turnover of dopamine (Bainbridge et al. 2008; Heinonen and Lammintausta 1991); (c) and "An intriguing mechanism of action of MAOB inhibition in ADHD may lie in the increased level of beta-phenylethylamine, which is assumed to act as an endogenous amphetamine." (Bortolato et al. 2008).

Both isoenzymes of MAO may play a role in ADHD, contingent on environmental factors. Individuals with ADHD typically have lower blood platelets levels of MAO type B (Coccini et al. 2009; Kiive et al. 2007; Nedic et al. 2010); higher levels, in association with anomalies in MAOA genes, exacerbate ADHD symptoms (Wargelius et al. 2010). Kim-Cohen and associates found that boys with the genotype conferring high MAOA activity (and therefore lower vesicular stores of dopamine) had more ADHD symptoms than boys with low activity genotype. But in children exposed to physical abuse, mental problems were prevalent, especially for those with the genotype conferring *low* MAOA activity (Kim-Cohen et al. 2006). Thus, a role for MAO in the etiology of ADHD is certain, and it is a role that is conditioned on the level of stress in the child's environment. Because females have a redundant copy of the X-chromosome, they are somewhat buffered against polymorphisms of the MAO genes, and thus against some of the symptoms of ADHD.

In light of the above analyses, it is interesting to note that tobacco contains strong MAO inhibitors of both types (Killeen 2011). This may explain why individuals with ADHD start smoking earlier and find it harder to quit than do non-ADHD individuals (Bainbridge et al. 2008; Upadhyaya et al. 2002).

4.2.2 Energetics

A number of other strands of evidence are converging on energy insufficiency as a potential cause of ADHD. The BOLD response, which is the indicator of glucose metabolism measured by fMRI, indicates decreased functionality of the superior frontal, premotor, and somatosensory cortices in ADHD (Castellanos et al. 2001; Paloyelis et al. 2007; Rubia et al. 1999; Vaidya et al. 2005; Yu-Feng et al. 2007). It is notable that astrocytes mediate and modify this function (Carmignoto and Gömez-Gonzalo 2010; Wang et al. 2009; Yu et al. 2010); the hypoenergetic response may play a causal, as much as indexical, role in the hypofunctional behavior associated with ADHD (Russell et al. 2006). Children with ADHD employ learning and memorial strategies that involve less effort than normals (Egeland et al. 2010), possibly out of necessity. This energetics hypothesis is consistent with the analysis of the previous section, as MPH, amphetamine, and the MAOIs all increase the levels of adrenaline in circulation. These higher levels, acting through adrenergic receptors on astroglia, stimulate the glia to increase lactate production, speeding the astrocyte–neuron lactate shuttle, and thereby increase the energy available to the neurons.

Because ADHD is a polygenetic condition, with involvement of monoamine receptors, transporters, and oxidizers, it is likely that some inauspicious alignment of these factors, along with stressful pre- and postnatal environments, all play a role in creating the brains of children with ADHD. The vicissitudes of brain development may contribute to the character, extremity, and comorbidities of the individual (Table 1).

5 Final Causes; What Is ADHD Good for?

Why do ADHDers still run among us? Dictionaries define pathology as a "deviation from the normal structure and function that causes discomfort and behavioral dysfunction." The relative rarity of ADHD makes it a "deviation from the normal"; "discomfort" is a personal evaluation, and may not always apply to ADHD; "behavioral dysfunction" is guaranteed by the DSM-IV diagnostic criterion 2c ("clear evidence of clinically significant impairment in social, academic, or occupational functioning.") By filtering out those who cope successfully with the other symptoms of ADHD, diagnosis foreordains dysfunction. The significantly greater presence of individuals with ADHD in emergency rooms, drug rehabilitation clinics, compensatory education classes, and prisons (Klein and Mannuzza 1991; Weiss et al. 1985) qualifies "ADHD" as a dysfunction in society's eyes. Less-dramatic forms of deviance, such as obesity, are also correlated with ADHD (Cortese and Vincenzi 2011; Campbell and Eisenberg 2007; Pagoto et al. 2009). Ability to communicate, a critical skill in the modern world, is impaired (Baird

Table 1 Examples of the four causal types involved in ADHD

Proximate	Ultimate
Formal causes	
Descriptions, diagnostic criteria, simple models	Theories, molar accounts
Clinical models (DSM, ICD)	• *Cognitive theories*
• *Symptomatological*	• *Neurobiological theories*
• *Decision theoretic*	• *Contextual theories*
Category/taxon vs. dimension/continuum	
Animal models	
• *Spontaneously hypertensive rat*	
Efficient causes	
Triggers of symptoms	Origins of syndrome
• *Inadequate reinforcement*	Prenatal
• *Processing demand overload (speed, duration, complexity)*	• *Maternal smoking, alcohol*
	• *Low birth weight, prematurity*
• *Inadequate control of context (chaotic, stressful, unpredictable)*	• *Exposure to environmental toxins – lead, PCPs*
	Perinatal
• *Boring environments*	• *Head injury*
• *Repeated tasks*	• *Malnutrition*
• *Imposed delays to action*	• *Stressful environments*
Material causes	
Concurrent dynamic brain events	Static brain structure
Neuromodulatory systems	Genetic differences
• *Dopaminergic deficit*	• *Rare variants*
• *Noradrenergic deficit*	• *Alterations, mutations*
• *Monoamine oxidase hyperfunction*	Brain differences
• *Astrocyte hypofuntion*	• *Regional volumes*
	• *White matter*
	• *Connectivity*
Final causes	
Immediate consequents of behavior	Historic environmental consequents of phenotype
Negative reinforcement	
• *Escape from boredom*	Evolutionary advantages
• *Escape from mental fatigue*	• *Alertness to novel stimuli*
Positive reinforcement	• *Escape from local environments causing pre- and peri-natal stress*
• *Approach novel stimuli*	
• *Achieve goals more quickly*	• *Founder effect in new environments*
• *Peer approval*	• *Exploitation of opportunities*

et al. 2000; Purvis and Tannock 1997). The costs attendant on ADHD have been estimated to be in excess of 30 billion dollars in the United States in 2000 (Birnbaum et al. 2005). The benefits are not tallied.

These handicaps, and the compromised function they entail, constitute strong evolutionary pressures that should have eliminated ADHD from the population within a few score generations. But, like some other genetic traits (Masel and Siegal 2009), ADHD is surprisingly robust; its high heritability and worldwide distribution attests that it cannot be explained solely by novel evolutionary forces such as industrial toxins or food additives. The factors which make ADHD a dysfunction compromise the individual's health, employment, and longevity, and therefore their ability to see their children through to their own place in society. What are the ultimate evolutionary factors that could have led to this syndrome? What are the proximate factors that buffer the syndrome against regression to a healthier mean of function?

5.1 Ultimate Function: Evolutionary Utility

One hypothesis of compensatory factors is found in the evolutionary history of *Homo sapiens*. *Response readiness* is a character of ADHD – characterized by readiness to move, seen in out-of-seat behavior; in the focus on actionable outcomes seen in impulsiveness (Urcelay and Dalley 2011); in the unreadiness to inhibit action in favor of contemplation, seen in attention dysfunction. These traits may have played a positive role in environments that were unsafe or resource scarce, quotidian conditions for our ancestors. If they did, they would have selected for the phenotype of ADHD (Jensen et al. 1997). Fugacious attention is better attuned to survival when every change in the environment could mean threat or opportunity; "shoot first" may have left few opportunities for questioning later, but has undeniable survival value in desperate environments.

Not all environments are so desperate, though, and impulsive behavior would have disadvantaged an individual, even in early hominid groups, where cooperation and communication were essential for survival (Hill and Hurtado 1996, 2009). It is in such situations that epigenetics may play a crucial role in buffering the more extreme aspects of a genetic suasion. Early experience interacts with genetic variants to predispose different phenotypes (Bennett et al. 2002). Just as pathologies of parents exacerbate those of the child (Hill et al. 1999; Maestripieri et al. 2006), a safe affectionate environment can render normal those individuals who otherwise would classify as deviant (Suomi 2006). A polymorphism in the serotonin transporter gene, which is implicated among other genes in ADHD (Bobb et al. 2006; Comings 2006), is associated with "deficits in early neurobehavioral functioning and serotonin metabolism, extreme aggression, and excessive alcohol consumption among monkeys who experienced insecure early attachment relationships, but not in monkeys who developed secure attachment relationships with their mothers during infancy." (Suomi 2005). Genes that predispose externalizing behavior could lead to antisocial tendencies in one context, drug or alcohol abuse in another, risky sports and adventurous careers in a third. Gene-by-environment interactions (Barry and Gill 2007; Kieling et al. 2008) and epigenetics (Elia et al. 2011) are instruments by which nature hedges her bets.

In a fresh perspective on the evolutionary origins of ADHD, Williams and Taylor (2006) parsed costs and benefits into those incurred by the individual, and those incurred by his community. Some actions whose costs are largely borne by the individual, such as adventurous exploitation of new sources of food, exploration of new territories and niches, and heroism in battle, benefit the family and community. The potential benefits to kin of a few young men of action in the extended family are obvious; it is kin selection. Williams and Taylor provided a simulation that demonstrated the conditions under which such risky behavior would be maintained in a population. How much of a bad thing for one might be good for his community is a delicate and dynamic balance (Dall 2004), suggesting that ADHD may be best thought of as an evolutionary stable strategy (Smith 1982). If so, it opens the door to the use of powerful quantitative tools, such as the Price Equation (Gardner 2008), to understand its population genetics. For further discussion of the evolutionary costs and benefits of ADHD, see Adriani and Laviola (this volume).

5.2 Proximate Function: Immediate Reinforcement

Klimkeit and Bradshaw (2007) provide a more nuanced treatment of relevance to the proximate causes of the maintenance of ADHD. Industrialized countries are higher in resources and lower in risk than were the ice ages. But this security is of recent times; many of today's population have grandparents who survived calamitous exigencies of war or famine. Modern societies support many trades, from linesmen, arborists, and high-steel workers to firemen and entrepreneurs, in which risk takers have some advantage. Nigg and associates (Nigg et al. 2002) emphasize the continuity of many of the symptoms of ADHD with traditional personality measures. A diagnosis of ADHD has not interfered with the success of many individuals in sports, entertainment, and business, and in many cases has helped those careers. The dashing personality and risky sexual practices of some individuals with ADHD may further increase their reproductive fitness.

5.2.1 The Heterozygote Advantage

Because ADHD results from the additive effect of many genes, traits may be selected independently, and each may be highly adaptive in moderation. Thus, even with strong pressures against the most extreme forms of ADHD and its comorbidities, there will always be pressure favoring energy, pleasure in novelty, and readiness for action. There are many examples of heterozygote advantage, in which a mixture of alleles has greater fitness than either the dominant or the recessive types by themselves. This is one of the mechanisms that maintain polymorphisms, and the rich diversity of human personalities they give rise to, including the traits that characterize ADHD. The disabilities of the extremes are a price for the excellence of the moderates.

6 Recurrent Causes: How Feedback Exacerbates or Meliorates the Syndrome and Symptoms

6.1 Ultimate Causes: the Baldwin Effect

As the genotype provides the score for the music of life, early experience informs the conductor on how to interpret those inchoate notes. Stress in utero is a cause of ADHD and, perhaps surprisingly, so also are developmental situations remote from the construction of the genotype, such as dysfunctional families, low socioeconomic status, poor personal economy, less schooling, and poorer occupational stability (Sagvolden et al. 2005). Parents who raise children in such environments are both cause and effect of those environments, and grace their children not only with their genes but also with the differential realization of them as they are switched on or off by the environment created or suffered by the parents. Daughters of some nonhuman primates "tend to develop the same type of attachment relations with their own offspring that they experienced with their mothers early in life; such early experiences provide a possible nongenetic mechanism for transmitting these patterns to subsequent generations." (Suomi 2005, p. 216) This is an intergenerational positive feedback loop, also seen in the intergenerational transmission of infant abuse in monkeys (Maestripieri 2005; Maestripieri et al. 2006). The effects are associated with long-term developmental alterations in neurotransmitters such as serotonin, dopamine, and norepinephrine (Cirulli et al. 2009; Pryce et al. 2005; Sagvolden et al. 2005). The bootstrapping nature of behavior and genetics, in which differences in learning or attention create new niches for organisms, within which those traits flourish, is called the Baldwin effect (Simpson 1953).

6.2 Proximate Causes: Reinforcement

A central problem of ADHD is positive feedback for the symptoms that characterize it. Lapses of attention are often reinforced by the discovery of more interesting objects of attention; hyperactivity uncovers new vistas for kids bored by the quotidian; inability to hold in mind the ultimate rewards of good behavior precludes their being operative in the press of life. Eventually the teacher's, principal's, and parent's displeasure will be felt, but often too late to guide action in the heat of the moment; and too remote to be remembered at the next opportunity.

The most prominent kind of proximate feedback is reinforcement, the strengthening of the associations between an act and its positive (or punitive) consequence. The reinforcement mechanisms that normally keep behavior on track seem to be deficient in ADHD (Johansen et al. 2009a). All problems of self-control involve difficulties in the control of current behavior by delayed consequences. But for children with ADHD, those delayed consequences are at a special disadvantage against immediate and easily realized diversions.

7 Therapy: Controlling Causes

An advantage to using Fig. 1 to parse this amorphous field is that it alerts us to the different kinds of intervention that may decrease the incidence of the condition known as "ADHD." Each of the causes offers unique opportunities for intervention, some of which are discussed in detail by Heal et al. (2011).

7.1 Formal Causes: Redefinition

Each new edition of the diagnostic manuals continually adjusts who has "ADHD" (Doyle 2004, provides a review, emphasizing the implications for adults with ADHD). As seen in Fig. 3, the prevalence of the dysfunction covaries with the various diagnostic tactics employed. Should reading and learning disabilities, or other comorbidities be required as part of the criteria? If so, then prevalence goes down. Should autistic spectrum disorder and other psychoses allow a co-diagnosis of "ADHD"? If so, then prevalence goes up. Will there be more symptoms to select from and fewer required in some populations? Prevalence goes up again, as there are more paths over the criterion. Can symptoms be manifest at a later age and still qualify? Then prevalence goes up yet again. The preceding questions are the focus of discussion among those involved with revising diagnostic criteria of ADHD for DSM5; in the end, we predict that the prevalence of "ADHD" will increase.

How to diagnose is fundamentally entangled with why we diagnose; if there are effective therapies contingent on the correct diagnosis, then such diagnosis is centrally important (Jensen et al. 2001). But medical therapies are often more functional than strategic: "Take two of these every day and call me if you're not feeling better in a week." If hypoactivity is hypothesized to be due to poor diet, diet is consequently improved and activity returns, we have witnessed a successful intervention. If hyperactivity is hypothesized to be due to "ADHD," and parental coaching, or Ritalin works, we have witnessed a successful intervention. But what if such conventional treatments tame the behavior, but do nothing to improve other outcomes, such as cognitive function (Currie and Stabile 2006)? Does that constitute a successful intervention?

Diagnosis of "ADHD" may be counterproductive if it distracts attention from hyperactive individuals who fall below diagnostic criteria. Consistent with more recent analysis (Bussing et al. 2010), Currie and Stabile (2006) poignantly noted that "children whose relatively low level of symptoms make them unlikely candidates for diagnosis will suffer significant ill effects. ... Efforts to find better ways to teach the relatively small number of children diagnosed with ADHD could have a large payoff in terms of improving the academic outcomes of many children with milder symptoms. (p. 1114)" And, we would add, payoff in the teaching of children with undetected learning disabilities; and in teaching parenting skills in general.

The issues with which our field must deal are hyperactivity and inattention and learning disabilities; not "ADHD." Attempts to find endophenotypes have, to date, failed. Diagnostic criteria formulated in light of all associated costs and benefits for classification are essential. But as Currie and Stabile (2006) showed, and Bussing and associates reaffirmed (Bussing et al. 2010), the cost in human capital does not start when those criteria are crossed; nor should our efforts to arrange environments that minimize those costs, environments in which the appropriate characterization of the behavior can change from dysfunctional to heterofunctional.

7.2 Efficient Causes: Context Modification

Because ADHD is highly heritable, young adults and their parents can make informed choices among potential mates. Such long-term considerations are, of course, rarely first in the minds of young adults in love. Therefore, good prenatal care, minimal stress during pregnancy, and good child-rearing environments will optimize the prognosis for the child by reducing the likelihood of expression of the problematic genes. Chapters by Asherson and Gurling (2011), and by Elia et al. (2011) remind us of the roles that both genetic and epigenetic factors play in exacerbating or ameliorating this condition.

Interventions such as parent training and classroom-based interventions aim to eliminate or minimize the triggers of ADHD symptoms in the home and school contexts. For instance, most parent training programs aim to increase parental use of explicit and positive contingencies to promote the child's learning. The hope is that the concomitant increase in socially desirable behavior will lead to improved habit formation, better family- and peer-relationships, as well as improved academic functioning.

7.3 Material Causes: Brain Modification

Current pharmacotherapies use alternate means of enhancing monoaminergic functions, in particular those of dopamine and noradrenaline (see Heal et al. 2011). Because ADHD is a polygenetic disorder, different physiological dysfunctions may underlie different subsets of the afflicted population (Burden et al. 2010; Comings et al. 2000), who may show differential benefit from different therapeutic approaches (Kegel et al. 2011).

To the extent that an energetic dysfunction underlies the condition, there may be means of re-energizing the weakened processes other than by increasing the saturation of monoamines. If overactive MAO plays a role in ADHD, inhibitors of MAO should ameliorate its symptoms, as they will slow the deamination of MAO (Bortolato et al. 2008). Indeed, selegiline is just as effective as methylphenidate in treating the symptoms of ADHD (Akhondzadeh et al. 2003; Mohammadi et al. 2004), as is another MAO inhibitor, moclobemide (Trott et al. 1992). Whereas the advantage of these drugs over more traditional pharmacotherapies is not established, new MAOIs (Youdim and Weinstock 2004) and routes of administration (Azzaro et al. 2007) may lead to more effective pharmacotherapies. On the other hand, nutritional approaches, such as restriction or supplementation of diets, have unfortunately led to no definitive improvements (Rucklidge et al. 2009). It is likely that different agents may work better for some individuals than others (Kegel et al. 2011); careful experimentation with behavioral- and pharmacotherapies, in concert with genetic analysis, will put a finer point on the varieties of ADHD and their responsivity to different therapies (Taubert et al. 2011).

7.4 Final Causes: The Uses of ADHD

Ultimate final causes – the evolutionary pressures that made ADHD part of the birthright of humanity – are beyond intervention. This is just as well, as the rapidly evolving political/technological/ecological fronts we witness may soon throw a few of our descendants back into conditions not dissimilar to those that birthed humanity.

To address the proximate final causes requires that we understand what gives ADHDers a reproductive advantage – or at least parity. For instance, it might be possible to target certain deleterious interactions between genes and environment, in which low birth rate or perinatal stress activates genetic predispositions. However, this would require multinational collaboration and willingness to commit the necessary resources to maternal and perinatal health, and to the mitigation of environmental toxins.

Final causes are functional descriptions and focus our attentions on the consequences of ADHD. Many of those consequences are in fact good. Finding ways to channel affected youths into careers that put their abilities to best use, and training them how to work around their deficits, will be to everyone's immediate advantage. Finding a place in our schools and work places and societies for individuals with these dispositions is our challenge. Routine medicalization of the condition is our failure.

Acknowledgments We thank Jonathon Williams and Vivienne Russell for comments on this manuscript. This work would not have been conceived without support from the Centre for Advanced Studies in Oslo. This work was also partly supported by the Canada Research Chairs Program (RT).

References

Aase H, Sagvolden T (2005) Moment-to-moment dynamics of ADHD behaviour. Behav Brain Funct 1:12
Adriani W, Zoratto F, Laviola G (2011) Brain processes in discounting: consequences of adolescent methylphenidate exposure. Curr Topics Behav Neurosci. doi:10.1007/7854_2011_156

Agrawal A, Scherrer JF, Grant JD, Sartor CE, Pergadia ML, Duncan AE, Madden PAF, Haber JR, Jacob T, Bucholz KK (2010) The effects of maternal smoking during pregnancy on offspring outcomes. Prev Med 50:13–18

Akhondzadeh S, Tavakolian R, Davari-Ashtiani R, Arabgol F, Amini H (2003) Selegiline in the treatment of attention deficit hyperactivity disorder in children: a double blind and randomized trial. Prog Neuropsychopharmacol Biol Psychiatry 27:841–845

Asherson P, Gurling H (2011) Quantitative and molecular genetics of ADHD. Curr Topics Behav Neurosci. doi:10.1007/7854_2011_155

Azzaro AJ, Ziemniak J, Kemper E, Campbell BJ, VanDenBerg C (2007) Pharmacokinetics and absolute bioavailability of selegiline following treatment of healthy subjects with the selegiline transdermal system (6 mg/24 h): a comparison with oral selegiline capsules. J Clin Pharmacol 47:1256

Bainbridge JL, Page RI, Ruscin J (2008) Elucidating the mechanism of action and potential interactions of MAO-B inhibitors. Neurol Clin 26:85–96

Baird J, Stevenson JC, Williams DC (2000) The evolution of ADHD: a disorder of communication? Q Rev Biol 75:17–35

Banaschewski T, Becker K, Scherag S, Franke B, Coghill D (2010) Molecular genetics of attention-deficit/hyperactivity disorder: an overview. Eur Child Adolesc Psychiatry 19:237–257

Barkley RA (1997) Behavioral inhibition, sustained attention, and executive functions: constructing a unifying theory of ADHD. Psychol Bull 121:65–94

Barkley RA, Murphy KR, Fischer M (2007) ADHD in adults: what the science says. Guilford, New York

Barry E, Gill M (2007) Environmental risk factors and gene-environment interaction in attention deficit hyperactivity disorder. In: Fitzgerald M, Bellgrove M, Gill M (eds) Handbook of attention deficit hyperactivity disorder. Wiley, West Sussex, Eng, pp 149–182

Beck JR, Shultz EK (1986) The use of relative operating characteristic (ROC) curves in test performance evaluation. Arch Pathol Lab Med 110:13–20

Bennett AJ, Lesch KP, Heils A, Long JC, Lorenz JG, Shoaf SE, Champoux M, Suomi SJ, Linnoila MV, Higley JD (2002) Early experience and serotonin transporter gene variation interact to influence primate CNS function. Mol Psychiatry 7:118–122

Birnbaum HG, Kessler RC, Lowe SW, Secnik K, Greenberg PE, Leong SA, Swensen AR (2005) Costs of attention deficit-hyperactivity disorder (ADHD) in the US: excess costs of persons with ADHD and their family members in 2000. Curr Med Res Opin 21:195–206

Bobb AJ, Castellanos FX, Addington AM, Rapoport JL (2006) Molecular genetic studies of ADHD: 1991 to 2004. Am J Med Genet B Neuropsychiatr Genet 141B:551–565

Bortolato M, Chen K, Shih JC (2008) Monoamine oxidase inactivation: from pathophysiology to therapeutics. Adv Drug Deliv Rev 60:1527–1533

Bouchard TJ (2004) Genetic influence on human psychological traits. Curr Dir Psychol Sci 13:148–151

Brunner HG, Nelen M, Breakefield X, Ropers H, Van Oost B (1993) Abnormal behavior associated with a point mutation in the structural gene for monoamine oxidase A. Science 262:578–580

Burden MJ, Jacobson JL, Westerlund A, Lundahl LH, Morrison A, Dodge NC, Klorman R, Nelson CA, Avison MJ, Jacobson SW (2010) An event-related potential study of response inhibition in ADHD with and without prenatal alcohol exposure. Alcohol Clin Exper Res 34:617–627

Bussing R, Fernandez M, Harwood M, Hou W, Wilson-Garvan C, Eyeberg SM, Swanson JM (2008) Parent and teacher SNAP-IV ratings of attention deficit/hyperactivity disorder symptoms. Assessment 15:317–328

Bussing R, Mason DM, Bell L, Porter P, Garvan C (2010) Adolescent outcomes of childhood attention-deficit/hyperactivity disorder in a diverse community sample. J Am Acad Child Adolesc Psychiatry 49:595–605

Campbell BC, Eisenberg D (2007) Obesity, attention deficit-hyperactivity disorder and the dopaminergic reward system. Coll Antropol 31:33–38

Carmignoto G, Gömez-Gonzalo M (2010) The contribution of astrocyte signalling to neurovascular coupling. Brain Res Rev 63:138–148

Carroll A, Houghton S, Taylor M, West J, List-Kerz M (2006) Responses to interpersonal and physically provoking situations: The utility and application of an observation schedule for school-aged students with and without Attention. Educ Psychol 26:483–498

Castellanos FX, Giedd JN, Berquin PC, Walter JM, Sharp W, Tran T, Vaituzis AC, Blumenthal JD, Nelson J, Bastain TM (2001) Quantitative brain magnetic resonance imaging in girls with attention-deficit/hyperactivity disorder. Arch Gen Psychiatry 58:289

Castellanos FX, Sonuga-Barke EJS, Scheres A, Di Martino A, Hyde C, Walters JR (2005) Varieties of attention-deficit/hyperactivity disorder-related intra-individual variability. Biol Psychiatry 57:1416–1423

Catania AC (1971) Reinforcement schedules: the role of responses preceding the one that produces the reinforcer. J Exp Anal Behav 15:271–287

Catania AC (2005) Attention-deficit/hyperactivity disorder (ADHD): delay-of-reinforcement gradients and other behavioral mechanisms. Behav Brain Sci 28:419–424

Chen Z, Powell J, Hsu Y, Breakefield X, Craig I (1992) Organization of the human monoamine oxidase genes and long-range physical mapping around them. Genomics 14:75–82

Chen WJ, Faraone SV, Biederman J, Tsuang MT (1994) Diagnostic accuracy of the Child Behavior Checklist scales for attention-deficit hyperactivity disorder: a receiver-operating characteristic analysis. J Consult Clin Psychol 62:1017–1025

Cirulli F, Francia N, Berry A, Aloe L, Alleva E, Suomi SJ (2009) Early life stress as a risk factor for mental health: role of neurotrophins from rodents to non-human primates. Neurosci Biobehav Rev 33:573–585

Coccini T, Crevani A, Rossi G, Assandri F, Balottin U, Nardo RD, Manzo L (2009) Reduced platelet monoamine oxidase type B activity and lymphocyte muscarinic receptor binding in unmedicated children with attention deficit hyperactivity disorder. Biomarkers 14:513–522

Coghill D, Nigg JT, Rothenberger A, Sonuga-Barke EJ, Tannock R (2005) Whither causal models in the neuroscience of ADHD? Dev Sci 8:105–114

Comings DE (2006) Clinical and molecular genetics of ADHD and Tourette syndrome: two related polygenic disorders. Ann N Y Acad Sci 931:50–83

Comings DE, Gade Andavolu R, Gonzalez N, Wu S, Muhleman D, Blake H, Dietz G, Saucier G (2000) Comparison of the role of dopamine, serotonin, and noradrenaline genes in ADHD, ODD and conduct disorder: multivariate regression analysis of 20 genes. Clin Genet 57:178–196

Cortese S, Vincenzi B (2011) Obesity and ADHD: clinical and neurobiological implications. Curr Topics Behav Neurosci. doi:10.1007/7854_2011_154

Currie J, Stabile M (2006) Child mental health and human capital accumulation: the case of ADHD. J Health Econ 25:1094–1118

Dall SRX (2004) Behavioural biology: fortune favours bold and shy personalities. Curr Biol 14: R470–R472

Domschke K, Sheehan K, Lowe N, Kirley A, Mullins C, O'Sullivan R, Freitag C, Becker T, Conroy J, Fitzgerald M (2005) Association analysis of the monoamine oxidase A and B genes with attention deficit hyperactivity disorder (ADHD) in an Irish sample: preferential transmission of the MAO-A 941 G allele to affected children. Am J Med Genet B Neuropsychiat Genet 134:110–114

Doyle R (2004) The history of adult attention-deficit/hyperactivity disorder. Psychiatr Clin North Am 27:203–214

Egeland J, Johansen SN, Ueland T (2010) Do low-effort learning strategies mediate impaired memory in ADHD? J Learn Disabil 50:347–354

Elia J, Laracy S, Allen J, Nissley-Tsiopinis J, Borgmann-Winter K (2011) Epigenetics: genetics versus life experiences. Curr Topics Behav Neurosci. doi:10.1007/7854_2011_144

Fan X, Bruno K, Hess E (2011) Rodent Models of ADHD. Curr Top Behav Neurosci

Faraone SV, Mick E (2010) Molecular genetics of attention deficit hyperactivity disorder. Psychiatr Clin North Am 33:159–180

Faraone SV, Biederman J, Sprich-Buckminster S, Chen W, Tsuang MT (1993) Efficiency of diagnostic criteria for attention deficit disorder: toward an empirical approach to designing and validating diagnostic algorithms. J Am Acad Child Adolesc Psychiatr 32:166–174

Fodor JA (1983) The modularity of mind. Bradford Books, Cambridge, MA

Franke B, Neale BM, Faraone SV (2009) Genome-wide association studies in ADHD. Hum Genet 126:13–50

Froehlich TE, Lanphear BP, Auinger P, Hornung R, Epstein JN, Braun J, Kahn RS (2009) Association of tobacco and lead exposures with attention-deficit/hyperactivity disorder. Pediatrics 124:e1054–e1063

Gardner A (2008) The Price equation. Curr Biol 18:R198–R202

Gizer IR, Ficks C, Waldman ID (2009) Candidate gene studies of ADHD: a meta-analytic review. Hum Genet 126:51–90

Gordon M, Antshel K, Faraone S, Barkley R, Lewandowski L, Hudziak JJ, Biederman J, Cunningham C (2006) Symptoms versus impairment: the case for respecting DSM-IV's Criterion D. J Atten Disord 9:465–475

Gottesman II, Gould TD (2003) The endophenotype concept in psychiatry: etymology and strategic intentions. Am J Psychiatry 160:636–645

Guimarães AP, Zeni C, Polanczyk G, Genro JP, Roman T, Rohde LA, Hutz MH (2009) MAOA is associated with methylphenidate improvement of oppositional symptoms in boys with attention deficit hyperactivity disorder. Int J Neuropsychopharmacol 12:709–714

Heal DJ, Smith SL, Findling RL (2011) ADHD: current and future therapeutics. Curr Topics Behav Neurosci. doi:10.1007/7854_2011_125

Heinonen E, Lammintausta R (1991) A review of the pharmacology of selegiline. Acta Neurol Scand 84:44–59

Heptinstall E, Taylor E (2002) Sex differences and their significance Hyperact Attn Disord Child. Cambridge University Press, Cambridge, pp 99–125

Hill K, Hurtado AM (1996) Ache life history: the ecology and demography of a foraging people. Aldine de Gruyter, New York

Hill K, Hurtado AM (2009) Cooperative breeding in South American hunter-gatherers. Proceed Royal Soc B Biol Sci 276:3863–3870

Hill SY, Locke J, Lowers L, Connolly J (1999) Psychopathology and achievement in children at high risk for developing alcoholism. J Am Acad Child Adolesc Psychiatry 38:883–891

Hocutt M (1974) Aristotle's four becauses. Philos 49:385–399

Hudziak JJ, Heath AC, Madden PF, Reich W, Bucholz KK, Slutske W, Bierut LJ, Neuman RJ, Todd RD (1998) Latent class and factor analysis of DSM-IV ADHD: a twin study of female adolescents. J Am Acad Child Adolesc Psychiatry 37:848–857

Huizink AC, Mulder EJH (2006) Maternal smoking, drinking or cannabis use during pregnancy and neurobehavioral and cognitive functioning in human offspring. Neurosci Biobehav Rev 30:24–41

Irwin RJ (2009) Equivalence of the statistics for replicability and area under the ROC curve. Br J Math Stat Psychol 62:485–487

Jacobi-Polishook T, Shorer Z, Melzer I (2009) The effect of methylphenidate on postural stability under single and dual task conditions in children with attention deficit hyperactivity disorder – a double blind randomized control trial. J Neurol Sci 280:15–21

Jensen AL, Poulsen JS (1992) Evaluation of diagnostic tests using relative operating characteristic (ROC) curves and the differential positive rate. An example using the total serum bile acid concentration and the alanine aminotransferase activity in the diagnosis of canine hepatobiliary diseases. Zentralbl Veterinarmed A 39:656–668

Jensen PS, Mrazek D, Knapp PK, Steinberg L, Pfeffer C, Schowalter J, Shapiro T (1997) Evolution and revolution in child psychiatry: ADHD as a disorder of adaptation. J Am Acad Child Adolesc Psychiatr 36:1672–1681

Jensen PS, Hinshaw SP, Kraemer HC, Lenora N, Newcorn JH, Abikoff HB, March JS, Arnold LE, Cantwell DP, Conners CK (2001) ADHD comorbidity findings from the MTA study: comparing comorbid subgroups. J Am Acad Child Adolesc Psychiatry 40:147–158

Johansen EB, Killeen PR, Russell VA, Tripp G, Wickens JR, Tannock R, Williams J, Sagvolden T (2009a) Origins of altered reinforcement effects in ADHD. Behav Brain Funct 5:7

Johansen EB, Killeen PR, Sagvolden T (2009b) Behavioral variability, elimination of responses, and delay-of-reinforcement gradients in SHR and WKY rats. Behav Brain Funct 3:60

Kahn RS, Khoury J, Nichols WC, Lanphear BP (2003) Role of dopamine transporter genotype and maternal prenatal smoking in childhood hyperactive-impulsive, inattentive, and oppositional behaviors. J Pediatr 143:104–110

Kapoor A, Petropoulos S, Matthews SG (2008) Fetal programming of hypothalamic-pituitary-adrenal (HPA) axis function and behavior by synthetic glucocorticoids. Brain Res Rev 57:586–595

Kegel CAT, Bus AG, van IJzendoorn MH (2011) Differential susceptibility in early literacy instruction through computer games: the role of the dopamine D4 receptor gene (DRD4). Mind Brain Educ 5:71–78

Kieling C, Goncalves RRF, Tannock R, Castellanos FX (2008) Neurobiology of attention deficit hyperactivity disorder. Child Adolesc Psychiatr Clin N Am 17:285–307

Kieling C, Kieling RR, Rohde L, Frick PJ, Moffitt TE, Nigg JT, Tannock R, Castellanos FX (2010) The age at onset of attention deficit hyperactivity disorder. Am J Psychiatry 167:14–16

Kiive E, Fischer K, Harro M, Harro J (2007) Platelet monoamine oxidase activity in association with adolescent inattentive and hyperactive behaviour: a prospective longitudinal study. Personality Indiv Diff 43:155–166

Killeen PR (1999) Modeling modeling. J Exp Anal Behav 71:275–280

Killeen PR (2003) The yins/yangs of science. Behav Phil 31:251–258

Killeen PR (2006) Beyond statistical inference: a decision theory for science. Psychon Bull Rev 13:549–562

Killeen PR (2011) Markov model of smoking cessation. Proc Natl Acad Sci USA

Kim-Cohen J, Caspi A, Taylor A, Williams B, Newcombe R, Craig IW, Moffitt TE (2006) MAOA, maltreatment, and gene-environment interaction predicting children's mental health: new evidence and a meta-analysis. Mol Psychiatry 11:903–913

Klein RG, Mannuzza S (1991) Long-term outcome of hyperactive children: a review. J Am Acad Child Adolesc Psychiatr 30:383–387

Klimkeit EI, Bradshaw JL (2007) Evolutionary aspects of ADHD. In: Fitzgerald M, Bellgrove M, Gill M (eds) Handbook of attention deficit hyperactivity disorder. Wiley, Chichester, England, pp 467–479

Knopik VS, Heath AC, Jacob T, Slutske WS, Bucholz KK, Madden PAF, Waldron M, Martin NG (2006) Maternal alcohol use disorder and offspring ADHD: disentangling genetic and environmental effects using a children-of-twins design. Psychol Med 36:1461–1471

Kochersperger L, Parker E, Siciliano M, Darlington G, Denney R (1986) Assignment of genes for human monoamine oxidases A and B to the X chromosome. J Neurosci Res 16:601–616

Kofler MJ, Rapport MD, Bolden J, Sarver DE, Raiker JS (2010) ADHD and working memory: the impact of central executive deficits and exceeding storage/rehearsal capacity on observed inattentive behavior. J Abnorm Child Psychol 38:149–161

Kooistra L, Ramage B, Crawford S, Cantell M, Wormsbecker S, Gibbard B, Kaplan BJ (2009) Can attention deficit hyperactivity disorder and fetal alcohol spectrum disorder be differentiated by motor and balance deficits? Hum Mov Sci 28:529–542

Kuntsi J, Andreou P, Ma J, Börger NA, van der Meere JJ (2005) Testing assumptions for endophenotype studies in ADHD: reliability and validity of tasks in a general population sample. BMC Psychiatry 5:40

Lambek R, Tannock R, Dalsgaard S, Trillingsgaard A, Damm D, Thomsen PH (2010) Executive dysfunction in school-age children with ADHD. J Atten Disord

Langley K, Rice F, Van den Bree MB, Thapar A (2005) Maternal smoking during pregnancy as an environmental risk factor for attention deficit hyperactivity disorder behaviour. A review. Minerva Pediatr 57:359–371

Lauth G, Heubeck B, Mackowiak K (2006) Observation of children with attention deficit hyperactivity (ADHD) problems in three natural classroom contexts. Br J Educ Psychol 76:385–404

Leth-Steensen C, King-Elbaz Z, Douglas VI (2000) Mean response times, variability, and skew in the responding of ADHD children: a response time distributional approach. Acta Psychol (Amst) 104:167–190

Levy F, Swanson JM (2001) Timing, space and ADHD: the dopamine theory revisited. Aust N Z J Psychiatry 35:504–511

Levy ER, Powell JF, Buckle VJ, Hsu YPP, Breakefield XO, Craig IW (1989) Localization of human monoamine oxidase-A gene to Xp11. 23–11.4 by in situ hybridization: implications for Norrie disease. Genomics 5:368–370

Levy F, Hay DA, McStephen M, Wood C, Waldman I (1997) Attention-deficit hyperactivity disorder: a category or a continuum? Genetic analysis of a large-scale twin study. J Am Acad Child Adolesc Psychiatr 36:737–744

Lindblad F, Hjern A (2010) ADHD after fetal exposure to maternal smoking. Nicotine Tobacco Res 12:408–415

Liu L, Guan LL, Chen Y, Ji N, Li HM, Li ZH, Qian QJ, Yang L, Glatt SJ, Faraone SV, Wang YF (2011) Association analyses of MAOA in Chinese Han subjects with attention deficit/hyperactivity disorder: Family based association test, case-control study, and quantitative traits of impulsivity. Am J Med Genet B 156:737–748

Loo SK, Humphrey LA, Tapio T, Moilanen IK, McGough JJ, McCracken JT, Yang MH, Dang J, Taanila A, Ebeling H (2007) Executive functioning among Finnish adolescents with attention-deficit/hyperactivity disorder. J Am Acad Child Adolesc Psychiatry 46:1594–1604

Maestripieri D (2005) Early experience affects the intergenerational transmission of infant abuse in rhesus monkeys. Proc Natl Acad Sci U S A 102:9726–9729

Maestripieri D, Higley JD, Lindell SG, McCormack KM, Sanchez MM, Newman TK (2006) Early maternal rejection affects the development of monoaminergic systems and adult abusive parenting in rhesus macaques (*Macaca mulatta*). Behav Neurosci 120:1017–1024

Mahone E, Crocetti D, Ranta M, Gaddis A, Cataldo M, Slifer K, Denckla M, Mostofsky S (2011) A Preliminary Neuroimaging Study of Preschool Children with ADHD. Clin Neuropsychol 1–20.

Malmberg K, Edbom T, Wargelius HL, Larsson JO (2011) Psychiatric problems associated with subthreshold ADHD and disruptive behaviour diagnoses in teenagers. Acta Paediatr

Masel J, Siegal ML (2009) Robustness: mechanisms and consequences. Trends Genet 25:395–403

Matza LS, Paramore C, Prasad M (2005) A review of the economic burden of ADHD. Cost Eff Resour Alloc 3:5

Meyer A, Sagvolden T (2006) Fine motor skills in South African children with symptoms of ADHD: influence of subtype, gender, age, and hand dominance. Behav Brain Funct 2:33

Mill J, Petronis A (2008) Pre-and peri-natal environmental risks for attention-deficit hyperactivity disorder (ADHD): the potential role of epigenetic processes in mediating susceptibility. J Child Psychol Psychiatry 49:1020–1030

Mohammadi MR, Ghanizadeh A, Alaghband-Rad J, Tehranidoost M, Mesgarpour B, Soori H (2004) Selegiline in comparison with methylphenidate in attention deficit hyperactivity disorder children and adolescents in a double-blind, randomized clinical trial. J Child Adolesc Psychopharmacol 14:418–425

Mota VL, Schachar RJ (2000) Reformulating attention-deficit/hyperactivity disorder according to signal detection theory. J Am Acad Child Adolesc Psychiatry 39:1144–1151

Muir T, Zegarac M (2001) Societal costs of exposure to toxic substances: economic and health costs of four case studies that are candidates for environmental causation. Environ Health Perspect 109:885–903

Nedic G, Pivac N, Hercigonja DK, Jovancevic M, Curkovic KD, Muck-Seler D (2010) Platelet monoamine oxidase activity in children with attention-deficit/hyperactivity disorder. Psychiatry Res 175:252–255

Neuman RJ, Lobos E, Reich W, Henderson CA, Sun LW, Todd RD (2007) Prenatal smoking exposure and dopaminergic genotypes interact to cause a severe ADHD subtype. Biol Psychiatry 61:1320–1328

Nigg JT, Breslau N (2007) Prenatal smoking exposure, low birth weight, and disruptive behavior disorders. J Am Acad Child Adolesc Psychiatry 46:362–369

Nigg JT, Casey B (2005) An integrative theory of attention-deficit/hyperactivity disorder based on the cognitive and affective neurosciences. Dev Psychopathol 17:785–806

Nigg JT, John OP, Blaskey LG, Huang-Pollock CL, Willicut EG, Hinshaw SP, Pennington B (2002) Big Five dimensions and ADHD symptoms: links between personality traits and clinical symptoms. J Pers Soc Psychol 83:451–469

Nigg JT, Willcutt EG, Doyle AE, Sonuga-Barke EJ (2005) Causal heterogeneity in attention-deficit/hyperactivity disorder: do we need neuropsychologically impaired subtypes? Biol Psychiatry 57:1224–1230

Ohashi K, Vitaliano G, Polcari A, Teicher MH (2010) Unraveling the nature of hyperactivity in children with attention-deficit/hyperactivity disorder. Arch Gen Psychiatry 67:388–396

Pagoto SL, Curtin C, Lemon SC, Bandini LG, Schneider KL, Bodenlos JS, Ma Y (2009) Association between adult attention deficit/hyperactivity disorder and obesity in the US population. Obesity 17:539–544

Paloyelis Y, Mehta MA, Kuntsi J, Asherson P (2007) Functional MRI in ADHD: a systematic literature review. Expert Rev Neurother 7:1337–1356

Pelham WE, Milich R, Walker JL (1986) Effects of continuous and partial reinforcement and methylphenidate on learning in children with attention deficit disorder. J Abnorm Psychol 95:319–325

Pelham WE Jr, Gnagy EM, Greenslade KE, Milich R (1992) Teacher ratings of DSM-III-R symptoms for the disruptive behavior disorders. J Am Acad Child Adolesc Psychiatry 31:210–218

Perdan K, Lipnik-Stangelj M, Krzan M (2009) The impact of astrocytes in the clearance of neurotransmitters by uptake and inactivation. Advances in Planar Lipid Bilayers and Liposomes 9:211–235

Pillow DR, Pelham WE Jr, Hoza B, Molina BSG, Stultz CH (1998) Confirmatory factor analyses examining attention deficit hyperactivity disorder symptoms and other childhood disruptive behaviors. J Abnorm Child Psychol 26:293–309

Pliszka SR (2005) The neuropsychopharmacology of attention-deficit/hyperactivity disorder. Biol Psychiatry 57:1385–1390

Polanczyk G, Jensen P (2008) Epidemiologic considerations in attention deficit hyperactivity disorder: a review and update. Child Adolesc Psychiatr Clin N Am 17:245–260

Posner ML, Petersen SE (1990) The attention system of the human brain. Ann Rev Neurosci 13:25–42

Pryce CR, Rüedi-Bettschen D, Dettling AC, Weston A, Russig H, Ferger B, Feldon J (2005) Long-term effects of early-life environmental manipulations in rodents and primates: potential animal models in depression research. Neurosci Biobehav Rev 29:649–674

Purvis KL, Tannock R (1997) Language abilities in children with attention deficit hyperactivity disorder, reading disabilities, and normal controls. J Abnorm Child Psychol 25:133–144

Rapport MD, Kofler MJ, Alderson RM, Timko TM, DuPaul GJ (2009) Variability of attention processes in ADHD. J Atten Disord 12:563–573

Rommelse NNJ, Altink ME, Arias Vasquez A, Buschgens CJM, Fliers E, Faraone SV, Buitelaar JK, Sergeant JA, Oosterlaan J, Franke B (2008) Differential association between MAOA, ADHD and neuropsychological functioning in boys and girls. Am J Med Genet B Neuropsychiatr Genet 147:1524–1530

Ronk MJ, Hund AM, Landau S (2011) Assessment of social competence of boys with attention-deficit/hyperactivity disorder: problematic peer entry, host responses, and evaluations. J Abnorm Child Psychol 39:829–840

Ross W (1936) Aristotle's physics. Oxford University Press, Oxford

Rowland AS, Skipper B, Rabiner DL, Umbach DM, Stallone L, Campbell RA, Hough RL, Naftel A, Sandler DP (2008) The shifting subtypes of ADHD: classification depends on how symptom reports are combined. J Abnorm Child Psychol 36:731–743

Rubia K, Overmeyer S, Taylor E, Brammer M, Williams SCR, Simmons A, Bullmore ET (1999) Hypofrontality in attention deficit hyperactivity disorder during higher-order motor control: a study with functional MRI. Am J Psychiatry 156:891

Rubinstein S, Malone MA, Roberts W, Logan WJ (2006) Placebo-controlled study examining effects of selegiline in children with attention-deficit/hyperactivity disorder. J Child Adolesc Psychopharmacol 16:404–415

Rucklidge JJ, Johnstone J, Kaplan BJ (2009) Nutrient supplementation approaches in the treatment of ADHD. Expert Rev Neurother 9:461–476

Russell V, Sagvolden T, Johansen E (2005) Animal models of attention-deficit hyperactivity disorder. Behav Brain Funct 1:9

Russell VA, Oades RD, Tannock R, Killeen PR, Auerbach JG, Johansen EB, Sagvolden T (2006) Response variability in Attention-Deficit/Hyperactivity Disorder: a neuronal and glial energetics hypothesis. Behav Brain Funct 2:30

Sagiv SK, Thurston SW, Bellinger DC, Tolbert PE, Altshul LM, Korrick SA (2010) Prenatal organochlorine exposure and behaviors associated with attention deficit hyperactivity disorder in school-aged children. Am J Epidemiol 171:593–601

Sagvolden T, Johansen E (2011) Rat Models of ADHD. Curr Top Behav Neurosci

Sagvolden T, Johansen EB, Aase H, Russell VA (2005) A dynamic developmental theory of attention-deficit/hyperactivity disorder (ADHD) predominantly hyperactive/impulsive and combined subtypes. Behav Brain Sci 28:397–419, discussion 419–468

Santayana G (1957) Dialogs in limbo. University of Michigan Press, Ann Arbor

Saricoban HE, Ozen AO, Harmanci K, Razi C, Zahmacioglu O, Cengizlier MR (2011) Common behavioral problems among children with asthma: is there a role of asthma treatment? Ann Allergy Asthma Immunol 106:200–204

Schlander M (2010) The pharmaceutical economics of child psychiatric drug treatment. Curr Pharm Des 16:2443–2461

Schneider H, Eisenberg D (2006) Who receives a diagnosis of attention-deficit/hyperactivity disorder in the United States elementary school population? Pediatrics 117:e601–e609

Sciberras E, Ukoumunne OC, Efron D (2011) Predictors of Parent-Reported Attention-Deficit/Hyperactivity Disorder in Children Aged 6–7 years: A National Longitudinal Study. J Abnorm Child Psychol 39:1–10

Secnik K, Swensen A, Lage MJ (2005) Comorbidities and costs of adult patients diagnosed with attention-deficit hyperactivity disorder. Pharmacoeconomics 23:93–102

Sergeant JA (2000) The cognitive-energetic model: an empirical approach to attention-deficit hyperactivity disorder. Neurosci Biobehav Rev 24:7–12

Sergeant JA (2005) Modeling attention-deficit/hyperactivity disorder: a critical appraisal of the cognitive-energetic model. Biol Psychiatry 57:1248–1255

Simpson GG (1953) The baldwin effect. Evolution 7:110–117

Singh I (2008) Beyond polemics: science and ethics of ADHD. Nat Rev Neurosci 9:957–964

Smith JM (1982) Evolution and the theory of games. Cambridge University Press, Cambridge, England

Solanto MV, Abikoff H, Sonuga-Barke EJ, Schachar R, Logan GD, Wigal T, Hechtman L, Hinshaw S, Turkel E (2001) The ecological validity of delay aversion and response inhibition as measures of impulsivity in AD/HD: a supplement to the NIMH multimodal treatment study of AD/HD. J Abnorm Child Psychol 29:215–228

Sonuga-Barke EJ (2003) The dual pathway model of AD/HD: an elaboration of neurodevelopmental characteristics. Neurosci Biobehav Rev 27:593–604

Sonuga-Barke EJ, Sergeant JA, Nigg JT, Willcutt E (2008) Executive dysfunction and delay aversion in attention deficit hyperactivity disorder: nosologic and diagnostic implications. Child Adolesc Psychiatr Clin N Am 17:367–384

Spitzer RL, Siegel B (1990) The DSM-II-R field trial of pervasive developmental disorders. J Am Acad Child Adolesc Psychiatry 29:855–862

Suomi SJ (2005) Aggression and social behaviour in rhesus monkeys. Novartis Found Symp 268:216–222

Suomi SJ (2006) Risk, resilience, and gene x environment interactions in Rhesus monkeys. Ann N Y Acad Sci 1094:52–62

Swanson J, Flodman P, Kennedy J, Spence MA, Moyzis R, Schuck S, Murias M, Moriarity J, Barr C, Smith M (2000) Dopamine genes and ADHD. Neurosci Biobehav Rev 24:21–25

Swets JA (1973) The relative operating characteristic in psychology. Science 182:990–1000

Swets JA, Dawes RM, Monahan J (2000) Psychological science can improve diagnostic decisions. Psychol Sci Publ Int 1:1–26

Szasz TS (2003) Pharmacracy: medicine and politics in America. Syracuse University Press, Syracuse, NY

Taubert M, Lohmann G, Margulies DS, Villringer A, Ragert P (2011) Long-term effects of motor training on resting-state networks and underlying brain structure. Neuroimage 57:1492–1498

Timimi S, Taylor E (2004) ADHD is best understood as a cultural construct. Br J Psychiatry 184:8

Tipton K (1973) Biochemical aspects of monoamine oxidase. Br Med Bull 29:116

Todd RD, Botteron KN (2001) Is attention-deficit/hyperactivity disorder an energy deficiency syndrome? Biol Psychiatry 50:151–158

Tripp G, Wickens JR (2008) Research review: dopamine transfer deficit: a neurobiological theory of altered reinforcement mechanisms in ADHD. J Child Psychol Psychiatry 49:691–704

Trott GE, Friese HJ, Menzel M, Nissen G (1992) Use of moclobemide in children with attention deficit hyperactivity disorder. Psychopharmacology (Berl) 106:134–136

Upadhyaya HP, Deas D, Brady KT, Kruesi M (2002) Cigarette smoking and psychiatric comorbidity in children and adolescents. J Am Acad Child Adolesc Psychiatry 41:1294–1305

Urcelay GP, Dalley JW (2011) Linking ADHD, impulsivity and drug abuse: a neuropsychological perspective. Curr Topics Behav Neurosci. doi:10.1007/7854_2011_119

Uttal WR (2008) Distributed neural systems: beyond the new phrenology. Sloan Educational Publishing, Cambridge, MA

Vaidya CJ (2011) Neurodevelopmental abnormalities in ADHD. Curr Top Behav Neurosci

Vaidya CJ, Bunge SA, Dudukovic NM, Zalecki CA, Elliott GR, Gabrieli JDE (2005) Altered neural substrates of cognitive control in childhood ADHD: evidence from functional magnetic resonance imaging. Am J Psychiatry 162:1605–1613

Valo S, Tannock R (2010) Diagnostic instability of DSM-IV ADHD subtypes: Effects of informant source, instrumentation, and methods for combining symptom reports. J Clin Child Adolesc Psychol 39:749–760

Waldman ID (2005) Statistical approaches to complex phenotypes: evaluating neuropsychological endophenotypes for attention-deficit/hyperactivity disorder. Biol Psychiatry 57:1347–1356

Wang X, Takano T, Nedergaard M (2009) Astrocytic calcium signaling: mechanism and implications for functional brain imaging. Methods Mol Biol 489:93–109

Wargelius HL, Fahlke C, Suomi SJ, Oreland L, Higley JD (2010) Platelet monoamine oxidase activity predicts alcohol sensitivity and voluntary alcohol intake in rhesus monkeys. Ups J Med Sci 115:49–55

Weiss G, Hechtman L, Milroy T, Perlman T (1985) Psychiatric status of hyperactives as adults: a controlled prospective 15-year follow-up of 63 hyperactive children. J Am Acad Child Adolesc Psychiatry 24:211–220

Whitaker R (2010) Anatomy of an epidemic. Crown Publishers, New York

Willcutt EG, Doyle AE, Nigg JT, Faraone SV, Pennington BF (2005) Validity of the executive function theory of attention-deficit/hyperactivity disorder: a meta-analytic review. Biol Psychiatry 57:1336–1346

Williams J (2010) Attention-deficit/hyperactivity disorder and discounting: multiple minor traits and states. In: Madden GJ, Bickel WK (eds) Impulsivity: the behavioral and neurological science of discounting. American Psychological Association, Washington, DC, pp 323–357

Williams JHG, Ross L (2007) Consequences of prenatal toxin exposure for mental health in children and adolescents. Eur Child Adolesc Psychiatry 16:243–253

Williams J, Taylor E (2006) The evolution of hyperactivity, impulsivity and cognitive diversity. J R Soc Interface 3:399–413

Wolraich ML, Lambert EW, Bickman L, Simmons T, Doffing MA, Worley KA (2004) Assessing the impact of parent and teacher agreement on diagnosing attention-deficit hyperactivity disorder. J Dev Behav Pediatr 25:41–47

Youdim MBH, Weinstock M (2004) Therapeutic applications of selective and non-selective inhibitors of monoamine oxidase A and B that do not cause significant tyramine potentiation. Neurotoxicology 25:243–250

Yu H, Schummers J, Sur M (2010) The Influence of Astrocyte Activation on Hemodynamic Signals for Functional Brain Imaging. In: Roe AW (ed) Image Brain Opt Meth. Springer, New York, pp 45–64

Yu-Feng Z, Yong H, Chao-Zhe Z, Qing-Jiu C, Man-Qiu S, Meng L, Li-Xia T, Tian-Zi J, Yu-Feng W (2007) Altered baseline brain activity in children with ADHD revealed by resting-state functional MRI. Brain Dev 29:83–91

Zametkin A, Rapoport JL, Murphy DL, Linnoila M, Ismond D (1985) Treatment of hyperactive children with monoamine oxidase inhibitors: I. Clinical efficacy. Arch Gen Psychiatry 42:962

Index

A
Addiction. *See* Drugs
ADHD. *See* Attention-deficit hyperactivity disorder (ADHD)
Adolescent methylphenidate exposure, brain processes
  ADHD, behavioral features
    adolescence and modeling, 117–119
    MPH use, 116–117
  consequences, MPH exposure
    persistent brain metabolic outcomes, 128–130
    striatal gene expression, 130–131
  discounting processes, MPH
    methodological remarks, 125–127
    self-control behavior, 123–125
  dopaminergic peculiarities, rodent models
    neurobiological basis, 119–120
    pharmacological imaging, MPH, 120–122
Adolescents
  emotional and contextual visual processing, 229
  face processing developments, 227
  and modeling, 117–119
  negative emotional expressions deficits, 230
α2-Adrenoceptor agonists, 376–377
Altered reinforcement mechanisms, 209–210
Amphetamine
  action mode, 367
  animal model, 275
  clinical trials, 367, 369
  D-and L-isomers, 365–366
  monoamine release, 365
  monoamine reuptake transporters, 367
  noradrenaline and dopamine efflux, 365
  pharmacological characteristics, 368

  striatum and prefrontal cortex, 366
Amygdala, 226–227
Anteroventral periventricular nucleus of the hypothalamus (AVPV), 349
Arousal, 222
Attention-deficit hyperactivity disorder (ADHD)
  in adolescents, 9–11
  in adults, 11–12
  controversial issues, 5–7
  diagnostic criteria, 3–5
  DSM-IV-TR, 3
  heterogeneity, executive dysfunction and, 178–180
  vs. HPA axis function, 96–97, 106–107
  in preschool children, 7–8
  in school-age children, 8–9
  smoking and, 151
  treatment, substance use disorders (SUD)
    clinical populations studies, 158–159
    potential mechanisms, 156–158

B
Baldwin effect, 413
Basal cortisol secretion, 97–100
Basal ganglia, 22, 176
Behavioral and neurobiological mechanisms, obesity
  altered reinforcement mechanisms, 209–210
  binge eating, 208
  brain-derived neurotropic factor, 210
  excessive daytime sleepiness, 208–209
  impulsivity role, 211–212
  inattention role, 212
  melanocortin-4-receptor (MC4-R) deficiency, 210–211

Binge eating, 208
Bipolar disorder (BD), 232–233
Body mass index (BMI).
  See Obesity and ADHD
Brain development.
  See Neurodevelopmental
    abnormalities
Bupropion
  clinical trial, 375–376
  microdialysis experiments, 375

C

Cadherin 13 gene (CDH13), 258
Catechol-*O*-methyl transferase (COMT) gene, 254–255
Caudate anomaly, 21–23
Causal framework
  efficient causes
    behavioral characters, 406
    examples, 410
    intervention, 415
    prenatal causal factors, 405–406
    state changes and trigger, 394, 396
  final causes
    behavioral dysfunction, 409
    biological function, 395, 396
    evolutionary history, 411–412
    examples, 410
    genetic traits, 410
    immediate reinforcement, 412
    intervention, 416
  formal causes
    characterization, 393, 396
    examples, 410
    intervention, 414–415
    proximate, 396–402
    symbolic models, 394
    ultimate level, 402–405
  material causes
    energetics, 409
    examples, 410
    genetics and epigenetics, 407
    intervention, 415–416
    neurobiological mechanisms, 394, 396, 407
    sex, cigarettes, and monoamine oxidase, 407–408
  recurrent causes
    Baldwin effect, 413
    closed-loop systems, 395, 396
    reinforcement, 413
Cerebellar anomaly, 23–24

Coloboma mutant mouse model
  monoaminergic regulation
    dopamine, 280
    norepinephrine, 280–281
    serotonin, 281
  therapeutic mechanisms, 281–282
  validity, 279–280
COMT. *See* Catechol-*O*-methyl transferase (COMT) gene
Conduct disorder (CD), 232. *See also*
  Substance use disorders (SUD)
Copy number variant (CNV), 259–260
Cortical morphology, 24–26
Cortisol hyporeactivity, 100–104
Cyclin-dependent kinase inhibitor 1C
  (Cdkn1c), 324
Cyclooxygenase-2 (COX-2), 347

D

Diagnostic and Statistical Manual, Fourth
  Edition (DSM-IV), 146, 149
Discounting. *See* Adolescent methylphenidate
  exposure, brain processes
Disruptive behavior disorder (DBD), 103, 400–401
DNA methylation, 320–321
Dopamine. *See also* Adolescent
  methylphenidate exposure, brain
  processes
  monoaminergic regulation, 282–283
  SUD–ADHD relationship
    genetics, 154–156
    heritability, 153–154
    neuropsychological functioning, 151–153
Dopamine D4 receptor gene (DRD4), 252
Dopamine transporter gene (DAT1/SLC6A3), 249–250
Dopamine transporter knockout (DAT-KO)
  mouse model
  monoaminergic regulation
    dopamine, 282–283
    norepinephrine, 283
    serotonin, 283
  therapeutic mechanisms, 283–284
  validity, 282
Drugs
  addiction
    evidence, executive dysfunction, 175–176
    modeling vulnerability, 182–183
  characteristics, 381

classification, mechanism of action and
neurotransmitter target, 364
combined interventions, 380
new developments
compounds efficacy and safety, 381–382
histamine H3 antagonists, 383
knowledge gain mechanism, 382
nicotinic receptor agonists, 383
reuptake inhibitors, 383
SHR model, 384
stimulants, 382
stimulants, 380
translational pharmacology evaluation
amphetamines, 365–369
bupropion, 375–376
guanfacine and α2-adrenoceptor agonists, 376–377
methylphenidate, 369–375
modafinil, 377–379
stimulant reuptake inhibitor, 364
DSM definition, 396–397
DSM-IV classification, 275

E

Emotional impairments, 221
Endophenotypes, 246, 260–263
Epigenetics
ADHD
dopamine transporter gene, 322
imprinting, 323
inattention and hyperactivity, 322
SLC6A3, role in, 322
twin studies, 321
chromosome organization, 319–320
DNA methylation
dietary methionine, 321
neuropsychiatric disorders, 321
transcriptional repression, 320
tumor suppression, 320
environmental factors, 319
gene–environmental interactions
alcohol exposure, 325
comorbidity, 325
cyclin-dependent kinase inhibitor 1C, 324
dopaminergic and serotonergic deficiencies, 323
maternal mutations, 324
psychosocial adversity, 325
smoking, 324–325
synthetic glucocorticoids, 323
medications and gene expression, ADHD
adverse effects and, 331
d-amphetamine, 326
gene/protein expression, 327–329
immediate early genes, 330
methylphenidate, 326, 330
noradrenaline, 331
paramutation and bookmarking, 319
transcription factors, 321
Epistatic interactions, 249
Equal environment assumption (EEA), 242
Estradiol, 345
Executive dysfunction, 178–180

F

Face processing
ADHD
aberrant face processing, 231
behavioral intermediate phenotype, 231
contextual and emotional visual processing, 229
emotional recognition and intensity, 229
facial affect task, 229
negative emotional expressions deficits, 230
neurocognitive deficits, prototypical domains, 228
reactive *vs.* executive inhibition, 230
attentional mechanisms
cortical regulation, 226
neural activation, 227
subcortical circuit, 226–227
attention's role
automatic processes, 225
brain regions variation, 225
dynamic ecological validity, 225–226
process disruption, 225
speed of, 223–224
unconscious, 224
bipolar disorder, 232–233
conduct disorder, 232
developmental effects, 227
emotional stimuli
appraisal and judgment, 222
body's response, visual stimuli, 222–223
motivational salience, 222
techniques, 223
neural network, 222
oppositional defiant disorder, 232
1/f noise, 70
Frontostriatal anomaly, 19–21
Functional imaging, neurodevelopmental abnormalities

Functional imaging, neurodevelopmental
abnormalities (cont.)
described, 55–56
frontal-striatal-cerebellar circuitry, 56–57
mesolimbic circuitry, 57
motor-premotor regions, 58–59
visual-spatial attention, 58
Functional magnetic resonance imaging
(fMRI), 51
Fusiform face area, 226
Fusiform gyrus, 226

G
Gene–environmental interactions
alcohol exposure, 325
comorbidity, 325
cyclin-dependent kinase inhibitor 1C, 324
dopaminergic and serotonergic deficiencies, 323
maternal mutations, 324
psychosocial adversity, 325
smoking, 324–325
synthetic glucocorticoids, 323
Genetic studies. See also Molecular genetic
studies; Quantitative genetic
studies; copy number variants
(CNVs), 259–260
endophenotypes, 260–263
genome-wide association studies (GWAS)
cadherin 13 gene, 258
GFOD1 gene, 259
SNP arrays, 257–258
intermediate phenotypes, 260–263
linkage analysis, 256–257
spontaneously hypertensive rat model, 305–307
Genome-wide association studies (GWAS)
cadherin 13 gene, 258
GFOD1 gene, 259
SNP arrays, 257–258
Glucose-fructose oxidoreductase-domain
containing 1 gene (GFOD1), 259
Gonadal steroids
apoptosis, 349
brain development and pathology, 344
Guanfacine, 376–377
GWAS. See Genome-wide association studies
(GWAS)

H
Hippocampus projects, 185–190
5-HT serotonin receptors, 278

Hunter–Farmer theory, 132
Hypothalamic–pituitary–adrenocortical (HPA)
axis function
vs. ADHD, 96–97, 106–107
advantages, cortisol levels, 105
assess cortisol, waking time, 104–105
basal cortisol secretion, 97–100
cortisol hyporeactivity, 100–104
described, 95–96
psychoneuroendocrine research, 105–106

I
Impulsivity
and choice behavior, 183–185
dimensions of, 176–178
neural substrates, 180–182
role, obesity and ADHD, 211–212
Intermediate phenotypes, 246, 260–263
Intraindividual variability (IIV)
measurement, 69–71
neural basis, 77–81
nosological specificity, 84–85
strength of
genetic association, 75–77
phenotypic association, 72–75
theories, 81–84
Involuntary movements, 31

L
Limbic system, 176
Linkage analysis, 256–257
Linking ADHD, drug addiction, 190–191

M
Magnetic resonance imaging. See
Neurodevelopmental
abnormalities, 51–62
Melanocortin-4-receptor (MC4-R)
deficiency, 210–211
Methylphenidate. See Adolescent
methylphenidate exposure, brain
processes
in adolescents, 326
animal model, 275
atomoxetine
adjunctive therapy, 374
adverse effects, 374
catecholamine efflux, 373
clinical trials, 374–375
dopamine reuptake and catabolism, 372
noradrenaline and dopamine efflux, 374

Index   431

catecholamine efflux, 370–371
clinical trials, 371
D-enantiomer, 369
gene expression, 330
noradrenaline/dopamine reuptake inhibitor, 370
structure, 369
Modafinil
  behavioral effects, 378
  clinical trials, 378–379
  disadvantages, 377
  microdialysis, 378
Molecular genetic studies
  ADHD
    adults, 255–256
    COMT gene, 254–255
    conduct problems, 254
    dopamine D4 receptor gene (DRD4), 252
    dopamine transporter gene (DAT1/SLC6A3), 249–250
    gene association
      meta-analysis, 251
      replicated participants, effect sizes, 252
    history, 248
    serotonin 1B receptor gene (HTR1B), 253
    serotonin transporter gene (SLC6A4/5-HTT), 253
    synaptosomal-associated protein 25 isoform gene, 254
    tachykinin receptor 1 (TACR1), 252–253
Monoamine oxidase (MAO), 407–408
Motivational salience, 222
Motor function, ADHD
  approaches, 33
  described, 30–31
  overflow, 31–33
  subtle signs, 31
Motor inhibition, 182

N

Neonatal 6-hydroxydopamine-lesioned rat model
  monoaminergic regulation
    dopamine, 276–277
    norepinephrine, 277
    serotonin, 277–278
  therapeutic mechanisms, 278–279
  validity, 276
Neurodevelopmental abnormalities
  functional imaging
    described, 55–56

  frontal-striatal-cerebellar circuitry, 56–57
  mesolimbic circuitry, 57
  motor-premotor regions, 58–59
  visual-spatial attention, 58
resting-state imaging, 61
  described, 60
  key findings, 61
structural imaging
  described, 51–52
  key findings, 52–54
Neurokinin 1 receptor knockout mice (NK1$^{-/-}$)
  monoaminergic regulation, 287–288
  therapeutic mechanisms, 288
  validity, 287
Nonfrontal cortical anomaly, 24
Noradrenaline, 331

O

Obesity and ADHD
  behavioral and neurobiological mechanisms
    altered reinforcement mechanisms, 209–210
    binge eating, 208
    brain-derived neurotropic factor, 210
    excessive daytime sleepiness, 208–209
    impulsivity role, 211–212
    inattention role, 212
    melanocortin-4-receptor (MC4-R) deficiency, 210–211
  clinical management
    problematic eating behavior, 212–213
    therapeutic strategies, 213–214
  prevalence, 202–204
  weight status, 204–205
Oculomotor functions, ADHD
  assessment
    response inhibition, 37
    response preparation, 36–37
    working memory, 37–38
  described, 34–35
  experimental assessment, eye movements, 35–36
Oppositional defiant disorder (ODD), 232

P

Paramutation, 319
Pharmacology. *See* Drugs
Pleiotropy, 246
Prefrontal cortex (PFC), 120

Q

Quantitative genetic studies
  adults
    familial effects, 243–244
    twin studies, 244
  clinical feature, 243
  cognitive performance and brain function
    aetiological relationships, 246
    family design, 247
    putative endophenotype, 246
    sibling pair studies, 247–248
  combined type ADHD, 243
  environmental risk measures, 242
  inattention and hyperactivity-impulsivity, 244–246
  monozygotic and dizygotic twin pairs, 242
  twin pairs, 242

R

Rat models, ADHD
  age and development, 309–310
  behavioral differences, 304–305
  defining features and situational factors, 308–309
  genetic differences, 305–307
  implications, 310
  WKY heterogeneity, 307–308
Reaction time (RT) variability, 69
Relative operating characteristic (ROC), 399
Response readiness, 411
Resting-state imaging
  described, 60
  key findings, 61
Reward deficiency syndrome, 209–210
Rodent models
  coloboma mutant mouse
    monoaminergic regulation, 280–281
    therapeutic mechanisms, 281–282
    validity, 279–280
  dopamine and serotonin receptors, 291–292
  dopamine transporter knockout mouse
    monoaminergic regulation, 282–283
    therapeutic mechanisms, 283–284
    validity, 282
  hyperdopaminergic neurotransmission, 288–290
  neonatal 6-hydroxydopamine-lesioned rat
    monoaminergic regulation, 276–278
    therapeutic mechanisms, 278–279
    validity, 276
  neurokinin 1 receptor knockout mice
    monoaminergic regulation, 287–288
    therapeutic mechanisms, 288
    validity, 287
  norepinephrine transmission, 290–291
  serotonin transmission, 291
  spontaneously hypertensive rat
    monoaminergic regulation, 284–286
    therapeutic mechanisms, 287
    validity, 284

S

SDN. *See* Sexually dimorphic nucleus (SDN)
  Sensorimotor, 25
Serotonin 1B receptor gene (HTR1B), 253
Serotonin transporter gene (SLC6A4/5-HTT), 253
Sexual differentiation, brain and ADHD
  apoptosis
    gonadal steroids, 349
    hormones, 348
  development and pathology
    animal models, 344–345
    estradiol, 345
    gonadal steroids, 344
    learning disabilities and hyperactivity, 343
    reproduction, 343
    testosterone, 344, 345
  dopaminergic excitation, 352–353
  GABA-mediated excitation
    cell membrane depolarization, 350
    chloride transporter, 351
    developmental progression, 350
    hippocampus development, 351
    substantia nigra, 352
  hormones, 348–349
  hypothalamus
    arcuate nucleus, 346–347
    astrocytes, 347
    dendritic spines, 347
    estradiol, 347, 348
    medial zone, 346
    periventricular zone, 346
    preoptic area, 346
    testosterone administration, 348
  neurogenesis, 349
  ventral tegmental area, 352
Sexually dimorphic nucleus (SDN), 348–349
Sign trackers, 183
Smoking and ADHD, 151
Spontaneously hypertensive rat (SHR) model.
  See also Rat models, ADHD
  monoaminergic regulation

dopamine, 284–286
norepinephrine, 286
serotonin, 286
therapeutic mechanisms, 287
validity, 284
Stimulant reuptake inhibitor, 364
Stress. *See* Hypothalamic-pituitary-adrenocortical (HPA) axis function
Substance use disorders (SUD)
ADHD treatment
clinical populations studies, 158–159
potential mechanisms, 156–158
dopamine relationship
genetics, 154–156
heritability, 153–154
neuropsychological functioning, 151–153
management guidelines, 162–164
pharmacological treatment effect, 160–162
prevalence, 147–148
psychostimulants misuse, 159–160
risk, ADHD and, 148–150
smoking and ADHD, 151
terminology, 146–147
Synaptosomal-associated protein 25 isoform gene, 254

T
Tachykinin receptor 1 (TACR1), 252–253
Testosterone, 344, 345
Therapeutical evaluation. *See* Drugs

V
Valence, 222
Validation, animal model. *See* Rat models, ADHD
Ventral medial prefrontal cortex (VMPFC), 179, 180
Volumetry function, ADHD
caudate anomaly, 21–23
cerebellar anomaly, 23–24
cortical morphology, 24–26
frontostriatal anomaly, 19–21
genes in, 28–29
longitudinal volumetric studies, 26–27
nonfrontal cortical anomaly, 24
shape analysis, 27
treatment, stimulant medication, 27–28

Printed by Publishers' Graphics LLC USA
MO20120327-119
2012